LINEAR ALGEBRA
GEOMETRY AND
TRANSFORMATION

TEXTBOOKS in MATHEMATICS

Series Editors: Al Boggess and Ken Rosen

PUBLISHED TITLES CONTINUED

TEXTBOOKS in MATHEMATICS

LINEAR ALGEBRA
GEOMETRY AND
TRANSFORMATION

BRUCE SOLOMON
INDIANA UNIVERSITY, BLOOMINGTON

USA

CRC Press
Taylor & Francis Group
Boca Raton   London   New York

CRC Press is an imprint of the
Taylor & Francis Group an **informa** business
A CHAPMAN & HALL BOOK

CRC Press
Taylor & Francis Group
6000 Broken Sound Parkway NW, Suite 300
Boca Raton, FL 33487-2742

Printed on acid-free paper
Version Date: 20140717

International Standard Book Number-13: 978-1-4822-9928-1 (Pack - Book and Ebook)

Library of Congress Cataloging-in-Publication Data

Solomon, Bruce, 1953- author.
    Linear algebra, geometry and transformation / Bruce Solomon.
        pages cm. -- (Textbooks in mathematics)
    Includes bibliographical references and index.
    ISBN 978-1-4822-9928-1
    1. Algebras, Linear--Textbooks. I. Title.

QA184.2.S65 2015
512.5--dc23   2014027909

Visit the Taylor & Francis Web site at
http://www.taylorandfrancis.com

and the CRC Press Web site at
http://www.crcpress.com

To my teachers. . .
. . . and to my students

Contents

Preface

*"The eyes of the mind, by which it sees and observes
things, are none other than proofs."*

Baruch Spinoza

The organizing concept of this book is this: every topic should bring
students closer to a solid geometric grasp of *linear transformations*.

Even more specifically, we aim to build a strong foundation for two
enormously important results that no undergraduate math student
should miss:

- The Spectral Theorem for symmetric transformations, and

- The Inverse/Implicit Function Theorem for differentiable mappings, or even better, the strong form of that result, sometimes called the Rank Theorem.

Every student who continues in math or its applications will encounter
both these results in many contexts. The Spectral Theorem belongs
to Linear Algebra proper; a course in the subject is simply remiss if it
fails to get there. The Rank Theorem actually belongs to multivariable
calculus, so we don't state or prove it here. Roughly, it says that a
differentiable map of constant rank can be locally approximated by—
and indeed, behaves geometrically just like—a *linear* map of the same
rank. A student *cannot* understand this without a solid grasp of the
linear case, which we do formulate and prove here as the *Linear Rank
Theorem* in Chapter 7, making it, and the Spectral Theorem, key goals
of our text.

The primacy we give those results motivates an unconventional start
to our book, one that moves quickly to a first encounter with multivariable mappings and to the basic questions they raise about images,
pre-images, injectivity, surjectivity, and distortion. While these are
fundamental concerns throughout mathematics, they can be frustrat-
ingly difficult to analyze in general. The beauty and power of Linear

Algebra stem in large part from the utter transparency of these problems in the linear setting. A student who follows our discussion will apprehend them with a satisfying depth, and find them easy to apply in other areas of mathematical pursuit.

Of course, we cover all the standard topics of a first course in Linear Algebra—linear systems, vector geometry, matrix algebra, subspaces, independence, dimension, orthogonality, eigenvectors, and diagonalization. In our view, however, these topics mean more when they are directed toward the motivating results listed above.

We therefore introduce linear mappings and the basic questions they raise in our very first chapter, and aim the rest of our book toward answering those questions.

Key secondary themes emerge along the way. One is the centrality of the homogeneous system and the version of Gauss-Jordan we teach for solving it—and for expressing its solution as the span of independent "homogeneous generators." The number of such generators, for instance, gives the nullity of the system's coefficient matrix $\mathbf{A}$, which in turn answers basic questions about the structure of solutions to inhomogeneous systems having $\mathbf{A}$ as coefficient matrix, and about the linear transformation represented by $\mathbf{A}$.

Throughout, we celebrate the beautiful dualities that illuminate the subject:

- An $n \times m$ matrix $\mathbf{A}$ is both a list of rows, acting as linear functions on $\mathbf{R}^m$, and a list of columns, representing vectors in $\mathbf{R}^n$. Accordingly, we can interpret matrix/vector multiplication in dual ways: As a transformation of the input vector, or as a linear combination of the matrix columns. We stress the latter viewpoint more than many other authors, for it often delivers surprisingly clear insights.

- Similarly, an $n \times m$ system $\mathbf{Ax} = \mathbf{b}$ asks for the intersection of certain hyperplanes in $\mathbf{R}^m$, while simultaneously asking for ways to represent $\mathbf{b} \in \mathbf{R}^n$ as a linear combination of the columns of $\mathbf{A}$.

- The solution set of a homogeneous system can be alternatively expressed as the image (column-space) of one linear map, or as the pre-image (kernel) of another.

- The ubiquitous operations of addition and scalar multiplication manifest as pure algebra in the numeric vectorspaces $\mathbf{R}^n$, while simultaneously representing pure geometry in 2- and 3-dimensional Euclidean space.

- Every subspace of $\mathbf{R}^n$ can be described in essentially just two dual ways: as a *span*—the span of a generating set, or as an intersection of hyperplanes—what we call a *perp*.

We emphasize the computational and conceptual skills that let students navigate easily back and forth along any of these dualities, since problems posed from one perspective can often be solved with less effort from the dual viewpoint.

Finally, we strive to make all this material a ramp, lifting students from the computational mathematics that dominates their experience before this course, to the conceptual reasoning that often dominates after it. We move very consciously from simple "identity verification" proofs early on (where students check, using the definitions, for instance, that vector addition commutes, or that it distributes over dot products) to constructive and contrapositive arguments—e.g., the proof that the usual algorithm for inverting a matrix fulfills its mission. One can base many such arguments on reasoning about the outcome of the Gauss-Jordan algorithm—i.e., row-reduction and reduced row-echelon form—which students easily master. Linear algebra thus forms an ideal context for fostering and growing students' mathematical sophistication.

Our treatment omits abstract vector spaces, preferring to spend the limited time available in one academic term focusing on $\mathbf{R}^n$ and its subspaces, orthogonality and diagonalization. We feel that when students develop familiarity and the ability to reason well with $\mathbf{R}^n$ and—especially—its subspaces, the transition to abstract vector spaces, if and when they encounter it, will pose no difficulty.

Most of my students have been sophomores or juniors, typically majoring in math, informatics, one of the sciences, or business. The lack of an engineering school here has given my approach more of a liberal arts flavor, and allowed me to focus on the mathematics and omit applications. I know that for these very reasons, my book will not satisfy everyone. Still, I hope that all who read it will find themselves sharing the pleasure I always feel in learning, teaching, and writing about linear algebra.

Acknowledgments. This book springs from decades of teaching linear algebra, usually using other texts. I learned from each of those books, and from every group of students. About 10 years ago, Gilbert Strang's lively and unique introductory text inspired many ideas and syntheses of my own, and I began to transition away from his book toward my own notes. These eventually took the course over, evolving into the present text. I thank all the authors, teachers, and students with whom I have learned to think about this beautiful subject, starting with the late Prof. Richard F. Arens, my undergraduate linear algebra teacher at UCLA.

Sincere thanks also go to CRC Press for publishing this work, and especially editor Bob Ross, who believed in the project and advocated for me within CRC.

I could not have reached this point without the unflagging support of my wife, family, and friends. I owe them more than I can express.

Indiana University and its math department have allowed me a life of continuous mathematical exploration and communication. A greater privilege is hard to imagine, and I am deeply grateful.

On a more technical note, I was lucky to have excellent software tools: TeXShop and LaTeX for writing and typesetting, along with Wolfram *Mathematica*®,[1] which I used to create all figures except Figure 28 in Chapter 3. The latter image of M.C. Escher's striking 1938 woodcut *Day and Night* (which also graces the cover) comes from the Official M.C. Escher website (www.mcescher.com).

Bruce Solomon
Indiana University
Bloomington, Indiana

---

[1]Wolfram *Mathematica*® is a registered trademark of Wolfram Research, Inc.

CHAPTER 1

Vectors, Mappings, and Linearity

1. Numeric Vectors

The overarching goal of this book is to impart a sure grasp of the numeric vector functions known as *linear transformations*. Students will have encountered *functions* before. We review and expand that familiarity in Section 2 below, and we define *linearity* in Section 4. Before we can properly discuss these matters though, we must introduce *numeric vectors* and their basic arithmetic.

DEFINITION 1.1 (Vectors and scalars). A **numeric vector** (or just **vector** for short) is an ordered n-tuple of the form $(x_1, x_2, \ldots, x_n)$. Here, each x_i—the i**th entry** (or ith **coordinate**) of the vector—is a real number.

The (x, y) pairs often used to label points in the plane are familiar examples of vectors with $n = 2$, but we allow more than two entries as well. For instance, the triple $(3, -1/2, 2)$, and the 7-tuple $(1, 0, 2, 0, -2, 0, -1)$ are also numeric vectors.

In the linear algebraic setting, we usually call single numbers **scalars**. This helps highlight the difference between numeric vectors and individual numbers. □

Vectors can have many entries, so to clarify and save space, we often label them with single bold letters instead of writing out all their entries. For example, we might define

$$\begin{aligned}
\mathbf{x} &:= (x_1, x_2, \ldots, x_n) \\
\mathbf{a} &:= (a_1, a_2, a_3, a_4) \\
\mathbf{b} &:= (-5, 0, 1)
\end{aligned}$$

and then use $\mathbf{x}$, $\mathbf{a}$, or $\mathbf{b}$ to indicate the associated vector. We use boldface to distinguish vectors from scalars. For instance, the same letters, without boldface, would typically represent scalars, as in $x = 5$, $a = -4.2$, or $b = \pi$.

Often, we write numeric vectors vertically instead of horizontally, in which case $\mathbf{x}$, $\mathbf{a}$, and $\mathbf{b}$ above would look like this:

$$\mathbf{x} = \begin{pmatrix} x_1 \\ x_2 \\ \vdots \\ x_m \end{pmatrix}, \quad \mathbf{a} = \begin{pmatrix} a_1 \\ a_2 \\ a_3 \\ a_4 \end{pmatrix}, \quad \mathbf{b} = \begin{pmatrix} -5 \\ 0 \\ 1 \end{pmatrix}$$

In our approach to the subject (unlike some others) we draw absolutely no distinction between

$$(x_1, x_2, \ldots, x_n) \quad \text{and} \quad \begin{pmatrix} x_1 \\ x_2 \\ \vdots \\ x_n \end{pmatrix}$$

These are merely different notations for the same vector—the very same mathematical object.

DEFINITION 1.2. We denote the set of all scalars—also known as the real number line—by $\mathbf{R}^1$ or simply $\mathbf{R}$.

Similarly, $\mathbf{R}^n$ denotes the collection of **all** numeric vectors with n entries; that is, all $(x_1, x_2, \ldots, x_n)$. The "all zero" vector $(0, 0, \ldots, 0) \in \mathbf{R}^n$ is called the **origin**, and denoted by $\mathbf{0}$.   □

As examples, the vectors $\mathbf{x}$, $\mathbf{a}$, and $\mathbf{b}$ above belong to $\mathbf{R}^m$, $\mathbf{R}^4$, and $\mathbf{R}^3$, respectively. We express this symbolically with the "element of" symbol "$\in$":

$$\mathbf{x} \in \mathbf{R}^m, \ \mathbf{a} \in \mathbf{R}^4, \ \text{and } \mathbf{b} \in \mathbf{R}^3$$

If $\mathbf{a}$ does **not** lie in $\mathbf{R}^5$, we can write $\mathbf{a} \notin \mathbf{R}^5$.

$\mathbf{R}^m$ is more than just a *set*, though, because it supports two important algebraic operations: *vector addition* and *scalar multiplication*.

1.3. Vector addition. To add (or subtract) vectors in $\mathbf{R}^m$, we simply add (or subtract) coordinates, entry-by-entry. This is best depicted vertically. Here are two examples, one numeric and one symbolic:

$$\begin{pmatrix} 1 \\ 2 \\ 3 \end{pmatrix} + \begin{pmatrix} 4 \\ -5 \\ 6 \end{pmatrix} = \begin{pmatrix} 1+4 \\ 2-5 \\ 3+6 \end{pmatrix} = \begin{pmatrix} 5 \\ -3 \\ 9 \end{pmatrix}$$

$$\begin{pmatrix} a_1 \\ a_2 \\ a_3 \\ a_4 \end{pmatrix} + \begin{pmatrix} b_1 \\ b_2 \\ b_3 \\ b_4 \end{pmatrix} - \begin{pmatrix} c_1 \\ c_2 \\ c_3 \\ c_4 \end{pmatrix} = \begin{pmatrix} a_1 + b_1 - c_1 \\ a_2 + b_2 - c_2 \\ a_3 + b_3 - c_3 \\ a_4 + b_4 - c_4 \end{pmatrix}$$

Adding the origin $\mathbf{0} \in \mathbf{R}^m$ to any vector obviously leaves it unchanged: $\mathbf{0} + \mathbf{x} = \mathbf{x}$ for any $\mathbf{x} \in \mathbf{R}^m$. For this reason, $\mathbf{0}$ is called the *additive identity* in $\mathbf{R}^m$.

Recall that addition of *scalars* is commutative and associative. That is, for any scalars x, y, and z we have

$$x + y = y + x \qquad \text{(Commutativity)}$$
$$(x + y) + z = x + (y + z) \qquad \text{(Associativity)}$$

It follows easily that *vector* addition has these properties too:

PROPOSITION 1.4. *Given any three vectors* $\mathbf{x}, \mathbf{y}, \mathbf{z} \in \mathbf{R}^m$, *we have*

$$\mathbf{x} + \mathbf{y} = \mathbf{y} + \mathbf{x} \qquad \text{(Commutativity)}$$
$$(\mathbf{x} + \mathbf{y}) + \mathbf{z} = \mathbf{x} + (\mathbf{y} + \mathbf{z}) \qquad \text{(Associativity)}$$

PROOF. We prove associativity, and leave commutativity as an exercise.

The associativity statement is an *identity*: it asserts that two things are equal. Our approach is a basic and useful one for proving such assertions: *Expand both sides of the identity to show individual entries, then simplify using the familiar algebra of scalars. If the simplified expressions can be made equal using legal algebraic moves, we have a proof.*

Here, we start with the left-hand side, labeling the coordinates of $\mathbf{x}$, $\mathbf{y}$, and $\mathbf{z}$ using x_i, y_i, and z_i, and then using the definition of vector addition twice:

$$(\mathbf{x} + \mathbf{y}) + \mathbf{z} = \left[\begin{pmatrix} x_1 \\ x_2 \\ \vdots \\ x_m \end{pmatrix} + \begin{pmatrix} y_1 \\ y_2 \\ \vdots \\ y_m \end{pmatrix} \right] + \begin{pmatrix} z_1 \\ z_2 \\ \vdots \\ z_m \end{pmatrix}$$

$$= \begin{pmatrix} x_1 + y_1 \\ x_2 + y_2 \\ \vdots \\ x_m + y_m \end{pmatrix} + \begin{pmatrix} z_1 \\ z_2 \\ \vdots \\ z_m \end{pmatrix}$$

$$= \begin{pmatrix} (x_1 + y_1) + z_1 \\ (x_2 + y_2) + z_2 \\ \vdots \\ (x_m + y_m) + z_m \end{pmatrix}$$

Similarly, for the right-hand side of the identity, we get

$$\mathbf{x} + (\mathbf{y} + \mathbf{z}) = \begin{pmatrix} x_1 \\ x_2 \\ \vdots \\ x_m \end{pmatrix} + \left[\begin{pmatrix} y_1 \\ y_2 \\ \vdots \\ y_m \end{pmatrix} + \begin{pmatrix} z_1 \\ z_2 \\ \vdots \\ z_m \end{pmatrix} \right]$$

$$= \begin{pmatrix} x_1 \\ x_2 \\ \vdots \\ x_m \end{pmatrix} + \begin{pmatrix} y_1 + z_1 \\ y_2 + z_2 \\ \vdots \\ y_m + z_m \end{pmatrix}$$

$$= \begin{pmatrix} x_1 + (y_1 + z_1) \\ x_2 + (y_2 + z_2) \\ \vdots \\ x_m + (y_m + z_m) \end{pmatrix}$$

The simplified expressions for the two sides are now very similar. The parentheses don't line up the same way on both sides, but we can fix that by using the associative law for *scalars*. The two sides then agree, exactly, and we have a proof.

In short, the associative law for vectors boils down, after simplification, to the associative law for scalars, which we already know. $\square$

1.5. Scalar multiplication. The second fundamental operation in $\mathbf{R}^n$ is even simpler than vector addition. *Scalar multiplication* lets us multiply any vector $\mathbf{x} \in \mathbf{R}^m$ by an arbitrary scalar t to get a new vector $t\,\mathbf{x}$. As with vector addition, we execute it entry-by-entry:

$$t\,\mathbf{x} = t \begin{pmatrix} x_1 \\ x_2 \\ \vdots \\ x_m \end{pmatrix} := \begin{pmatrix} t\,x_1 \\ t\,x_2 \\ \vdots \\ t\,x_m \end{pmatrix}$$

For instance, $2\,(1,3,5) = (2,6,10)$ and $-3\,(1,1,0,1) = (-3,-3,0,-3)$, while $0\,\mathbf{x} = (0,0,\ldots,0)$ no matter what $\mathbf{x}$ is.

Recall that for *scalars*, multiplication distributes over addition. This means that for any scalars t, x, and y, we have

$$t(x + y) = tx + ty$$

Since scalar multiplication and vector addition both operate entry-by-entry, scalar multiplication distributes over *vector* addition too. This simple relationship between the two operations is truly fundamental in linear algebra. Indeed, we shall see in Section 4 below, that it models the concept of *linearity*.

PROPOSITION 1.6. *Scalar multiplication distributes over vector addition. That is, if t is any scalar and $\mathbf{x}_1$, $\mathbf{x}_2$, ..., $\mathbf{x}_k$ are arbitrary vectors in $\mathbf{R}^m$, we have*

$$t\left(\mathbf{x}_1 + \mathbf{x}_2 + \cdots + \mathbf{x}_k\right) = t\,\mathbf{x}_1 + t\,\mathbf{x}_2 + \cdots + t\,\mathbf{x}_k$$

PROOF. To keep things simple, we prove this for just two vectors $\mathbf{x}, \mathbf{y} \in \mathbf{R}^m$. The argument for k vectors works exactly the same way.

Using the same approach we used in proving the associativity identity in Proposition 1.4, we expand both sides of the identity in individual entries, simplify, and observe that we get the same result either way.

Let $\mathbf{x} = (x_1, x_2, \ldots, x_m)$ and $\mathbf{y} = (y_1, y_2, \ldots, y_m)$ be any two vectors in $\mathbf{R}^m$. Then for each scalar t, the left-hand side of the identity expands like this:

$$t\left(\mathbf{x} + \mathbf{y}\right)$$

$$= t\begin{pmatrix} x_1 + y_1 \\ x_2 + y_2 \\ \vdots \\ x_m + y_m \end{pmatrix} = \begin{pmatrix} t(x_1 + y_1) \\ t(x_2 + y_2) \\ \vdots \\ t(x_m + y_m) \end{pmatrix} = \begin{pmatrix} tx_1 + ty_1 \\ tx_2 + ty_2 \\ \vdots \\ tx_m + ty_m \end{pmatrix}$$

While the right-hand side expands thus:

$$t\mathbf{x} + t\mathbf{y} = t\begin{pmatrix} x_1 \\ x_2 \\ \vdots \\ x_n \end{pmatrix} + t\begin{pmatrix} y_1 \\ y_2 \\ \vdots \\ y_n \end{pmatrix} = \begin{pmatrix} tx_1 \\ tx_2 \\ \vdots \\ tx_n \end{pmatrix} + \begin{pmatrix} ty_1 \\ ty_2 \\ \vdots \\ ty_n \end{pmatrix}$$

$$= \begin{pmatrix} tx_1 + ty_1 \\ tx_2 + ty_2 \\ \vdots \\ tx_m + ty_m \end{pmatrix}$$

We get the same result either way, so the identity holds.   □

1.7. Linear combination. We now define a third operation that combines scalar multiplication and vector addition. Actually, scalar multiplication and vector addition can be seen as mere special cases of this new operation:

DEFINITION 1.8. Given vectors $\mathbf{a_1}, \mathbf{a_2}, \ldots, \mathbf{a_m} \in \mathbf{R}^n$ and equally many scalars $x_1, x_2, \ldots, x_m$, the "weighted sum"

$$x_1 \mathbf{a_1} + x_2 \mathbf{a_2} + \cdots + x_m \mathbf{a_m}$$

is again a vector in $\mathbf{R}^n$. We call it a **linear combination** of the $\mathbf{a_i}$'s. We say that x_i is the **coefficient** of $\mathbf{a_i}$ in the linear combination. $\square$

EXAMPLE 1.9. Suppose $\mathbf{a_1} = (1, -1, 0)$, $\mathbf{a_2} = (0, 1, -1)$ and $\mathbf{a_3} = (1, 0, -1)$. If we multiply these by the scalar coefficients $x_1 = 2$, $x_2 = -3$, and $x_3 = 4$, respectively and then add, we get the linear combination

$$\begin{aligned} 2\mathbf{a_1} - 3\mathbf{a_2} + 4\mathbf{a_3} &= 2\begin{pmatrix} 1 \\ -1 \\ 0 \end{pmatrix} - 3\begin{pmatrix} 0 \\ 1 \\ -1 \end{pmatrix} + 4\begin{pmatrix} 1 \\ 0 \\ -1 \end{pmatrix} \\ &= \begin{pmatrix} 2 - 0 + 4 \\ -2 - 3 + 0 \\ 0 + 3 - 4 \end{pmatrix} \\ &= \begin{pmatrix} 6 \\ -5 \\ -1 \end{pmatrix} \end{aligned}$$

$\square$

Ultimately, many (perhaps most!) problems in linear algebra reduce to that of finding coefficients that linearly combine several given vectors to make a specified target vector. Here's an example. Because it involves just two vectors in $\mathbf{R}^2$, we can solve it by elementary methods.

EXAMPLE 1.10. Does some linear combination of $(2, 1)$ and $(-1, 2)$ add up to $(8, -1)$?

This is equivalent to asking if we can find coefficients x and y such that

$$x\begin{pmatrix} 2 \\ 1 \end{pmatrix} + y\begin{pmatrix} -1 \\ 2 \end{pmatrix} = \begin{pmatrix} 8 \\ -1 \end{pmatrix}$$

After performing the indicated scalar multiplications and vector addition, this becomes

$$\begin{pmatrix} 2x - y \\ x + 2y \end{pmatrix} = \begin{pmatrix} 8 \\ -1 \end{pmatrix}$$

Solving this for x and y is now clearly the same as simultaneously solving

$$
\begin{aligned}
2x - y &= 8 \\
x + 2y &= 1
\end{aligned}
$$

To do so, we can multiply the second equation by 2 and subtract it from the first to get

$$-3y = 6 \quad \text{hence} \quad y = -2$$

Setting $y = -2$ now reduces the first equation to $2x + 2 = 8$, so $x = 3$. This solves our problem: With $x = 3$ and $y = -2$, we get a linear combination of the given vectors that adds up to $(8, -1)$:

$$3 \begin{pmatrix} 2 \\ 1 \end{pmatrix} - 2 \begin{pmatrix} -1 \\ 2 \end{pmatrix} = \begin{pmatrix} 8 \\ -1 \end{pmatrix}$$

$\square$

We end our introductory discussion of linear combination by introducing the *standard basis vectors* of $\mathbf{R}^n$. They play key roles later on.

DEFINITION 1.11. The **standard basis vectors** in $\mathbf{R}^n$ are the n numeric vectors

$$
\begin{aligned}
\mathbf{e}_1 &= (1,\ 0,\ 0,\ \ldots,\ 0,\ 0) \\
\mathbf{e}_2 &= (0,\ 1,\ 0,\ \ldots,\ 0,\ 0) \\
\mathbf{e}_3 &= (0,\ 0,\ 1,\ \ldots,\ 0,\ 0) \\
&\ \ \vdots \quad \vdots \qquad \vdots \\
\mathbf{e}_n &= (0,\ 0,\ 0,\ \ldots,\ 0,\ 1)
\end{aligned}
$$

$\square$

Simple as they are, these vectors are central to our subject. We introduce them here partly because problems like Example 1.10 and Exercises 6 and 7 become *trivial* when we're combining standard basis vectors, thanks to the following:

OBSERVATION 1.12. *We can express any numeric vector*

$$\mathbf{x} = (x_1, x_2, \ldots, x_n)$$

as a linear combination of standard basis vectors in an obvious way:

$$\mathbf{x} = x_1\,\mathbf{e}_1 + x_2\,\mathbf{e}_2 + x_3\,\mathbf{e}_3 + \cdots + x_n\,\mathbf{e}_n$$

PROOF. Since $x_1\,\mathbf{e}_1 = (x_1, 0, 0, \ldots, 0)$, $x_2\,\mathbf{e}_2 = (0, x_2, 0, \ldots, 0)$ and so forth, the identity is easy to verify. $\square$

1.13. Matrices. One of the most fundamental insights in linear algebra is simply this: *We can view any linear combination as the result of multiplying a vector by a* matrix:

DEFINITION 1.14 (Matrix). An $n \times m$ **matrix** is a rectangular array of scalars, with n horizontal **rows** (each in $\mathbf{R}^m$), and m vertical **columns** (each in $\mathbf{R}^n$). For instance:

$$\mathbf{A} = \begin{bmatrix} 1 & 0 & -2.5 \\ \pi & 4 & 1/2 \end{bmatrix} \qquad \mathbf{B} = \begin{bmatrix} 0 & 1 \\ 2 & 0 \\ 3 & -3 \end{bmatrix}$$

Here $\mathbf{A}$ has 2 rows and 3 columns, while $\mathbf{B}$ has 3 rows, 2 columns.

We generally label matrices with bold uppercase letters, as with $\mathbf{A}$ and $\mathbf{B}$ above. We *double-subscript* the corresponding *lowercase* letter to address the **entries**—the individual scalars—in the matrix. So if we call a matrix $\mathbf{X}$, then x_{34} names the entry in row 3 and column 4 of $\mathbf{X}$.

With regard to $\mathbf{A}$ and $\mathbf{B}$ above, for example, we have

$$a_{21} = \pi, \quad a_{12} = 0, \quad a_{13} = -2.5, \quad \text{and} \quad b_{11} = b_{22} = 0.$$

We sometimes label a matrix $\mathbf{X}$ by $[x_{ij}]$ or write $\mathbf{X} = [x_{ij}]$ to emphasize that the entries of $\mathbf{X}$ will be called x_{ij}.

Finally, if we want to clarify that a matrix $\mathbf{C}$ has, say, 4 rows and 5 columns, we can call it $\mathbf{C}_{4\times 5}$. Just as with entries, the first index refers to rows, while the second refers to columns. $\qquad \square$

1.15. Matrix addition and scalar multiplication. Matrices, like numeric vectors, can be scalar multiplied: When k is a scalar and $\mathbf{A}$ is a matrix, we simply multiply each entry in $\mathbf{A}$ by k to get $k\mathbf{A}$.

EXAMPLE 1.16. Suppose

$$\mathbf{A} = \begin{bmatrix} 1 & 2 & 3 & 4 \\ -4 & -3 & -2 & -1 \end{bmatrix} \quad \text{and} \quad \mathbf{B} = \begin{bmatrix} 1 & -1 & 0 \\ 0 & 1 & -1 \\ -1 & 0 & 1 \end{bmatrix}$$

Then

$$\pi\mathbf{A} = \begin{bmatrix} \pi & 2\pi & 3\pi & 4\pi \\ -4\pi & -3\pi & -2\pi & -\pi \end{bmatrix} \quad \text{while} \quad 5\mathbf{B} = \begin{bmatrix} 5 & -5 & 0 \\ 0 & 5 & -5 \\ -5 & 0 & 5 \end{bmatrix}$$

$\square$

Similarly, matrices *of the same size* can be added together. Again, just as with numeric vectors, we do this entry-by-entry:

EXAMPLE 1.17. If

$$\mathbf{A} = \begin{bmatrix} 1 & 0 & 0 \\ 0 & 2 & 0 \\ 0 & 0 & 3 \end{bmatrix} \quad \text{and} \quad \mathbf{B} = \begin{bmatrix} 0 & 0 & 1 \\ 0 & 2 & 0 \\ 3 & 0 & 0 \end{bmatrix}$$

then

$$\mathbf{A} + \mathbf{A} = \begin{bmatrix} 2 & 0 & 0 \\ 0 & 4 & 0 \\ 0 & 0 & 6 \end{bmatrix} \quad \text{while} \quad \mathbf{A} + \mathbf{B} = \begin{bmatrix} 1 & 0 & 1 \\ 0 & 4 & 0 \\ 3 & 0 & 3 \end{bmatrix}$$

□

1.18. Matrix/vector products. The matrix/vector product we describe next is an operation much richer than either matrix addition or scalar multiplication. In particular, the matrix/vector product gives us a new and useful way to handle linear combination. The rule is very simple:

We can express any linear combination

$$x_1 \mathbf{v}_1 + x_2 \mathbf{v}_2 + \cdots + x_m \mathbf{v}_m$$

as a matrix/vector product, as follows:

Write the vectors $\mathbf{v}_i$ as the columns of a matrix $\mathbf{A}$, and stack the coefficients x_i up as a vector $\mathbf{x}$. The given linear combination then agrees with the product $\mathbf{A}\mathbf{x}$.

EXAMPLE 1.19. To write the linear combination

$$x \begin{pmatrix} 7 \\ -3 \end{pmatrix} + y \begin{pmatrix} -5 \\ 2 \end{pmatrix} + z \begin{pmatrix} 1 \\ -4 \end{pmatrix}$$

as a matrix/vector product, we then take the *vectors* in the linear combination, namely

$$\begin{pmatrix} 7 \\ -3 \end{pmatrix}, \begin{pmatrix} -5 \\ 2 \end{pmatrix}, \quad \text{and} \quad \begin{pmatrix} 1 \\ -4 \end{pmatrix}$$

and line them up as columns in a matrix

$$\mathbf{A} = \begin{bmatrix} 7 & -5 & 1 \\ -3 & 2 & -4 \end{bmatrix}$$

We then stack the coefficients x, y, and z up as the vector

$$\mathbf{x} = \begin{pmatrix} x \\ y \\ z \end{pmatrix}$$

In short, we can now write the original linear combination, which was

$$x \begin{pmatrix} 7 \\ -3 \end{pmatrix} + y \begin{pmatrix} -5 \\ 2 \end{pmatrix} + z \begin{pmatrix} 1 \\ -4 \end{pmatrix}$$

as the matrix/vector product

$$\mathbf{Ax} = \begin{bmatrix} 7 & -5 & 1 \\ -3 & 2 & -4 \end{bmatrix} \begin{pmatrix} x \\ y \\ z \end{pmatrix}$$

Note that the coefficient vector $\mathbf{x} = (x, y, z)$ here lies in $\mathbf{R}^3$, while $\mathbf{Ax}$ lies in $\mathbf{R}^2$. Indeed, if we actually compute it, we get

$$\mathbf{Ax} = \begin{pmatrix} 7x - 5y + z \\ -3x + 2y - 4z \end{pmatrix} \in \mathbf{R}^2$$

$\square$

With this example in mind, we carefully state the general rule:

DEFINITION 1.20 (Matrix/vector multiplication). If a matrix $\mathbf{A}$ has n rows and m columns, we can multiply it by any vector $\mathbf{x} \in \mathbf{R}^m$ to produce a result $\mathbf{Ax}$ in $\mathbf{R}^n$.

To compute it, we linearly combine the columns of $\mathbf{A}$ (each a vector in $\mathbf{R}^n$), using the entries of $\mathbf{x} = (x_1, x_2, \ldots, x_m)$ as coefficients:

$$\mathbf{Ax} := x_1 \, \mathbf{c}_1(\mathbf{A}) + x_2 \, \mathbf{c}_2(\mathbf{A}) + \cdots + x_m \, \mathbf{c}_m(\mathbf{A})$$

where $\mathbf{c}_j(\mathbf{A})$ signifies column j of $\mathbf{A}$.

Conversely, any linear combination

$$x_1 \, \mathbf{v}_1 + x_2 \mathbf{v}_2 + \cdots + x_m \mathbf{v}_m$$

can be written as the product $\mathbf{Ax}$, where $\mathbf{A}$ is the matrix with columns $\mathbf{v}_1, \mathbf{v}_2, \ldots, \mathbf{v}_m$ (in that order) and $\mathbf{x} = (x_1, x_2, \ldots, x_m)$. Symbolically,

$$\mathbf{A} = \begin{bmatrix} \mathbf{v}_1 & \mathbf{v}_2 & \cdots & \mathbf{v}_m \\ | & | & \cdots & | \end{bmatrix}, \qquad \mathbf{x} = (x_1, x_2, \ldots, x_m)$$

and then

$$\mathbf{Ax} = x_1 \, \mathbf{v}_1 + x_2 \mathbf{v}_2 + \cdots + x_m \mathbf{v}_m$$

$\square$

REMARK 1.21 (Warning!). We can **only** multiply $\mathbf{A}$ by $\mathbf{x}$ when *the number of columns in* $\mathbf{A}$ *equals the number of entries in* $\mathbf{x}$. When the vector $\mathbf{x}$ lies in $\mathbf{R}^m$, the matrix $\mathbf{A}$ must have exactly m columns.

On the other hand, $\mathbf{A}$ can have any number n of rows. The product $\mathbf{Ax}$ will then lie in $\mathbf{R}^n$.

REMARK 1.22. It is useful to conceptualize matrix/vector multiplication via the following mnemonic "mantra":

$$Matrix/vector\ multiplication\ =\ Linear\ combination$$

Commit this phrase to memory—we will have many opportunities to invoke it. □

EXAMPLE 1.23. If

$$\mathbf{A} = \begin{bmatrix} 1 & 2 \\ 3 & 4 \\ -4 & -3 \\ -1 & -2 \end{bmatrix} \quad \text{and} \quad \mathbf{x} = (-1, 5)$$

then

$$\mathbf{Ax} = -1 \begin{pmatrix} 1 \\ 3 \\ -4 \\ -1 \end{pmatrix} + 5 \begin{pmatrix} 2 \\ 4 \\ -3 \\ -2 \end{pmatrix} = \begin{pmatrix} -1 + 10 \\ -3 + 20 \\ 4 - 15 \\ 1 - 10 \end{pmatrix} = \begin{pmatrix} 9 \\ 17 \\ -11 \\ -9 \end{pmatrix}$$

More generally, if $\mathbf{x} = (x, y)$, then

$$\mathbf{Ax} = x \begin{pmatrix} 1 \\ 3 \\ -4 \\ -1 \end{pmatrix} + y \begin{pmatrix} 2 \\ 4 \\ -3 \\ -2 \end{pmatrix} = \begin{pmatrix} x + 2y \\ 3x + 4y \\ -4x - 3y \\ -x - 2y \end{pmatrix}$$

Note how dramatically we abbreviate the expression on the right above when we write it as simply $\mathbf{Ax}$. □

1.24. Properties of matrix/vector multiplication. To continue our discussion of matrix/vector multiplication we record two crucial properties:

PROPOSITION 1.25. *Matrix/vector multiplication commutes with scalar multiplication, and distributes over vector addition. More precisely, if* $\mathbf{A}$ *is any* $n \times m$ *matrix, the following two facts always hold:*

i) *If k is any scalar and $\mathbf{x} \in \mathbf{R}^m$, then*

$$\mathbf{A}(k\mathbf{x}) = k(\mathbf{A}\mathbf{x}) = (k\mathbf{A})\mathbf{x}.$$

ii) *For any two vectors $\mathbf{x}, \mathbf{y} \in \mathbf{R}^m$, we have*

$$\mathbf{A}(\mathbf{x} + \mathbf{y}) = \mathbf{A}\mathbf{x} + \mathbf{A}\mathbf{y}.$$

PROOF. For simplicity here, we denote the columns of $\mathbf{A}$, respectively by $\mathbf{a}_1, \mathbf{a}_2, \ldots, \mathbf{a}_m$. We then prove (i) and (ii) in the usual way: we simplify each side of the equation separately and show that they agree.

Start with the first equality in (i). Expanding $\mathbf{x}$ as $\mathbf{x} = (x_1, x_2, \ldots, x_m)$ we know that $k\mathbf{x} = k(x_1, x_2, \ldots, x_m) = (kx_1, kx_2, \ldots, kx_m)$. The definition of matrix/vector multiplication (Definition 1.20) then gives

$$\mathbf{A}(k\mathbf{x}) = kx_1\mathbf{a}_1 + kx_2\mathbf{a}_2 + \cdots + kx_m\mathbf{a}_m$$

Similarly, we can rewrite the middle expression in (i) as

$$\begin{aligned}
k(\mathbf{A}\mathbf{x}) &= k(x_1\mathbf{a}_1 + x_2\mathbf{a}_2 + \cdots + x_m\mathbf{a}_m) \\
&= kx_1\mathbf{a}_1 + kx_2\mathbf{a}_2 + \cdots + kx_m\mathbf{a}_m
\end{aligned}$$

because scalar multiplication distributes over vector addition (Proposition 1.6). This expression matches exactly with what we got before. Since $\mathbf{A}$, k, and $\mathbf{x}$ were completely arbitrary, this proves the first equality in (i). We leave the reader to expand out $(k\mathbf{A})\mathbf{x}$ and show that it takes the same form.

A similar left/right comparison confirms (ii). Given arbitrary vectors $\mathbf{x} = (x_1, x_2, \ldots, x_m)$ and $\mathbf{y} = (y_1, y_2, \ldots, y_m)$ in $\mathbf{R}^m$, we have

$$\mathbf{x} + \mathbf{y} = (x_1 + y_1, x_2 + y_2, \cdots, x_m + y_m)$$

and hence

$$\begin{aligned}
\mathbf{A}(\mathbf{x} + \mathbf{y}) &= (x_1 + y_1)\mathbf{a}_1 + (x_2 + y_2)\mathbf{a}_2 + \cdots + (x_m + y_m)\mathbf{a}_m \\
&= x_1\mathbf{a}_1 + y_1\mathbf{a}_1 + x_2\mathbf{a}_2 + y_2\mathbf{a}_2 + \cdots + x_m\mathbf{a}_m + y_m\mathbf{a}_m
\end{aligned}$$

by the definition of matrix/vector multiplication, and the distributive property (Proposition 1.6). When we simplify the right side of (ii), namely $\mathbf{A}\mathbf{x} + \mathbf{A}\mathbf{y}$, we get the same thing. (The summands come in a different order, but that's allowed, since vector addition is commutative, by Proposition 1.4). We leave this to the reader. □

1.26. The dot product. As we have noted, the matrix/vector product $\mathbf{Ax}$ makes sense only when the number of columns in $\mathbf{A}$ matches the number of entries in $\mathbf{x}$.

The number of *rows* in $\mathbf{A}$ will then match the number of entries in $\mathbf{Ax}$. So any number of rows is permissible—even just one.

In that case $\mathbf{Ax} \in \mathbf{R}^1 = \mathbf{R}$. So when $\mathbf{A}$ has just one row, $\mathbf{Ax}$ reduces to a single scalar.

EXAMPLE 1.27. Suppose we have

$$\mathbf{A} = \begin{bmatrix} -4 & 1 & 3 & -2 \end{bmatrix} \quad \text{and} \quad \mathbf{x} = \begin{pmatrix} 1 \\ 1 \\ -1 \\ -1 \end{pmatrix}$$

Then

$$\mathbf{Ax} = \begin{bmatrix} -4 & 1 & 3 & -2 \end{bmatrix} \begin{pmatrix} 1 \\ 1 \\ -1 \\ -1 \end{pmatrix}$$

$$= 1(-4) + 1(1) - 1(3) - 1(-2)$$
$$= -3$$

$\square$

Note, however, that a $1 \times m$ matrix corresponds in an obvious way to a vector in $\mathbf{R}^m$. Seen in that light, matrix/vector multiplication provides a way to multiply two *vectors* $\mathbf{a}$ and $\mathbf{x}$ in $\mathbf{R}^m$: we just regard the first vector $\mathbf{a}$ as a $1 \times m$ matrix, and multiply it by $\mathbf{x}$ using matrix/vector multiplication. As noted above, this produces a scalar result.

Multiplying two vectors in $\mathbf{R}^m$ this way—by regarding the first vector as a $1 \times m$ matrix—is therefore sometimes called a **scalar product**. We simply call it the **dot product** since we indicate it with a dot.

DEFINITION 1.28 (Dot product). Given any two vectors

$$\mathbf{u} = (u_1, u_2, \ldots, u_m) \quad \text{and} \quad \mathbf{v} = (v_1, v_2, \ldots, v_m)$$

in $\mathbf{R}^m$, we define the **dot product** $\mathbf{u} \cdot \mathbf{v}$ via

$$(1) \qquad \mathbf{u} \cdot \mathbf{v} := u_1 v_1 + u_2 v_2 + \cdots u_m v_m$$

bearing in mind that this is exactly what we get if we regard $\mathbf{u}$ as a $1 \times m$ matrix and multiply it by $\mathbf{v}$.

Effectively, however, this simply has us multiply the two vectors entry-by-entry, and then sum up the results.   □

EXAMPLE 1.29. In $\mathbf{R}^2$,

$$\begin{pmatrix} 2 \\ -1 \end{pmatrix} \cdot \begin{pmatrix} 3 \\ 2 \end{pmatrix} = 2 \cdot 3 + (-1) \cdot 2 = 6 - 2 = 4$$

while in $\mathbf{R}^4$,

$$\begin{pmatrix} 2 \\ -1 \\ 0 \\ -1 \end{pmatrix} \cdot \begin{pmatrix} 3 \\ 2 \\ -1 \\ 1 \end{pmatrix} = 2 \cdot 3 + (-1) \cdot 2 + 0 \cdot (-1) + (-1) \cdot 1 = 3$$

□

PROPOSITION 1.30. *The dot product is commutative. It also commutes with scalar multiplication and distributes over vector addition. Thus, for any vectors* $\mathbf{u}, \mathbf{v}, \mathbf{w} \in \mathbf{R}^n$, *we have*

$$\mathbf{v} \cdot \mathbf{w} = \mathbf{w} \cdot \mathbf{v}$$

$$\mathbf{u} \cdot (k\mathbf{v}) = k(\mathbf{u} \cdot \mathbf{v}) = (k\mathbf{u}) \cdot \mathbf{v}$$

$$\mathbf{u} \cdot (\mathbf{v} + \mathbf{w}) = \mathbf{u} \cdot \mathbf{v} + \mathbf{u} \cdot \mathbf{w}$$

PROOF. We leave the proof of the first identity to the reader (Exercise 18). The last two identities follow straight from the matrix identities in Proposition 1.25, since the dot product can be seen as the "$(1 \times n)$ times $(n \times 1)$" case of matrix/vector multiplication.   □

1.31. Fast matrix/vector multiplication via dot product. We have seen that the dot product (Definition 1.28) corresponds to matrix/vector multiplication with a one-rowed matrix. We now turn this around to see that the dot product gives an efficient way to compute matrix/vector products—without forming linear combinations.

To see how, take any matrix $\mathbf{A}$ and vector $\mathbf{v}$, like these:

$$\mathbf{A} = \begin{bmatrix} a_{11} & a_{12} & a_{13} & \cdots & a_{1m} \\ a_{21} & a_{22} & a_{23} & \cdots & a_{2m} \\ \vdots & \vdots & \vdots & \cdots & \vdots \\ a_{n1} & a_{n2} & a_{n3} & \cdots & a_{nm} \end{bmatrix}, \quad \text{and} \quad \mathbf{v} = \begin{pmatrix} v_1 \\ v_2 \\ \vdots \\ v_m \end{pmatrix}$$

By the definition of matrix/vector multiplication (as a linear combination) we get

$$\mathbf{A}\mathbf{v} = v_1 \begin{pmatrix} a_{11} \\ a_{21} \\ \vdots \\ a_{n1} \end{pmatrix} + v_2 \begin{pmatrix} a_{12} \\ a_{22} \\ \vdots \\ a_{n2} \end{pmatrix} + \cdots + v_m \begin{pmatrix} a_{1m} \\ a_{2m} \\ \vdots \\ a_{nm} \end{pmatrix}$$

Now carry out the scalar multiplications and vector additions to rewrite as a single vector:

$$\mathbf{A}\mathbf{v} = \begin{pmatrix} v_1 a_{11} + v_2 a_{12} + \cdots + v_m a_{1m} \\ v_1 a_{21} + v_2 a_{22} + \cdots + v_m a_{2m} \\ \vdots \quad \vdots \quad \vdots \\ v_1 a_{n1} + v_2 a_{n2} + \cdots + v_m a_{nm} \end{pmatrix}$$

Each entry is now a dot product! The first entry dots $\mathbf{v}$ with the first row of $\mathbf{A}$, the second entry dots $\mathbf{v}$ with the second row of $\mathbf{A}$, and so forth. In other words, we have:

OBSERVATION 1.32 (Dot-product formula for matrix/vector multiplication). *We can compute the product of any $n \times m$ matrix $\mathbf{A}$ with any vector $\mathbf{v} = (v_1, v_2, \ldots, v_m) \in \mathbf{R}^m$ as a vector of dot products:*

$$\mathbf{A}\mathbf{v} = \begin{pmatrix} \mathbf{r}_1(\mathbf{A}) \cdot \mathbf{v} \\ \mathbf{r}_2(\mathbf{A}) \cdot \mathbf{v} \\ \mathbf{r}_3(\mathbf{A}) \cdot \mathbf{v} \\ \vdots \\ \mathbf{r}_n(\mathbf{A}) \cdot \mathbf{v} \end{pmatrix}$$

where $\mathbf{r}_i(\mathbf{A})$ denotes row i of $\mathbf{A}$.

EXAMPLE 1.33. Given

$$\mathbf{A} = \begin{bmatrix} 2 & -1 & 3 \\ 1 & 4 & -5 \end{bmatrix} \quad \text{and} \quad \mathbf{v} = \begin{pmatrix} 3 \\ -2 \\ -7 \end{pmatrix}$$

we compute $\mathbf{A}\mathbf{v}$ using dot products as follows:

$$\mathbf{A}\mathbf{v} = \begin{pmatrix} \mathbf{r}_1(\mathbf{A}) \cdot \mathbf{v} \\ \mathbf{r}_2(\mathbf{A}) \cdot \mathbf{v} \\ \mathbf{r}_3(\mathbf{A}) \cdot \mathbf{v} \\ \vdots \\ \mathbf{r}_n(\mathbf{A}) \cdot \mathbf{v} \end{pmatrix} = \begin{pmatrix} (2, -1, 3) \cdot (3, -2, -7) \\ (1, 4, -5) \cdot (3, -2, -7) \end{pmatrix} = \begin{pmatrix} -13 \\ 30 \end{pmatrix}$$

The reader will easily check that this against our definition of $\mathbf{Av}$, namely

$$3\begin{pmatrix} 2 \\ 1 \end{pmatrix} - 2\begin{pmatrix} -1 \\ 4 \end{pmatrix} - 7\begin{pmatrix} 3 \\ -5 \end{pmatrix}$$

□

EXAMPLE 1.34. Similarly, given

$$\mathbf{A} = \begin{bmatrix} 3 & -1 \\ 2 & 2 \\ -1 & 3 \end{bmatrix} \quad \text{and} \quad \mathbf{v} = \begin{pmatrix} 7 \\ -5 \end{pmatrix}$$

the dot-product approach gives

$$\mathbf{Av} = \begin{pmatrix} (3, -1) \cdot (7, -5) \\ (2, 2) \cdot (7, -5) \\ (-1, 3) \cdot (7, -5) \end{pmatrix} = \begin{pmatrix} 26 \\ 4 \\ -22 \end{pmatrix}$$

□

1.35. Eigenvectors. Among matrices, *square* matrices—matrices having the same number of rows and columns—are particularly interesting and important. One reason for their importance is this:

When we multiply a vector $\mathbf{x} \in \mathbf{R}^m$ *by a* ***square*** *matrix* $\mathbf{A}_{m \times m}$, *the product* $\mathbf{Ax}$ *lies in the* ***same*** *space as* $\mathbf{x}$ *itself:* $\mathbf{R}^m$.

This fact makes possible a phenomenon that unlocks some of the deepest ideas in linear algebra: *The product* $\mathbf{Ax}$ *may actually be a scalar multiple of the original vector* $\mathbf{x}$. That is, there may be certain "lucky" vectors $\mathbf{x} \in \mathbf{R}^m$ for which $\mathbf{Ax} = \lambda\mathbf{x}$, where λ (the Greek letter *lambda*) is some scalar.

DEFINITION 1.36 (Eigenvalues and eigenvectors). If $\mathbf{A}$ is an $m \times m$ matrix, and there exists a vector $\mathbf{x} \neq \mathbf{0}$ in $\mathbf{R}^m$ such that $\mathbf{Ax} = \lambda\mathbf{x}$ for some scalar $\lambda \in \mathbf{R}$, we call $\mathbf{x}$ an **eigenvector** of $\mathbf{A}$, and we call λ its **eigenvalue**.   □

EXAMPLE 1.37. The vectors $(1, 1)$ and $(-3, 3)$ in $\mathbf{R}^2$ are eigenvectors of the matrix

$$\mathbf{A} = \begin{bmatrix} 1 & 2 \\ 2 & 1 \end{bmatrix}$$

but the vector $(2, 1)$ is *not* an eigenvector. To verify these statements, we just multiply each vector by $\mathbf{A}$ and see whether the product is a scalar multiple of **A** or not. This is easy to verify using the dot-product method of matrix/vector multiplication:

$$\begin{bmatrix} 1 & 2 \\ 2 & 1 \end{bmatrix} \begin{pmatrix} 1 \\ 1 \end{pmatrix} = \begin{pmatrix} 1+2 \\ 2+1 \end{pmatrix} = \begin{pmatrix} 3 \\ 3 \end{pmatrix} = 3 \begin{pmatrix} 1 \\ 1 \end{pmatrix}$$

Thus, when $\mathbf{x} = (1,1)$, we have $\mathbf{Ax} = 3\mathbf{x}$. This makes $\mathbf{x}$ an eigenvector of **A** with eigenvalue $\lambda = 3$.

Similarly, when $\mathbf{x} = (-3, 3)$, we have

$$\begin{bmatrix} 1 & 2 \\ 2 & 1 \end{bmatrix} \begin{pmatrix} -3 \\ 3 \end{pmatrix} = \begin{pmatrix} -3+6 \\ -6+3 \end{pmatrix} = \begin{pmatrix} 3 \\ -3 \end{pmatrix} = -1 \begin{pmatrix} -3 \\ 3 \end{pmatrix}$$

Thus, when $\mathbf{x} = (-3, 3)$, we again have $\mathbf{Ax} = -\mathbf{x}$, which makes $\mathbf{x}$ an eigenvector of **A**, this time with eigenvalue $\lambda = -1$.

On the other hand, when we multiply **A** by $\mathbf{x} = (2, 1)$, we get

$$\begin{bmatrix} 1 & 2 \\ 2 & 1 \end{bmatrix} \begin{pmatrix} 2 \\ 1 \end{pmatrix} = \begin{pmatrix} 2+2 \\ 4+1 \end{pmatrix} = \begin{pmatrix} 4 \\ 5 \end{pmatrix}$$

Since $(4, 5)$ is *not* a scalar multiple of $(2, 1)$, it is *not* an eigenvector of **A**. $\qquad\square$

EXAMPLE 1.38. The vector $\mathbf{x} = (2, 3, 0)$ is an eigenvector of the matrix

$$\mathbf{B} = \begin{bmatrix} 1 & 2 & 3 \\ 0 & 4 & 5 \\ 0 & 0 & 6 \end{bmatrix}$$

since (again by the dot-product method of matrix/vector multiplication)

$$\mathbf{Bx} = \begin{bmatrix} 1 & 2 & 3 \\ 0 & 4 & 5 \\ 0 & 0 & 6 \end{bmatrix} \begin{pmatrix} 2 \\ 3 \\ 0 \end{pmatrix} = \begin{pmatrix} 2+6+0 \\ 0+12+0 \\ 0+0+0 \end{pmatrix} = \begin{pmatrix} 8 \\ 12 \\ 0 \end{pmatrix} = 4 \begin{pmatrix} 2 \\ 3 \\ 0 \end{pmatrix}$$

In short, we have $\mathbf{Bx} = 4\mathbf{x}$, and hence $\mathbf{x} = (2, 3, 0)$ is an eigenvector of **B** with eigenvalue $\lambda = 4$.

In an exercise below, we ask the reader to verify that $(1, 0, 0)$ and $(16, 25, 10)$ are also eigenvectors of **B**, and to discover their eigenvalues. Most vectors in $\mathbf{R}^3$, however, are *not* eigenvectors of **B**. For instance, if we multiply **B** by $\mathbf{x} = (1, 2, 1)$, we get $\mathbf{Bx} = (8, 13, 6)$ which is clearly not a scalar multiple of $(1, 2, 1)$. (Scalar multiples of $(1, 2, 1)$ always have the same first and third coordinates.) $\qquad\square$

Eigenvectors and eigenvalues play an truly fundamental role in linear algebra. We won't be prepared to grasp their full importance until Chapter 7, where our explorations all coalesce. We have introduced them here, however, so they can begin to take root in students' minds. We will revisit them off and on throughout the course so that when we reach Chapter 7, they will already be familiar.

– Practice –

1. Find the vector sum and difference $\mathbf{a} \pm \mathbf{b}$, if

    a) $\mathbf{a} = (2, -3, 1)$ and $\mathbf{b} = (0, 0, 0)$

    b) $\mathbf{a} = (1, -2, 0)$ and $\mathbf{b} = (0, 1, -2)$

    c) $\mathbf{a} = (1, 1, 1, 1)$ and $\mathbf{b} = (1, 1, -1, -1)$

2. Guided by the proof of associativity for Proposition 1.4, prove that Proposition's claim that vector addition is also *commutative*.

3. Compute

    a) $5 \, (0, 1, 2, 1, 0)$

    b) $-1 \, (2, -2)$

    c) $\frac{1}{10} \, (10, 25, 40)$

4. Rework the proof of Proposition 1.6 for the case of three vectors $\mathbf{x}$, $\mathbf{y}$, and $\mathbf{z}$ instead of just two vectors $\mathbf{x}$ and $\mathbf{y}$.

5. Compute these additional linear combinations of the vectors $\mathbf{a}_1$, $\mathbf{a}_2$, and $\mathbf{a}_3$ in Example 1.9.

    a) $\mathbf{a}_1 + 2\mathbf{a}_2 + \mathbf{a}_3$ b) $-2\mathbf{a}_1 + \mathbf{a}_2 - 2\mathbf{a}_3$ c) $x \, \mathbf{a}_1 + y \, \mathbf{a}_2 + z \, \mathbf{a}_3$

(In part (c), treat $x, y,$ and z as unevaluated scalars, and leave them that way in your answer.)

6. Find a linear combination of the vectors $\mathbf{v} = (1, 2, 3)$ and $\mathbf{w} = (-2, 3, -1)$ in $\mathbf{R}^3$ that adds up to $(8, -5, 9)$.

7. Find 3 *different* linear combinations of $\mathbf{a} = (1, -2)$, $\mathbf{b} = (2, 3)$, and $\mathbf{c} = (3, -1)$ that add up to $(0, 0)$ in $\mathbf{R}^2$.

8. Without setting the scalars x and y *both* equal to zero, find a linear combination $x(1,1)+y(1,-1)$ that adds up to $(0,0) \in \mathbf{R}^2$, or explain why this cannot be done.

9. Express each vector below as a linear combination of the standard basis vectors:

    a) $(1,2,-1)$

    b) $(1,-1,-1,1)$

    c) $(0,3,0,-4,0)$

10. Write each linear combination below as a matrix/ vector product **Ax**.

    a) $\quad 2\begin{pmatrix} -1 \\ 1 \end{pmatrix} - 3\begin{pmatrix} 1 \\ -1 \end{pmatrix}$

    b) $\quad \frac{1}{2}\begin{pmatrix} 1 \\ 1 \\ 0 \end{pmatrix} + 0.9\begin{pmatrix} 0 \\ 1 \\ 1 \end{pmatrix} + \pi\begin{pmatrix} 1 \\ 0 \\ 1 \end{pmatrix}$

    c) $\quad x_1\begin{pmatrix} 1 \\ 0 \end{pmatrix} - x_2\begin{pmatrix} 0 \\ 1 \end{pmatrix} + x_3\begin{pmatrix} 1 \\ 3 \end{pmatrix} - x_4\begin{pmatrix} 2 \\ 4 \end{pmatrix}$

    d) $\quad z\begin{pmatrix} 1 \\ 2 \\ 3 \\ 4 \end{pmatrix} - w\begin{pmatrix} 4 \\ 3 \\ 2 \\ 1 \end{pmatrix}$

11. Expand each matrix/vector product below as a linear combination, then simplify as far as possible, writing each product as a single vector.

    a)

$$\begin{bmatrix} 1 & -2 & 3 \\ -4 & 5 & -6 \end{bmatrix}\begin{pmatrix} 1 \\ 1 \\ 1 \end{pmatrix}$$

    b)

$$\begin{bmatrix} 1 & 0 & 0 \\ 0 & 1 & 0 \\ 0 & 0 & 1 \end{bmatrix}\begin{pmatrix} a \\ b \\ c \end{pmatrix}$$

c)

$$\begin{bmatrix} 1 & 2 & 0 \\ 2 & 1 & 2 \\ 0 & 2 & 1 \\ 1 & 1 & 1 \end{bmatrix} \begin{pmatrix} x \\ -1 \\ \pi \end{pmatrix}$$

d)

$$\begin{bmatrix} 1 & 2 & 3 & 4 & 5 \\ 2 & 3 & 4 & 5 & 1 \\ 3 & 4 & 5 & 1 & 2 \end{bmatrix} \begin{pmatrix} 1 \\ -2 \\ 4 \\ -2 \\ 1 \end{pmatrix}$$

12. Complete the proof of Proposition 1.25 by showing that for any $\mathbf{x} = (x_1, x_2, \ldots, x_m)$ and $\mathbf{y} = (y_1, y_2, \ldots, y_m)$ in $\mathbf{R}^m$, we have

$$\mathbf{Ax} + \mathbf{Ay} = x_1\mathbf{a}_1 + y_1\mathbf{a}_1 + x_2\mathbf{a}_2 + y_2\mathbf{a}_2 + \cdots + x_m\mathbf{a}_m + y_m\mathbf{a}_m \,,$$

the same result we got there for $\mathbf{A}(\mathbf{x} + \mathbf{y})$.

13. Compute each matrix/vector product below using dot products, as in Examples 1.33 and 1.34 above.

a)

$$\begin{bmatrix} 1 & -2 & 3 \\ -4 & 5 & -6 \end{bmatrix} \begin{pmatrix} 1 \\ 1 \\ 1 \end{pmatrix}$$

b)

$$\begin{bmatrix} 1 & 0 & 0 \\ 0 & 1 & 0 \\ 0 & 0 & 1 \end{bmatrix} \begin{pmatrix} a \\ b \\ c \end{pmatrix}$$

c)

$$\begin{bmatrix} 1 & 2 & 0 \\ 2 & 1 & 2 \\ 0 & 2 & 1 \\ 1 & 1 & 1 \end{bmatrix} \begin{pmatrix} x \\ -1 \\ \pi \end{pmatrix}$$

d)

$$\begin{bmatrix} 1 & 2 & 3 & 4 & 5 \\ 2 & 3 & 4 & 5 & 1 \\ 3 & 4 & 5 & 1 & 2 \end{bmatrix} \begin{pmatrix} 1 \\ -2 \\ 4 \\ -2 \\ 1 \end{pmatrix}$$

14. Show that $(1, 0, 0)$ and $(16, 25, 10)$ are both eigenvectors of the matrix $\mathbf{B}$ of Example 1.38. What are the corresponding eigenvalues? Is $(0, 2, 3)$ an eigenvector? How about $(0, -3, 2)$?

15. A 3-by-3 diagonal matrix is a matrix of the form

$$\begin{bmatrix} a & 0 & 0 \\ 0 & b & 0 \\ 0 & 0 & c \end{bmatrix}$$

where $a, b,$ and c are any (fixed) scalars. Show that the standard basis vectors $\mathbf{e}_1, \mathbf{e}_2, \mathbf{e}_3 \in \mathbf{R}^3$ are always eigenvectors of a diagonal matrix. What are the corresponding eigenvalues? Do the analogous statements hold for 2×2 diagonal matrices? How about $n \times n$ diagonal matrices?

16. Consider the matrices

$$\mathbf{Y} = \begin{bmatrix} 2 & -1 & 0 & 3 \\ 0 & -2 & -3 & 1 \end{bmatrix} \qquad \mathbf{Z} = \begin{bmatrix} 1 & z & y \\ 0 & 1 & x \\ 0 & 0 & 1 \end{bmatrix}$$

   a) How many rows and columns does each matrix have?

   b) What are y_{21}, y_{14}, and y_{23}? Why is there no y_{32}?

   c) What are z_{11}, z_{22}, and z_{33}? What is z_{13}? z_{31}?

17. Compute the dot product $\mathbf{x} \cdot \mathbf{y}$ for:

   a) $\mathbf{x} = (1, 2, 3)$, $\mathbf{y} = (4, -5, 6)$

   b) $\mathbf{x} = (-1, 1, -1, 1)$, $\mathbf{y} = (2, 2, 2, 2)$

   c) $\mathbf{x} = (\pi, \pi)$, $\mathbf{y} = (\frac{1}{4}, \frac{3}{4})$

18. Prove commutativity of the dot product (i.e., the first identity in Proposition 1.30 of the text).

19. Prove the third identity of Proposition 1.30 (the distributive law) in $\mathbf{R}^2$ and $\mathbf{R}^4$ directly:

   a) In $\mathbf{R}^2$, consider arbitrary vectors $\mathbf{u} = (u_1, u_2)$, $\mathbf{v} = (v_1, v_2)$ and $\mathbf{w} = (w_1, w_2)$, and expand out both

$$\mathbf{u} \cdot (\mathbf{v} + \mathbf{w}) \quad \text{and} \quad \mathbf{u} \cdot \mathbf{v} + \mathbf{u} \cdot \mathbf{w}$$

   to show that they are equal.

   b) In $\mathbf{R}^4$, carry out the same argument for vectors $\mathbf{u}, \mathbf{v}, \mathbf{w} \in \mathbf{R}^4$. Do you see that it would work for any $\mathbf{R}^n$?

20. Suppose $\mathbf{x} \in \mathbf{R}^m$ is an eigenvector of an $m \times m$ matrix $\mathbf{A}$. Show that if $k \in \mathbf{R}$ is any scalar, then $k\mathbf{x}$ is *also* an eigenvector of $\mathbf{A}$, and has the same eigenvalue as $\mathbf{x}$.

Similarly, if both $\mathbf{v}$ and $\mathbf{w}$ are eigenvectors of $\mathbf{A}$, and both have the same eigenvalue λ, show that any linear combination $a\mathbf{v} + b\mathbf{w}$ is *also* an eigenvector of $\mathbf{A}$, again with the same eigenvalue λ.

— ⋆ —

2. Functions

Now that we're familiar with numeric vectors and matrices, we can consider *vector functions*—functions that take numeric vectors as inputs and produce them as outputs. The ultimate goal of this book is to give students a detailed understanding of *linear vector functions*, both algebraically, and geometrically. Here and in Section 3, we lay out the basic vocabulary for the kinds of questions one seeks to answer for any vector function, linear or not. Then, in Section 4, we introduce linearity, and with these building blocks all in place, we can at least *state* the main questions we'll be answering in later chapters.

2.1. Domain, image, and range.
Roughly speaking, a **function** is an input-output rule. Here is is a more precise formal definition.

DEFINITION 2.2. A **function** is an input/output relation specified by three data:

   i) A **domain** set X containing all allowed inputs,

   ii) A **range** set Y containing all allowed outputs, and

   iii) A **rule** f that assigns exactly one output $f(x)$ to every input x in the domain.

We typically signal all three of these at once with a simple diagram like this:

$$f\colon X \to Y$$

For instance, if we apply the rule $T(x,y) = x + y$ to any input pair $(x,y) \in \mathbf{R}^2$, we get a scalar output in $\mathbf{R}$, and we can summarize this situation by writing $T\colon \mathbf{R}^2 \to \mathbf{R}$.   □

Technically, *function* and *mapping* are synonyms, but we will soon reserve the term *function* for the situation where (as with T above) the range is just $\mathbf{R}$. When the range is $\mathbf{R}^n$ for some $n > 1$, we typically prefer the term *mapping* or *transformation*.

2.3. Image. Suppose S is a subset of the domain X of a function. Notationally, we express this by writing $S \subset X$. This subset S may consist of one point, the entire domain X, or anything in between.

Whatever S is, if we apply f to every $x \in S$, the resulting outputs $f(\mathbf{x})$ form a subset of the range Y called the **image of S under f**, denoted $f(S)$. In particular,

- The image of a domain point $\mathbf{x} \in X$ is the single point $f(\mathbf{x})$ in the range.

- The image of the *entire domain* X, written $f(X)$, is called the **image** of the mapping f.

The image of any subset $S \subset X$ lies *in* the range, of course. But even when $S = X$ (the entire domain), *its image may not fill the entire range.*

EXAMPLE 2.4. Consider the familiar *squaring* rule $f(x) = x^2$. If we take its domain to be $\mathbf{R}$ (the set of all real numbers), what is its image? What is its range?

Since x^2 cannot be negative, $f(x)$ has no negative outputs. On the other hand, every *non-negative* number $y \geq 0$ *is* an output, since $y = f(\sqrt{y})$. Note that $f(-\sqrt{y}) = y$ too, a fact showing that in general, *different inputs* may produce the *same output*.

In any case, we see that with $\mathbf{R}$ as domain, the squaring function has the half-line $[0,\infty)$ (all $0 \leq y < \infty$) as its image.

We may take the image—*or any larger set*—to serve as the **range** of f. One often takes the range to be all of $\mathbf{R}$, for instance. We would write

$$f\colon \mathbf{R} \to [0,\infty) \qquad \text{or} \qquad f\colon \mathbf{R} \to \mathbf{R}$$

to indicate that we have a rule named f with domain $\mathbf{R}$, and range either $[0, \infty)$ or $\mathbf{R}$, depending on our choice. Technically speaking, each choice yields a different function, since the domain is one of the three data that define the function.

Now consider the subset $S = [-1, 1]$ in the domain $\mathbf{R}$. What is the image of this subset? That is, what is $f(S)$? The answer is $f(S) = [0, 1]$, which the reader may verify as an exercise.   □

We thus associate three basic sets with any function:

- **Domain:** The set of all allowed *inputs* to the function f.

- **Range:** The set of all allowed *outputs* to the function.

- **Image:** The collection of all *actual* outputs $f(\mathbf{x})$ as $\mathbf{x}$ runs over the entire domain. It is always *contained* in the range, and *may or may not* fill the entire range.

REMARK 2.5. It may seem pointless—perhaps even perverse—to make the range larger than the image. Why should the range include points that never actually arise as outputs?

A simple example illustrates at least part of the reason. Indeed, suppose we have a function given by a somewhat complicated formula like

$$h(t) = 2.7\,t^6 - 1.3\,t^5 + \pi\,t^3 - \sin|t|$$

Determining the exact image of h would be difficult at best. But we can easily see that every output $h(x)$ will be a real number. So we can take $\mathbf{R}$ as the range, and then describe the situation correctly, albeit roughly, by writing

$$h \colon \mathbf{R} \to \mathbf{R}$$

We don't know the image of h, because we can't say exactly which numbers are actual outputs—but we *can* be sure that all outputs are real numbers. So we can't easily specify the image, but we *can* make a valid choice of range.   □

   2.6. Onto. As emphasized above, the image of a function is always a *subset* of the range, but it may not fill the entire range. When the image *does* equal the entire range, we say the function is *onto*:

DEFINITION 2.7 (Onto). We call a function **onto** if every point in the range also lies in the image—that is, the image fills the entire range. Figures 1 and 2 illustrate the concept.   □

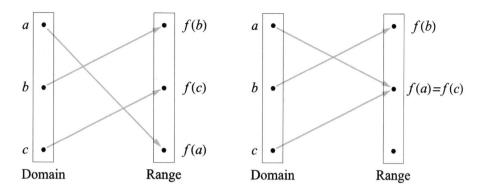

Figure 1. A function sends each point in the domain to
one in the range. The function on the left is **onto**—each
point in the range also lies in the image. The function on the
right is *not* onto: the lowest point in the range does *not* lie
in the image. That point has no *pre-image* (Definition 2.9).

EXAMPLE 2.8. The squaring function with domain **R** is *onto* if we
take its range to be *just* the interval $[0, \infty)$ of non-negative numbers.
If we take its range to be all of **R**, however, it is *not* onto, because **R**
contains negative numbers, which do not lie in the squaring function's
image.   □

There is a useful counterpart to the term *image* which, among other
things, makes it easier to discuss the difference between the image and
range of a function.

DEFINITION 2.9 (Pre-image). Suppose we have a function $f \colon X \to Y$,
and a subset S of the range Y. Notionally, $S \subset Y$. The **pre-image**
(or **inverse image**) of S consists of all points x in the domain X
that f sends into S—all points whose images lie in S. We denote the
pre-image of S by $f^{-1}(S)$ (pronounced "f *inverse of* S").   □

REMARK 2.10. We are *not* claiming here that f has an inverse *function*
f^{-1}. It may or may not—this is a topic we take up later. In general,
"f^{-1}" by itself means *nothing* unless it occurs with a subset of the
range, as in $f^{-1}(S)$ or $f^{-1}(x)$, in which case it means the pre-image
of that subset, as defined above.

In certain cases, we *can* define an inverse mapping called f^{-1} (see
Section 2.17). Then the pre-image $f^{-1}(S)$ equals the *image* of S
under the *inverse* mapping f^{-1}, so our notation $f^{-1}(S)$ is consistent.
□

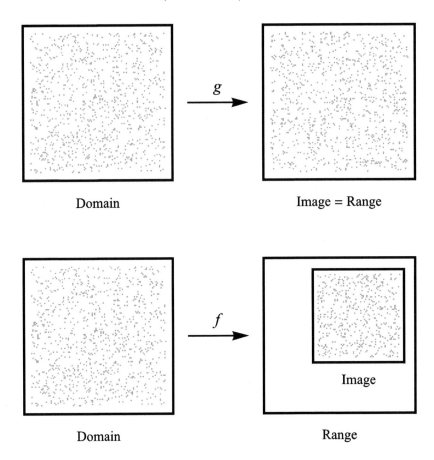

Figure 2. The image of g (above) fills its entire range, so g is *onto*. The image of f (below) does not fill the range, so f is *not* onto.

We can use the concept of *pre-image* to offer an alternate definition of the term *onto*:

OBSERVATION 2.11. *A function is **onto** exactly when every point in the range has **at least** one pre-image.*

Indeed, if every point in the range has a pre-image, then every point in the range is the image of some point in the domain. In this case, the image fills the entire range, and our function is indeed onto.

Pre-image is also a useful term because it gives an alternate name for something quite familiar, and very central to mathematics: *the solution of an equation.* The most basic question we ask about any equation $f(x) = y$ is whether we can solve it for x, given y. But this is the same as *finding a pre-image* $f^{-1}(y)$. Solving an equation and finding a pre-image are exactly the same thing.

EXAMPLE 2.12. Again consider the squaring function $f\colon \mathbf{R} \to \mathbf{R}$ given by $f(x) = x^2$. What is the pre-image of 4? Of 0? Of -4?

According to Definition 2.9 above, the pre-image of 4 consists of all x such that $f(x) = 4$. Here f is the squaring function, so that means we seek all x for which $x^2 = 4$. Clearly, this means

$$f^{-1}(4) = \{-2, 2\}$$

Similarly, we get the pre-image of $y = 0$ by solving $x^2 = 0$. Here there is only one solution—only one point in the pre-image:

$$f^{-1}(0) = \{0\}$$

The pre-image of -4, on the other hand, consists of all solutions to $x^2 = -4$. Since this equation has *no* solutions in the domain $\mathbf{R}$ we specified here, -4 has no pre-image; its pre-image is the *empty* set:

$$f^{-1}(-4) = \emptyset$$

Finally, we might ask for the pre-image of a set *larger* than just one point; say the pre-image of the interval $[0, 1]$. Since every number in $[0, 1]$ has a square root in $[0, 1]$, and also a square-root in $[-1, 0]$, it is easy to see that

$$f^{-1}([0, 1]) = [-1, 1]$$

Note that $f^{-1}[-1, 1]$ is *also* $[-1, 1]$. $\quad\square$

2.13. One-to-one. By definition, a *function* $f\colon X \to Y$ sends each point x in the domain X to a point $y = f(x)$ in the range Y. In that case, x belongs to the pre-image of y. But the pre-image of y may contain *other* inputs beside x. This happens, for instance, with the squaring function $f(x) = x^2$. The pre-image of any positive number contains *two* inputs. For example, $f^{-1}(4)$ contains both 2 and -2.

A nicer situation arises with the function $g\colon \mathbf{R} \to \mathbf{R}$ given by $g(x) = 2x + 3$. Here, the pre-image of an output y always contains exactly one point—no more, and no less. We know this because we can easily solve for $2x + 3 = y$ for x, and we always get exactly one solution, namely $x = (y - 3)/2$ (try it).

Since each point in the range of this function *has* a pre-image, it is *onto*. But we also know that pre-images never contain *more* than one point. This makes g *one-to-one*:

DEFINITION 2.14. A function $f\colon X \to Y$ is **one-to-one** if each point in the range has at **most** one point in its pre-image. Examples show (see below) that a function may be one-to-one, onto, both, or neither. □

EXAMPLE 2.15. The function depicted on the left in Figure 1 (above, not below) is both onto and one-to-one, because each point in the range has one—and only one—point in its pre-image. The function on the *right* in that figure, however, is *neither* one-to-one *nor* onto. It's not one-to-one because the pre-image of the middle point in the range contains *two* points (a and c). Neither is it onto, since the lowest point in the range has no pre-image at all.

The functions in Figure 3 below, on the other hand, each have one of the properties, but not the other. The function on the left is one-to-one but not onto. The one on the right is onto, but not one-to-one. (Make sure you see why.)

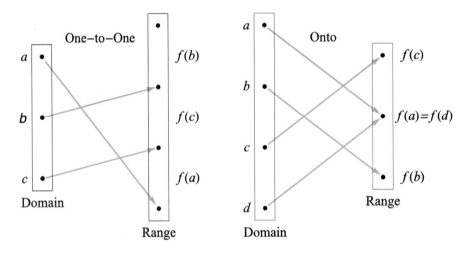

Figure 3. The function on the left is one-to-one, but not onto. The function on the right is onto, but not one-to-one.

REMARK 2.16 (One-to-one vs. Onto). The definitions of *one-to-one* and *onto* compare and contrast very nicely if we summarize them like this:

- A function is **one-to-one** if every point in the range has **at most** one pre-image.

- A function is **onto** if every point in the range has **at least** one pre-image. □

2.17. Inverse functions. When a function $f\colon X \to Y$ is one-to-one, each y in the image has exactly one pre-image $x = f^{-1}(y)$, hence exactly one solution of $f(x) = y$. If f is also *onto*, then *every* y in the range has a unique pre-image x in this way. In this case, the assignment $y \to x$ defines a new function that "undoes" f:

DEFINITION 2.18 (Inverse mapping). If $f\colon X \to Y$ is *both* one-to-one *and* onto, the mapping that sends each $y \in Y$ to its unique pre-image $x = f^{-1}(y)$, is called the **inverse** of f. We denote it by f^{-1}. $\qquad\square$

EXAMPLE 2.19. Consider the mapping $f\colon \mathbf{R} \to \mathbf{R}$ given by $f(x) = x + 1$. The solutions of $f(x) = y$ constitute the pre-image of y, and here that means solving $y = x + 1$. Doing so, we get $x = y - 1$. There's no restriction on y here—this gives a solution for *every* y. In fact, it gives *exactly one* solution for every y, so the function is both one-to-one and onto, and hence has an inverse. The inverse maps each y in the range to its unique pre-image in the domain, and our solution gives a formula for it: $f^{-1}(y) := y - 1$. $\qquad\square$

EXAMPLE 2.20. The identity mapping $I(\mathbf{x}) = \mathbf{x}$ on $\mathbf{R}^m$ is its *own* inverse, since each $\mathbf{x}$ is obviously its own pre-image. Thus, $I^{-1} = I$. A slightly less trivial example is given by the *doubling* map on $\mathbf{R}^m$, given by $D(\mathbf{x}) = 2\mathbf{x}$. To solve $D(\mathbf{x}) = \mathbf{y}$, we write $2\mathbf{x} = \mathbf{y}$, which implies $\mathbf{x} = \mathbf{y}/2$. It follows that the inverse of D is the "halving" map: $D^{-1}(\mathbf{y}) = \mathbf{y}/2$. $\qquad\square$

– Practice –

21. Define a function $f : X \to Y$ whose domain and range both contain just the first five letters of the alphabet: $X = Y = \{a, b, c, d, e\}$. Define the "rule" f for this function by setting

$$f(a) = b, \; f(b) = c, \; f(c) = a, \; f(d) = b, \text{ and } f(e) = c$$

a) Find the images of these sets: $\{a, b, c\}$, $\{a, b, d\}$, $\{a, b, e\}$.

b) Find the pre-images of these sets: $\{a\}$, $\{a, b\}$, $\{a, b, c\}$, $\{c\}$, and $\{d, e\}$.

c) What is the image of f?

d) Is this function one-to-one? Is it onto? Explain.

e) Define a function $g : X \to Y$ (same X and Y as above) which is one-to-one, onto, *and* satisfies $g(a) = c$.

22. Suppose X and Y are the sets of young men and young women at a dance where the protocol is that each $x \in X$ chooses a dance partner $y \in Y$. Let $f : X \to Y$ be the "choosing" function, so that for each young man x, $y = f(x)$ is the partner he chooses.

a) What does it mean for f to be *onto*?

b) What does it mean for f to be *one-to-one*?

c) What is the image of f?

d) If $S \subset Y$ is a subset of the young woman, what does $f^{-1}(S)$ correspond to?

e) If $S \subset X$ is a subset of the young men, what is $f(S)$?

23. The following questions refer to Figure 4.

a) If we take $\mathbf{R}$ as the domain of the constant function $c(x) \equiv 1$, what is the image of c? (The triple equal sign emphasizes that $c(x) = 1$ for *all* inputs x in the domain.)

b) Assuming 1, 0 and -1 are in the range of c, what are the pre-images $c^{-1}(1)$, $c^{-1}(0)$, and $c^{-1}(-1)$?

c) Could the interval $[0, \infty)$ serve as the range of c? How about the interval $(-\infty, 0]$? How about the entire real line $\mathbf{R}$? Is c *onto* in any of these cases?

24. The following questions refer to Figure 4.

a) Suppose we take $\mathbf{R}$ as both the domain and range of the function $g(x) = 2x + 1$. What is the image of g? Is g onto?

b) What are the pre-images $g^{-1}(1)$, $g^{-1}(0)$, and $g^{-1}(-1)$?

c) If $\mathbf{R}$ is the domain of g, could the interval $[0, \infty)$ serve as its range? How about the interval $(-\infty, 0]$? Why or why not?

d) If we changed the domain of g from $\mathbf{R}$ to just the the interval $[-1, 1]$ (all $-1 \leq x \leq 1$), what would the image be?

25. Devise examples of:

    a) A function $N : \mathbf{Z} \to \mathbf{Z}$ that is one-to-one, but not onto.

    b) A function $F : \mathbf{Z} \to \mathbf{Z}$ that is onto, but not one-to-one.

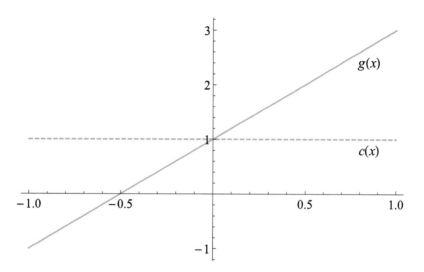

Figure 4. Graphs of the functions in Exercises 23 and 24.

26. Let $\mathbf{Z}$ be the set of all integers (positive and negative), and let $E \subset \mathbf{Z}$ be the subset of *even* integers. Show that the map $H : E \to \mathbf{Z}$ given by the rule $H(n) = n/2$ is *onto*.

Thus, even though E is a *proper* subset of $\mathbf{Z}$ (E is not *all* of $\mathbf{Z}$), we can map it *onto* $\mathbf{Z}$. This is only possible because $\mathbf{Z}$ contains infinitely many elements (integers). A proper subset of a *finite* set Y can never map *onto* Y. Can you give a reason for this?

27. Let A, B, C be scalars with $A \neq 0$ and consider the quadratic function

$$Q(x) = Ax^2 + Bx + C$$

    a) Give a precise description of the image of Q (in terms of the coefficients A, B, C). Thinking about the shape of the graph of Q should help you answer.

    b) What is the pre-image $f^{-1}(0)$? How about $f^{-1}(y)$ if $y \neq 0$? (Again, your answers will be functions of the coefficients.)

28. Take both the domain and range of each function below to be the integers $\ldots, -3, -2, -1, 0, 1, 2, 3, \ldots$. In each case, say whether the function is one-to-one and/or onto. If it isn't one-to-one, give an example of an integer with *multiple* pre-images. If it isn't onto, give an example of an integer with *no* pre-image.

   a) $f(n) = n + 2$

   b) $g(n) = 2n$

   c) $h(n) = |n|$

   d) $k(n) = (n + 1)^2$

   e)

$$p(n) = \begin{cases} n/2, & \text{if } n \text{ is even} \\ (1 - n)/2, & \text{if } n \text{ is odd} \end{cases}$$

29. Show that each map below is *both* one-to-one *and* onto. Then find its inverse function.

   a) Domain and range both equal to all integers, $f(n) = 2 - n$.

   b) Domain and range both equal to **R**, with $f(x) = 7 + x/10$.

30. When $f \colon X \to Y$ is both one-to-one and onto, will its *inverse* $f^{-1} \colon Y \to X$ also be one-to-one and onto? Explain why or why not.

31. When X has only finitely many members, an invertible function $f \colon X \to X$ is called a **permutation**. For instance, when $X = \{a, b, c\}$, the rule

$$f(a) = b, \quad f(b) = c, \quad f(c) = a$$

defines a permutation.

   a) Make a table (with 3 columns, $f(a), f(b)$, and $f(c)$) listing all six permutations of $X = \{a, b, c\}$.

   b) Are any of these permutations their *own* inverses (so that $f(f(x)) = x$ for every $x \in X$)? Such a function is called an **involution**.

   c) If a set has n members, how many permutations will it have?

   d) If a set has n members, how many of its permutations will be involutions?

— ★ —

3. Mappings and Transformations

The functions we study in Linear Algebra usually have domains and/or ranges in one of the numeric vector spaces $\mathbf{R}^n$ we introduced in Section 1.

DEFINITION 3.1. A function with numeric vector inputs or outputs is called a **mapping** or **transformation**—synonymous terms. A mapping, or transformation is thus simply a function described by a diagram of the form

$$F \colon \mathbf{R}^n \to \mathbf{R}^m$$

where $n > 1$ and/or $m > 1$. Typically, we use uppercase letters like F, G, or H to label mappings, and from now on, we try to reserve the word **function** for the case of scalar outputs $(m = 1)$. □

EXAMPLE 3.2. A simple mapping

$$J \colon \mathbf{R}^2 \to \mathbf{R}^2$$

is given by the rule

$$J(x, y) = (-y, x)$$

This formula makes it easy to compute $J(x, y)$ for any specific input $(x, y) \in \mathbf{R}^2$. For instance, we have

$$J(1, 2) = (-2, 1), \quad J(-3, 5) = (-5, -3), \quad \text{and} \quad J(0, 0) = (0, 0)$$

Is J one-to-one and/or onto? We leave that as part of Exercise 32 below.

While the domain and range of J are the same, other mappings often have domains and ranges that differ, as the following examples illustrate. □

EXAMPLE 3.3. The rule

$$F(x, y, z, w) = (x - y,\ z + w)$$

has four scalar entries in its input, but only two in its output. It therefore defines a mapping $F \colon \mathbf{R}^4 \to \mathbf{R}^2$. Contrastingly,

$$G(x, y) = \left(x^2,\ 2y,\ 2x,\ y^2 \right)$$

defines a mapping $G \colon \mathbf{R}^2 \to \mathbf{R}^4$. □

Mappings in general display an enormous variety of behaviors. We present a few more examples here to suggest the almost endless possibilities.

EXAMPLES 3.4.

- The formula
$$F(x, y, z) = xyz$$
has three inputs, but only one output, so it defines a transformation $F: \mathbf{R}^3 \to \mathbf{R}$. It is onto, since every $t \in \mathbf{R}$ has a pre-image (for example, $(t, 1, 1)$), but it is certainly not one-to-one. For example, $(1, t, 1)$ and $(1, 1, t)$ and infinitely other inputs (can you find any?) are all pre-images of t.

- The formula
$$G(t) = (t, t^2, t^3)$$
defines a transformation $G: \mathbf{R} \to \mathbf{R}^3$. Its image can be visualized as an infinitely long curve in (x, y, z)-space. The mapping is one-to-one, but not onto. For instance $(1, 2, 3)$ has no pre-image. Do you see why?

- The formula
$$Q(x_1, x_2, x_3, x_4, x_5) = (x_1, x_2, x_3, x_4, x_5, 0, 0, 0)$$
defines a transformation $P: \mathbf{R}^5 \to \mathbf{R}^8$. It "inserts" a copy of $\mathbf{R}^5$ into $\mathbf{R}^8$. It is far from being onto, since most points in $\mathbf{R}^8$ have no pre-image: the only points that do are those whose last three coordinates vanish.

- The formula
$$P(x_1, x_2, x_3, x_4, x_5) = (x_1, x_2, x_3)$$
defines a transformation $P: \mathbf{R}^5 \to \mathbf{R}^3$. It "projects" $\mathbf{R}^5$ onto $\mathbf{R}^3$ by dropping the last two coordinates. It is onto, but not one-to-one, since every point $(x, y, z) \in \mathbf{R}^3$ has multiple pre-images. For instance, any vector of the form $(1, 2, 3, s, t) \in \mathbf{R}^5$ is a pre-image of $(1, 2, 3) \in \mathbf{R}^3$.

$\square$

EXAMPLE 3.5. Every matrix $\mathbf{A}_{n \times m}$ determines a mapping $F: \mathbf{R}^m \to \mathbf{R}^n$ via matrix/vector multiplication. The rule is simple:
$$F(\mathbf{x}) = \mathbf{A}\mathbf{x}$$
In other words, to each input $\mathbf{x} \in \mathbf{R}^m$, we assign the vector $\mathbf{A}\mathbf{x} \in \mathbf{R}^n$ as output. For instance, the mapping J in Example 3.2 above fits this description. The rule we gave there was $J(x, y) = (-y, x)$. This is the same as multiplying the input (x, y) by the 2×2 matrix

$$\mathbf{J} = \begin{bmatrix} 0 & -1 \\ 1 & 0 \end{bmatrix}$$

For when $\mathbf{x} = (x, y)$, we have

$$\mathbf{Jx} = \begin{bmatrix} 0 & -1 \\ 1 & 0 \end{bmatrix} \begin{pmatrix} x \\ y \end{pmatrix} = \begin{pmatrix} -y \\ x \end{pmatrix} = J(x, y)$$

as claimed. All the questions we wish to ask about such a mapping (see Section 3.6 below) can be answered by analyzing the matrix that defines it. The primary goal of our course is to learning how to do such analyses.   □

3.6. Basic mapping questions. Suppose we have a mapping

$$F \colon \mathbf{R}^n \to \mathbf{R}^m$$

Given a vector $\mathbf{y}$ in the range $\mathbf{R}^m$ of F, we often want to answer the most basic mapping question:

- *Does the equation*

$$F(\mathbf{x}) = \mathbf{y}$$

  have one or more solutions? If so, how might we find them?

In general this question can be difficult or impossible to answer—it's just a sad fact of life that most equations are very hard to solve. Fortunately, however, one can *always* solve this equation when F is *linear*, as defined in Section 4 below. We will learn how in Chapter 2.

Whether we can answer that question or not for a given specific vector $\mathbf{y}$ in the range, however, we often hope to get more general information. For instance we frequently ask

- *Does $F(\mathbf{x}) = \mathbf{y}$ have a solution for **every** $\mathbf{y}$ in the range? In other words, is F **onto**?*

Of course, this is the same as asking whether every $\mathbf{y} \in \mathbf{R}^m$ has a pre-image.

If F is not onto, we often seek a way to describe those $\mathbf{y}$ in the range $\mathbf{R}^m$ that *do* have pre-images. In other words, we want to know

- *Can we (geometrically or algebraically) describe the **image** of F as a subset of the range?*

When a vector $\mathbf{y}$ in the range does belong to the image, we often want to know *how many* points lie in its pre-image $f^{-1}(\mathbf{y})$. The simplest situation occurs when each $\mathbf{y}$ in the image has exactly one pre-image. Determining whether this is the case is the same as asking

- *Is F **one-to-one**?*

If F is *not* one-to-one, some pre-images contain multiple points—perhaps infinitely many. In this case, we often want good descriptions of typical pre-images.

- *Can we (geometrically or algebraically)* describe *the **pre-image** of a typical* $\mathbf{y} \in \mathbf{R}^m$ *as a subset of the domain* $\mathbf{R}^n$?

Finally, we often want information about how F "distorts" sets of inputs. When the outputs of F tend to be farther apart than the corresponding inputs, we say the mapping *expands*. If then outputs tend to be closer than the corresponding inputs, we say that mapping *contracts*. The simplest question about distortion would be this like to know:

- *Does F expand or contract? Can we understand the nature of its distortion?*

To answer this, we must first define *distances in* $\mathbf{R}^n$. We will do that using the dot product in Chapter 3. Mappings usually distort in ways more complicated than simple expansion or contraction, but for *linear* maps, which we'll define in Section 4 below, we will eventually be able to analyze distortion with satisfying precision. Eigenvalues and eigenvectors of square matrices play a large part in that analysis, as we shall see in Chapter 7.

First, however, to get a better feeling for some of the questions above, we consider them with reference to a few very basic examples.

EXAMPLE 3.7. We start with a recognizable example that is one-to-one, but not onto, namely the mapping

$$F \colon \mathbf{R} \to \mathbf{R}^2$$

given by the rule

$$F(x) = (x, x - 1)$$

Here the first coordinate of each output can be any real number x, and the second coordinate is then simply $x - 1$. If we regard $\mathbf{R}^2$ as the (x, y)-plane in the usual way, then the image of F is simply the line $y = x - 1$ (Figure 5). This line misses "most" points in $\mathbf{R}^2$, so the image is much smaller than the range, and F badly fails to be onto.

On the other hand, F *is* one-to-one, since each point of the image has just one pre-image, given by its x-coordinate.

As for distortion, we can be very precise: F expands all distances by exactly $\sqrt{2}$. For the distance between two outputs $F(a)$ and $F(b)$ is exactly $\sqrt{2}$ times the distance between the inputs a and b. This is easy to see because, as Figure 5 illustrates (with $a = 2$ $b = 4$), $F(a)$ and $F(b)$ bound the hypotenuse of an isoceles right triangle whose short sides both have length exactly $b - a$.

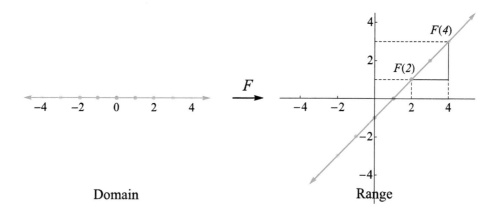

Figure 5. Here $F(x) = (x, x - 1)$ sends $\mathbf{R}$ into $\mathbf{R}^2$. Its image is the diagonal line $y = x - 1$ on the right, not all of $\mathbf{R}^2$, so F is not onto. It *is* one-to-one, however, since each point $(x, x - 1)$ in the image has exactly one pre-image, namely x, on the left.

$\square$

EXAMPLE 3.8. Next, consider the rule

$$P(x, y) = (x, 0)$$

This is another simple linear mapping. Domain and range are both $\mathbf{R}^2$, and we now propose to verify that it is *neither* onto, nor one-to-one.

Since $P(x, y) = (x, 0)$, which lies on the x-axis, namely $(x, 0)$, the x-axis contains the entire image of P. Conversely, every point $(a, 0)$ on the x-axis lies in the image, since $P(a, 0) = (a, 0)$. In fact, however, since $P(a, y) = (a, 0)$ for any $y \in \mathbf{R}$, the entire vertical line through $(a, 0)$ lies in the pre-image of $(a, 0)$ (Figure 6).

From these facts, we easily see that P is *not* onto. The image (the x-axis) does *not* fill the entire range $\mathbf{R}^2$. On the other hand, points that *do* lie on the x-axis—have an entire vertical line of pre-images, which means P isn't one-to-one either.

Clearly, this map contracts (or at least, doesn't expand) distances, since it doesn't change the horizontal distance between points, while collapsing vertical differences to zero. $\square$

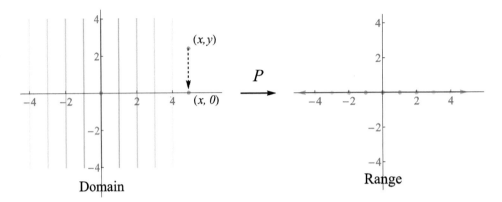

Figure 6. P maps each vertical line in the plane to the point where it meets the x-axis (right). The image of P is thus the entire x-axis, and the pre-image of any point on the x-axis is the vertical line through that point.

EXAMPLE 3.9. We get a somewhat similar, though more complicated mapping from the function $d\colon \mathbf{R}^2 \to \mathbf{R}$ depicted in Figure 7 and given by

$$d(x, y) = \sqrt{x^2 + y^2}$$

Again, the mapping is *not* onto. Sums of squares are always non-negative, so outputs never lie in the negative half of the number line. The image fills out that entire *non*-negative half-line, however, since every $r \geq 0$ has a pre-image: for instance, we have $r = d(r, 0)$.

The mapping is not one-to-one either, since the pre-image of a positive number $r > 0$ always contains infinitely many points. In fact, the pre-image $d^{-1}(r)$ forms a circle of radius r centered at the origin of the (x, y)-plane. This follows from the Pythagorean theorem, since any point (x, y) on that circle satisfies $x^2 + y^2 = r^2$ (Figure 7).

Finally, we mention (omitting the easy proof) that d *contracts*: When we apply it to a pair of points in the domain, we always get two points in the image that are closer (or at least not farther) than the original points.   □

General mappings can be far (far!) more complicated than the simple examples above, and that can make it truly difficult to address the questions we have raised—in general. Fortunately for us, when it comes to the *linear* mappings we study in this course, one can answer them all to a very satisfying degree. Doing so is the main goal of our book.

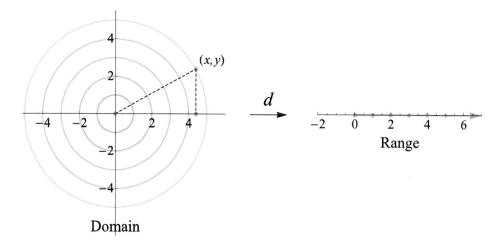

Figure 7. d maps each circle on the left to the same-colored dot on the right. Pre-images of positive numbers are thus circles. Negative numbers on the right have no pre-image. The hypotenuse of the right triangle associated to any (x, y) on one of these circles has length $d(x, y) = \sqrt{x^2 + y^2}$ by the Pythagorean theorem.

That ability, by the way, makes it possible to answer the same sorts of questions—at least in part—for the mappings of multivariable calculus. Those mappings are called *differentiable*, which simply means they can be well-approximated, at least near any point in their domain, by a *linear* mapping. So besides being useful themselves, linear mappings turn out to be crucial for understanding differentiable mappings. Indeed, virtually every field of mathematics requires a good understanding of linear mappings.

– Practice –

32. Determine which of the mappings J, F, and G in Examples 3.2 and 3.3 (if any) is onto. Which is one-to-one? Justify your answers.

33. Determine which of the following mappings is one-to-one, and which is onto. Justify your answers.

    a) The mapping $P\colon \mathbf{R}^3 \to \mathbf{R}^2$ given by $S(x, y, z) = (x, y)$.

    b) The mapping $S\colon \mathbf{R}^3 \to \mathbf{R}^3$ given by $S(x, y, z) = (0, x, y)$.

    c) The mapping $T\colon \mathbf{R}^3 \to \mathbf{R}^4$ given by $S(x, y, z) = (0, x, y, z)$.

34. Consider the mapping $A: \mathbf{R}^2 \to \mathbf{R}^2$ given by $A(\mathbf{x}) = -\mathbf{x}$. Show that A is both one-to-one and onto, hence invertible. Does the same hold if we make the domain and range equal to $\mathbf{R}^n$ instead of $\mathbf{R}^2$?

35. Consider the mapping $T: \mathbf{R}^2 \to \mathbf{R}^3$ given by $T(x, y) = (x, 2, y)$. Which points in $\mathbf{R}^3$ have non-empty pre-images? How would you describe the image of T? Is T one-to-one? Is it onto?

36. Consider the mapping $F: \mathbf{R}^3 \to \mathbf{R}^2$ given by $S(x, y, z) = (z, x)$. Which points in $\mathbf{R}^2$ have non-empty pre-images? How would you describe the pre-image of a typical point in the range? Is F onto? Is it one-to-one?

37. Consider the mapping $E: \mathbf{R}^2 \to \mathbf{R}^2$ given by
$$E(x, y) = (x + y, \ x - y)$$
a) Find the images and pre-images of the points $(1, 0)$, $(0, 1)$, and $(1, 1)$ with respect to E.

b) Find the (possibly empty) pre-image $E^{-1}(b, c)$ of a general point $(b, c) \in \mathbf{R}^2$.

c) Is E one-to-one? Is E onto? Justify your answers.

38. Consider the mapping $E: \mathbf{R}^2 \to \mathbf{R}^2$ given by
$$W(x, y) = (x + y, \ xy)$$
a) Find the images and pre-images of the points $(1, 0)$, $(0, 1)$, and $(1, 1)$ with respect to W.

b) Find the (possibly empty) pre-image $W^{-1}(b, c)$ of a general point $(b, c) \in \mathbf{R}^2$.

c) Is W one-to-one? Is W onto? Justify your answers.

The next exercise treats a situation where the "points" in the domain and range of a mapping are themselves functions. This perspective on functions became common in mathematics during the 20th century.

39. (Requires basic calculus) A polynomial of degree k in the variable x is a function of the form
$$p(x) = a_k x^k + a_{k-1} x^{k-1} + \cdots + a_1 x + a_0$$
where the a_i's are scalars. Let $\mathcal{P}^k$ denote the set of all such polynomials, and let $D: \mathcal{P}^k \to \mathcal{P}^{k-1}$ be the *differentiation mapping* given

$D(p) = p'$. For instance, $D(x^n) = nx^{n-1}$. Show that D is onto, but not one-to-one. What is the pre-image of x^2? of x^3? If two different polynomials lie in $D^{-1}(p)$ for some $p \in \mathcal{P}^{k-1}$, what can we say about their difference?

4. Linearity

Recall that both scalar multiplication and matrix/vector multiplication distribute over vector addition (Propositions 1.6 and 1.25). The definition of *linearity* generalizes those distributivity rules:

DEFINITION 4.1. A mapping $F: \mathbf{R}^n \to \mathbf{R}^m$ is **linear** if it has *both* these properties:

    i) *F commutes with vector addition*, meaning that for any two inputs $\mathbf{x}, \mathbf{y} \in \mathbf{R}^m$, we have

$$F(\mathbf{x} + \mathbf{y}) = F(\mathbf{x}) + F(\mathbf{y})$$

    ii) *F commutes with scalar multiplication*, meaning that for any input $\mathbf{x} \in \mathbf{R}^m$ and any scalar $c \in \mathbf{R}$, we have

$$F(c\,\mathbf{x}) = c\,F(\mathbf{x})$$

Linear mappings are often called **linear transformations**, and for this reason the favorite symbol for a linear mapping is the letter T. $\square$

EXAMPLE 4.2. The mapping $T: \mathbf{R}^2 \to \mathbf{R}^2$ given by

(2) $$T(a, b) = (2b,\, 3a)$$

is linear.

To verify this, we have to show that T has both properties in Definition 4.1 above.

First property: T commutes with addition: We have to show that for any two vectors $\mathbf{x} = (x_1, x_2)$, and $\mathbf{y} = (y_1, y_2)$, we have

(3) $$T(\mathbf{x} + \mathbf{y}) = T(\mathbf{x}) + T(\mathbf{y})$$

We do so by expanding each side of the equation separately in coordinates, and checking that they give the same result. On the left, we have

$$T(\mathbf{x} + \mathbf{y}) = T\left(\begin{pmatrix} x_1 \\ x_2 \end{pmatrix} + \begin{pmatrix} y_1 \\ y_2 \end{pmatrix} \right) = T\left(x_1 + y_1,\ x_2 + y_2 \right)$$

and now the rule for T, namely (2), reduces this to

$$T(\mathbf{x} + \mathbf{y}) = (2(x_2 + y_2), 3(x_1 + y_1)) = (2x_2 + 2y_2, 3x_1 + 3y_1)$$

We have to compare this with the result of the right-hand side of (3), namely $T(\mathbf{x}) + T(\mathbf{y})$. For that we get

$$
\begin{aligned}
T(\mathbf{x}) + T(\mathbf{y}) &= T(x_1, x_2) + T(y_1, y_2) \\
&= (2x_2, 3x_1) + (2y_2, 3y_1) \\
&= (2x_2 + 2y_2, 3x_1 + 3y_1)
\end{aligned}
$$

Both sides of (3) give exactly the same result, and we calculated with completely arbitrary coordinates x_1, x_2, y_1 and y_2, so we can be sure that the first linearity property always holds for T.

Second property: T commutes with scalar multiplication: Now we must show that for any vector $\mathbf{x} = (x_1, x_2)$ and any *scalar* c, we have

(4) $$T(k\mathbf{x}) = k\,T(\mathbf{x})$$

As before, we can check this by computing both sides of the identity and seeing whether they agree or not. We start with

$$T(k\mathbf{x}) = T\left(k(x_1, x_2)\right) = T\left(kx_1, kx_2\right) = (2k\,x_2, 3k\,x_1)$$

and compare that with

$$k\,T(\mathbf{x}) = k\,T(x_1, x_2) = k\,(2x_2, 3x_1) = (2k\,x_2, 3k\,x_1)$$

Again, we get the same result both ways, and since k and the input vector (x_1, x_2) were completely arbitrary, T obeys the identity for all possible scalars and inputs.

In short, both the first and second linearity properties hold for T, and this makes it linear, as claimed.   □

In combination, the two linearity rules, simple as they are, have truly far-reaching consequences—indeed, our whole subject is based on them. Most mappings (by far) do *not* enjoy these properties, however. Even very simple mappings can lack linearity, as the next two examples show.

EXAMPLE 4.3. The square-root function is not linear, because

$$\sqrt{a + b} \quad \text{does } \textbf{not} \text{ typically equal} \quad \sqrt{a} + \sqrt{b}$$

For instance,

$$\sqrt{1 + 1} \neq \sqrt{1} + \sqrt{1}$$

This violates the first linearity rule, so the square-root function is not linear.   □

EXAMPLE 4.4. Suppose $T \colon \mathbf{R}^m \to \mathbf{R}^n$ is the *constant* mapping given by $T(\mathbf{v}) \equiv \mathbf{b}$, where $\mathbf{b} \in \mathbf{R}^n$ is some fixed vector. No matter what input $\mathbf{v}$ we feed to T, the output is $\mathbf{b}$. The constant mapping is extremely simple, but is it linear?

Answer: Not unless $\mathbf{b} = \mathbf{0}$.

To see this, we test T with regard to scalar multiplication. The second rule of linearity demands that $T(2\mathbf{x})$ and $2T(\mathbf{x})$ be the *same* for any $\mathbf{x} \in \mathbf{R}^m$. But $T(\mathbf{x})$ and $T(2\mathbf{x})$ (like T of anything) both equal $\mathbf{b}$, so that

$$T(2\mathbf{x}) = \mathbf{b} \quad \text{and} \quad 2T(\mathbf{x}) = 2\mathbf{b}$$

Since $\mathbf{b} \neq 2\mathbf{b}$ (unless $\mathbf{b} = \mathbf{0}$), we see that the constant mapping fails the second linearity rule except when $\mathbf{b} = \mathbf{0}$. In short, T is definitely **not** linear when $\mathbf{b} \neq \mathbf{0}$.

Is it linear when $\mathbf{b} = \mathbf{0}$? The answer is easily seen to be *yes*, but we leave that as an exercise. $\square$

EXAMPLE 4.5. The transformation $F \colon \mathbf{R}^3 \to \mathbf{R}^2$ given by

$$F(x, y, z) = (x + 1, y - z)$$

is **not** linear. For one thing, it doesn't commute with vector addition. To see this, take two input vectors

$$\mathbf{x} = (x, y, z) \quad \text{and} \quad \mathbf{x}' = (x', y', z')$$

and check both sides of the addition rule in Definition 4.1. On the left side, we get

$$
\begin{aligned}
F\left(\mathbf{x} + \mathbf{x}'\right) \\
&= F\left(x + x',\, y + y',\, z + z'\right) \\
&= \left(x + x' + 1,\, y + y' - z - z'\right)
\end{aligned}
$$

while on the right,

$$
\begin{aligned}
F(\mathbf{x}) + F(\mathbf{x}') \\
&= (x + 1, y - z) + (x' + 1, y' - z') \\
&= (x + x' + 2,\, y + y' - z - z')
\end{aligned}
$$

Although the second coordinates of $F(\mathbf{x} + \mathbf{x}')$ and $F(\mathbf{x}) + F(\mathbf{x}')$ come out the same either way, the first coordinates do *not*. To commute with addition, $F(\mathbf{x} + \mathbf{x}')$ and $F(\mathbf{x}) + F(\mathbf{x}')$ must be *exactly* the same in all coordinates, and they are not. So F fails to commute with addition, and is therefore *not* linear.

It doesn't commute with scalar multiplication either, by the way. We may again show this by checking both sides of the condition with an arbitrary input $\mathbf{x} \in \mathbf{R}^3$ and scalar $s \in \mathbf{R}$. On the left, we get

$$
\begin{aligned}
F(s\,\mathbf{x}) &= F(sx, sy, sz) \\
&= (sx + 1, sy - sz)
\end{aligned}
$$

while on the right,

$$
\begin{aligned}
s\,F(\mathbf{x}) &= s(x + 1, y - z) \\
&= (sx + s, sy - sz)
\end{aligned}
$$

The two sides are close, but not equal, except when $s = 1$. But the definition requires equality for *all* scalars s, and we don't have that. So we have an alternative proof that F is *not* linear.

Note that failing *either* condition (i) *or* condition (ii) from Definition 4.1 suffices to make F non-linear. We showed that both fail here just to demonstrate the way one uses each. $\qquad\square$

REMARK 4.6. Perhaps we should stress that while functions $f\colon \mathbf{R} \to \mathbf{R}$ of the form $f(x) = mx + b$ are often called "linear" in high school (because their graphs are straight lines), they do *not* satisfy the linearity conditions in Definition 4.1 unless $b = 0$, and hence are *not* linear for our purposes here. We leave the reader to verify this in Exercise 43. $\qquad\square$

EXAMPLE 4.7 (Projection). When $n > m$, the map $P\colon \mathbf{R}^n \to \mathbf{R}^m$ given by

$$
P(x_1, x_2, \ldots, x_n) = (x_1, x_2, \ldots, x_m)
$$

(*projection* onto the first m-coordinates) is linear.

As always, we prove this by verifying both conditions from Definition (4.1). To verify that P commutes with addition, we take an arbitrary pair of vectors $\mathbf{x} = (x_1, x_2, \ldots, x_n)$ and $\mathbf{x}' = (x_1', x_2', \ldots, x_n')$, compute $P(\mathbf{x} + \mathbf{x}')$ and $P(\mathbf{x}) + P(\mathbf{x}')$, and compare them:

$$
P(\mathbf{x} + \mathbf{x}') = P(x_1 + x_1', x_2 + x_2', \ldots, x_n + x_n')
$$

The formula for P above says that we get P by simply keeping the first $m < n$ coordinates of its input. So

$$
P(\mathbf{x} + \mathbf{x}') = (x_1 + x_1', x_2 + x_2', \ldots, x_m + x_m')
$$

To determine whether P commutes with addition, we have to compare this with

$$
\begin{aligned}
P(\mathbf{x}) + P(\mathbf{x}') &= (x_1, x_2, \ldots, x_m) + (x_1', x_2', \ldots, x_m') \\
&= (x_1 + x_1', x_2 + x_2', \ldots, x_m + x_m')
\end{aligned}
$$

This is the same result we got by computing $P(\mathbf{x}+\mathbf{x}')$, so P *does* commute with addition. The first condition for linearity therefore holds.

To show that P also commutes with scalar multiplication, we have to compare $P(c\mathbf{x})$ and $c\,P(\mathbf{x})$ for an arbitrary scalar c and vector $\mathbf{x} \in \mathbf{R}^n$. We have

$$
\begin{aligned}
P(c\mathbf{x}) &= P(c\,x_1, c\,x_2, \ldots, c\,x_n) \\
&= (c\,x_1, c\,x_2, \ldots, c\,x_m)
\end{aligned}
$$

while

$$
\begin{aligned}
c\,P(\mathbf{x}) &= c\,(x_1, x_2, \ldots, x_m) \\
&= (c\,x_1, c\,x_2, \ldots, c\,x_m)
\end{aligned}
$$

Same result both ways, and the calculation applies no matter what $c \in \mathbf{R}$ and $\mathbf{x} \in \mathbf{R}^n$ we choose. So P commutes with scalar multiplication as well as addition. In short, P is linear, as claimed. $\qquad\square$

EXAMPLE 4.8 (Matrix/vector multiplication). The most important examples of linear transformations are given by matrix/vector multiplication. As we first noted in Example 3.5 above, every $n \times m$ matrix $\mathbf{A}$ produces a linear mapping

$$
T \colon \mathbf{R}^m \to \mathbf{R}^n
$$

via the simple formula

$$
T(\mathbf{x}) = \mathbf{A}\mathbf{x}
$$

The input vector can be any $\mathbf{x} \in \mathbf{R}^m$, since $\mathbf{A}$ has m columns. Since $\mathbf{A}$ has n rows, each column lies in $\mathbf{R}^n$. Each output $\mathbf{A}\mathbf{x}$, as a linear combination of those columns, therefore lies in $\mathbf{R}^n$.

The *linearity* of such a transformation follows immediately from the distributive property of matrix/vector multiplication stated in Proposition 1.25. By combining the definition of T with that Proposition, for instance, we get

$$
T(\mathbf{x} + \mathbf{y}) = \mathbf{A}\,(\mathbf{x} + \mathbf{y}) = \mathbf{A}\mathbf{x} + \mathbf{A}\mathbf{y} = T(\mathbf{x}) + T(\mathbf{y})
$$

which is the first property of linearity. We leave verification of the second property as an exercise. $\qquad\square$

– **Practice** –

40. The *zero transformation* $Z: \mathbf{R}^m \to \mathbf{R}^n$ maps every input $\mathbf{x} \in \mathbf{R}^m$ to the origin $\mathbf{0} \in \mathbf{R}^n$. Show that Z is linear.

41. Suppose k is a scalar, and $T_k: \mathbf{R}^m \to \mathbf{R}^m$ is scalar multiplication by k. Thus, $T_k(\mathbf{x}) = k\,\mathbf{x}$ for any $\mathbf{x} \in \mathbf{R}^m$. Show that T_k is linear.

42. Show that the transformation T defined by matrix/vector multiplication in Example 4.8 above obeys the second linearity property, namely

$$T(k\mathbf{x}) = k\,T(\mathbf{x})$$

for all scalars k and all vectors $\mathbf{x} \in \mathbf{R}^m$ (where m is the number of columns in $\mathbf{A}$).

43. Show that a function $f: \mathbf{R} \to \mathbf{R}$ of the form $f(x) = mx+b$, though often called "linear" in other contexts, does not meet the linearity criteria of Definition 4.1 (and hence is not linear in our terminology) unless $b = 0$.

44. Show that a linear function $f: \mathbf{R} \to \mathbf{R}$ *must* take the form $f(x) = ax$ for some scalar $a \in \mathbf{R}$. [*Hint: Any scalar x can be written as $x \cdot 1$, so if we apply the second linearity rule and define $a := f(1) \ldots$*]

45 (Homothety). Show that in $\mathbf{R}^n$, multiplication by a scalar m produces a linear transformation $H: \mathbf{R}^n \to \mathbf{R}^n$. Mappings of this simple type are called **homotheties**.

46. Consider the transformations $\mathbf{R}^3 \to \mathbf{R}^2$ given by

    a) $T(x, y, z) = (|x|, |z|)$

    b) $P(x, y, z) = (x, z)$

Prove that P is linear, but T is not.

47. Consider the transformations $\mathbf{R}^2 \to \mathbf{R}^2$ given by

    a) $S(x, y) = (x + 2y, 2x - y)$

    b) $F(x, y) = (1 - 2y, 2x - 1)$

Prove that S is linear, but F is not.

48. Consider the "shift-right" mapping $S \colon \mathbf{R}^n \to \mathbf{R}^n$ that shifts each coordinate one slot to the right (and rotates the last around to the first). In symbols,

$$S(x_1, x_2, x_3, \ldots, x_{n-1}, x_n) := (x_n, x_1, x_2, x_3, \cdots, x_{n-1})$$

Is S linear? [*Suggestion: Warm up with the case $n = 3$.*]

49. Supposing $n < m$, show that the *injection* map $P \colon \mathbf{R}^n \to \mathbf{R}^m$ given by

$$P(x_1, x_2, \ldots, x_n) = (x_1, x_2, \ldots, x_n, 0, 0, \ldots, 0)$$

(and so-called because it "injects" $\mathbf{R}^n$ into $\mathbf{R}^m$ in a one-to-one way) is linear.

— ⋆ —

5. The Matrix of a Linear Transformation

We end this chapter on a point of great importance: that *every* linear transformation $T \colon \mathbf{R}^m \to \mathbf{R}^n$ amounts to multiplication by a matrix $\mathbf{A}$. In this case, we say that $\mathbf{A}$ *represents* T:

DEFINITION 5.1. A linear transformation $T \colon \mathbf{R}^m \to \mathbf{R}^n$ is **represented by a matrix $\mathbf{A}$** when we can compute T using multiplication by $\mathbf{A}$. In other words, $\mathbf{A}$ represents T when we have

$$T(\mathbf{x}) = \mathbf{A}\mathbf{x}$$

for all inputs $\mathbf{x} \in \mathbf{R}^m$.   □

As the course proceeds, we'll learn how to answer almost any question about a linear transformation—like the basic mapping questions listed in Section 3.6 above—by analyzing the matrix that represents it. We'll begin acquiring tools for that kind of analysis in Chapter 2. First though, we want to show how to *find* the matrix that represents a given linear map.

We start with Observation 1.12, which shows how to expand any vector $\mathbf{x} := (x_1, x_2, \ldots, x_n) \in \mathbf{R}^m$ as a linear combination of standard basis vectors in a simple way:

$$(x_1, x_2, \ldots, x_n) = x_1\, \mathbf{e}_1 + x_2\, \mathbf{e}_2 + \cdots + x_m\, \mathbf{e}_m$$

If we expand a vector $\mathbf{x}$ this way, and then map it into $\mathbf{R}^n$ using a *linear* transformation $T \colon \mathbf{R}^m \to \mathbf{R}^n$, the linearity rules (Definition 4.1) yield

$$
\begin{aligned}
T(\mathbf{x}) &= T(x_1\,\mathbf{e}_1 + x_2\,\mathbf{e}_2 + \cdots + x_m\,\mathbf{e}_m) \\
&= T(x_1\,\mathbf{e}_1) + T(x_2\,\mathbf{e}_2) + \cdots + T(x_m\,\mathbf{e}_m) \\
&= x_1\,T(\mathbf{e}_1) + x_2\,T(\mathbf{e}_2) + \cdots + x_m\,T(\mathbf{e}_m)
\end{aligned}
$$

This reveals a powerful fact:

OBSERVATION 5.2. *If we know where a linear transformation* $T\colon \mathbf{R}^m \to$ $\mathbf{R}^n$ *sends the standard basis vectors, we can find its effect on* any *input vector* $\mathbf{x} = (x_1, x_2, \ldots, x_m)$. *For,*

$$
T(x_1, x_2, \ldots, x_m) = x_1\,T(\mathbf{e}_1) + x_2\,T(\mathbf{e}_2) + \cdots + x_m\,T(\mathbf{e}_m)
$$

In short: *A linear transformation on* $\mathbf{R}^m$ *is completely determined by its effect on the standard basis vectors.*

EXAMPLE 5.3. Suppose we know that a linear transformation $T\colon \mathbf{R}^2 \to$ $\mathbf{R}^2$ affects the standard basis vectors $\mathbf{e}_1$ and $\mathbf{e}_2$ in $\mathbf{R}^2$ as follows:

$$
T(\mathbf{e}_1) = \begin{pmatrix} 7 \\ -3 \end{pmatrix} \quad \text{and} \quad T(\mathbf{e}_2) = \begin{pmatrix} -5 \\ 2 \end{pmatrix}
$$

Then Observation 5.2 tells us that for *any* input $\mathbf{x} = (x, y)$, we have

$$
\begin{aligned}
T(x, y) &= x\,T(\mathbf{e}_1) + y\,T(\mathbf{e}_2) \\[2mm]
&= x \begin{pmatrix} 7 \\ -3 \end{pmatrix} + y \begin{pmatrix} -5 \\ 2 \end{pmatrix} \\[2mm]
&= \begin{pmatrix} 7x - 5y \\ -3x + 2y \end{pmatrix}
\end{aligned}
$$

So for instance,

$$
T(-2, 1) = \begin{pmatrix} 7 \cdot (-2) - 5 \cdot 1 \\ -3 \cdot (-2) + 2 \cdot 1 \end{pmatrix} = \begin{pmatrix} -19 \\ 8 \end{pmatrix}
$$

The power of linearity shows up clearly here: From just the two data points $T(\mathbf{e}_1)$ and $T(\mathbf{e}_2)$, we produced a general formula expressing $T(x, y)$ for all inputs (x, y). $\qquad\square$

EXAMPLE 5.4. Suppose we have a linear mapping $F\colon \mathbf{R}^3 \to \mathbf{R}$ (typically called a linear *function* in a case like this where the outputs are mere scalars). Then Observation 5.2 says that if we know where F sends the standard basis vectors $\mathbf{e}_i$, we can compute $F(x, y, z)$ for *any* (x, y, z):

$$
F(x, y, z) = x\,F(\mathbf{e}_1) + y\,F(\mathbf{e}_2) + z\,F(\mathbf{e}_3)
$$

For example, suppose

$$F(\mathbf{e}_1) = 2, \quad F(\mathbf{e}_2) = -1, \quad \text{and} \quad F(\mathbf{e}_3) = 3$$

Then

$$F(1,1,1) = 1 \cdot 2 + 1 \cdot (-1) + 1 \cdot 3 = 4$$

and

$$F(-3,2,5) = -3 \cdot 2 + 2 \cdot (-1) + 5 \cdot 3 = 7$$

□

These examples illustrate the power of Observation 5.2, but there is more. For it expresses $T(\mathbf{x})$, given any input $\mathbf{x}$, as a *linear combination* of the $T(\mathbf{e}_i)$'s, and by our mantra (Remark 1.22), *linear combination = matrix/vector multiplication*. So we can actually rewrite Observation 5.2 like this:

OBSERVATION 5.5 (Matrix construction rule). *If we know where a linear transformation* $T\colon \mathbf{R}^m \to \mathbf{R}^n$ *sends the standard basis vectors, then its effect on* any *input vector* $\mathbf{x} = (x_1, x_2, \ldots, x_m)$ *is given by*

$$T(\mathbf{x}) = \mathbf{A}\mathbf{x}$$

where $\mathbf{A}$ *is the matrix whose columns are the* $T(\mathbf{e}_j)$'s:

$$\mathbf{A} = \begin{bmatrix} T(\mathbf{e}_1) & T(\mathbf{e}_2) & \cdots & T(\mathbf{e}_m) \\ | & | & \cdots & | \\ \vdots & \vdots & \cdots & \vdots \end{bmatrix}$$

This way of stating the Observation immediately yields a sweeping fact:

THEOREM 5.6. *Every linear transformation* $T\colon \mathbf{R}^m \to \mathbf{R}^n$ *amounts to multiplication by a matrix—the matrix whose columns are the* $T(\mathbf{e}_j)$'s.

In particular, this means: **Every linear transformation is represented by a matrix.**

EXAMPLE 5.7. In Example 5.3, our linear mapping $T\colon \mathbf{R}^2 \to \mathbf{R}^2$ acted on the standard basis vectors of $\mathbf{R}^2$ as follows:

$$T(\mathbf{e}_1) = \begin{pmatrix} 7 \\ -3 \end{pmatrix} \quad \text{and} \quad T(\mathbf{e}_2) = \begin{pmatrix} -5 \\ 2 \end{pmatrix}$$

Theorem 5.6 immediately yields a matrix that represents T. In fact, it tells us that for any input vector (x, y), we have

$$T(x, y) = \begin{bmatrix} 7 & -5 \\ -3 & 2 \end{bmatrix} \begin{pmatrix} x \\ y \end{pmatrix}$$

It's easy to see that this gives the same formula for T we developed in Example 5.3:

$$T(x,y) = \begin{pmatrix} 7x - 5y \\ -3x + 2y \end{pmatrix}$$

□

EXAMPLE 5.8. Similarly, the linear *function* $F \colon \mathbf{R}^3 \to \mathbf{R}$ of Example 5.4, which satisfied

$$F(\mathbf{e}_1) = 2, \quad F(\mathbf{e}_2) = -1, \quad \text{and} \quad F(\mathbf{e}_3) = 3$$

can be represented by the one-rowed matrix $\mathbf{A} = \begin{bmatrix} 2 & -1 & 3 \end{bmatrix}$:

$$F(x,y,z) = \begin{bmatrix} 2 & -1 & 3 \end{bmatrix} \begin{pmatrix} x \\ y \\ z \end{pmatrix}$$

Note that because $\mathbf{A}$ here has only one row, we could write this even more compactly as a dot product:

$$F(x, y, z) = (2, -1, 3) \cdot (x, y, z)$$

Whenever the range of a linear mapping is just $\mathbf{R}$, that map will be represented by a one-rowed matrix, and hence by dotting, in this same way.

□

EXAMPLE 5.9. What 2×2 matrix represents the linear transformation $J \colon \mathbf{R}^2 \to \mathbf{R}^2$ given by the formula

$$J(x, y) = (-y, x) \ ?$$

According to Theorem 5.6, J is the same as multiplication by the matrix $\mathbf{J}$ whose columns are the $J(e_i)$'s. The formula for J above immediately gives

$$J(\mathbf{e}_1) = J(1, 0) = \begin{pmatrix} 0 \\ 1 \end{pmatrix} \quad \text{and} \quad J(\mathbf{e}_2) = J(0, 1) = \begin{pmatrix} -1 \\ 0 \end{pmatrix}$$

So J is represented by the matrix

$$\mathbf{J} = \begin{bmatrix} 0 & -1 \\ 1 & 0 \end{bmatrix}$$

It's easy to check that when we multiply this matrix by (x, y) we do indeed get $(-y, x)$.

□

EXAMPLE 5.10. What 3×3 matrix represents the linear transformation $R \colon \mathbf{R}^3 \to \mathbf{R}^3$ given by the formula

$$R(x, y, z) = (z, 2x, 3y) \ ?$$

By Theorem 5.6, R agrees with multiplication by the matrix $\mathbf{R}$ whose columns are the $R(\mathbf{e}_i)$'s. Using the formula for R above, we easily compute

$$R(\mathbf{e}_1) = R(1, 0, 0) = \begin{pmatrix} 0 \\ 2 \\ 0 \end{pmatrix}, \quad R(\mathbf{e}_2) = R(0, 1, 0) = \begin{pmatrix} 0 \\ 0 \\ 3 \end{pmatrix}$$

and

$$R(\mathbf{e}_3) = R(0, 0, 1) = \begin{pmatrix} 1 \\ 0 \\ 0 \end{pmatrix}$$

So R is represented by the matrix

$$\mathbf{R} = \begin{bmatrix} 0 & 0 & 1 \\ 2 & 0 & 0 \\ 0 & 3 & 0 \end{bmatrix}$$

Again, it's easy to check that when we multiply this matrix by an input vector (x, y, z), we get $(z, 2x, 3y)$, as expected. $\qquad\square$

5.11. Numeric vectors as functions. We conclude this chapter by noting that we can view any numeric vector

$$\mathbf{a} = (a_1, a_2, \ldots, a_m) \in \mathbf{R}^m$$

as a one-rowed matrix

$$\begin{bmatrix} a_1 & a_2 & \cdots & a_m \end{bmatrix}$$

A matrix with only one row, however, represents a linear *function* via matrix/vector multiplication, as we saw in Example 5.8. Calling this function $\mathbf{a}^*$ to emphasize its close relationship to the numeric vector $\mathbf{a}$, we then have, given any $\mathbf{x} = (x_1, x_2, \ldots, x_m) \in \mathbf{R}^m$,

$$\mathbf{a}^*(\mathbf{x}) := \begin{bmatrix} a_1 & a_2 & \cdots & a_m \end{bmatrix} \begin{pmatrix} x_1 \\ x_2 \\ \vdots \\ x_m \end{pmatrix}$$

Alternatively, using the dot-product formulation of matrix/vector multiplication, we can say the same thing like this:

$$\mathbf{a}^*(\mathbf{x}) \; := \; \mathbf{a} \cdot \mathbf{x}$$
$$= \; a_1 x_1 + a_2 x_2 + \cdots + a_m x_m$$

In short, $\mathbf{a}^*$ *is the linear function we get by dotting with* $\mathbf{a}$, and we call it the *dual* of $\mathbf{a}$. Formally:

DEFINITION 5.12 (Duals). Every numeric vector $\mathbf{a} \in \mathbf{R}^m$ determines a linear function $\mathbf{a}^* \colon \mathbf{R}^m \to \mathbf{R}$ via the simple rule

$$\mathbf{a}^*(\mathbf{x}) = \mathbf{a} \cdot x$$

We say that the linear function $\mathbf{a}^*$, and the numeric vector $\mathbf{a}$, are **duals** of each other. The function $\mathbf{a}^*$ is the dual of the vector $\mathbf{a}$, and vice-versa.

EXAMPLE 5.13. The numeric vector $\mathbf{a} := (1, 2, 3) \in \mathbf{R}^3$ is dual to the linear function given by $\mathbf{a}^*(x, y, z) = x + 2y + 3z$. So for instance,

$$\mathbf{a}^*(-2, 0, 5) \; = \; \mathbf{a} \cdot (-2, 0, 5) = 1 \cdot (-2) + 2 \cdot 0 + 3 \cdot 5 = 13$$
$$\mathbf{a}^*(-1, -1, 1) \; = \; \mathbf{a} \cdot (-1, -1, 1) = 1 \cdot (-1) + 2 \cdot (-1) + 3 \cdot 1 = 0$$

Similarly, the linear function on $\mathbf{R}^4$ given by

$$f(x, y, z, w) = x - y + 2z - 2w$$

is dual to the numeric vector $(1, -1, 2, -2) \in \mathbf{R}^4$. $\square$

Lastly, we note that our two ways to understand matrix/vector multiplication—as linear combinations of columns, and using dot products with rows—correspond precisely to the vector/function duality of Definition 5.12.

For we saw that when we multiply an $n \times m$ matrix $\mathbf{A}$ by a vector $\mathbf{x} = (x_1, x_2, \ldots, x_m) \in \mathbf{R}^m$, we can compute the product $\mathbf{Ax}$ by either:

- *Forming a linear combination:*

$$\mathbf{Ax} = x_1 \, \mathbf{c}_1(\mathbf{A}) + x_2 \, \mathbf{c}_2(\mathbf{A}) + \cdots + x_m \mathbf{c}_m(\mathbf{A})$$

  *Here the **columns** of $\mathbf{A}$ are numeric **vectors** in $\mathbf{R}^n$, or*

- *Forming a vector of dot products:*

$$\mathbf{A}\mathbf{x} = \begin{pmatrix} \mathbf{r}_1(\mathbf{A}) \cdot \mathbf{x} \\ \mathbf{r}_2(\mathbf{A}) \cdot \mathbf{x} \\ \vdots \\ \mathbf{r}_n(\mathbf{A}) \cdot \mathbf{x} \end{pmatrix}$$

Here the **rows** of $\mathbf{A}$ operate as **linear functions** on $\mathbf{R}^m$, since $\mathbf{r}_i(\mathbf{A}) \cdot \mathbf{x}$ is the same as $\mathbf{r}_i(\mathbf{A})^*(\mathbf{x})$.

We get the same result either way, of course, but we have two entirely different "dual" ways of getting there.

REMARK 5.14. (Duality) The row/column duality just described shows up throughout our subject, one of many beautiful dualities we highlight in this book.

In Linear Algebra, the word **duality** traditionally refers to the dual roles of *vector* and *function* associated with a given numeric vector, as in Definition 5.12.

We use the word more broadly, however, invoking it whenever we meet two fundamentally different ways of interpreting the same mathematical object. We can regard an $n \times n$ matrix, for instance, as a list of n (row) vectors in $\mathbf{R}^m$, or as a list of m (column) vectors in $\mathbf{R}^n$, and the difference is not superficial, as we shall come to see. Here and with each of the dualities we shall encounter—vector/function, row/column, image/pre-image, geometric/numeric, and perp/span—problems that seem difficult from one viewpoint often turn out to be much easier from the dual viewpoint. Each of the two perspectives illuminates the other in a beautiful and useful way, and we shall make good use of this principle many times. □

– Practice –

50. What matrix represents the projection $P\colon \mathbf{R}^4 \to \mathbf{R}^2$ given by $P(x_1, x_2, x_3, x_4) = (x_2, x_3)$?

51. What matrix represents the injection $Q\colon \mathbf{R}^2 \to \mathbf{R}^4$ given by $Q(x, y) = (0, x, y, 0)$?

52. In Exercise 41 we saw that multiplication by a scalar k defines a linear map $T_k\colon \mathbf{R}^m \to \mathbf{R}^m$ called a *homothety*. What matrix represents T_k?

53. What matrix represents the shift mapping $S : \mathbf{R}^n \to \mathbf{R}^n$ given by

$$S(x_1, x_2, x_3, \ldots, x_{n-1}, x_n) := (x_n, x_1, x_2, x_3, \cdots, x_{n-1})?$$

(Compare with Exercise 48.)

54. What matrix represents the mapping $R\colon \mathbf{R}^n \to \mathbf{R}^n$ we get by reversing the order of the coordinates? That is,

$$R(x_1, x_2, \ldots, x_{n-1}, x_n) = (x_n, x_{n-1}, \ldots, x_2, x_1)$$

55. Show that the mapping $T\colon \mathbf{R}^3 \to \mathbf{R}^3$ given by

$$T(x, y, z) = (2x, -y, 3z)$$

is linear. What matrix represents this mapping?

56. What matrix represents the linear transformation $S\colon \mathbf{R}^3 \to \mathbf{R}^3$ given by

$$S(x, y, z) = (y - z, \ z - x, \ x - y) \ ?$$

57. Show that the **identity** mapping $I\colon \mathbf{R}^n \to \mathbf{R}^n$ given by $I(\mathbf{x}) = \mathbf{x}$ is linear, no matter what n is. What matrix represents the identity transformation on $\mathbf{R}^n$?

58. Let $\mathbf{e}_i$ and $\mathbf{e}_j$ denote the ith and jth standard basis vectors in $\mathbf{R}^m$ and $\mathbf{R}^m$, respectively. Verify the linearity of the mapping $E_{ij}\colon \mathbf{R}^m \to \mathbf{R}^n$ given by

$$E_{ij}(\mathbf{x}) := (\mathbf{x} \cdot \mathbf{e}_i)\, \mathbf{e}_j$$

What matrix represents E_{ij}?

59. If $\mathbf{u} = (u_1, u_2, \ldots u_m) \in \mathbf{R}^m$ and $\mathbf{v} = (v_1, v_2, \ldots, v_n) \in \mathbf{R}^n$, are non-zero, we get a linear mapping $T\colon \mathbf{R}^m \to \mathbf{R}^n$ given by

$$T(\mathbf{x}) = \mathbf{u}^*(\mathbf{x})\mathbf{v}$$

   a) Show that this formula always defines a *linear* mapping.

   b) What is the entry in row i, column j of the matrix representing T ?

Mappings like the ones in Exercises 58 and 59 are called rank-1 *mappings for reasons that will become clear in later chapters. All linear mappings can be expressed as linear combinations of rank-1 mappings.*

60. Show that when a mapping is linear, it must send the origin in the domain to the origin in the range.

61. Find $\mathbf{a}^*(-2, 1, -4, 3)$ and $\mathbf{a}^*(0, 3, -5, 0)$ for each choice of $\mathbf{a} \in \mathbf{R}^4$ below.

    a) $\mathbf{a} = (1, 1, 1, 1)$

    b) $\mathbf{a} = (1, 2, -2, -1)$

    c) $\mathbf{a} = (0, 0, 1, 0)$

Similarly, find the numeric vector $\mathbf{a}$ dual to each linear function below:

    d) $f(x, y, z) = -3x + y - 5z$

    e) $g(u, v) = 16u + 26v$

    f) $h(x_1, x_2, x_3, x_4, x_5) = 5x_1 - 4x_2 + 3x_3 - 2x_4 + x_5$

62. What do we get when we apply the dual of $\mathbf{a} = (a_1, a_2, \ldots, a_n)$ to $\mathbf{a}$ itself? In other words, what is the formula for $\mathbf{a}^*(\mathbf{a})$?

63. (Cross product) Given any two vectors $\mathbf{a} = (a_1, a_2, a_3)$ and $\mathbf{x} = (x, y, z) \in \mathbf{R}^3$, their *cross-product* $\mathbf{a} \times \mathbf{x}$ (a new vector in $\mathbf{R}^3$) is given by the formula

$$\mathbf{a} \times \mathbf{x} = (a_2 z - a_3 y, \, a_3 x - a_1 z, \, a_1 y - a_2 x)$$

For any *fixed* $\mathbf{a}$, the formula $C_{\mathbf{a}}(\mathbf{x}) := \mathbf{a} \times \mathbf{x}$ defines a linear transformation $C_{\mathbf{a}} : \mathbf{R}^3 \to \mathbf{R}^3$. What matrix represents $C_{\mathbf{a}}$?

CHAPTER 2

Solving Linear Systems

1. The Linear System

We now begin to focus on answering the basic mapping questions for linear transformations; that is, for linear mappings

$$T : \mathbf{R}^m \to \mathbf{R}^n$$

As we observed in Theorem 5.6, every linear transformation is represented by a matrix, via matrix/vector multiplication. Specifically, we have the formula

$$T(\mathbf{x}) = \mathbf{A}\mathbf{x}$$

where $\mathbf{A}$ is the matrix whose columns are given by the $T(\mathbf{e}_j)$'s. For this reason, we can usually reduce questions about the mapping T to calculations involving the matrix $\mathbf{A}$.

In this chapter, we focus on the question of *pre-image*:

Problem: *Given a linear transformation* $T : \mathbf{R}^m \to \mathbf{R}^n$, *and a point* $\mathbf{b}$ *in the range of* T *how can we find all points in the pre-image* $T^{-1}(\mathbf{b})$.

Since every linear map amounts to multiplication by a matrix (and conversely, multiplication by any matrix $\mathbf{A}$ defines a linear map), finding $T^{-1}(\mathbf{b})$ is the same as *solving* $T(\mathbf{x}) = \mathbf{b}$ for $\mathbf{x}$. When T is represented by $\mathbf{A}$, we have $T(\mathbf{x}) = \mathbf{A}\mathbf{x}$, so the Problem above is exactly the same as this equivalent problem: *Given an* $n \times m$ *matrix* $\mathbf{A}$, *and a vector* $\mathbf{b} \in \mathbf{R}^n$, *how can we find every* $\mathbf{x} \in \mathbf{R}^m$ *that solves the matrix/vector equation*

$$(5) \qquad\qquad \mathbf{A}\mathbf{x} = \mathbf{b}$$

This statement of the problem is nice and terse, but to solve it, we first need to expand its symbols in terms of matrix entries and coordinates. We start with $\mathbf{A}$.

As an $n \times m$ matrix, $\mathbf{A}$ has n rows and m columns. Double-subscripting its entries in the usual way, we have

$$\mathbf{A} = \begin{bmatrix} a_{11} & a_{12} & a_{13} & \cdots & a_{1m} \\ a_{21} & a_{22} & a_{23} & \cdots & a_{2m} \\ \vdots & \vdots & \vdots & \cdots & \vdots \\ a_{n1} & a_{n2} & a_{n3} & \cdots & a_{nm} \end{bmatrix}$$

while

$$\mathbf{x} = \begin{pmatrix} x_1 \\ x_2 \\ \vdots \\ x_m \end{pmatrix} \quad \text{and} \quad \mathbf{b} = \begin{pmatrix} b_1 \\ b_2 \\ \vdots \\ b_n \end{pmatrix}$$

When we multiply (5) out using this notation, the single matrix/vector equation we want to solve, namely $\mathbf{Ax} = \mathbf{b}$, expands into the following system of n equations in m variables:

$$(6) \qquad \begin{array}{rcl} a_{11}x_1 + a_{12}x_2 + a_{13}x_3 + \cdots + a_{1m}x_m &=& b_1 \\ a_{21}x_1 + a_{22}x_2 + a_{23}x_3 + \cdots + a_{2m}x_m &=& b_2 \\ \vdots \qquad\quad \vdots \qquad\qquad\quad \vdots &=& \vdots \\ a_{n1}x_1 + a_{n2}x_2 + a_{n3}x_3 + \cdots + a_{nm}x_m &=& b_n \end{array}$$

DEFINITION 1.1 (Linear system). Any system having the form (6) above is called a *linear system in m variables*, or an $\mathbf{n} \times \mathbf{m}$ *linear system*. The a_{ij}'s are called the system **coefficients**, and form the entries of the **coefficient matrix A**. The **variables** x_j and the **right-hand sides** b_i form the variable vector $\mathbf{x}$ and the right-hand side vector $\mathbf{b}$, respectively.

By a **solution** of such a system, we mean *any numeric vector* $\mathbf{x} = (x_1, x_2, \ldots, x_m)$ *that simultaneously solves every equation in the system.* Such a vector then solves $\mathbf{Ax} = \mathbf{b}$, and is also a pre-image of $\mathbf{b}$ under the mapping $T(\mathbf{x}) = \mathbf{Ax}$. □

EXAMPLE 1.2. Suppose

$$\mathbf{A} = \begin{bmatrix} 1 & -1 & 0 \\ 0 & 1 & -1 \\ -1 & 0 & 1 \end{bmatrix}, \quad \mathbf{x} = (x, y, z), \quad \text{and} \quad \mathbf{b} = (-1, -1, 2)$$

Then the matrix/vector equation $\mathbf{Ax} = \mathbf{b}$ expands as

$$\begin{array}{rcl} x - y + 0z &=& -1 \\ 0x + y - z &=& -1 \\ -x + 0y + z &=& 2 \end{array} \quad \text{or more simply,} \quad \begin{array}{rcl} x - y &=& -1 \\ y - z &=& -1 \\ z - x &=& 2 \end{array}$$

□

Linear mappings can be onto, one-to-one, both, or neither. It follows that linear *systems* may have one, many, or no solutions. Before tackling the general problem of finding these solutions (if any), however, let's get a sense of the problem by considering simple examples.

EXAMPLE 1.3. This 2×2 system has exactly one solution, given by $(x, y) = (-2, 5)$:

$$
\begin{aligned}
3x + 2y &= 4 \\
5x + 3y &= 5
\end{aligned}
$$

Leaving aside the question of how to *find* this solution, it is easy to check that when $x = -2$ and $y = 5$, we do solve both equations. □

EXAMPLE 1.4. This 3×3 system has **no** solution:

$$
\begin{aligned}
x - y &= 1 \\
y - z &= 1 \\
z - x &= 1
\end{aligned}
$$

For clearly, the first equation implies $x > y$, the second implies $y > z$, and the third implies $z > x$. If there were a solution, it would therefore have to satisfy $x > y > z > x$, which is impossible, since x can't be greater than itself. □

EXAMPLE 1.5. This 3×3 system, though differing from the last one only on the right-hand side of the second equation, has infinitely many solutions:

$$
\begin{aligned}
x - y &= 1 \\
y - z &= -2 \\
z - x &= 1
\end{aligned}
$$

For the triple $(x, y, z) = (1, 0, 2)$ solves it, and it is easy to check that so does $(x, y, z) = (1, 0, 2) + (t, t, t)$, for *any* value of t. □

Amazingly, there's a fairly simple procedure, called the **Gauss–Jordan algorithm**—that finds all solutions of **any** $n \times m$ linear system, no matter how big m and n are. Admittedly, it can require many steps. In such cases humans get tired, bored, and tend to make mistakes. Computers, however, have none of those problems, and can solve systems with millions of equations and millions of unknowns. Our main goal in this chapter will be to discover and understand the Gauss–Jordan algorithm.

Before we get there, however, we want to point out that linear systems have not just a *dual,* but a *triple* nature. Every linear system can be seen as posing three completely different problems.

1.6. Three ways to interpret systems. We have introduced linear systems as a way of solving the "pre-image" problem: When $\mathbf{A}$ is $n \times m$, the solutions of $\mathbf{Ax} = \mathbf{b}$ are the pre-images (in $\mathbf{R}^m$) of $\mathbf{b} \in \mathbf{R}^n$ under the linear map T represented by $\mathbf{A}$, so that for every input $\mathbf{x} \in \mathbf{R}^m$,

$$T(\mathbf{x}) = \mathbf{Ax}$$

Along with this *pre-image* problem, however, there are two other important ways to interpret the same system. We call them the *column* and *row* problems, respectively.

Column problem. The *column problem* is about linear combinations in the *range* $\mathbf{R}^n$.

EXAMPLE 1.7. Consider the simple 2×3 system

$$\begin{aligned} 3x - 4y + z &= 0 \\ 5x + 3y - 2z &= 1 \end{aligned}$$

This has the form $\mathbf{Ax} = \mathbf{b}$, where

$$\mathbf{A} = \begin{bmatrix} 3 & -4 & 1 \\ 5 & 3 & -2 \end{bmatrix}, \qquad \mathbf{x} = \begin{pmatrix} x \\ y \\ z \end{pmatrix}, \quad \text{and} \quad \mathbf{b} = \begin{pmatrix} 0 \\ 1 \end{pmatrix}$$

As we emphasized in Remark 1.22, however, *matrix/vector multiplication equals linear combination,* so we can *also* read the system as a problem about linear combinations of the columns of $\mathbf{A}$. In this case, for instance, $\mathbf{Ax} = \mathbf{b}$ becomes

$$x \begin{pmatrix} 3 \\ 5 \end{pmatrix} + y \begin{pmatrix} -4 \\ 3 \end{pmatrix} + z \begin{pmatrix} 1 \\ -2 \end{pmatrix} = \begin{pmatrix} 0 \\ 1 \end{pmatrix}$$

or

$$x \, \mathbf{a}_1 + y \, \mathbf{a}_2 + z \, \mathbf{a}_3 = \mathbf{e}_2$$

where $\mathbf{a}_j$ denotes column j of $\mathbf{A}$. Seen in this light, solving the linear system means *finding a linear combination of the* $\mathbf{a}_i$ *vectors that adds up to* $(0, 1)$. □

The same reasoning applies to *every* linear system. Just as in the example above, we can rewrite the matrix/vector equation $\mathbf{Ax} = \mathbf{b}$ as the vector equation

$$(7) \qquad x_1\, \mathbf{c}_1(\mathbf{A}) + x_2\, \mathbf{c}_2(\mathbf{A}) + \cdots + x_m\, \mathbf{c}_m(\mathbf{A}) = \mathbf{b}$$

where the $\mathbf{c}_i(\mathbf{A})$ denotes column i of the coefficient matrix $\mathbf{A}$. This reformulates the problem entirely: instead of seeking all pre-images of $\mathbf{b}$ under the linear mapping represented by $\mathbf{A}$, we now have a question about linear combinations:

What linear combination of the columns of $\mathbf{A}$ (if any) will add up to the "target" vector $\mathbf{b}$?

We call this the *column problem* because it focuses on the columns of $\mathbf{A}$. It is the second of three basic ways of interpreting linear systems (the first being the *pre-image* problem).

Row problem. The third way to interpret a linear system is called the *row problem*. It asks us to solve simultaneous linear equations in the *domain* $\mathbf{R}^n$. Specifically, it sees $\mathbf{Ax} = \mathbf{b}$ as n distinct equations, one for each *row* of $\mathbf{A}$, that we must solve simultaneously. For instance, when

$$\mathbf{A} = \begin{bmatrix} 3 & 2 \\ 5 & 3 \end{bmatrix} \quad \text{and} \quad \mathbf{b} = \begin{pmatrix} 4 \\ 5 \end{pmatrix}$$

we get the system of Example (1.3):

$$\begin{aligned} 3x + 2y &= 4 \\ 5x + 3y &= 5 \end{aligned}$$

Individually, there are two equations here—and each has infinitely many solutions. Indeed, given *any* value of x we can solve $3x + 2y = 4$ for y and get a solution, and the same technique gives infinitely many solutions for the second equation. The *system* however, asks us to find all (x, y) pairs that solve *both* equations. In this case there is exactly one such pair: $(x, y) = (-2, 5)$.

There is another way to formulate this. Each individual equation in the system $\mathbf{Ax} = \mathbf{b}$ has the form $\mathbf{r}_i(\mathbf{A}) \cdot \mathbf{x} = b_i$, where $\mathbf{r}_i(\mathbf{A})$ denotes row i of $\mathbf{A}$ (and b_i is the ith *entry* in $\mathbf{b}$). We could write the 2-by-2 system above, for example, as

$$\begin{aligned} (3, 2) \cdot (x, y) &= 4 \\ (5, 3) \cdot (x, y) &= 5 \end{aligned}$$

Here the ith row of $\mathbf{A}$ is really acting in its *dual* capacity as a *linear function*, and to emphasize that, we could replace $\mathbf{r}_i(\mathbf{A}) \cdot \mathbf{x}$ with $\mathbf{r}_i(\mathbf{A})^*(\mathbf{x})$. In any case, the *rows* of $\mathbf{A}$ take center stage in this interpretation, each row yielding a single equation that we have to solve. Each of these equations takes the form

$$\mathbf{r}_i(\mathbf{A}) \cdot \mathbf{x} = b_i$$

So from the row standpoint, the problem $\mathbf{Ax} = \mathbf{b}$ becomes:

Find all vectors in $\mathbf{R}^m$ (if any) that dot with each row of $\mathbf{A}$ to give the corresponding entry in $\mathbf{b}$.

This is the row problem. It is dual to the column problem in exactly the sense of Definition 5.12: In the *column* problem, we view the columns of $\mathbf{A}$ as numeric vectors, and seek a linear combination that adds up to the right-hand side $\mathbf{b}$. In the *row* problem, each row of $\mathbf{A}$ acts in its dual capacity as a linear function, and we seek an input $\mathbf{x}$ for which each of these linear functions returns the value given by the right-hand side of its row.

In Chapter 3 we will develop a vivid *geometric* visualization for the row problem.

1.8. Summary. There are three fundamentally different ways of interpreting any linear system $\mathbf{Ax} = \mathbf{b}$. We can view it as either a

- **Pre-image problem:** Find all pre-images $\mathbf{b} \in \mathbf{R}^n$ under the linear map $T : \mathbf{R}^m \to \mathbf{R}^n$ represented by the coefficient matrix $\mathbf{A}$.

- **Column problem:** Find all ways to linearly combine the columns of $\mathbf{A}$ to produce $\mathbf{b} \in \mathbf{R}^n$. This is a problem about linear combinations in the range, $\mathbf{R}^n$.

- **Row problem:** Find all vectors in the domain $\mathbf{R}^m$ that simultaneously solve each of the linear equations $\mathbf{r}_i(\mathbf{A})^*(\mathbf{x}) = b_i$ for $i = 1, 2, \ldots, n$. Equivalently, find all vectors in the *intersection* of the solution sets of all these equations. These sets all lie in $\mathbf{R}^m$.

By moving between these three interpretations of a linear system, we will see into $\mathbf{Ax} = \mathbf{b}$ much more deeply than we could using any one or two of these perspectives alone.

– Practice –

64. Find the coefficient matrix $\mathbf{A}$, the variable vector $\mathbf{x}$, and the right-hand side vector $\mathbf{b}$ for this system:

$$
\begin{array}{rcl}
2x_1 - 3x_2 + x_4 &=& 12 \\
2x_2 - 7x_3 - x_4 &=& -5 \\
-4x_1 + x_2 - 5x_3 - 2x_4 &=& 0
\end{array}
$$

65. Find the coefficient matrix $\mathbf{A}$, the variable vector $\mathbf{x}$, and the right-hand side vector $\mathbf{b}$ for these systems:

a)
$$
\begin{array}{rcl}
x_1 - x_2 &=& 0 \\
x_2 - x_3 &=& 0 \\
x_3 - x_4 &=& 0 \\
x_4 - x_4 &=& 0
\end{array}
$$

b)
$$
x_1 + x_2 - x_3 + x_4 - x_5 = 0
$$

c)
$$
\begin{array}{rcl}
x_1 - x_2 + x_3 - x_4 &=& 0 \\
x_2 - x_3 + x_4 &=& 0 \\
x_3 - x_4 &=& 0
\end{array}
$$

66. If $\mathbf{A}$ and $\mathbf{b}$ are the matrix and vector given by

$$
\mathbf{A} = \begin{bmatrix} -8 & -5 & 9 \\ 6 & 3 & -1 \\ 9 & -5 & 7 \\ -8 & -3 & -8 \end{bmatrix} \quad \text{and} \quad \mathbf{b} = \begin{pmatrix} -3 \\ 8 \\ 2 \\ -5 \end{pmatrix}
$$

What linear system (written out in full) corresponds to the vector equation $\mathbf{Ax} = \mathbf{b}$?

67. What linear systems (written-out in full) correspond to the vector equation $\mathbf{Ax} = \mathbf{b}$ when $\mathbf{A}$ and $\mathbf{b}$ are given, respectively, by

a)
$$
\mathbf{A} = \begin{bmatrix} 2 & 0 & -1 & 3 \\ 0 & 2 & -3 & 1 \end{bmatrix} \quad \text{and} \quad \mathbf{b} = \begin{pmatrix} 1.2 \\ 2.1 \end{pmatrix}
$$

b)
$$
\mathbf{A} = \begin{bmatrix} 2 & 0 \\ 0 & 2 \\ -1 & -3 \\ -3 & 1 \end{bmatrix} \quad \text{and} \quad \mathbf{b} = \begin{pmatrix} 5 \\ -3 \\ 1 \\ 4 \end{pmatrix}
$$

68. Let
$$\mathbf{A} = \begin{bmatrix} 3 & 0 & -1 \\ 2 & 1 & 0 \end{bmatrix} \qquad \mathbf{b} = \begin{pmatrix} 2 \\ 1 \end{pmatrix}$$

Which of the vectors $\mathbf{x}$ below solve $\mathbf{Ax} = \mathbf{b}$? (Show your work.)

a) $\mathbf{x} = \begin{pmatrix} 2 \\ -3 \\ 4 \end{pmatrix}$ b) $\mathbf{x} = \begin{pmatrix} -3 \\ 2 \\ 1 \end{pmatrix}$ c) $\mathbf{x} = \begin{pmatrix} 1 \\ -1 \\ 1 \end{pmatrix}$

d) $\mathbf{x} = \begin{pmatrix} 3 \\ -5 \\ 7 \end{pmatrix}$ e) $\mathbf{x} = \begin{pmatrix} 2 \\ -2 \\ 2 \end{pmatrix}$

69. a) Let $\mathbf{x} = (x, y, z)$. Express the "row problem"
$$\begin{aligned} \mathbf{a}_1^*(\mathbf{x}) &= 0 \\ \mathbf{a}_2^*(\mathbf{x}) &= -1 \\ \mathbf{a}_3^*(\mathbf{x}) &= 2 \end{aligned}$$

as a linear system, assuming $\mathbf{a}_1 = (1, -2, 3)$, $\mathbf{a}_2 = (-2, 1, -2)$, and $\mathbf{a}_3 = (3, -2, 1)$.

  b) Express the linear system below in "row" form like the system in part (a):
$$\begin{aligned} 3x - 2y &= 0 \\ 2x - 3y &= 1 \\ 2x + 3y &= 0 \\ 3x + 2y &= -1 \end{aligned}$$

70. a) Express the "column problem"
$$u\,\mathbf{a}_1 + v\,\mathbf{a}_2 + w\,\mathbf{a}_3 = \mathbf{b}$$

as a linear system in the variables u, v, w, assuming $\mathbf{a}_1 = (1, -2, 3)$, $\mathbf{a}_2 = (-2, 1, -2)$, and $\mathbf{a}_3 = (3, -2, 1)$.

  b) Express the linear system below in "column" form $x\,\mathbf{a}_1 + y\,\mathbf{a}_2 = \mathbf{b}$:
$$\begin{aligned} 3x - 2y &= 0 \\ 2x - 3y &= 1 \\ 2x + 3y &= 0 \\ 3x + 2y &= -1 \end{aligned}$$

71.   a) Express the "column problem"

$$u\,\mathbf{a}_1 + v\,\mathbf{a}_2 + w\,\mathbf{a}_3 = \mathbf{b}\,,$$

as a linear system in the variables u, v, w, assuming $\mathbf{a}_1 = (1, -2, 3)$, $\mathbf{a}_2 = (-2, 1, -2)$, and $\mathbf{a}_3 = (3, -2, 1)$.

b) Express the linear system below in "column" form $x\,\mathbf{a}_1 + y\,\mathbf{a}_2 = \mathbf{b}$:

$$
\begin{aligned}
3x - 2y &= 0 \\
2x - 3y &= 1 \\
2x + 3y &= 0 \\
3x + 2y &= -1
\end{aligned}
$$

72. Consider this problem:

What vectors $\mathbf{x} \in \mathbf{R}^4$ *dot to zero with all three vectors* $(1, 1, 2, 3)$, $(1, 2, 3, 1)$, *and* $(2, 1, 1, 3)$ *simultaneously?*

What linear system expresses this problem? You needn't solve it.

73. Consider this problem:

What linear combinations (if any) of the vectors $(1, 1, 2, 3)$, $(1, 2, 3, 1)$, *and* $(2, 1, 1, 3)$ *add up to* $(5, -2, 3, 7)$?

What linear system expresses this problem? You needn't solve it.

2. The Augmented Matrix and RRE Form

We now turn to the problem of actually *solving* linear systems. There is a standard procedure for doing so, and the solutions it provides (if any) *always* arise as the *image* of a new mapping.

We formalize this fact as a definition:

DEFINITION 2.1. We **solve an** $n \times m$ **linear system** $\mathbf{Ax} = \mathbf{b}$ by *finding a mapping* $\mathbf{X} : \mathbf{R}^k \to \mathbf{R}^m$ *(for some* $0 \le k \le m$ *) whose outputs provide all solutions to the system—and no others.* In short, we solve a system by finding a mapping $\mathbf{X}$ whose *image* forms the solution set.

EXAMPLE 2.2. For instance, in Example 1.5, we got all solutions of the system

$$\begin{aligned} x - y &= 1 \\ y - z &= -2 \\ z - x &= 1 \end{aligned}$$

by plugging any real number t into the formula

$$(x, y, z) = (1, 0, 2) + (t, t, t)$$

This means the solutions of the system above are the *same as the image of the mapping* $\mathbf{X} : \mathbf{R} \to \mathbf{R}^3$ *given by*

$$\mathbf{X}(t) = \begin{pmatrix} 1 \\ 0 \\ 2 \end{pmatrix} + t \begin{pmatrix} 1 \\ 1 \\ 1 \end{pmatrix}$$

$\square$

In general, when $\mathbf{A}$ represents a linear map $T : \mathbf{R}^m \to \mathbf{R}^n$, and we're given a "target" vector $\mathbf{b} \in \mathbf{R}^n$, we seek a solution mapping X that fits into a simple diagram like this:

$$\mathbf{R}^k \xrightarrow{X} \mathbf{R}^m \xrightarrow{T} \mathbf{R}^n$$

and *we want the image of* $\mathbf{X}$ *to equal the pre-image of* $\mathbf{b}$, *so that* $\mathbf{X}(\mathbf{R}^k) = T^{-1}(\mathbf{b})$.

The crucial question is, *How can we find a solution mapping* $\mathbf{X}$ *with this property for any linear system we encounter?*

2.3. The augmented matrix of a system. The first step toward answering that question is purely notational: We form the *augmented matrix* of the system:

DEFINITION 2.4. We get the **augmented matrix** of the $n \times m$ system $\mathbf{Ax} = \mathbf{b}$ by taking the coefficient matrix $\mathbf{A}$ and adjoining $\mathbf{b}$ as an additional $(m + 1)$th column: We indicate the augmented matrix like this: $[\mathbf{A} \,|\, \mathbf{b}]$.

Written out entry-wise, the augmented matrix of the system

$$\begin{aligned} a_{11}x_1 + a_{12}x_2 + a_{13}x_3 + \cdots + a_{1m} &= b_1 \\ a_{21}x_1 + a_{22}x_2 + a_{23}x_3 + \cdots + a_{2m} &= b_2 \\ \vdots \qquad\qquad\qquad \vdots \; &= \; \vdots \\ a_{n1}x_1 + a_{n2}x_2 + a_{n3}x_3 + \cdots + a_{nm} &= b_n \end{aligned}$$

is

$$\begin{bmatrix} a_{11} & a_{12} & a_{13} & \cdots & a_{1m} & b_1 \\ a_{21} & a_{22} & a_{23} & \cdots & a_{2m} & b_2 \\ \vdots & \vdots & & \cdots & \vdots & \\ a_{n1} & a_{n2} & a_{n3} & \cdots & a_{nm} & b_n \end{bmatrix}$$

EXAMPLE 2.5. The augmented matrix of the system

$$\begin{aligned} 2x + 3y &= -1 \\ 3x - 2y &= 0 \end{aligned}$$

is

$$[\mathbf{A} \,|\, \mathbf{b}] = \begin{bmatrix} 2 & 3 & -1 \\ 3 & -2 & 0 \end{bmatrix}$$

$\square$

Note that the augmented matrix contains all the information we need to reconstruct the original system.

EXAMPLE 2.6. Suppose we know that the augmented matrix of a certain system is

$$\begin{bmatrix} 3 & -2 & 1 & 4 \\ 5 & -3 & 1 & 5 \end{bmatrix}$$

then all we have to do to recover the original system is to put the variables and $+, -$ and $=$ signs in the appropriate locations:

$$\begin{aligned} 3x_1 - 2x_2 + x_3 &= 4 \\ 5x_1 - 3x_2 + x_3 &= 5 \end{aligned}$$

$\square$

2.7. Systems with obvious solutions: RRE form. Starting with the augmented matrix of a system—any system—we will solve the system using the simple strategy discovered two centuries ago by mathematical giant C.F. Gauss.[1]

---

[1]The procedure was later modified toward the precise algorithm we'll use by scientist W. Jordan—and others.

Gauss–Jordan Strategy: *Simplify the augmented matrix of the original system **without changing its solution set**, until it represents a system so simple that we can read off the solution by inspection. The solution of the final, simple system will then solve the original system too.*

We'll discover the Gauss–Jordan algorithm for ourselves by working backwards from a crucial question:

When is a system so simple that one can deduce its solution by inspection?

Some basic examples will guide us toward the answer.

EXAMPLE 2.8. Consider the system whose augmented matrix is

$$\begin{bmatrix} 1 & 0 & 0 & -2 \\ 0 & 1 & 0 & 3 \\ 0 & 0 & 1 & 0 \end{bmatrix}$$

It is easy to see that this system has exactly one solution, and we can write it down immediately. First note that the matrix has four *columns*, so there are three unknowns. Further, each *row* represents an equation that completely determines the value of one unknown: the first row says $x_1 = -2$, the second says $x_2 = 3$, and the third says $x_3 = 0$.

That makes $(x_1, x_2, x_3) = (-2, 3, 0)$ a solution—and the only possible solution.   □

EXAMPLE 2.9. Drop the third equation from the previous example. The augmented matrix will then be

$$\begin{bmatrix} 1 & 0 & 0 & -2 \\ 0 & 1 & 0 & 3 \end{bmatrix}$$

Again the first equation says $x_1 = -2$, and the second says $x_2 = 3$. There are still four columns, and thus three variables, but we no longer have an equation that sets x_3 to any particular value. So what value can we assign it?

The answer is... *any* scalar! The third variable gets multiplied by zero in both equations of the system, so its value affects nothing.

We therefore have a system with *infinitely many solutions*. They all have $x_1 = -2$ and $x_2 = 3$, but x_3 can be any number t, and we can express this very simply by writing the general solution as

$$(x_1, x_2, x_3) = (-2, 3, t)$$

Alternatively we can see this as the image of the mapping $\mathbf{X} : \mathbf{R} \to \mathbf{R}^3$ given by

$$\mathbf{X}(t) = \begin{pmatrix} -2 \\ 3 \\ 0 \end{pmatrix} + t \begin{pmatrix} 0 \\ 0 \\ 1 \end{pmatrix}$$

$\square$

EXAMPLE 2.10. If we switch the last two columns of the matrix in the previous example, we get a more interesting situation, but the solution is only slightly harder to tease out. The augmented matrix becomes

$$\begin{bmatrix} 1 & 0 & -2 & 0 \\ 0 & 1 & 3 & 0 \end{bmatrix}$$

We can *still* assign any value to x_3 here, and write $x_3 = t$, where t can be any scalar, just as in our last example. Then write out the two equations: they are $x_1 - 2x_3 = 0$ and $x_2 + 3x_3 = 0$. Moving the x_3-terms to the right-hand sides and using $x_3 = t$ we then get

$$\begin{aligned} x_1 &= 2t \\ x_2 &= -3t \end{aligned}$$

If we now view our assignment

$$x_3 = t$$

as a third equation, we have three equations that clearly describe all possible solutions. The t-value we give to x_3 completely determines the other two variables, and this again gives us infinitely many solutions— one for each value t. We can express the result by the simple formula

$$(x_1, x_2, x_3) = (2t, -3t, t)$$

Again, we may interpret the solutions as the image of a mapping $\mathbf{X} : \mathbf{R} \to \mathbf{R}^3$, namely

$$\mathbf{X}(t) = t \begin{pmatrix} 2 \\ -3 \\ 1 \end{pmatrix}$$

$\square$

Each system above was easy to solve. What made that true? Before we highlight the key feature they share, we should consider one more situation:

EXAMPLE 2.11. Modify the last example by adding the row $(0\ 0\ 0\ 1)$ at the bottom, to get the augmented matrix

$$\begin{bmatrix} 1 & 0 & -2 & 0 \\ 0 & 1 & 3 & 0 \\ 0 & 0 & 0 & 1 \end{bmatrix}$$

This matrix describes a system with **no** solution. Indeed, whenever the augmented matrix has a row of the form $(0\ 0\ \cdots\ 0\ c)$ (where $c \neq 0$), that row represents an absurd equation of the form

$$0x_1 + 0x_2 + 0x_3 = c$$

Since $c \neq 0$, this equation is preposterous—it obviously isn't solved by *any* values of x_1, x_2, and x_3. □

This last situation may seem bizarre or even silly: it's hard to imagine an equation like $0x_1 + 0x_2 + 0x_3 = 1$ arising in a "real-world" system. And in truth, we seldom encounter such an equation when we *start* solving a system. But such equations often *do* arise *during* the Gauss–Jordan algorithm. When they do, they tell us the original system has no solution. So we really must include them in our thinking.

2.12. Reduced row-echelon form. Consideration of examples like the above eventually yields an answer to the question posed earlier:

Question: *When is a system so simple that we can read off its solution by inspection?*

Answer: When its augmented matrix has *reduced row-echelon* form.

DEFINITION 2.13 (RRE form). A matrix has **reduced row-echelon** (RRE) form when it satisfies these four conditions:

- **Leading 1's:** *The leftmost non-zero entry in every row is a one (called **leading ones**, or **pivot**).*

- **Southeast:** *Each pivot lies to the right of all pivots above it ("Pivots head southeast").*

- **Pivot columns:** *If a column has a pivot, all its other entries are zeros.*

- **0-rows sink:** *Rows populated entirely by zeros lie below all rows with non-zero entries.*

EXAMPLE 2.14. Here are three simple, but rather different matrices in RRE form. You should check that all of the four conditions above apply in each case.

$$\begin{bmatrix} 1 & 2 & 0 & 3 \\ 0 & 0 & 1 & -1 \end{bmatrix} \qquad \begin{bmatrix} 0 & 1 \\ 0 & 0 \\ 0 & 0 \\ 0 & 0 \end{bmatrix} \qquad \begin{bmatrix} 1 & 0 & 0 \\ 0 & 1 & 0 \\ 0 & 0 & 1 \end{bmatrix}$$

□

EXAMPLE 2.15. Every 2×3 in RRE form looks like one of the seven matrices below, where a and b can be any scalars (including 0 or 1):

$$\begin{bmatrix} 0 & 0 & 0 \\ 0 & 0 & 0 \end{bmatrix}$$

$$\begin{bmatrix} 1 & a & b \\ 0 & 0 & 0 \end{bmatrix} \qquad \begin{bmatrix} 0 & 1 & a \\ 0 & 0 & 0 \end{bmatrix} \qquad \begin{bmatrix} 1 & 0 & a \\ 0 & 1 & b \end{bmatrix}$$

$$\begin{bmatrix} 0 & 0 & 1 \\ 0 & 0 & 0 \end{bmatrix} \qquad \begin{bmatrix} 1 & a & 0 \\ 0 & 0 & 1 \end{bmatrix} \qquad \begin{bmatrix} 0 & 1 & 0 \\ 0 & 0 & 1 \end{bmatrix}$$

The reader should check that all seven of these satisfy each of the four conditions in Definition 2.13.

Let's work out the **solutions** to the systems corresponding to each 2×3 RRE form in Example 2.15 above.

• The top one (the zero matrix) is easy: both its rows represent the trivial equation $0x + 0y = 0$. *Every* $(x, y) \in \mathbf{R}^2$ solves this equation, so that's the solution: x and y can be anything. Both variables can take any value. We therefore set $x = s$, $y = t$, and get the solution mapping

$$\mathbf{X}(s, t) = (s, t) = s\, \mathbf{e}_1 + t\, \mathbf{e}_2$$

• The matrices on the last line of Example 2.15 are easy too, because each has a row of the form $(0 \ 0 \ 1)$. That row stands for the absurd equation $0x + 0y = 1$, as discussed in Example 2.11 above. When one equation has *no* solution, the system certainly can't have one either. So any system with an augmented matrix like the ones in the third row of our list above has **no** solution.

• The middle row of Example 2.15 is more interesting. Consider the first matrix in that row, for example:

$$\begin{bmatrix} 1 & a & b \\ 0 & 0 & 0 \end{bmatrix}$$

The second row here is all zeros, which we know represents an equation solved by *any* (x, y) pair. So only the first row constrains the variables, and it stands for the equation

$$x + ay = b$$

We can set $y = s$ for any scalar s, and then solve the first equation for x to get $x = b - as$. We then have both x and y in terms of s, and the solution set is thus the image of the mapping

$$\mathbf{X}(s) = \begin{pmatrix} b \\ 0 \end{pmatrix} + s \begin{pmatrix} -a \\ 1 \end{pmatrix}$$

$\square$

The number of possible $n \times m$ RRE forms grows quickly as n and m increase. Still, there is a straightforward way to read off the solution of a system, no matter how large, *when its augmented matrix is in* RRE *form*. Before we analyze that general problem, however, we note that it becomes truly trivial in two cases, independent of size:

Trivial case #1. ("Anything goes.") *The augmented matrix contains only zeros.*

In this case, *all* vectors $(x_1, x_2, \ldots, x_m) \in \mathbf{R}^m$ solve the system, just as we saw in the first 2×3 case above. For the equation

$$0x_1 + 0x_2 + \cdots + 0x_m = 0$$

obviously holds no matter what values we give to $x_1, x_2, \ldots, x_m$.

Trivial case #2. ("Nothing goes.") *The augmented matrix has a pivot in the last column.*

In this case, the system has *no* solution, as illustrated by the matrices in the last row of Example 2.15. A pivot in the last column always stands for an equation of the form

$$0x_1 + 0x_2 + \cdots + 0x_m = 1$$

Obviously, no vector solves this equation, and it follows that no vector can solve a system containing such an equation.

– **Practice** –

74. What are the augmented and coefficient matrices for these systems?

a) $\begin{aligned} 2x + 3y &= 7 \\ 3x - 2y &= 4 \end{aligned}$ b) $\begin{aligned} y &= 12 \\ x &= -1 \end{aligned}$ c) $\begin{aligned} 7x - 2y &= 5 \\ 6x - 3y &= 0 \end{aligned}$

75. What are the augmented and coefficient matrices for these systems?

a) $\begin{aligned} x + y + z &= 3 \\ x - 2y + z &= -9 \\ x + y - z &= 3 \end{aligned}$ b) $\begin{aligned} 2x_1 + x_2 &= -1 \\ x_2 + 2x_3 + x_4 &= -1 \\ x_3 + 2x_4 + x_5 &= -1 \\ x_4 + 2x_5 &= -1 \end{aligned}$

76. What systems have the augmented matrices below?

a) $\begin{bmatrix} 1 & 2 & 3 \\ 4 & 5 & 6 \end{bmatrix}$ b) $\begin{bmatrix} 0 & 3 & 0 \\ 2 & 0 & 1 \end{bmatrix}$ c) $\begin{bmatrix} 1 & 1 & -1 \\ 1 & -1 & 1 \end{bmatrix}$

77. What systems have these augmented matrices?

a) $\begin{bmatrix} 1 & 2 & -3 \\ 2 & 1 & 4 \end{bmatrix}$ b) $\begin{bmatrix} 1 & 2 & -2 & 1 & 3 \\ 2 & 1 & 3 & -5 & 0 \\ -2 & 3 & 2 & 1 & 3 \end{bmatrix}$

c) $\begin{bmatrix} 1 & -1 & 1 & -1 & 1 & -1 \end{bmatrix}$

78. Find all solutions to the systems represented by the remaining two augmented matrices in Example 2.15, namely

$$\begin{bmatrix} 0 & 1 & a \\ 0 & 0 & 0 \end{bmatrix} \quad \text{and} \quad \begin{bmatrix} 1 & 0 & a \\ 0 & 1 & b \end{bmatrix}$$

79. The four matrices below are **not** in reduced row-echelon (RRE) form. Each fails exactly one of the four necessary conditions. Which matrix fails which condition?

a) $\begin{bmatrix} 1 & 2 & 0 & 3 \\ 0 & 0 & 2 & -1 \end{bmatrix}$ b) $\begin{bmatrix} 0 & 1 \\ 1 & 0 \\ 0 & 0 \\ 0 & 0 \end{bmatrix}$ c) $\begin{bmatrix} 1 & 0 & 1 \\ 0 & 1 & 0 \\ 0 & 0 & 1 \end{bmatrix}$

$$\text{d)} \quad \begin{bmatrix} 1 & 5 & 0 \\ 0 & 0 & 0 \\ 0 & 0 & 1 \end{bmatrix}$$

80. Which matrices below are in reduced row-echelon (RRE) form?

$$\text{a)} \quad \begin{bmatrix} 1 & 2 & 0 & -1 \\ 0 & 1 & 0 & -1 \\ 0 & 0 & 1 & 0 \end{bmatrix} \qquad \text{b)} \quad \begin{bmatrix} 0 & 1 & 0 & -1 \\ 0 & 0 & 1 & -1 \\ 0 & 0 & 0 & 0 \end{bmatrix}$$

$$\text{c)} \quad \begin{bmatrix} 0 & 1 & 0 & -1 \\ 0 & 0 & 1 & -1 \\ 0 & 0 & 1 & -1 \end{bmatrix} \qquad \text{d)} \quad \begin{bmatrix} 0 & 0 & 1 & -1 \\ 0 & 1 & 0 & -1 \\ 1 & 0 & 0 & 0 \end{bmatrix}$$

$$\text{e)} \quad \begin{bmatrix} 1 & 0 & 0 \\ 0 & 1 & 0 \\ 0 & 0 & 0 \\ 0 & 0 & 1 \end{bmatrix} \qquad \text{f)} \quad \begin{bmatrix} 1 & 1 & 1 \\ 0 & 0 & 0 \\ 0 & 0 & 0 \\ 0 & 0 & 0 \end{bmatrix}$$

81. Every 3×3 matrix in RRE form falls into one of eight different classes. The **zero** and **identity** matrices

$$\mathbf{0} = \begin{bmatrix} 0 & 0 & 0 \\ 0 & 0 & 0 \\ 0 & 0 & 0 \end{bmatrix} \qquad \mathbf{I} = \begin{bmatrix} 1 & 0 & 0 \\ 0 & 1 & 0 \\ 0 & 0 & 1 \end{bmatrix}$$

are two of them. What are the other six? Use a and b to represent arbitrary numbers as in Example 2.15.

82. Write down all 16 different 4-by-4 RRE forms. Use a and b to represent arbitrary numbers as in Example 2.15.

83. Suppose a matrix has RRE form and we remove a row, the new matrix will still have RRE form.

    a) Explain why each of the RRE conditions still holds for the new matrix.

    b) Give an example of a 3-by-4 matrix in RRE form that, upon removal of column three, no longer has RRE form.

c) If a matrix has RRE form and we delete the *last* column, the new matrix will still have RRE form. Explain why each of the RRE conditions still holds for the new matrix.

84. If an $n \times n$ matrix in RRE form has n pivots, what matrix must it be?

85. Show that for every n, there are 2^n different RRE forms of size $n \times n$. [*Hint: Row n of Pascal's triangle always adds up to 2^n.*]

3. Homogeneous Systems in RRE Form

Most of the work required to solve an arbitrary system in RRE form arises in solving the related system we get by setting all right-hand sides to zero.

DEFINITION 3.1. A linear system is called **homogeneous** if the right-hand side of each equation is zero. Otherwise, the system is **inhomogeneous**. Given an **in**homogenous system $\mathbf{Ax} = \mathbf{b}$, we call $\mathbf{Ax} = \mathbf{0}$ its **homogeneous version**. Both systems have the same *coefficient* matrix $\mathbf{A}$, but their *augmented* matrices differ in the last column.

EXAMPLE 3.2. The system on the right below is the homogeneous version of the inhomogeneous system on the left. Both have the same coefficient matrices.

$$
\begin{aligned}
2x - 3y + z &= 6 \\
x + 2y - 3z &= 4 \\
3x - y - 2z &= 0
\end{aligned}
\qquad\qquad
\begin{aligned}
2x - 3y + z &= 0 \\
x + 2y - 3z &= 0 \\
3x - y - 2z &= 0
\end{aligned}
$$

Inhomogeneous system   Homogeneous version

□

When the right-hand side of a single linear *equation* is zero, we get an obvious solution by setting all variables equal to zero. This is called the **trivial solution**. It solves every homogeneous *system* too:

OBSERVATION 3.3 (Trivial solution). *A homogeneous system $\mathbf{Ax} = \mathbf{0}$ always has at least one solution, namely the trivial one $\mathbf{x} = \mathbf{0}$. In other words, the origin in the range of a linear map always has at least one pre-image: the origin of the domain.*

REMARK 3.4 (*Non*-trivial solutions!). We cannot emphasize strongly enough:

Homogeneous systems often have *non*-trivial solutions too.

In Example 2.10 above, for instance, we studied a homogeneous system solved by

$$\mathbf{X}(t) = t(2, -3, 1)$$

where t is any scalar. When $t = 0$, we get the trivial solution, but every $t \neq 0$ gives a **non**-trivial solution.

The existence of non-trivial solutions is a fundamental difference between scalar equations and vector equations.

Indeed, every student knows that the *scalar* equation $ax = 0$ forces either $a = 0$ or $x = 0$. This basic fact about the algebra of scalars does **not** hold for vector equations:

*We **can** and often **do** have $\mathbf{Ax} = \mathbf{0}$ even though $\mathbf{A} \neq \mathbf{0}$ **and** $\mathbf{x} \neq \mathbf{0}$.*

This is not just a curiosity: it is a key feature of our subject. It occurs often, and it will affect almost everything we do. It makes the algebra of vectors and matrices more rich and interesting, than the algebra of scalars.

So *please* don't fall into the trap of thinking that $\mathbf{Ax} = \mathbf{0}$ means $\mathbf{A} = \mathbf{0}$ or $\mathbf{x} = \mathbf{0}$. In the world of matrices and vectors, it does not. Review Example 2.10 if you have trouble seeing how this can happen. It happens often! □

Indeed, the phenomenon just described spawns one of the most basic procedures in our entire subject because, among other things, solving a homogeneous system constitutes the main step in solving an *in*homogeneous system.

Before we present the general algorithm for finding *all* solutions to a homogeneous system in RRE form, we make a simplifying observation: *When a system is homogeneous, the last column of its augmented matrix is all zeros. So when solving a homogeneous system, we can work with the coefficient matrix alone—we needn't carry the last column.*

3.5. Pivot variables and free variables. Each *column* of the coefficient matrix corresponds to a particular *variable*: the first matrix column lists the coefficients of x_1 in each equation, the second column gives the coefficients of x_2, and so forth. In a system with m variables, the last column of the coefficient matrix goes with x_m.

DEFINITION 3.6. Suppose the coefficient matrix of a system has RRE form. If column j contains a pivot, we call it a **pivot column**, and we call x_j a **pivot variable**. Otherwise we say the column or variable is **free**.

Each column of the coefficient matrix, and each variable, is one or the other: *pivot* or *free*. □

EXAMPLE 3.7. Consider the system whose coefficient matrix has this RRE form:

$$\begin{bmatrix} 1 & 2 & 0 & 0 \\ 0 & 0 & 0 & 1 \end{bmatrix}$$

Columns 1 and 4 have pivots, which makes x_1 and x_4 pivot variables. Columns 2 and 3—and the variables x_2 and x_3—are free. □

EXAMPLE 3.8. This coefficient matrix has RRE form too:

$$\begin{bmatrix} 1 & 0 & 0 & 1 \\ 0 & 1 & 0 & 1 \\ 0 & 0 & 1 & 1 \end{bmatrix}$$

The first three columns have leading 1's, so x_1, x_2 and x_3 are pivot variables. The remaining variable x_4 is consequently free. □

EXAMPLE 3.9. Here is a coefficient matrix in RRE form with **no** free variables—only pivot variables. Do you see why?

$$\begin{bmatrix} 1 & 0 & 0 \\ 0 & 1 & 0 \\ 0 & 0 & 1 \\ 0 & 0 & 0 \end{bmatrix}$$

□

EXAMPLE 3.10. This coefficient matrix represents a system with only one equation:

$$\begin{bmatrix} 0 & 1 & 3 & 1 & 2 \end{bmatrix}$$

There's a leading 1 in column two, so x_2 is a pivot variable. The other four variables, x_1, x_3, x_4, and x_5 are all free. There's a one in column 4, but it's not a *leading* one, so it's not a pivot. □

3.11. The algorithm. We now give a simple five-step procedure for finding all solutions to a homogeneous system in RRE form. After listing the steps, we illustrate with some typical examples.

Important: This procedure applies only to systems that are both *homogeneous* and *have RRE form*.

- **Step 1.** *Identify the free columns*—let's say there are k of them—and delete (or cross out) the rows of zeros.

- **Step 2.** *Write down k blank **generating** vectors $\mathbf{h}_1, \mathbf{h}_2, \ldots, \mathbf{h}_k$ in $\mathbf{R}^m$.* (Write them vertically to make Step 4 a bit easier.)

- **Step 3.** *Set all free variables equal to 1 or 0:* For each $j = 1, 2, \ldots, k$, select $\mathbf{h}_j$, and set the jth *free* variable in it equal to 1. Set all other free variables in $\mathbf{h}_j$ equal to zero.

- **Step 4.** *Fill in the pivot variables.* For each $j = 1, 2, \ldots, k$, identify the jth free column in the matrix. Set the pivot variables of $\mathbf{h}_j$ equal to the **opposites** of the entries in that column, in order of their appearance, top to bottom.

- **Step 5.** *Express the complete solution as all linear combinations of the $\mathbf{h}_j$'s:*

$$\mathbf{x} = \mathbf{Hs} = s_1\mathbf{h}_1 + s_2\mathbf{h}_2 + \cdots + s_k\mathbf{h}_k \,.$$

Here $\mathbf{H}$ is the matrix having the $\mathbf{h}_i$'s as columns

$$\mathbf{H} = \begin{bmatrix} \mathbf{h}_1 & \mathbf{h}_2 & \cdots & \mathbf{h}_k \\ | & | & \vdots & | \end{bmatrix}$$

while we allow $\mathbf{s} = (s_1, s_2, \ldots, s_k)$ to be any vector in $\mathbf{R}^k$. The product $\mathbf{Hs}$ then gives the linear combination above by our "mantra": Linear combination is matrix/vector multiplication.

As we let $\mathbf{s}$ vary over all vectors in $\mathbf{R}^k$, the products $\mathbf{Hs}$ give all solutions $\mathbf{x} \in \mathbf{R}^m$ for the system. Note too that this allows us to see the solution set as the *image* of a mapping $\mathbf{R}^k \to \mathbf{R}^n$.

We illustrate the algorithm by working through two numeric examples.

EXAMPLE 3.12. Consider the homogeneous system with this coefficient matrix:

$$\begin{bmatrix} 0 & 1 & 3 & 0 & -2 \\ 0 & 0 & 0 & 1 & 4 \\ 0 & 0 & 0 & 0 & 0 \end{bmatrix}$$

The matrix has RRE form (check!), so we may apply the 5-step procedure above.

Step 1. *Identify the free variables and delete the rows of zeros.*

Columns 2 and 4 have pivots, so x_1, x_3, and x_5 are free. Three free
variables makes $k = 3$. There is one row of zeros, so we discard it.
Our matrix now looks like this:

$$\begin{bmatrix} 0 & 1 & 3 & 0 & -2 \\ 0 & 0 & 0 & 1 & 4 \end{bmatrix}$$

Step 2. *Set up* k *"blank" vertical **generating** vectors* $\mathbf{h}_1, \mathbf{h}_2, \ldots,$
$\mathbf{h}_k \in \mathbf{R}^m$.

Here $k = 3$, so we get three generators with blank entries, entries
represented here by p's (for pivot variables) and f's (free variables):

$$\mathbf{h}_1 = \begin{pmatrix} f \\ p \\ f \\ p \\ f \end{pmatrix}, \quad \mathbf{h}_2 = \begin{pmatrix} f \\ p \\ f \\ p \\ f \end{pmatrix}, \quad \mathbf{h}_3 = \begin{pmatrix} f \\ p \\ f \\ p \\ f \end{pmatrix}$$

Step 3. *Fill in the free variables.* Taking each generating vector $\mathbf{h}_j$
in turn, set the jth free variable equal to 1 and all other free variables
equal to zero. That is, the first free variable (x_1) in $\mathbf{h}_1$ is a 1, the
second free variable (x_3) in $\mathbf{h}_2$ is a 1, the third free variable (x_5) in $\mathbf{h}_3$
is a 1. All other free variables in all three generators are zeros. The
pivot variables x_2 and x_4 remain undetermined.

This fills in the entries for x_1, x_3, and x_5 in each generator:

$$\mathbf{h}_1 = \begin{pmatrix} 1 \\ p \\ 0 \\ p \\ 0 \end{pmatrix}, \quad \mathbf{h}_2 = \begin{pmatrix} 0 \\ p \\ 1 \\ p \\ 0 \end{pmatrix}, \quad \mathbf{h}_3 = \begin{pmatrix} 0 \\ p \\ 0 \\ p \\ 1 \end{pmatrix}$$

Step 4. *Set the pivot variables of each* $\mathbf{h}_j$ *equal to the **opposites** of
the values in the jth free variable column of the non-zero rows of the
matrix—in order of their appearance, top to bottom.*

We start with $j = 1$: The first free variable is x_1, so we take the
entries in the two remaining rows of column 1. In this example both
those entries are zero. We then use their opposites (still zero) to set
the pivot variables x_2 and x_4 of $\mathbf{h}_1$. We thus get

$$\mathbf{h}_1 = \begin{pmatrix} 1 \\ -0 \\ 0 \\ -0 \\ 0 \end{pmatrix}$$

(The minus signs are irrelevant here since $-0 = 0$, but they do show which entries we just filled in.) It's easy to check that $\mathbf{h}_1$ now solves the homogeneous system—just dot it with each row of the coefficient matrix: We get zero every time.

Now set $j = 2$. The second free variable is x_3, so we take the entries in the two rows of column three: 3 and 0. We then put **minus** 3 and **minus** 0 in the pivot positions of the second generator to get

$$\mathbf{h}_2 = \begin{pmatrix} 0 \\ -3 \\ 1 \\ -0 \\ 0 \end{pmatrix}$$

Again, this checks: it dots to zero with each row of the coefficient matrix.

Finally we complete $\mathbf{h}_3$. The third free variable is x_5, so we take the entries in column 5, namely -2 and 4, **flip their signs** to get 2 and -4, and put these in the pivot positions to get

$$\mathbf{h}_3 = \begin{pmatrix} 0 \\ 2 \\ 0 \\ -4 \\ 1 \end{pmatrix}$$

Again it is easy to check that $\mathbf{h}_3$ solves the homogeneous system, and this completes Step 4.

Step 5. *Express the homogeneous solution map* $\mathbf{H}$ *as all linear combinations of the generating vectors* $\mathbf{h}_j$.

Here, the full homogeneous solution set is the set of all products $\mathbf{Hs}$, where

$$(8) \qquad \mathbf{H} = \begin{bmatrix} 1 & 0 & 0 \\ 0 & -3 & 2 \\ 0 & 1 & 0 \\ 0 & 0 & -4 \\ 0 & 0 & 1 \end{bmatrix}, \quad \mathbf{s} = (s_1, s_2, s_3)$$

If we multiply it out, the solution takes the compact form

$$\mathbf{Hs} = (s_1, \; -3s_2 + 2s_3, \; s_2, \; -4s_3, \; s_3)$$

$\square$

EXAMPLE 3.13. Consider the homogeneous system whose coefficient matrix is given by

$$\begin{bmatrix} 1 & 0 & 0 & 2 \\ 0 & 1 & 0 & 3 \\ 0 & 0 & 1 & 4 \end{bmatrix}$$

Checking that this matrix does indeed have RRE form, we can apply our procedure. The only free variable here is x_4, and there are no rows of zeros.

Step 2 therefore gives us just one generating vector for the homogeneous solution set, namely

$$\mathbf{h}_1 = \begin{pmatrix} p \\ p \\ p \\ 1 \end{pmatrix}$$

The free column is column four, so we take the values in that column of the matrix and flip their signs to get -2, -3, and -4, respectively. Replacing the $*$'s in $\mathbf{h}_1$ with these values, top to bottom, we conclude that the homogeneous solution is given by all products $s\mathbf{H}$, where

$$\mathbf{H} = \begin{bmatrix} -2 \\ -3 \\ -4 \\ 1 \end{bmatrix}$$

Here $\mathbf{s} = s$ is just a single scalar, and as usual, it is easy to check that the single column of $\mathbf{H}$ does dot to zero with each row of the coefficient matrix. So its multiples really do solve the homogeneous system. $\square$

REMARK 3.14. It can certainly happen that a matrix in RRE form has *no* free columns. In this case, our procedure yields *no* homogeneous generators $\mathbf{h}_j$, which means the system has *only* the trivial solution $\mathbf{x} = \mathbf{0}$. $\qquad\qquad\square$

– **Practice** –

86. Each matrix below is in RRE form. Label each column as pivot (P) or free (F).

a) $\begin{bmatrix} 1 & 0 & 0 & -1 \\ 0 & 1 & 0 & -1 \\ 0 & 0 & 1 & 0 \end{bmatrix}$ b) $\begin{bmatrix} 0 & 1 & 0 & -1 \\ 0 & 0 & 1 & -1 \\ 0 & 0 & 0 & 0 \end{bmatrix}$

c) $\begin{bmatrix} 0 & 1 & 2 & 0 \\ 0 & 0 & 0 & 1 \\ 0 & 0 & 0 & 0 \end{bmatrix}$ d) $\begin{bmatrix} 0 & 0 & 1 & 1 \\ 0 & 0 & 0 & 0 \\ 0 & 0 & 0 & 0 \end{bmatrix}$

e) $\begin{bmatrix} 1 & 0 & 0 \\ 0 & 1 & 0 \\ 0 & 0 & 1 \\ 0 & 0 & 0 \end{bmatrix}$ f) $\begin{bmatrix} 1 & 1 & 0 \\ 0 & 0 & 0 \\ 0 & 0 & 0 \\ 0 & 0 & 0 \end{bmatrix}$

87. Verify that the coefficient matrix of each homogeneous system has RRE form, and find the corresponding solution map whose image gives all its solutions.

a) $\begin{aligned} x + 2y &= 0 \\ z &= 0 \end{aligned}$ b) $\begin{aligned} x + 2z &= 0 \\ y + 2z &= 0 \end{aligned}$ c) $x + 2y + 3z = 0$

88. Consider this homogeneous system:

$$x_1 - \tfrac{1}{2}x_3 - \tfrac{1}{4}x_4 = 0$$
$$x_2 + 2x_3 + x_4 = 0$$

Verify that its coefficient matrix has RRE form, and find the solution map whose image gives all its solutions.

89. Each matrix below is the coefficient matrix of a homogeneous system. Which system has only the trivial solution? Find all solutions of the other system.

a) $\begin{bmatrix} 1 & 0 \\ 0 & 1 \\ 0 & 0 \end{bmatrix}$ b) $\begin{bmatrix} 1 & 0 & 0 \\ 0 & 1 & 0 \end{bmatrix}$

90. Verify that the coefficient matrix of this system has RRE form, and find its solutions. Then check that each of your generators h_i solves the system.

$$\begin{aligned} x_1 + 2x_3 - x_5 &= 0 \\ x_2 - x_3 + 2x_5 &= 0 \\ x_4 - x_5 &= 0 \end{aligned}$$

91. Treat the homogeneous equation below as a system—a system of one equation in four variables. Write its 1-by-4 coefficient matrix, and find three generators $h_1, h_2,$ and h_3 for the solution set. Then find the matrix H that generates all its solutions.

$$x_1 + 2x_2 - 3x_3 + 2x_4 = 0$$

92. Verify that the coefficient matrix of the system below has RRE form, and find its solutions. Check that each of your generators h_i solves the system.

$$\begin{aligned} x_1 - 2x_3 + 3x_4 &= 0 \\ x_2 + 2x_5 &= 0 \end{aligned}$$

93. Suppose $\mathbf{a}_1 = (1, 0, 2, -3, 0)$ and $\mathbf{a}_2 = (0, 1, 0, 0, \pi)$ in $\mathbf{R}^5$. Find all vectors $\mathbf{x} \in \mathbf{R}^5$ for which $\mathbf{a}_1^*(\mathbf{x}) = \mathbf{a}_2^*(\mathbf{x}) = 0$.

94. Suppose an $n \times m$ matrix A has RRE form.

a) If A has r leading 1's, how many generators h_i will show up in the solution of $A\mathbf{x} = 0$? Explain.

b) What is the smallest number of free columns A can have if $m > n$? How about it $n > m$? Explain.

95. What 2×2 matrix $\mathbf{A}$ in RRE form yields

   a) The single homogeneous generator $\mathbf{h} = (-5, 1)$?

   b) The single homogeneous generator $\mathbf{h} = (1, 0)$?

96. What 3×3 matrix $\mathbf{A}$ in RRE form yields

   a) The single homogeneous generator $\mathbf{h} = (3, 1, 0)$?

   b) These two homogenous generators: $\mathbf{h}_1 = (1, 0, 0)$, and $\mathbf{h}_2 = (0, -2, 1)$?

97. What 2×4 matrix $\mathbf{A}$ in RRE form yields

   a) These two homogeneous generators: $\mathbf{h}_1 = (3, 1, 0, 0)$, and $\mathbf{h}_2 = (-3, 0, 1, 0)$?

   b) These two homogenous generators: $\mathbf{h}_1 = (1, 0, 0, 0)$, and $\mathbf{h}_2 = (4, 0, 1, 0)$?

— ⋆ —

4. Inhomogeneous Systems in RRE Form

Continuing to assume our system has RRE form, we now turn to the **in**homogeneous case. We clearly need to work with the full *augmented* matrix here: inhomogeneous systems always have one or more non-zero entries on the right-hand side, so the last column of the augmented matrix contains critical data.

As we noted at the end of Section 2, a *pivot* in that column rules out any solution at all. So the first rule of solving an inhomogeneous system in RRE form is this:

• *A pivot in the last column of the augmented matrix means the system has **no solution**.*

If the last column of an augmented matrix in RRE form does *not* have a pivot, however, the system *will* have a solution. And if we can find *just one solution*, we are essentially done, thanks to this crucial fact:

THEOREM 4.1 (Structure of the pre-image). *Given just one solution* $\mathbf{x}_p$ *of a linear system* $\mathbf{A}\mathbf{x} = \mathbf{b}$, *every other solution* $\mathbf{x}'$ *is gotten by adding a solution of the associated homogeneous system to* $\mathbf{x}_p$. *That is,* $\mathbf{x}' = \mathbf{x}_p + \mathbf{h}$, *where* $\mathbf{h}$ *solves* $\mathbf{A}\mathbf{x} = \mathbf{0}$.

This theorem plays a huge role in our subject, and its proof is easy. It boils down to linearity—in this case, the distributivity of matrix/vector multiplication over vector addition.

PROOF. If we have one ("particular") solution $\mathbf{x}_p$ of $\mathbf{Ax} = \mathbf{b}$, and $\mathbf{x}'$ is any other solution, $\mathbf{x}'$ and $\mathbf{x}_p$ both solve the system, so the distributive law gives

$$\mathbf{A}\left(\mathbf{x}' - \mathbf{x}_p\right) = \mathbf{Ax}' - \mathbf{Ax}_p = \mathbf{b} - \mathbf{b} = \mathbf{0}$$

In short, $\left(\mathbf{x}' - \mathbf{x}_p\right)$ solves the *homogeneous* system:

$$\mathbf{A}\left(\mathbf{x}' - \mathbf{x}_p\right) = \mathbf{0}$$

If $\mathbf{h} = \mathbf{x}' - \mathbf{x}_p$ denotes this homogeneous solution, we then have

$$\mathbf{x}' = \mathbf{x}_p + \left(\mathbf{x}' - \mathbf{x}_p\right) = \mathbf{x}_p + \mathbf{h}$$

which expresses our arbitrary solution $\mathbf{x}'$ as "$\mathbf{x}_p$ plus a homogeneous solution $\mathbf{h}$," just as the Theorem asserts.   □

This Theorem tells us that once we have all solutions $\mathbf{h}$ of the *homogeneous version* of a system (as we learned to do, at least for RRE form systems, in the previous section) then we only have to find *one* solution of the *in*homogeneous system to get *all* other solutions. Any solution of the inhomogeneous system can serve as $\mathbf{x}_p$, which we often call a *particular solution* to the inhomogeneous system.

4.2. Finding a particular solution. Finding a particular solution to a system in RRE form is far simpler than finding the full homogeneous solution. Two quick steps are all it takes:

To produce a particular solution $\mathbf{x}_p$ of a system in RRE form, write down a blank vector $\mathbf{x}_p \in \mathbf{R}^m$, and then fill it in by

- *Setting all free variables equal to zero.*

- *Setting each pivot variable equal to the rightmost entry in the row of that pivot.*

We demonstrate with some examples.

EXAMPLE 4.3. Let us find a particular solution of the system with this augmented matrix:

$$\begin{bmatrix} 1 & 2 & 0 & 0 & | & 7 \\ 0 & 0 & 0 & 1 & | & 3 \end{bmatrix}$$

The matrix does have RRE form (the procedure doesn't apply otherwise!), and there are four variables, so we write down a blank vector $\mathbf{x}_p \in \mathbf{R}^4$:

$$\mathbf{x}_p = (\_, \_, \_, \_)$$

The free variables are x_2 and x_3, and the first instruction has us set those equal to zero:

$$\mathbf{x}_p = (\_, 0, 0, \_)$$

We fill in the pivot slots using the last column of the augmented matrix. We find the x_1 pivot in row 1, where the last entry is 7, so we set $x_1 = 7$. We find the x_4 pivot in row 2, where the last entry is 3, so we set $x_4 = 3$, and that's it—we have a particular solution:

$$\mathbf{x}_p = (7, 0, 0, 3)$$

It's easy to check that this really does solve both the equations in the system—just dot it with each row in the coefficient matrix and verify that we get the right-hand side of that row.   □

EXAMPLE 4.4. Consider the system with augmented matrix

$$\begin{bmatrix} 1 & 0 & 0 & 1 & 7 \\ 0 & 1 & 0 & 1 & -2 \\ 0 & 0 & 1 & 1 & 3 \end{bmatrix}$$

The only free variable is x_4, so we put a zero in that slot. The right-hand sides of the pivot rows are 7, -2, and 3, so we set the corresponding pivot variables equal to those values, and we're done:

$$\mathbf{x}_p = (7, -2, 3, 0)$$

Again, it's easy to check that this vector solves the system.   □

EXAMPLE 4.5. The system with this augmented matrix:

$$\begin{bmatrix} 1 & 0 & 0 & 7 \\ 0 & 1 & 0 & -2 \\ 0 & 0 & 1 & 3 \end{bmatrix}$$

has **no** free variables, and a pivot in every row. So the last column, as it is, gives a (in fact, the only) particular solution: $\mathbf{x}_p = (7, -2, 3)$. □

EXAMPLE 4.6. Our method gives $(0, 2, 0, 0)$ as the particular solution to the "system" (really just one equation) with this augmented matrix?

$$\begin{bmatrix} 0 & 1 & 3 & 1 & 2 \end{bmatrix}$$

Can you see why? $\qquad \square$

Before going any further, we emphasize that one can easily see *why* our procedure always works.

It works because each row of an RRE matrix corresponds to an equation where the coefficients of all pivot variables except one—the "leading" variable—are zero. The value we assign to every free variable is *also* zero. So when we plug our particular solution into equation i, say, of a system with RRE augmented matrix, we end up with only one non-zero term on the left: the leading 1 times the value we gave to its pivot variable. But the value we gave to that variable is exactly b_i, the last entry in that row. So plugging our particular solution into equation i always reduces it to $b_i = b_i$, which is certainly true. Since this happens for every equation in the system, we have a solution.

Our procedure for solving the *homogeneous system* in RRE form is a bit more complicated, but it actually works for very similar reasons. With a little persistence, the interested student should be able to confirm the effectiveness of that algorithm too.

4.7. Complete solution of a system in RRE form. We can now find *all* solutions of *any* linear system in RRE form. For by Theorem 4.1, we get every solution by finding one particular solution $\mathbf{x}_p$—which we now know how to do—and then adding a homogeneous solution $\mathbf{h}$. We know how to find all homogeneous solutions too, so by putting the skills together, we can always get a complete solution of the form

$$\mathbf{x}_p + \mathbf{Hs} = \mathbf{x}_p + s_1 \mathbf{h}_1 + s_2 \mathbf{h}_2 + \cdots + s_k \mathbf{h}_k$$

The $\mathbf{h}_i$'s are the homogeneous generators introduced in Section 3, and k is the number of free variables.

EXAMPLE 4.8. In Example 4.3, we found a particular solution, namely $\mathbf{x}_p = (7, 0, 0, 3)$, to the system with augmented matrix

$$\begin{bmatrix} 1 & 2 & 0 & 0 & 7 \\ 0 & 0 & 0 & 1 & 3 \end{bmatrix}$$

To get *all* solutions, we need to solve the homogeneous version of the system. By now, this is a familiar drill: There are two free variables (x_2 and x_3), so we prepare two homogeneous generators, of the form

$$\mathbf{h}_1 = \begin{pmatrix} p \\ 1 \\ 0 \\ p \end{pmatrix} \quad \text{and} \quad \mathbf{h}_2 = \begin{pmatrix} p \\ 0 \\ 1 \\ p \end{pmatrix}$$

The first free variable is x_2, and its column in the coefficient matrix contains a 2 and a 0 (top to bottom) so we replace the p's in $\mathbf{h}_1$ with -2 and -0, respectively. The second free variable is x_3, but its column contains only zeros. So we replace the p in $\mathbf{h}_2$ with zeros. Our complete solution thus consists of all vectors

$$\mathbf{x}_p + s_1 \mathbf{h}_1 + s_2 \mathbf{h}_2 = \begin{pmatrix} 7 \\ 0 \\ 0 \\ 3 \end{pmatrix} + s_1 \begin{pmatrix} -2 \\ 1 \\ 0 \\ 0 \end{pmatrix} + s_2 \begin{pmatrix} 0 \\ 0 \\ 1 \\ 0 \end{pmatrix}$$

We can simplify this to express its as the image of a single mapping $\mathbf{x} : \mathbf{R}^2 \to \mathbf{R}^4$:

$$\mathbf{x}(s_1, s_2) = \begin{pmatrix} 7 - 2s_1 \\ s_1 \\ s_2 \\ 3 \end{pmatrix}$$

$\square$

EXAMPLE 4.9. In Example 4.4, we found that the system with augmented matrix

$$\begin{bmatrix} 1 & 0 & 0 & 1 & 7 \\ 0 & 1 & 0 & 1 & -2 \\ 0 & 0 & 1 & 1 & 3 \end{bmatrix}$$

has $\mathbf{x}_p = (7, -2, 3, 0)$ as a particular solution. We find only one free variable, namely x_4, so we get only one homogeneous generator, namely $\mathbf{h} = (-1, -1, -1, 1)$. (Verify!) The full solution of the system is consequently given by the image of a mapping $\mathbf{x} : \mathbf{R} \to \mathbf{R}^4$, namely

$$\mathbf{x}(s) = \begin{pmatrix} 7 \\ -2 \\ 3 \\ 0 \end{pmatrix} + s \begin{pmatrix} -1 \\ -1 \\ -1 \\ 1 \end{pmatrix} = \begin{pmatrix} 7-s \\ -2-s \\ 3-s \\ s \end{pmatrix}$$

Easy!   □

EXAMPLE 4.10. This augmented matrix

$$\begin{bmatrix} 1 & 0 & 0 & 7 \\ 0 & 1 & 0 & -2 \\ 0 & 0 & 1 & 3 \end{bmatrix}$$

has **no** free variables. So its particular solution is its *only* solution:
$\mathbf{x} = \mathbf{x}_p = (7, -2, 3)$.   □

EXAMPLE 4.11. In Example 4.5, we found the particular solution $\mathbf{x}_p = (0, 2, 0, 0)$ to the single equation with augmented matrix

$$\begin{bmatrix} 0 & 1 & 3 & 1 & 2 \end{bmatrix}$$

What is the complete solution?

There are three free variables: $x_1, x_3,$ and x_4, which give rise to three homogeneous generators

$$\mathbf{h}_1 = \begin{pmatrix} 1 \\ 0 \\ 0 \\ 0 \end{pmatrix}, \quad \mathbf{h}_2 = \begin{pmatrix} 0 \\ -3 \\ 1 \\ 0 \end{pmatrix}, \quad \text{and} \quad \mathbf{h}_3 = \begin{pmatrix} 0 \\ -1 \\ 0 \\ 1 \end{pmatrix}$$

Adding all linear combinations of these to our particular solution, we get the complete solution to our "system" of one equation. It is the image of this mapping $\mathbf{x} : \mathbf{R}^3 \to \mathbf{R}^4$:

$$\mathbf{x}(s_1, s_2, s_3) = \mathbf{x}_p + s_1\mathbf{h}_1 + s_2\mathbf{h}_2 + s_3\mathbf{h}_3 = \begin{pmatrix} s_1 \\ 2 - 3s_2 - s_3 \\ s_2 \\ s_3 \end{pmatrix}$$

                                                                        □

– Practice –

98. Verify that the augmented matrix for this system is in RRE form. Then find a particular solution for it.

$$x_2 + 2x_3 + 3x_5 = 4$$
$$x_4 + 5x_5 = 6$$

99. Determine whether each augmented matrix below has RRE form, then decide if if the corresponding system has a solution. Find one if it does.

a)

$$\begin{bmatrix} 1 & 0 & 3 \\ 0 & 1 & -3 \end{bmatrix}$$

b)

$$\begin{bmatrix} 1 & 3 & 0 & -2 \\ 0 & 0 & 1 & 4 \end{bmatrix}$$

c)

$$\begin{bmatrix} 0 & 0 & 1 & 0 & -1/2 \\ 0 & 0 & 0 & 1 & 1/3 \end{bmatrix}$$

d)

$$\begin{bmatrix} 0 & 0 & 1 & 2 & 0 \\ 0 & 0 & 0 & 0 & 1 \end{bmatrix}$$

e)

$$\begin{bmatrix} 0 & 1 & 0 & 0 & 1 \\ 0 & 0 & 1 & 0 & 1 \\ 0 & 0 & 0 & 1 & 1 \end{bmatrix}$$

f)

$$\begin{bmatrix} 0 & 1 & 0 & 0 & 0 \\ 0 & 0 & 1 & 0 & 0 \\ 0 & 0 & 0 & 1 & 0 \end{bmatrix}$$

g)

$$\begin{bmatrix} 1 & 0 & 0 & 0 \\ 0 & 1 & 0 & 0 \\ 0 & 0 & 0 & 1 \\ 0 & 0 & 0 & 0 \end{bmatrix}$$

100. This system has two equations in three unknowns:

$$\begin{aligned} x + 2y &= 1 \\ z &= 2 \end{aligned}$$

Verify that its augmented matrix has RRE form and find all solutions.

101. Verify that the augmented matrix for this system has RRE form and find all solutions:

$$\begin{aligned} x_1 + 2x_3 &= 4 \\ x_2 + 3x_3 &= -5 \\ x_4 &= 6 \end{aligned}$$

102. Verify that the augmented matrix for this system has RRE form and find all solutions:

$$\begin{aligned} x_2 - 2x_3 + x_5 &= 0 \\ x_4 - 2x_5 &= 3 \end{aligned}$$

103. Verify that each system below has RRE form and find all solutions.

a)

$$\begin{bmatrix} 1 & 0 & 3 \\ 0 & 1 & -3 \end{bmatrix}$$

b)

$$\begin{bmatrix} 1 & 3 & 0 & -2 \\ 0 & 0 & 1 & 4 \end{bmatrix}$$

c)

$$\begin{bmatrix} 0 & 0 & 1 & 0 & -1/2 \\ 0 & 0 & 0 & 1 & 1/3 \end{bmatrix}$$

d)

$$\begin{bmatrix} 0 & 0 & 1 & 0 & -1/2 \\ 0 & 0 & 0 & 0 & 1 \end{bmatrix}$$

e)

$$\begin{bmatrix} 0 & 1 & 0 & 0 & 1 \\ 0 & 0 & 1 & 0 & 1 \\ 0 & 0 & 0 & 1 & 1 \end{bmatrix}$$

f)

$$\begin{bmatrix} 0 & 1 & 0 & 0 & 0 \\ 0 & 0 & 1 & 0 & 0 \\ 0 & 0 & 0 & 1 & 0 \end{bmatrix}$$

g)

$$\begin{bmatrix} 1 & 0 & 0 & 0 \\ 0 & 1 & 2 & 0 \\ 0 & 0 & 0 & 1 \\ 0 & 0 & 0 & 0 \end{bmatrix}$$

104. Suppose $\mathbf{A}$ is a 7×3 matrix, and $\mathbf{b} \in \mathbf{R}^7$. If $(2, -1, 5)$ solves the homogeneous system $\mathbf{Ax} = \mathbf{0}$, and $(8, -2, 9)$ solves the *inhomo-geneous* system $\mathbf{Ax} = \mathbf{b}$, give three other solutions of $\mathbf{Ax} = \mathbf{b}$.

105. Suppose $\mathbf{A}$ is a 4×5 matrix, and $\mathbf{b} \in \mathbf{R}^5$. Assume $(2, -1, 5, 3)$ and $(5, -1, 2, 3)$ both solve the inhomogeneous system $\mathbf{Ax} = \mathbf{b}$. Give three solutions of the *homogeneous* system $\mathbf{Ax} = \mathbf{0}$.

106. The vectors $(3, -1, 2, 5)$, $(0, 1, -1, 0)$ and $(7, -7, 0, 1) \in \mathbf{R}^4$ all solve $\mathbf{Ax} = \mathbf{b}$ for a certain matrix $\mathbf{A}$ and right-hand side $\mathbf{b}$. Find three solutions of the *homogeneous* system $\mathbf{Ax} = \mathbf{0}$, and three additional solutions of $\mathbf{Ax} = \mathbf{b}$.

107. Create an augmented matrix $[\mathbf{A} \,|\, \mathbf{b}]$ in RRE form with the property that $\mathbf{Ax} = \mathbf{0}$ has infinitely many homogeneous solutions, but $\mathbf{Ax} = \mathbf{b}$ has *no* solution.

108. Suppose an inhomogeneous system has...

    a) ...three equations and four variables. Can it have just one solution? If so, give an example. If not, explain why not.

    b) ...four equations and three variables. Can it have exactly one solution? If so, give an example. If not, explain why not.

    c) ...four equations and three variables. Can it have infinitely many solutions? If so, give an example. If not, explain why not.

— ⋆

5. The Gauss–Jordan Algorithm

So far, we have learned to find all solutions of any linear system *whose augmented matrix has* RRE *form*. But most systems do *not* have RRE form—so what about them?

The answer was first discovered by C.F. Gauss, who saw that any system can be successively modified, using simple moves, *without changing its solution set*, until it reaches RRE form. This lets us implement the strategy we sketched at the end of Section 2. Once the system is in RRE form, we know how to solve it; but the solution set of the reduced system still equals that of the original system, so we've solved that too.

This procedure, known as **Gaussian elimination**, or the **Gauss–Jordan algorithm**, reduces any linear system step by step, in such a way that

    • Each step of the procedure changes the system *without* changing its solution set, and

    • The augmented matrix of the system reaches RRE form—and does so reasonably fast.

5.1. The Elementary row operations. The Gauss–Jordan algorithm can be repetitious and, for large systems, tedious, but it is surprisingly simple: It uses just three basic "moves" on the system's augmented matrix, called **elementary row operations**, and it is easy to see that none of them change the solution set—even while they do change the system. They are:

- **Row op #1: Swap two rows.** We execute this move by exchanging two rows of the augmented matrix. It only changes the *order* of the corresponding equations, so it has no affect on their solutions. This is simplest, but also the least frequent move.

- **Row op #2: Multiply a row by a non-zero scalar.** In this move, we multiply one row of the augmented matrix by a **non-zero** scalar. This corresponds to multiplying both sides of an equation by the same non-zero scalar. It's easy to see that this too leaves the solution set unchanged.

- **Row op #3: Add a multiple of one row to another.** This operation is a bit more complex, and the most commonly used. We'll examine it more closely after an example.

EXAMPLE 5.2. Consider the system

$$\begin{aligned} x_2 + x_3 + x_4 &= 7 \\ 2x_1 - 4x_2 - 2x_3 &= -12 \end{aligned}$$

Its augmented matrix,

$$\left[\begin{array}{ccccc} 0 & 1 & 1 & 1 & 7 \\ 2 & -4 & -2 & 0 & -12 \end{array} \right]$$

does *not* have RRE form. We can bring it to RRE form, however, by using elementary row operations. First, swap the two rows (which swaps the corresponding equations) to get

$$\left[\begin{array}{ccccc} 2 & -4 & -2 & 0 & -12 \\ 0 & 1 & 1 & 1 & 7 \end{array} \right]$$

Now multiply the first row by $1/2$, effectively dividing the (new) first equation by 2:

$$\left[\begin{array}{ccccc} 1 & -2 & -1 & 0 & -6 \\ 0 & 1 & 1 & 1 & 7 \end{array} \right]$$

Finally, use the third row operation, adding twice the second row to the first, to reach RRE form:

$$\left[\begin{array}{ccccc} 1 & 0 & 1 & 2 & 8 \\ 0 & 1 & 1 & 1 & 7 \end{array} \right]$$

The augmented matrix now has RRE form, so we know how to extract the solution. The free variables are x_3 and x_4, which give rise to these two homogeneous generators:

$$\mathbf{h}_1 = \begin{pmatrix} -1 \\ -1 \\ 1 \\ 0 \end{pmatrix} \quad \text{and} \quad \mathbf{h}_2 = \begin{pmatrix} -2 \\ -1 \\ 0 \\ 1 \end{pmatrix}$$

We get a particular solution by setting the free variables equal to zero and looking at the last column of the final augmented matrix, obtaining

$$\mathbf{x}_p = \begin{pmatrix} 8 \\ 7 \\ 0 \\ 0 \end{pmatrix}$$

The solution of our original system is then given by the image of the mapping

$$\mathbf{x}(s,t) = \mathbf{x}_p + s\mathbf{h}_1 + t\mathbf{h}_2 = \begin{pmatrix} 8 - s - 2t \\ 7 - s - t \\ s \\ t \end{pmatrix}$$

For example, if we take $s = 3$, $t = 2$, we get

$$\mathbf{x}(3,2) = (1,2,3,2)$$

Plug this into the system to see that it works—it solves both equations. Of course we get many other solutions too: any choice of (s,t) will give one. The point is, we have discovered all solutions of the system, even though it did not originally have RRE form. $\square$

How did we know which row operations to use above, and when? The Gauss–Jordan algorithm gives a complete answer to that question. It brings any given matrix into RRE form in finitely many steps, just as we did above.

Before we present Gauss–Jordan, however, let us tie up a loose end. When we listed the elementary row operations above, we could see easily that swapping rows and multiplying a row by a scalar—the first two operations—left the solution set of a system unchanged. Let us now see that the third row operation—adding a multiple of one row to another—also leaves the solution set alone. Like the proof of the Structure Theorem 4.1, it boils down to linearity—in this case, distributivity of the dot product over vector addition (Corollary 1.30).

Suppose we have a linear system with coefficient matrix $\mathbf{A}$ and right-hand side $\mathbf{b} = (b_1, b_2, \ldots, b_n)$. Suppose we also have a solution $\mathbf{x}$ of the system. *If we add c times row i of the augmented matrix to row j, will the same vector $\mathbf{x}$ still solve the new system?*

The answer is *yes.* To see why, recall (Observation 1.32) that we can abbreviate the equations in question by using dot products

$$\text{Equation } i: \quad \mathbf{r}_i(\mathbf{A}) \cdot \mathbf{x} = b_i$$
$$\text{Equation } j: \quad \mathbf{r}_j(\mathbf{A}) \cdot \mathbf{x} = b_j$$

Both these equations are *true*, here, because we're assuming $\mathbf{x}$ solves them both. So if we add c times the first one to the second and factor out the $\mathbf{x}$ using the distributive law, equation j becomes

$$(c\,\mathbf{r}_i(\mathbf{A}) + \mathbf{r}_j(\mathbf{A})) \cdot \mathbf{x} = b_j + c\,b_i$$

The point is, *we added equals to equals, so we still have equals.* The left side of equation i equaled the right side. That remains true if we multiply both those sides by c. So if we add c times the left side to the left side of equation j, and c times the right side to the right side of equation j, we're adding equals to equals, and the equation still holds.

We have just shown that that we never *lose* a solution $\mathbf{x}$ when we do row operation #3: If $\mathbf{x}$ solves the system *before* the operation, it still solves the modified system. But we must also check that we never *gain* a solution that we didn't have *before* the operation.

Fortunately, this is just a matter of reversing the operation: We can get back to the original equation j by adding $-c$ times row i to it. In other words, we can "undo" a row operation of type #3, with another row operation of type #3—we just change the sign of c.

In particular, if we had *gained* a solution by doing the operation with $+c$, we would *keep* that solution by using the operation with $-c$, since we have shown that row operation #3 never loses a solution. But then equation j would have a solution it didn't have before, which is impossible, since the equation itself hasn't changed.

With this we have proven for row operation #3 what we already knew for row operations #1 and #2. Namely,

PROPOSITION 5.3. *Each of the elementary row operations change a system **without** changing any of its solutions.*

5.4. The algorithm. The Gauss–Jordan algorithm works on any $n \times m$ matrix $\mathbf{A} = [a_{ij}]$. It consists mainly of a five-step "loop" (steps 1 through 5 below) that we repeat. The matrix is guaranteed to reach RRE form within m repetitions of this loop.

Gauss–Jordan Algorithm

0) Start by setting the "current" row and column numbers, respectively to $i = 1$ and $j = 1$.

1) Inspect the i, j entry a_{ij}. If $a_{ij} \neq 0$, skip to Step (3). If $a_{ij} = 0$, proceed to Step (2).

2) **SCAN** column j, from row i down, for a non-zero entry. If you find one, **swap its row with row** i. This puts a non-zero entry in the i, j position. If you find **only** zeros, move to the next column (i.e., replace j by $j + 1$), and repeat Step 1.

3) **MULTIPLY row** i by $1/a_{ij}$ to create a leading one in the i, j position.

4) **ADD appropriate multiples of row** i to the rows above and below it to zero-out the remaining entries in column j.

5) If $i < n$ and $j < m$, **increment** $i \mapsto i + 1$ and $j \mapsto j + 1$ and repeat, starting with Step (1) above. If $i = n$ or $j = m$ (no more rows or no more columns), we're done. The matrix is now guaranteed to have RRE form.

With a little practice—and we will get plenty—the Gauss–Jordan algorithm is easy to master. To help remember it, here's a mantra that summarizes:

Scan, Swap, Normalize, Zero-out, Increment

EXAMPLE 5.5. Let us use Gauss–Jordan to solve the system

$$
\begin{aligned}
y - 3z &= 3 \\
x - y + 5z &= 1 \\
-x + 2y - 8z &= 2
\end{aligned}
$$

To start, write down its augmented matrix:

$$
\begin{bmatrix}
0 & 1 & -3 & 3 \\
1 & -1 & 5 & 1 \\
-1 & 2 & -8 & 2
\end{bmatrix}
$$

This does **not** yet have RRE form, so we must apply Gauss–Jordan.

For that, we start by setting $i = j = 1$. Step (1) then has us look for a non-zero entry in the $(i, j) = (1, 1)$ position. We find that entry *is* zero however, so we cannot skip to Step (3).

Step (2) now tells us to **scan** column 1 for a non-zero entry. We find one in row 2, so we **swap** rows 1 and 2:

$$
\begin{bmatrix} 0 & 1 & -3 & 3 \\ 1 & -1 & 5 & 1 \\ -1 & 2 & -8 & 2 \end{bmatrix} \xrightarrow[\text{---}]{\text{swap r1, r2}} \begin{bmatrix} 1 & -1 & 5 & 1 \\ 0 & 1 & -3 & 3 \\ -1 & 2 & -8 & 2 \end{bmatrix}
$$

Next we apply Step (3), which asks us to **normalize**: divide the current row—row one—by its leading entry to produce a pivot. Here, that entry is already a 1, so Step (3) doesn't really change anything.

The point of the first three steps is to get a pivot in the current column. We now have that pivot, and RRE form requires zeros in the rest of its column. Step (4) produces them: To **zero-out** the -1 in the $(3,1)$ position (row 3, column 1), we add 1 times row 1 to row 3:

$$
\begin{bmatrix} 1 & -1 & 5 & 1 \\ 0 & 1 & -3 & 3 \\ -1 & 2 & -8 & 2 \end{bmatrix} \xrightarrow[\text{---}]{\text{add r1 to r3}} \begin{bmatrix} 1 & -1 & 5 & 1 \\ 0 & 1 & -3 & 3 \\ 0 & 1 & -3 & 3 \end{bmatrix}
$$

The $(3,2)$ position already has a zero, so we're now done with column one: It has a pivot in the current $(\, i = 1, \ j = 1 \,)$ position, with zeros above and below it. We therefore **increment**: move to the next row and column $(\, i = 2, \ j = 2 \,)$ and begin again.

First we inspect that position for a non-zero entry—which it has. In fact, it already has a leading 1 there, so we don't need to *swap* or *multiply*. We go straight to Step (4), the zero-out step. It puts a zero *above* the pivot by adding row 2 to row 1:

$$
\begin{bmatrix} 1 & -1 & 5 & 1 \\ 0 & 1 & -3 & 3 \\ 0 & 1 & -3 & 3 \end{bmatrix} \xrightarrow[\text{---}]{\text{add r2 to r1}} \begin{bmatrix} 1 & 0 & 2 & 4 \\ 0 & 1 & -3 & 3 \\ 0 & 1 & -3 & 3 \end{bmatrix}
$$

Then we put a zero *below* the pivot by *subtracting* row 2 from row 3—which here cancels the entire third row:

$$
\begin{bmatrix} 1 & 0 & 0 & 4 \\ 0 & 1 & -3 & 3 \\ 0 & 1 & -3 & 3 \end{bmatrix} \xrightarrow[\text{---}]{\text{add } -\text{r2 to r3}} \begin{bmatrix} 1 & 0 & 2 & 4 \\ 0 & 1 & -3 & 3 \\ 0 & 0 & 0 & 0 \end{bmatrix}
$$

This finishes off column 2. We increment again (Step 5) and move to the next row and column (row 3, column 3). But when we scan for the next potential pivot, we find none—our augmented matrix already has RRE form.

Now the familiar method for systems in RRE from takes over. We observe that x and y are pivot variables, while z is free.

To get a particular solution, we set the free variable equal to zero and use the right-hand sides of the pivot rows for the pivot variables x and y to get

$$\mathbf{x}_p = \begin{pmatrix} 4 \\ 3 \\ 0 \end{pmatrix}$$

The usual check shows that this *does* solve the original system.

Meanwhile, since we have one free variable, the homogeneous solution is generated by a single vector $\mathbf{h}$. To get it, we set the free variable z equal to 1, and flip the signs of the non-zero entries (2 and -3) in the z column of the pivot rows of the final RRE matrix to get

$$\mathbf{h} = \begin{pmatrix} -2 \\ 3 \\ 1 \end{pmatrix}$$

This solves the *homogeneous* version of the original system, and we're ready to write the complete solution of the original system, which takes the standard form $\mathbf{x}_p + s\,\mathbf{h}$, where s can be any scalar. Explicitly, the solution set is the image of the map $\mathbf{x} : \mathbf{R} \to \mathbf{R}^3$ given by

$$\begin{aligned}
\mathbf{x}(s) &= \begin{pmatrix} 4 \\ 3 \\ 0 \end{pmatrix} + s \begin{pmatrix} -2 \\ 3 \\ 1 \end{pmatrix} \\
&= \begin{pmatrix} 4 - 2s \\ 3 + 3s \\ s \end{pmatrix}
\end{aligned}$$

This formula gives all solutions of the original system. □

EXAMPLE 5.6. As a second demonstration of Gauss–Jordan, consider the **homogeneous** system whose **coefficient** (not augmented) matrix is

$$\begin{bmatrix} 3 & 6 & 0 & 12 \\ -1 & -2 & 0 & -3 \\ 2 & 4 & 1 & 11 \\ -2 & -4 & 0 & -8 \end{bmatrix}$$

We immediately find a non-zero entry—namely a 3—in the $(1, 1)$ position so we can skip the *swap* step and go straight to creating a pivot, a common situation.

We do so by multiplying all of row 1 by the reciprocal of its leading entry, that is, $1/3$:

$$\begin{bmatrix} 3 & 6 & 0 & 12 \\ -1 & -2 & 0 & -3 \\ 2 & 4 & 1 & 11 \\ -2 & -4 & 0 & -8 \end{bmatrix} \xrightarrow[\text{- - - -}]{\text{multiply r1 by 1/3}} \begin{bmatrix} 1 & 2 & 0 & 4 \\ -1 & -2 & 0 & -3 \\ 2 & 4 & 1 & 11 \\ -2 & -4 & 0 & -8 \end{bmatrix}$$

Here and every time we acquire a new pivot, we use it to zero-out the rest of its column (Step 4 of the algorithm). In this case, there are three other non-zero entries in the current column, so we have to apply Step 4 three times. We

- Add row 1 to row 2,

- Add minus twice row 1 to row 3, and

- Add twice row 1 to row 4.

These moves actually produce zeros in other columns too, getting us much closer to RRE form:

$$\begin{bmatrix} 1 & 2 & 0 & 4 \\ -1 & -2 & 0 & -3 \\ 2 & 4 & 1 & 11 \\ -2 & -4 & 0 & -8 \end{bmatrix} \xrightarrow[\text{- - - -}]{\text{3 moves}} \begin{bmatrix} 1 & 2 & 0 & 4 \\ 0 & 0 & 0 & 1 \\ 0 & 0 & 1 & 3 \\ 0 & 0 & 0 & 0 \end{bmatrix}$$

With a pivot in the $(1, 1)$ position and having zeroed-out the rest of its column, we *increment*, moving to the next row and column. This takes us to the $(2, 2)$ entry.

Here however, when we scan downwards starting at position $(2, 2)$, we find only zeros! In this case, Step (2) has us move right one column— taking us to position $(2, 3)$—and retry the scan from there.

In column 3, we do find a non-zero entry when we scan down to row 3. Step (2) tells us to swap that row with the "current" row 2:

$$\begin{bmatrix} 1 & 2 & 0 & 4 \\ 0 & 0 & 0 & 1 \\ 0 & 0 & 1 & 3 \\ 0 & 0 & 0 & 0 \end{bmatrix} \xrightarrow[\text{- - - -}]{\text{swap r3 w/r2}} \begin{bmatrix} 1 & 2 & 0 & 4 \\ 0 & 0 & 1 & 3 \\ 0 & 0 & 0 & 1 \\ 0 & 0 & 0 & 0 \end{bmatrix}$$

We now have a leading 1 in row 2. Typically, we'd now use it to zero-out the rest of its column, but the rest of column 3 is already zero, so we can skip to the next step, moving to the next row and column.

This takes us to position $(3, 4)$. There is already a pivot there, so we needn't scan, swap, or multiply. If we were dealing with the *augmented* matrix of a system here, we would be done: A leading 1 in the last column would mean *no solution*.

Here, however, we have the coefficient matrix of a *homogeneous* system. So the last column plays the same role as every other column, and we continue with the Gauss–Jordan algorithm, using the leading 1 to zero out the rest of the column. Following Step 4, we do so by

- Adding -3 times row 3 to row 2, and

- Adding -4 times row 3 to row 1.

This takes us to RRE form:

$$\begin{bmatrix} 1 & 2 & 0 & 4 \\ 0 & 0 & 1 & 3 \\ 0 & 0 & 0 & 1 \\ 0 & 0 & 0 & 0 \end{bmatrix} \quad \underset{\text{2 moves}}{-\,-\,-\,-\,\rightarrow} \quad \begin{bmatrix} 1 & 2 & 0 & 0 \\ 0 & 0 & 1 & 0 \\ 0 & 0 & 0 & 1 \\ 0 & 0 & 0 & 0 \end{bmatrix}$$

We now see that x_2 is the only free variable. Since the system is homogeneous, the solution will consist of all multiples of the single homogeneous generator **h**. To get **h** we set the free variable equal to 1, flip the signs of the three entries in the non-zero rows of the x_2 column, and obtain

$$\mathbf{h} = \begin{pmatrix} -2 \\ 1 \\ 0 \\ 0 \end{pmatrix}$$

The complete solution of our homogeneous system is then given by

$$\mathbf{x}(s) = (-2s, \ s, \ 0, \ 0)$$

□

We emphasized that the Gauss–Jordan algorithm is fully *deterministic*: each step is completely determined by the instructions and the current state of the matrix. Moreover, since the *increment* step moves it inevitably column-by-column to the right and row-by-row downwards, leaving every column it passes in RRE form, it *never* ends until the entire matrix has RRE form.

DEFINITION 5.7. Let us agree to write RRE($\mathbf{A}$) for the reduced row-echelon form that we get by applying the Gauss–Jordan algorithm to $\mathbf{A}$. For example, if

$$\mathbf{A} = \begin{bmatrix} 1 & 2 \\ 2 & 4 \end{bmatrix}$$

then

$$\text{RRE}(\mathbf{A}) = \begin{bmatrix} 1 & 2 \\ 0 & 0 \end{bmatrix}$$

$\square$

REMARK 5.8. The reduced row-echelon form of $\mathbf{A}$ is fully determined quite apart from the Gauss–Jordan algorithm. *Any* sequence of elementary row operations that brings $\mathbf{A}$ into reduced row-echelon form will bring it to exactly the same matrix we get using Gauss–Jordan, namely RRE($\mathbf{A}$). We will not tackle the proof of this, but it means that one needn't follow the Gauss–Jordan algorithm step-by-step as listed above. Often one can get to RRE form faster by making a "smarter" choice of row operations. $\square$

– Practice –

109. Reduce to RRE form using Gauss–Jordan:

a) $\begin{bmatrix} 1 & 2 \\ 3 & 4 \end{bmatrix}$ b) $\begin{bmatrix} 3 & 9 \\ 1 & 3 \end{bmatrix}$ c) $\begin{bmatrix} 0 & 5 \\ 0 & 12 \end{bmatrix}$

110. Reduce to RRE form using Gauss–Jordan:

$$\begin{bmatrix} 1 & 1 & 0 & 0 \\ 1 & 2 & 1 & 0 \\ 1 & 2 & 2 & 1 \end{bmatrix}$$

111. Reduce to RRE form using Gauss–Jordan:

$$\begin{bmatrix} 3 & 9 & 3 & 3 & 1 \\ 4 & 12 & 0 & -8 & -1 \\ -1 & -3 & 1 & 5 & 1 \end{bmatrix}$$

112. Reduce to RRE form using Gauss–Jordan:

$$\begin{bmatrix} 3 & 4 & -1 \\ 9 & 12 & -3 \\ 3 & 0 & 1 \\ 3 & -8 & 5 \\ 1 & -1 & 1 \end{bmatrix}$$

113. Find all solutions of the system.

$$\begin{aligned} x - z &= 1 \\ x - y &= 1 \\ y - z &= 1 \end{aligned}$$

114. Find the complete solution of each system.

a) $\begin{aligned} x + y + z &= 1 \\ x - y + z &= -1 \\ x + y - z &= -1 \end{aligned}$
b) $\begin{aligned} x + y + z &= 1 \\ x - y + z &= -1 \\ x + z &= 1 \end{aligned}$

115. Find the complete solution of each system.

a) $x + y + z = 1$
b) $\begin{aligned} x + y + z &= 1 \\ x - y + z &= -1 \end{aligned}$

116. Solve the homogeneous system:

$$\begin{aligned} 8x_1 + 4x_2 + 4x_3 + 2x_4 &= 0 \\ x_2 + 2x_3 + x_4 &= 0 \\ 8x_1 + 6x_2 + 8x_3 + 4x_4 &= 0 \end{aligned}$$

117. Solve the homogeneous system:

$$x_1 - 2x_2 + 3x_3 - 4x_4 = 0$$

118. If we let $\mathbf{R}^{m \times n}$ denote the set of all $n \times m$ matrices, *each elementary row operation can be seen as a map from $\mathbf{R}^{m \times n}$ to itself.*

   a) Let $S_{ij} \colon \mathbf{R}^{m \times n} \to \mathbf{R}^{m \times n}$ be the mapping that swaps rows i and j of any $\mathbf{A} \in \mathbf{R}^{m \times n}$. Is this map one-to-one? Is it onto? What is its inverse?

b) Let $M_i(k)\colon \mathbf{R}^{m\times n} \to \mathbf{R}^{m\times n}$ be the mapping that multiplies row i of any $\mathbf{A} \in \mathbf{R}^{m\times n}$ by the scalar k. Is this map one-to-one? Is it onto? Your answers should depend on whether $k = 0$ or not. When $k \neq 0$, what is the inverse of $M_i(k)$?

c) Let $A_{i,j}(k)\colon \mathbf{R}^{m\times n} \to \mathbf{R}^{m\times n}$ be the mapping that adds k times row i of $\mathbf{A}$ to row j. Is this map one-to-one? Is it onto? What is its inverse?

119. The Gauss–Jordan algorithm gives us a way to reach RRE($\mathbf{A}$) from $\mathbf{A}$ using elementary row operations. Can we always reach $\mathbf{A}$ from RRE($\mathbf{A}$) using elementary row operations? If so, how? If not, why not?

120. Let $\mathbf{R}^{m\times n}$ denote the set of all $n \times m$ matrices as in Exercise 118. The rule $\mathbf{A} \mapsto$ RRE($\mathbf{A}$) defines a mapping GJ ("Gauss–Jordan") from $\mathbf{R}^{m\times n}$ to itself.

a) Is GJ onto? If not, what is its image?

b) If GJ one-to-one? If not, how many pre-images does a matrix in its image typically have?

121. (Quadratic interpolation) Find a quadratic polynomial $q(x) = Ax^2 + Bx + C$ whose graph goes through the points $(-2, 19)$, $(1, 7)$, and $(2, 15)$. [*Hint: For each (x, y) point on the graph, the equation $q(x) = y$ gives a linear equation for $A, B,$ and C.*]

122. (Cubic interpolation) Find a *cubic* polynomial $c(x) = a_3 x^3 + a_2 x^2 + a_1 x + a_0$ whose graph goes through the points $(-3, 60)$, $(-1, 0)$, $(1, 12)$, and $(3, 24)$.

123. (General polynomial interpolation) *Suppose we have $n+1$ points*

$$(x_0, y_0),\ (x_1, y_1),\ (x_2, y_2), \ldots,\ (x_n, y_n) \in \mathbf{R}^2$$

with $x_0 < x_1 < x_2 < \cdots < x_n$. Then there exists a degree-n polynomial

$$p(x) = a_n x^n + a_{n-1} x^{n-1} + \cdots + a_2 x^2 + a_1 x_1 + a_0$$

whose graph goes through all the given points by solving a linear system. What is the augmented matrix of this system? [*Hint: Review Exercises 121 and 122 and generalize.*]

— ⋆ —

6. Two Mapping Answers

We now understand how to solve *any* linear system $\mathbf{Ax} = \mathbf{b}$ for any $n \times m$ matrix $\mathbf{A}$ —no matter how many equations ("n") or how many variables ("m") it has. The method, to summarize, goes like this:

- First form the augmented matrix $[\mathbf{A} \,|\, \mathbf{b}]$.

- Use the Gauss–Jordan algorithm to put $[\mathbf{A} \,|\, \mathbf{b}]$ into RRE form without changing the solution set of the system.

- Then extract both parts of the solution:

  - The particular solution $\mathbf{x}_p$, and

  - The homogeneous solution $\mathbf{Hs} = s_1 \mathbf{h}_1 + \cdots + s_k \mathbf{h}_k$.

  Here k is the number of free variables in $\mathrm{RRE}(\mathbf{A})$, and the $\mathbf{h}_i$'s are its *homogeneous generators*. Each $\mathbf{h}_i$ solves $\mathbf{Ax} = \mathbf{0}$, and as $\mathbf{s}$ varies over $\mathbf{R}^k$, the matrix/vector product $\mathbf{Hs}$ gives *all* homogeneous solutions.

- If $\mathbf{Ax} = \mathbf{b}$ has *any* solution $\mathbf{x}_p$, its *complete* solution is the image of the map $\mathbf{x} : \mathbf{R}^k \to \mathbf{R}^m$ given by

$$
(9) \qquad \begin{aligned}
\mathbf{X}(s_1, \ldots, s_k) &= \mathbf{x}_p + \mathbf{Hs} \\
&= \mathbf{x}_p + s_1 \mathbf{h}_1 + \cdots + s_k \mathbf{h}_k
\end{aligned}
$$

Let us emphasize again (as we did in Remark 3.4) that in the linear algebra of *vectors*—as opposed to scalars—we can, and often *do* have $\mathbf{Ax} = \mathbf{0}$ when neither $\mathbf{A}$ nor $\mathbf{x}$ vanishes individually. In fact, this may happen even when neither $\mathbf{A}$ nor $\mathbf{x}$ has a single zero entry. For instance

$$
\begin{bmatrix} 1 & 2 \\ 2 & 4 \end{bmatrix} \begin{pmatrix} 2 \\ -1 \end{pmatrix} = \begin{pmatrix} 0 \\ 0 \end{pmatrix}
$$

In fact, *finding* such homogeneous solutions (finding the matrix $\mathbf{H}$ above), forms the *main* step in solving a typical *in*homogeneous system $\mathbf{Ax} = \mathbf{b}$. If we change $\mathbf{b}$, the particular solution $\mathbf{x}_p$ will change, but the *homogeneous* solution $\mathbf{Hs}$ remains the same for all $\mathbf{b}$.

Finally, since *"Solve $\mathbf{Ax} = \mathbf{b}$"* means the same as *"Find the pre-image of $\mathbf{b}$ under the linear transformation T represented by $\mathbf{A}$,"* we can view it this way:

OBSERVATION 6.1. *The pre-image of every vector* **b** *in the image of a linear map* T *has the same form:* $T^{-1}(\mathbf{b}) = \mathbf{x}_p + \mathbf{Hs}$, *where* **H** *is the matrix whose columns are the homogeneous generators for the system, and* **s** *varies over* $\mathbf{R}^k$.

Using this fact, we can answer two of our basic mapping questions.

6.2. The "one-to-one" question.

PROPOSITION 6.3. *Suppose* $T : \mathbf{R}^m \to \mathbf{R}^n$ *is a linear transformation, and* **A** *is the matrix that represents it. Then* T *is one-to-one if and only if* RRE(**A**) *has a pivot in every* ***column***.

PROOF. Recall that "T is one-to-one" means *every* $\mathbf{b} \in \mathbf{R}^n$ has at *most* one point in its pre-image. But the homogeneous system $\mathbf{Ax} = \mathbf{0}$ has *infinitely* many solutions if we have even one homogeneous generator $\mathbf{h}_1$. That means the pre-image $T^{-1}(\mathbf{0})$ contains infinitely many points. So T *fails* to be one-to-one if RRE(**A**) has any free columns, i.e., unless there's a pivot in every column.

Conversely, if the matrix representing T has *no* free columns, then (9) shows that $T^{-1}(\mathbf{b})$ will contain only the vector $\mathbf{x}_p$ if it contains any vectors at all. Each $\mathbf{b} \in \mathbf{R}^n$ has *at most* one pre-image, so T is one-to-one.

So T is one-to-one if and only if RRE(**A**) has no free columns, which again means a pivot in every column, as the Proposition asserts. □

EXAMPLE 6.4. Consider the transformation $T : \mathbf{R}^3 \to \mathbf{R}^3$ given by

$$T(x, y, z) = (x + 2y + 3z, \ 4x + 5y + 6z, \ 7x + 8y + 9z)$$

Is it one-to-one?

The reader will easily verify that T is represented by

$$\mathbf{A} = \begin{bmatrix} 1 & 2 & 3 \\ 4 & 5 & 6 \\ 7 & 8 & 9 \end{bmatrix}$$

which row reduces to

$$\mathrm{RRE}(\mathbf{A}) = \begin{bmatrix} 1 & 0 & -1 \\ 0 & 1 & 2 \\ 0 & 0 & 0 \end{bmatrix}$$

Since RRE(**A**) does *not* have a pivot in every column, Proposition 6.3 tells us that T is *not* one-to-one. □

COROLLARY 6.5. *A linear transformation from* $\mathbf{R}^m$ *to* $\mathbf{R}^n$ *cannot be one-to-one if* $m > n$.

PROOF. The matrix $\mathbf{A}$ representing such a transformation has m columns and n rows. If $m > n$, it has more columns than rows. Since RRE$(\mathbf{A})$ can have at most one pivot in each row, "more columns than rows" means there aren't enough pivots to go in every column. So when $m > n$, at least one column lacks a pivot. This prevents T from being one-to-one, by Proposition 6.3. □

For instance, a linear transformation $T : \mathbf{R}^4 \to \mathbf{R}^3$ is *never* one-to-one, since $4 > 3$.

6.6. The "onto" question. This question requires a subtler analysis, but we can answer it too.

PROPOSITION 6.7. *Suppose* $T : \mathbf{R}^m \to \mathbf{R}^n$ *is a linear transformation, and* $\mathbf{A}$ *is the matrix that represents it. Then* T *is onto if and only if* RRE$(\mathbf{A})$ *has a pivot in every* ***row***.

PROOF. Recall that "T is onto" means every $\mathbf{b} \in \mathbf{R}^n$ has at *least* one point in its pre-image, which is the same as saying that $\mathbf{Ax} = \mathbf{b}$ *has* a solution for every right-hand side $\mathbf{b}$. The only thing that can prevent $\mathbf{Ax} = \mathbf{b}$ from having a solution, however, is a pivot in the *last* column when we reduce the *augmented* matrix $[\mathbf{A}|\mathbf{b}]$. That cannot happen if RRE$(\mathbf{A})$ has a pivot in every row, however, since a pivot in the "$\mathbf{b}$" column would then have to come *after* the pivot in that same row of RRE$(\mathbf{A})$. Since the rules for RRE form forbid two pivots in one row, "a pivot in every row of RRE$(\mathbf{A})$" ensures that $\mathbf{Ax} = \mathbf{b}$ has a solution, which makes T onto.

Conversely, if RRE$(\mathbf{A})$ has a row of with *no* pivot, T *cannot* be onto. This is the subtler part of the argument; it goes like this.

Suppose RRE$(\mathbf{A})$ has a row with no pivot. Such a row must be all zeros, so RRE$(\mathbf{A})$ has at least one row of zeros at the bottom. With this in mind, consider the augmented matrix

(10) $$\left[\text{RRE}(\mathbf{A}) \,\middle|\, \mathbf{e}_n \right] ,$$

where $\mathbf{e}_n = (0, 0, \ldots, 0, 1)$ is the nth standard basis vector in $\mathbf{R}^n$. Since RRE$(\mathbf{A})$ has a row of zeros at the bottom, this system has *no* solution.

Now suppose we "undo," in reverse order, all the row operations we used in reducing $\mathbf{A}$ to RRE$(\mathbf{A})$. Then (10) will "un-reduce" to an augmented matrix of the form $[\mathbf{A} \,|\, \mathbf{b}]$, where $\mathbf{A}$ is again the matrix representing T, and $\mathbf{b}$ is *some* vector in $\mathbf{R}^n$. Having constructed $\mathbf{b}$ using "reverse row-reduction," however, we now know that forward row reduction will return us to $[\text{RRE}(\mathbf{A}) \mid \mathbf{b}]$:

$$[\mathbf{A} \mid \mathbf{b}] \longrightarrow [\text{RRE}(\mathbf{A}) \mid \mathbf{e}_n]$$

In short, with this choice of $\mathbf{b}$, $\mathbf{Ax} = \mathbf{b}$ has *no* solution. In other words, $\mathbf{b}$ has no pre-image under T, and T is not onto.

In sum, we have shown that when $\text{RRE}(\mathbf{A})$ *has* a pivot in every row, it *is* onto, while conversely, if any row of $\text{RRE}(\mathbf{A})$ lacks a pivot, T is *not* onto. This is precisely what the Proposition asserts. $\qquad\square$

EXAMPLE 6.8. In Example 6.4, we showed that the linear transformation $T : \mathbf{R}^3 \to \mathbf{R}^3$ given by

$$T(x, y, z) = (x + 2y + 3z, \ 4x + 5y + 6z, \ 7x + 8y + 9z)$$

was *not* one-to-one, because the matrix $\mathbf{A}$ representing it reduced to

$$\text{RRE}(\mathbf{A}) = \begin{bmatrix} 1 & 0 & -1 \\ 0 & 1 & 2 \\ 0 & 0 & 0 \end{bmatrix}$$

which has a free column.

We now observe that $\text{RRE}(\mathbf{A})$ also has a *row* with no pivot. By Proposition 6.7, that means that T is not onto either. $\qquad\square$

COROLLARY 6.9. *A linear transformation* $T : \mathbf{R}^m \to \mathbf{R}^n$ *cannot be* **onto** *if* $n > m$.

PROOF. The matrix $\mathbf{A}$ representing such a transformation has m columns and n rows. If $n > m$, this matrix has more rows than columns. Since $\text{RRE}(\mathbf{A})$ can have at most one pivot in each column, "more rows than columns" means there aren't enough pivots to go in every row, even when every *column* has a pivot. So when $n > m$, *some* row lacks a pivot, and this prevents T from being onto, by Proposition 6.7. $\qquad\square$

COROLLARY 6.10. *A linear transformation* $T : \mathbf{R}^m \to \mathbf{R}^n$ *cannot be* **both** *one-to-one* **and** *onto unless* $m = n$.

PROOF. T can't be one-to-one if $m > n$ (Proposition 6.3), and it can't be onto if $m < n$ (by Proposition 6.7). We can avoid both inequalities only when $m = n$. $\qquad\square$

REMARK 6.11 (Warning!). By Corollary 6.10, a linear transformation from $\mathbf{R}^m$ to $\mathbf{R}^n$ can't be both one-to-one *and* onto unless $m = n$. It does *not* work the other way, though: having $m = n$ doesn't make T both one-to-one and onto. In fact, it may be *neither*! Example 6.8 illustrates this by exhibiting a transformation with $m = n = 3$ which is *neither* one-to-one, *nor* onto.

COROLLARY 6.12. *A linear transformation* $T : \mathbf{R}^n \to \mathbf{R}^n$ *is either* **both** *one-to-one* **and** *onto, or it is neither.*

PROOF. The matrix $\mathbf{A}$ that represents a transformation from $\mathbf{R}^n$ to itself will be square, with n rows and n columns. If T is one-to-one, then by Proposition 6.3, RRE($\mathbf{A}$) has a pivot in every column. Since there are equally many rows and columns, and no row can have two pivots, it follows that there is a pivot in every row too, and hence T is also onto, by Proposition 6.7. In short, if T is one-to-one and $m = n$, it is also onto.

Similar reasoning shows that if T is onto, it is also one-to-one. We leave this as an exercise.

Together these facts show that if T is *either* one-to-one or onto, it is *both*. So it is either both, or neither. □

6.13. Isomorphism.
Linear maps that *are* both one-to-one and onto are especially important, so they have a special name:

DEFINITION 6.14. A linear map $T : \mathbf{R}^m \to \mathbf{R}^n$ which is both one-to-one and onto is called an **isomorphism**.

The most basic isomorphism is the *identity* map $I : \mathbf{R}^m \to \mathbf{R}^m$ given by $I(\mathbf{x}) = \mathbf{x}$ for all $\mathbf{x} \in \mathbf{R}^m$. The identity map has exactly one pre-image for every $\mathbf{b}$, namely $\mathbf{b}$ itself. This makes it both one-to-one and onto, hence an isomorphism.

In this language we can restate Corollary 6.10 above as follows:

A linear transformation from $\mathbf{R}^m$ *to* $\mathbf{R}^n$ *cannot be an isomorphism unless* $m = n$.

We emphasize again (as in Remark 6.11) that while a linear map $T : \mathbf{R}^m \to \mathbf{R}^n$ cannot be an isomorphism unless domain equals range, having domain equal to range doesn't *necessarily* make it an isomorphism. The **zero** map $Z : \mathbf{R}^m \to \mathbf{R}^m$, given by $Z(\mathbf{x}) = \mathbf{0}$ for every $\mathbf{x} \in \mathbf{R}^m$, provides another example. It badly fails to be either one-to-one or onto, even though it has the same domain and range.

EXAMPLE 6.15 (Another isomorphism). The map $P : \mathbf{R}^3 \to \mathbf{R}^3$ given by

$$P(x, y, z) = (x + y, \ y + z, \ z + x)$$

is represented by

$$\mathbf{P} = \begin{bmatrix} 1 & 1 & 0 \\ 0 & 1 & 1 \\ 1 & 0 & 1 \end{bmatrix}$$

The reader can easily check that

$$\mathrm{RRE}(\mathbf{P}) = \begin{bmatrix} 1 & 0 & 0 \\ 0 & 1 & 0 \\ 0 & 0 & 1 \end{bmatrix}$$

This has a pivot in every row and every column, so by Propositions 6.3 and 6.7, P is *both* one-to-one and onto—an isomorphism. □

REMARK 6.16. Because isomorphisms are (by definition) *both* one-to-one *and* onto, every isomorphism T has an *inverse* mapping T^{-1}, as discussed in Section 2.17 of Chapter 1. It is not hard to show that the inverse of an isomorphism is again a (linear) isomorphism. Rather than prove that here, however, we wait until Chapter 4. There we introduce inverse matrices, which reduce the proof to a routine exercise (Exercise 262 of Chapter 4). □

– Practice –

124. Give a short proof that when a "domain = range" linear transformation $T : \mathbf{R}^n \to \mathbf{R}^n$ is onto, it must also be one-to-one. (Imitate the reasoning used in the first paragraph of the proof of Corollary 6.12.)

125. For each linear mapping below, find the matrix representing it, and use it to decide if the mapping is one-to-one and/or onto.

a) The map $T : \mathbf{R}^3 \to \mathbf{R}^3$ given by
$$T(x, y, z) = (x - y - z, \ y - z - x, \ z - x - y)$$

b) The map $Z : \mathbf{R}^2 \to \mathbf{R}^4$ given by
$$Z(x, y) = (x - y, \ 2y - 2x, \ 3x - 3y, \ 4y - 4x)$$

c) The map $W : \mathbf{R}^2 \to \mathbf{R}^4$ given by
$$W(x, y) = (x - y, \ 2y - 2x, \ 3x + 3y, \ 4y + 4x)$$

d) The map $G : \mathbf{R}^4 \to \mathbf{R}^2$ given by
$$G(x_1, x_2, x_3, x_4) = (x_1 + 2x_2 + 2x_3 - x_4, \ 2x_1 + x_2 + x_3 - 2x_4)$$

e) The mapping $S : \mathbf{R}^4 \to \mathbf{R}^4$ given by
$$S(x_1, x_2, x_3, x_4) = (x_1 - x_2, \ x_2 + x_3, \ x_3 - x_4, \ x_4 - x_1)$$

Are any of these maps isomorphisms?

126. Give an example (like the ones in Exercise 125 above) of

a) A linear mapping $F : \mathbf{R}^3 \to \mathbf{R}^5$ that is one-to-one. Can you also make it onto? Why or why not?

b) A linear mapping $F : \mathbf{R}^5 \to \mathbf{R}^3$ that is onto. Can you also make it one-to-one? Why or why not?

127. Consider the linear transformation $T : \mathbf{R}^2 \to \mathbf{R}^2$ represented by
$$\mathbf{A} = \begin{bmatrix} 1 & -1 \\ -1 & 1 \end{bmatrix}$$

a) Show that T is neither one-to-one nor onto.

b) Find a $\mathbf{b} \in \mathbf{R}^2$ (other than $(0,0)$) whose pre-image contains more than one vector.

c) Find a $\mathbf{b} \in \mathbf{R}^2$ with *no* pre-image under T.

128. Show that when a linear map $F : \mathbf{R}^2 \to \mathbf{R}^2$ is represented by a matrix of the form
$$\begin{bmatrix} 1 & a \\ b & 1 \end{bmatrix}$$
with $ab \neq 1$, it has to be an isomorphism.

129. Show that a linear map $T : \mathbf{R}^2 \to \mathbf{R}^2$ represented by *any* 2-by-2 matrix
$$\begin{bmatrix} a & b \\ c & d \end{bmatrix}$$
will be an isomorphism *unless* $ad - bc = 0$. (Thus, linear maps from $\mathbf{R}^2$ to itself are "usually" isomorphisms.)

130. True or False: *An $n \times n$ matrix* **A** *represents an isomorphism from* $\mathbf{R}^n$ *to* $\mathbf{R}^n$ *if and only if* RRE($\mathbf{A}$) = $\mathbf{I}_n$ *(the $n \times n$ identity matrix)*. Explain your conclusion.

131. The linear map $T : \mathbf{R}^2 \to \mathbf{R}^2$ given by $T(x, y) = (x + y, \, x - y)$ is an isomorphism. By solving $T(x, y) = (b_1, b_2)$ for x and y in terms of b_1 and b_2, find a formula for the inverse of T.

132. Given any three scalars a, b, c, form the matrix

$$\begin{bmatrix} 0 & c & -b \\ -c & 0 & a \\ b & -a & 0 \end{bmatrix}$$

Show that the linear transformation it represents is never one-to-one, nor onto, no matter what a, b, and c are.

133. Given any three scalars a, b, c, consider the matrix

$$\begin{bmatrix} a^2 & ab & ac \\ ab & b^2 & bc \\ ac & bc & c^2 \end{bmatrix}$$

Show that the linear transformation it represents is never one-to-one, nor onto, no matter what a, b, and c are.

— ⋆ —

CHAPTER 3

Linear Geometry

Systems of equations and their solutions, the subjects of Chapter 2, are purely algebraic. Virtually all that material has a *geometric* interpretation, however, that will help us understand the algebra better. The algebra, in return provides a powerful tool for solving geometry problems. Linear algebra is in many respects a dance between algebra and geometry: each partner helps the other to do things they could not do alone.

1. Geometric Vectors

1.1. Arrows as vectors. Suppose we take the plane of Euclidean geometry, and select—at random—one point, as an **origin** $\vec{0}$. We make no choice of axes yet. Imagine the plane as completely blank except for one infinitesimal dot at $\vec{0}$.

DEFINITION 1.2 (Geometric vector). We will call any arrow in the plane—any directed line segment with its tail at the origin, a **geometric vector**. We often view the collection of *all* these arrows as the plane itself, since each point p in the plane corresponds to the arrow from $\vec{0}$ to p, and hence geometric vectors are in one-to-one correspondence with the points of the plane (Figure 1). We denote the geometric vector from $\vec{0}$ to p by $\vec{p}$, and we regard the "infinitesimal" arrow from $\vec{0}$ to itself as a geometric vector too: the **zero vector**.

Note that we have defined geometric vectors *without mentioning numeric vectors—or even numbers—at all*. It turns out that, even so, we can do vector arithmetic with geometric vectors, *adding and scalar multiplying* them just as we did with numeric vectors in Chapter 2.

1.3. Geometric vector addition. The procedure for adding geometric vectors in the plane is simple. To add $\vec{p}$ to $\vec{q}$, we just slide $\vec{q}$, without rotating, moving its tail from the origin to p, the tip of $\vec{p}$. The head of the *translated* vector $\vec{q}$ now locates the sum $\vec{s} := \vec{p} + \vec{q}$. (Figure 2). Note that the arrows $\vec{p}$ and $\vec{p} + q'$, along with the relocated

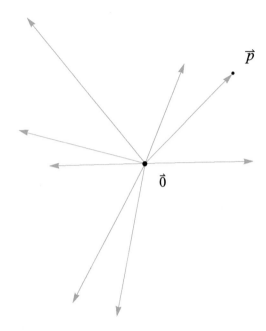

Figure 1. Geometric vectors. An arrow from the origin $\vec{0}$ to any point in the plane forms a *geometric vector*.

$\vec{q}$, now form a triangle. Accordingly, we call this the **triangle rule** for adding geometric vectors.

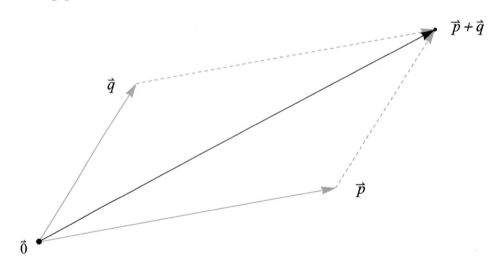

Figure 2. Addition. We form $\vec{p} + \vec{q}$ by sliding the tail of $\vec{q}$ to the tip of $\vec{p}$ (dashed arrow on the right) or vice-versa (dashed arrow above). Either way yields the black vector $\vec{p} + \vec{q}$.

Note a fact of key importance:

OBSERVATION 1.4. *Addition of geometric vectors is* **commutative:** $\vec{p} + \vec{q} = \vec{q} + \vec{p}$.

This is clear from Figure 2: Whether we translate $\vec{q}$ to the tip of $\vec{p}$ to get the dashed arrow on the right, or translate $\vec{p}$ to the tip of $\vec{q}$ to get the dashed arrow up above, we arrive at the same point s. Indeed, $\vec{p}$, $\vec{q}$, and their translated copies form a *parallelogram*[1] whose diagonal is exactly the sum of interest:

$$\vec{p} + \vec{q} = \vec{s} = \vec{q} + \vec{p}$$

We can add more than two vectors by repeating the process: simply translate each new vector to the tip of the previous sum. The last vector we add then points to the tip of the full sum. Figure 3 depicts an example.

Addition of geometric vectors is also **associative**:

OBSERVATION 1.5. *Vector addition is associative: Given any three geometric vectors $\vec{u}$, $\vec{v}$, $\vec{w}$, we have*

$$(\vec{u} + \vec{v}) + \vec{w} = \vec{u} + (\vec{v} + \vec{w})$$

We omit the proof of this fact, which is an exercise in plane geometry. Given any three specific geometric vectors, however, it is easy to verify (Exercise 139).

1.6. Scalar multiplication. We can also *scalar multiply* geometric vectors (Figure 4). This requires at most two steps, and takes place entirely within the line containing the vector.

To multiply a geometric vector $\vec{v}$ by a scalar s, we simply

- *Multiply its **length** by $|s|$, and,*
- *If the scalar s is negative, reverse its **direction**.*

In particular, multiplication by -1 reverses the direction of a geometric vector without changing its length.

A simple argument using similar triangles shows that *scalar multiplication distributes over geometric vector addition* (Exercise 140).

---

[1]Our "triangle rule" for vector addition is often called the "parallelogram rule" because of this parallelogram.

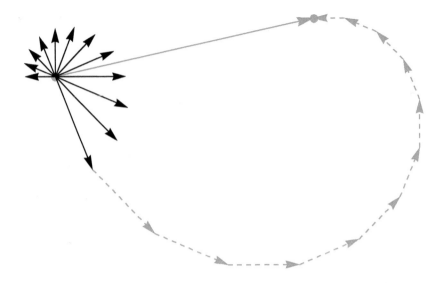

Figure 3. Vector summation. To sum the black vectors, we chain them tail-to-tip (dashed blue arrows). The solid blue arrow from the origin to the tip of the chain is the sum.

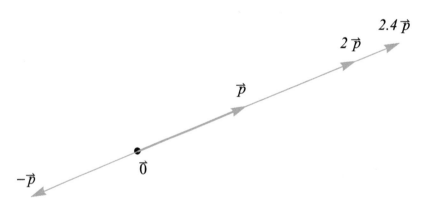

Figure 4. A geometric vector $\vec{p}$ and some of its scalar multiples.

1.7. Subtraction and linear combination. By combining addition and scalar multiplication, we can also *subtract* geometric vectors, or form *linear combinations*.

We subtract by *adding the opposite*:

$$\vec{p} - \vec{q} := \vec{p} + (-\vec{q})$$

The best way to visualize this is by forming the addition triangle for the sum $\vec{q} + (\vec{p} - \vec{q}) = \vec{p}$. If we do this by first drawing $\vec{q}$ and $\vec{p}$, then $\vec{p} - \vec{q}$ must be the vector that, when translated, goes from the tip of $\vec{q}$ to the tip of $\vec{p}$.

In short,

*$\vec{p} - \vec{q}$ is the geometric vector we get by drawing an arrow **from** the tip of $\vec{q}$ **to** the tip of $\vec{p}$, and then translating its tail back to the origin* (Figure 5).

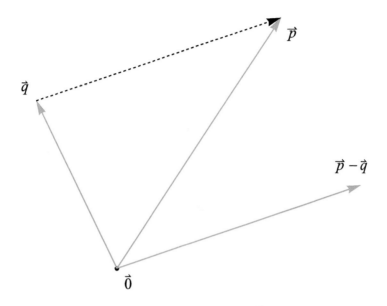

Figure 5. Vector subtraction. We find the difference $\vec{p} - \vec{q}$ by sketching an arrow from the tip of $\vec{q}$ to the tip of $\vec{p}$ (here dashed black), then translating its tail to the origin.

By combining our constructions for addition, subtraction, and scalar multiplication, we can form any linear combination of geometric vectors. Figure 6 depicts an example.

1.8. Geometric vectors in space. We have only discussed and depicted geometric vectors in the plane, but all we have said works equally well in space. Just as in the plane, once we choose an origin $\vec{0}$, geometric vectors are arrows emanating from it. We add/subtract by chaining vectors tail-to-tip, and we scalar multiply as before, within the line containing the vector. Figure 7 illustrates this with a linear combination in space.

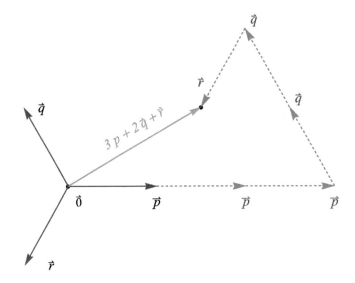

Figure 6. Linear combination. Multiplying $\vec{p}$, $\vec{q}$, and $\vec{r}$ by $3, 2$, and 1, respectively before summing, we get the linear combination $3\vec{p} + 2\vec{q} + \vec{r}$ (blue arrow).

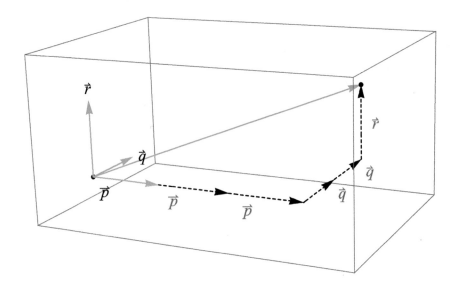

Figure 7. Linear combination in space works just as it does in the plane. Here the black dot marks the tip of $3\vec{p} + 2\vec{q} + \vec{r}$ (not shown).

– Practice –

134. Suppose $\vec{p}$ is a non-zero vector in the plane. How would you describe the set of

a) Tips of all scalar multiples of $\vec{p}$?

b) Tips of all *positive* scalar multiples of $\vec{p}$?

c) Tips of all *negative* scalar multiples of $\vec{p}$?

d) Tips of all *integer* scalar multiples of $\vec{p}$?

e) How (if at all) would your answers to (a)–(d) change if $\vec{p}$ were a vector in *space* instead of the plane?

f) How (if at all) would your answers to (a)–(e) change if $\vec{p}$ were the *zero* vector $\vec{0}$?

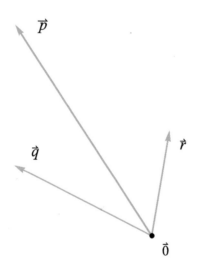

Figure 8. Template for Exercises 135 and 136.

135. Sketch or trace a fresh copy of Figure 8, then add and label these geometric vectors to your drawing:

a) $\vec{r} + \vec{p}$ b) $\vec{r} - \vec{p}$ c) $\vec{q} + \vec{p}$ d) $\vec{q} - \vec{p}$ e) $-\vec{q} - \vec{r}$

136. Sketch or trace a fresh copy of Figure 8, then add and label these geometric vectors to your drawing:

a) $\vec{r} + \vec{q} - \vec{p}$ b) $\vec{p} - 2\vec{r}$ c) $3\vec{r} - \vec{p} + 2\vec{q}$

137. In Figure 9, what linear combinations of $\vec{p}$ and $\vec{q}$ (if any) add up to the vectors with their tips at points (a), (b), and (c)? (A ruler may be helpful here.)

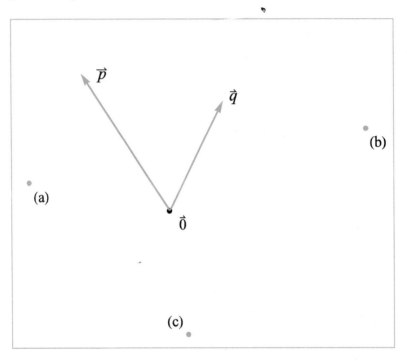

Figure 9. Template for Exercises 137 and 138.

138. Consider Figure 9.

  a) Can we reach *any* point in the plane with a linear combinations of the vectors $\vec{p}$ and $\vec{q}$ there? If so, how? If not, why not?

  b) Suppose we replace $\vec{p}$ and $\vec{q}$ there by some other pair of vectors $\vec{p}_*$ and $\vec{q}_*$. What must we assume about a pair of vectors to ensure that their linear combinations will (or will not) reach every point in the plane?

139. Verify the associative law $(\vec{u} + \vec{v}) + \vec{w} = \vec{u} + (\vec{v} + \vec{w})$ for the vectors shown in Figure 10 as follows:

  a) Sketch (or better, trace) a copy of Figure 10.

  b) Draw and label the addition triangle that constructs $(\vec{u} + \vec{v})$.

  c) Draw and label the addition triangle that adds $\vec{w}$ to that, giving $(\vec{u} + \vec{v}) + \vec{w}$.

d) Now go the other way: Draw and label the addition triangle that constructs $(\vec{v} + \vec{w})$.

e) Draw and label the addition triangle that adds $\vec{u}$ to that, giving $\vec{u} + (\vec{v} + \vec{w})$.

Done carefully, (use a ruler) both procedures should terminate at the blue dot.

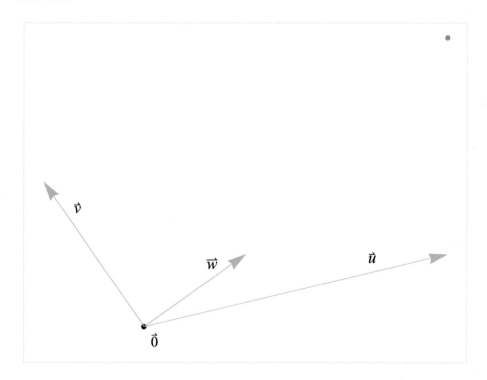

Figure 10. Template for Exercise 139.

140. *(Scalar multiplication distributes over geometric addition).* Draw and label two geometric vectors $\vec{a}$ and $\vec{b}$ with the same length, and forming an angle of roughly 45°.

a) Draw $2\vec{a} + \vec{b}$ and $2\vec{b} - \vec{a}$.

b) Is it true that $\vec{a} - \vec{a} = \vec{0}$? Will $\vec{v} - \vec{v}$ always equal $\vec{0}$? Explain why using the definitions of vector addition and scalar multiplication.

c) Redraw $\vec{a}$ and $\vec{b}$ on a fresh diagram and illustrate the identity $2(\vec{a} + \vec{b}) = 2\vec{a} + 2\vec{b}$ (a special case of the distributive law $k(\vec{a} + \vec{b}) = k\vec{a} + k\vec{b}$).

141. Suppose we travel around a closed (but not necessarily regular) polygon in the plane, turning each side we visit into a geometric vector by translating the first endpoint we reach to the origin, as in Figure 11. Explain why these vectors must sum to $\vec{0}$.

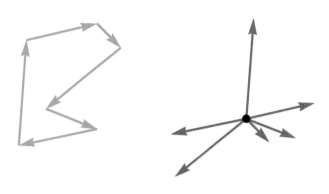

Figure 11. At left, a "tour" around a random polygon. At right, each edge is translated back to the origin to give a geometric vector. Why do the vectors at right sum to $\vec{0}$?

142. Does the converse of Exercise 141 hold? That is, if k non-zero geometric vectors sum to zero, will they form a closed, k-sided polygon when we join them end-to-tip? Does it matter what order we join them in? Why or why not?

143. a) Draw a 4-sided polygon that crosses itself (like a bow tie), but has no two sides parallel.

   b) Put arrows on one end of each side to indicate one complete tour around the polygon, as in the left side of Figure 11.

   c) Translate the arrows to the origin to get four geometric vectors, as in the right side of Figure 11.

   d) Can you arrange these same vectors to get a 4-sided polygon that does *not* cross itself?

— ★ —

2. Geometric/Numeric Duality

Numeric and geometric vectors don't just have similar properties—we can define an amazingly precise correspondence between them by introducing axes and units of measurement (Figure 12). The process should seem familiar:

To relate geometric vectors in the plane to numeric vectors in $\mathbf{R}^2$, we

i) *Choose x- and y-axes. Any two directed, non-parallel lines through the origin will serve.*

ii) *Choose a unit of length along each axis.*

Given these choices, we can assign a geometric vector $\vec{v}$ to any numeric vector $\mathbf{v} = (a, b) \in \mathbf{R}^2$ as follows:

- *Draw a line Y_a parallel to the y-axis through the point a on the x-axis.*

- *Draw a line X_b parallel to the x-axis through the point b on the y-axis.*

- *Define $\vec{v}$ as the arrow from $\vec{0}$ to the point where Y_a meets X_b.*

In essence, as a and b vary, the lines Y_a and X_b form a grid that assigns an (x, y) "address" to the tip of every geometric vector. We call this address the numeric **coordinate vector** $\mathbf{v}$ of the geometric vector $\vec{v}$. To summarize:

> *We can assign a numeric coordinate vector $\mathbf{v}$ to any geometric vector $\vec{v}$ in the plane by drawing two lines through the tip of $\vec{v}$—one parallel to each axis. The scalars a and b where these lines meet the x- and y-axes form the entries of $\mathbf{v}$. By reversing this process, we can assign a geometric vector $\vec{v}$ in the plane to any numeric vector $\mathbf{v} \in \mathbf{R}^2$.*

Our axes need *not* be perpendicular, and they can have different length units. If we choose different axes and/or units, the numeric coordinates we give to each geometric vector will also change. But once we fix axes and units, we get a one-to-one correspondence between geometric and numeric vectors.

DEFINITION 2.1. We call each choice of axes and units as described above a **coordinate system** for the plane. We use the term **geometric/numeric duality** to describe the correspondence a coordinate system creates between numeric and geometric vectors.

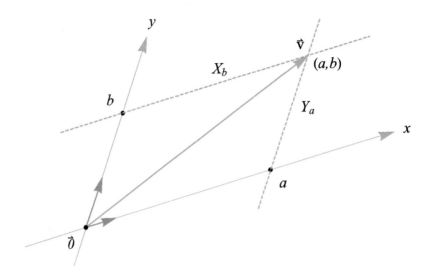

Figure 12. Coordinates. If we choose x and y axes, and introduce units (gray arrows), we can assign a numeric vector (a, b) to each geometric vector $\vec{v}$, and vice-versa.

The real power of this correspondence derives from an additional, wonderful property: it respects vector arithmetic.

Geometric/numeric duality respects vector arithmetic.

What does this mean? We explain carefully below, but in brief, it simply means that numeric/geometric duality is *linear*—it obeys the same two linearity rules that govern linear mappings from $\mathbf{R}^m \to \mathbf{R}^n$. We can add or scalar multiply either before, or after, we assign numeric addresses to geometric vectors (or vice-versa). Either order will produce the same result. The importance of this fact is hard to overstate, because:

*The linearity property of numeric/geometric duality lets us interpret vector **algebra** as **geometry** and vice-versa.*

Let us now be more precise. When we say that numeric/geometric duality *respects vector addition* we mean we get the same result whether we

– First add two numeric vectors $(x_1, y_1) + (x_2, y_2)$ and then interpret their sum as a geometric vector, or...

– First interpret (x_1, y_1) and (x_2, y_2) as geometric vectors, and add those via triangle addition.

Conversely, we get the same *numeric* result whether we

- First add two geometric vectors $\vec{v}_1 + \vec{v}_2$ and then assign a numeric coordinate vector to their sum, or...

- First assign numeric coordinates to $\vec{v}_1$ and $\vec{v}_2$ separately, then add those coordinates using numeric vector addition.

Figure 13 illustrates these facts.

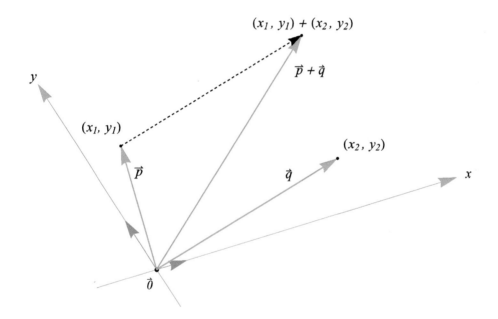

Figure 13. Geometric/numeric duality. We can add numeric vectors and then interpret geometrically, or add geometric vectors and interpret numerically—either way, we get the same result.

Precisely the same kind of statement holds for scalar multiplication: If we use numeric/geometric duality to assign a geometric vector $\vec{v}$ to a numeric vector $\mathbf{v} = (x, y)$, it will assign the geometric vector $s\vec{v}$ to the numeric vector $s\mathbf{v}$, and vice-versa. We can scalar multiply $\vec{v}$ and then read its coordinates, or read the coordinates of $\vec{v}$ and then scalar multiply. We get the same result either way.

Finally, because numeric/geometric duality respects both addition and scalar multiplication, it respects linear combinations.

In other words, if our coordinate system assigns numeric vectors $\mathbf{v}_i$ to a set of k geometric vectors $\vec{v}_i$, like this,

$$\mathbf{v}_1 \longleftrightarrow \vec{v}_1, \quad \mathbf{v}_2 \longleftrightarrow \vec{v}_2, \quad \cdots \quad \text{and} \quad \mathbf{v}_k \longleftrightarrow \vec{v}_k$$

then for any scalars $c_1, c_2, \ldots, c_k$, it assigns the same linear combination of numeric vectors to the geometric linear combination:

$$c_1\mathbf{v}_1 + c_2\mathbf{v}_2 + \cdots c_k\mathbf{v}_k \quad \longleftrightarrow \quad c_1\vec{v}_1 + c_2\vec{v}_2 + \cdots + c_k\vec{v}_k$$

We record all this as a fundamental proposition:

PROPOSITION 2.2. *Numeric/geometric duality respects vector arithmetic.*

PROOF. The proof is an exercise in similar triangles, easier to depict than to write out. We will describe part of it and supply the figure below, leaving the rest as an exercise.

Fix an arbitrary coordinate system. It will suffice to prove that the resulting numeric/geometric duality will respect addition and scalar multiplication, since linear combinations will then automatically be preserved too, since we get them by combining addition and scalar multiplication.

Consider scalar multiplication first. Suppose our coordinate system assigns a numeric vector $\mathbf{v} = (a, b)$ to a geometric vector $\vec{v}$, as depicted in Figure 14. Scalar multiply $\vec{v}$ by a scalar c, again as depicted in the figure. Which numeric vector (a', b') will the coordinate system assign to this new vector $c\vec{v}$? If we can show that $(a', b') = (ca, cb)$, then the Proposition holds for scalar multiplication. Figure 14 shows why this does in fact hold. The large triangle formed by $\vec{0}$, a', and $c\vec{v}$ is clearly similar to the smaller one formed by $\vec{0}$, a, and $\vec{v}$. And by definition of (geometric) scalar multiplication, side $c\vec{v}$ of the larger triangle is exactly c times as long as side $\vec{v}$ of the smaller one. It follows that the other sides must obey the same ratio. In particular, a' must be exactly c times a —in short $a' = ca$.

By drawing the dashed lines parallel to the x-axis instead of the y-axis, we could prove that $b' = cb$ in the same way. So $(a', b') = (ca, cb)$, and this proves that numeric/geometric duality respects scalar multiplication.

Similar arguments show that duality respects addition too—we leave the details as an exercise. $\qquad\square$

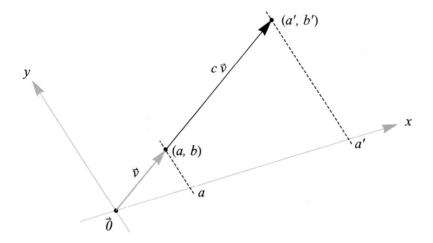

Figure 14. Scalar multiplication respects numeric/ geometric duality because the triangles cut off by the dashed lines are similar.

− **Practice** −

144. What numeric vectors **a**, **b**, **c**, and **d** correspond to the geometric vectors $\vec{a}$, $\vec{b}$, $\vec{c}$, and $\vec{d}$ in Figure 15, if $\vec{u}_x$ and $\vec{u}_y$ there mark *unit* vectors on the x- and y-axes, respectively?

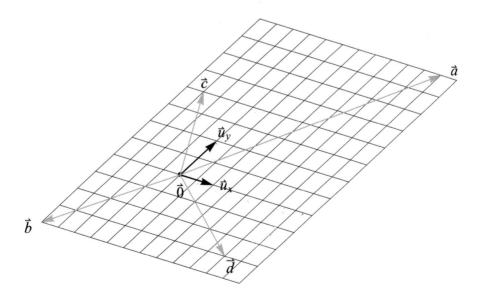

Figure 15. Graphic for Exercise 144.

145. Copy Figure 16, and draw the geometric vectors $\vec{a}$, $\vec{b}$, $\vec{c}$, and $\vec{d}$ that correspond respectively to the numeric "addresses" $\mathbf{a} = (1.5, 1.5)$, $\mathbf{b} = (3, -1.5)$, $\mathbf{c} = (-1.5, 3.5)$, and $\mathbf{d} = (-1.5, -1.5) = -\mathbf{a}$ in that Figure. (The geometric vectors $\vec{u}_x$ and $\vec{u}_y$ there mark the measurement units on the x- and y-axes, respectively.)

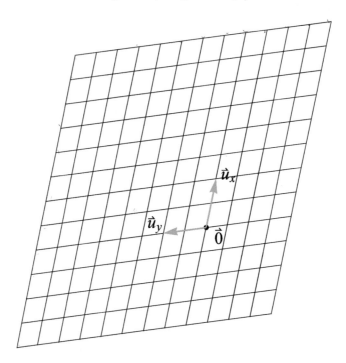

Figure 16. Template for Exercise 145.

146. Use a ruler to carefully draw any two axes and mark off units of measurement along them. [*Hint: Make a rough sketch to make sure you have room for everything before drawing a careful version with a ruler.*]

a) Carefully draw the geometric vectors $\vec{p}$ and $\vec{q}$ having coordinates $\mathbf{p} = (2, 1)$ and $\mathbf{q} = (1, -1)$.

b) Carefully draw the scalar multiples $-\vec{p}$, $-\vec{q}$, $2\vec{p}$, and $3\vec{q}$. Verify that their coordinates are $-\mathbf{p}$, $-\mathbf{q}$, $2\mathbf{p}$, and $3\mathbf{q}$, respectively, as predicted by numeric/geometric duality.

c) Draw the appropriate addition triangle to draw $\vec{r} := -\vec{p} - 3\vec{q}$, then verify that its numeric coordinate vector is indeed $-\mathbf{p} - 3\mathbf{q}$ as predicted by numeric/geometric duality.

147. Draw a coordinate system with geometric vectors $\vec{v}$ and $\vec{v}\,'$, respectively. Assign them numeric coordinates $\mathbf{v} = (a, b)$ and $\mathbf{v}' = (a', b')$. Draw the addition triangle for $\vec{v} + \vec{v}\,'$, and explain why numeric/geometric duality will assign $(a + a', b + b')$ to that vector sum. [*Hint: Use similar triangles.*]

— ⋆ —

3. Dot-Product Geometry

3.1. Entry-wise multiplication. We have seen that addition, subtraction, and scalar multiplication for numeric vectors in $\mathbf{R}^2$ mirror the corresponding operations on geometric vectors in the plane. The same goes for $\mathbf{R}^3$ and geometric vectors in space, by the way. And in both cases, this mirroring takes place no matter what coordinate system we choose.

It is natural to wonder why we don't define numeric vector *multiplication* entry-by-entry just like we define vector addition. Numeric/geometric duality (actually, the *lack* of it) is the reason.

Question: If we multiply numeric vectors in $\mathbf{R}^2$ entry-by-entry does that mirror some *geometric* operation the way vector addition and scalar multiplication are mirrored via numeric/geometric duality?

Answer: No. Entry-by-entry multiplication for numeric vectors is not respected by numeric/geometric duality—it doesn't correspond to any geometric operation under that duality. It is comparatively useless for that reason.

In other words, if we set up a coordinate system in the plane, the geometric vector corresponding to the entry-wise "product" of the numeric vectors *will change depending on the coordinate system we use*. It lacks the type of "coordinate-free" geometric interpretation that addition and scalar multiplication enjoy. We illustrate with an example.

EXAMPLE 3.2. Consider the numeric vector $\mathbf{v} = (1, 1)$ and its geometric counterpart $\vec{v}$ in a standard (x, y) coordinate system as shown in Figure 17. Let us use "$*$" to denote entry-wise multiplication, and multiply $\mathbf{v}$ by itself in this way, to get

$$\mathbf{v} * \mathbf{v} = \begin{pmatrix} 1 \\ 1 \end{pmatrix} * \begin{pmatrix} 1 \\ 1 \end{pmatrix} = \begin{pmatrix} 1 \cdot 1 \\ 1 \cdot 1 \end{pmatrix} = \begin{pmatrix} 1 \\ 1 \end{pmatrix} = \mathbf{v}$$

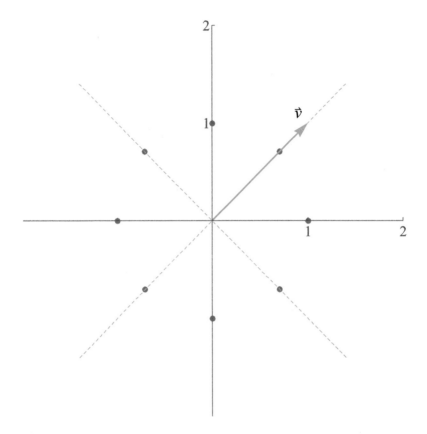

Figure 17. Using standard axes and units, the geometric vector $\vec{v}$ corresponds to the numeric vector $(1,1)$. What numeric vector corresponds to $\vec{v}$ if we use the dashed blue axes? (Gray dots mark "1 unit of length").

If numeric/geometric duality were to hold in this situation, we would apparently have $\vec{v}*\vec{v} = \vec{v}$—the "square" of this geometric vector would be itself.

Now, however, consider what happens if we tilt our axes by 45° along the blue dashed lines in Figure 17.

In this "blue" coordinate system, the same geometric vector $\vec{v}$ has a **different** numeric label, namely $\mathbf{v} = (\sqrt{2}, 0)$. Why? Firstly, it lies along the first blue coordinate axis, so its second blue coordinate is clearly 0. To see that its first blue coordinate is $\sqrt{2}$, note that it forms the hypotenuse of a right triangle with short sides 1 and 1, and hence has length $\sqrt{2}$. Since it lies on the positive first axis, its length and its first coordinate are the same.

When we try entry-wise multiplication in this blue coordinate system, the square of $\vec{v}$ will *not* be $\vec{v}$ itself:

$$\mathbf{v} * \mathbf{v} = \begin{pmatrix} \sqrt{2} \\ 0 \end{pmatrix} * \begin{pmatrix} \sqrt{2} \\ 0 \end{pmatrix} = \begin{pmatrix} \sqrt{2} \cdot \sqrt{2} \\ 0 \cdot 0 \end{pmatrix} = \begin{pmatrix} 2 \\ 0 \end{pmatrix} = \sqrt{2}\,\mathbf{v}$$

If we were to interpret that geometrically, it would tell us that $\vec{v} * \vec{v} = \sqrt{2}\vec{v}$—a significantly different result than the one we got with the gray axes.

Indeed, virtually every coordinate system we try will give us a different geometric result. We thus see that numeric/geometric duality does *not* respect entry-wise multiplication the way it respects entry-wise addition and scalar multiplication. □

REMARK 3.3. In $\mathbf{R}^2$ (and essentially **only** in $\mathbf{R}^2$), there **is** a way to multiply numeric vectors that **does** respect numeric/geometric duality, but **it is more complicated than entry-wise multiplication**. One multiplies (a, b) by (a', b') according to the surprising formula

$$(aa' - bb',\ ab' + ba')$$

Some students may recognize this formula: one gets it by regarding vectors in $\mathbf{R}^2$ as **complex numbers**. Though quite important, we won't pursue the consequences of this formula because it has no true analogue[2] in $\mathbf{R}^3$ or indeed, $\mathbf{R}^n$ for any $n > 2$. Our interest here lies with operations that respect numeric/geometric duality in $\mathbf{R}^n$ for **every** n.

3.4. Euclidean coordinates. We saw above that when we multiply numeric vectors entry-wise, we do *not* get a result that obeys numeric/geometric duality. If, however, we *sum* the entries of that result to get a single scalar (and use the right kind of coordinate system), that scalar *does* have a "coordinate-free" geometric interpretation—a very useful one.

Indeed, we have seen this scalar before: it is precisely the *dot product* we met in Section 1.26 of Chapter 1. We shall see that it has surprisingly rich geometric meaning in $\mathbf{R}^n$ for *every* n. The dot product is geometrically natural in an amazing way. To take advantage of this, however, we need to use certain coordinate systems only:

---

[2] There are partial analogues in $\mathbf{R}^4$ and $\mathbf{R}^8$, but they lack commutativity and/or associativity.

DEFINITION 3.5 (Euclidean coordinates). We call a coordinate system **euclidean** when it employs **perpendicular axes** measured by **a common unit of length**. When two euclidean coordinate systems use the same unit of length, we call them **equivalent**. $\square$

For instance, when we draw horizontal and vertical x- and y-axes in the usual way, with equally spaced tick marks, we get a euclidean coordinate system. If we then *rotate* the axes by a fixed angle without changing their length scales, we get a new, *equivalent* euclidean, coordinate system. The "gray" and "blue" coordinate systems in Figure 17, for example, are equivalent euclidean systems.

We now want to work our way toward Proposition 3.20 below, which roughly states: *Dot products don't change when we replace one euclidean coordinate system by another—equivalent—euclidean system.*

EXAMPLE 3.6. In Example 3.2 we saw that the blue vector $\vec{v}$ in Figure 17 had numeric coordinates $\mathbf{v}_{\text{gray}} = (1, 1)$ with respect to the gray axis/unit system, and coordinates $\mathbf{v}_{\text{red}} = (\sqrt{2}, 0)$ with respect to the equivalent, but rotated, red axes. These are completely different numeric vectors, but they represent the same geometric vector in different coordinate systems, and we have

$$\mathbf{v}_{\text{gray}} \cdot \mathbf{v}_{\text{gray}} = \mathbf{v}_{\text{blue}} \cdot \mathbf{v}_{\text{blue}} = 2$$

The numeric vectors representing $\vec{v}$ in equivalent coordinate systems have the same dot products with themselves. $\square$

This surprising invariance of dot product in equivalent euclidean coordinate systems holds precisely because *dot products compute a geometric quantity*—a function of length and angle, as we are about to see. If we use a fixed euclidean measurement scale, lengths of geometric vectors don't depend on our choice of axes, and angles between vectors don't either. So if we can interpret dot products in terms of length and angle, they should come out the same in all equivalent euclidean coordinate systems.

3.7. Length. Suppose we have a euclidean coordinate system, and a geometric vector $\vec{v}$ with numeric coordinate vector $\mathbf{v} = (a, b)$. Then $\vec{v}$ forms the hypotenuse of a right triangle with short sides parallel to the axes. The lengths of these short sides are the absolute values of the coordinates of $\mathbf{v}$, namely $|a|$ and $|b|$ (Figure 18). So by the Pythagorean theorem, the *length* of $\vec{v}$ is $\sqrt{a^2 + b^2} = \sqrt{\mathbf{v} \cdot \mathbf{v}}$, since

$$\mathbf{v} \cdot \mathbf{v} = (a, b) \cdot (a, b) = a^2 + b^2$$

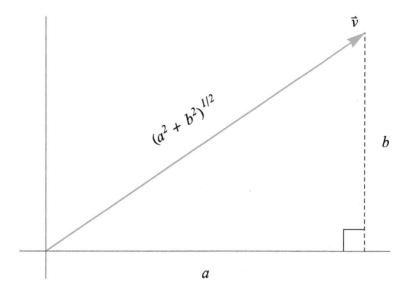

Figure 18. Dot products and lengths. In a euclidean coordinate system, a geometric vector $\vec{v}$ with numeric "address" $\mathbf{v} = (a, b)$ has length $|\vec{v}| = \sqrt{a^2 + b^2} = \sqrt{\mathbf{v} \cdot \mathbf{v}}$.

If we indicate lengths of geometric vectors with "absolute value" bars, we can summarize the calculation above like this:

$$|\vec{v}| = \sqrt{\mathbf{v} \cdot \mathbf{v}}$$

Note that $\vec{v}$, on the left, is geometric, while $\mathbf{v}$, on the right, is numeric. We are computing a geometric quantity—length—using algebra, via the dot product.

For instance, a geometric vector $\vec{v}$ with numeric coordinate vector $\mathbf{v} = (3, 4)$, has length

$$|\vec{v}| = \sqrt{\begin{pmatrix} 3 \\ 4 \end{pmatrix} \cdot \begin{pmatrix} 3 \\ 4 \end{pmatrix}} = \sqrt{3^2 + 4^2} = \sqrt{25} = 5$$

Thus, in any euclidean coordinate system, the numeric vector $\mathbf{v} = (a, b)$ labels a geometric vector with length $\sqrt{a^2 + b^2} = \sqrt{\mathbf{v} \cdot \mathbf{v}}$.

Using the Pythagorean theorem *twice*, it's easy to see that the same fact holds in space too, as depicted in Figure 19. In a 3D euclidean coordinate system (three mutually perpendicular axes with a common length unit), a numeric vector $\mathbf{v} = (a, b, c)$ labels a geometric vector $\vec{v}$ with length $\sqrt{a^2 + b^2 + c^2}$. Again, this equals $\sqrt{\mathbf{v} \cdot \mathbf{v}}$.

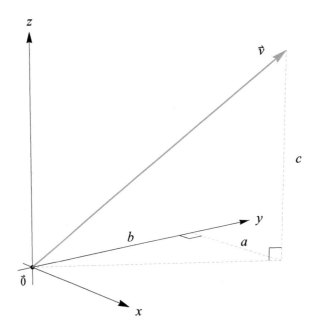

Figure 19. Lengths in 3D. In a 3D euclidean system that assigns numeric coordinates $\mathbf{v} = (a, b, c)$ to a geometric vector $\vec{v}$, the Pythagorean theorem, applied *twice*, gives $|\vec{v}| = \sqrt{a^2 + b^2 + c^2} = \sqrt{\mathbf{v} \cdot \mathbf{v}}$. Can you see why?

In sum, we have

PROPOSITION 3.8. *If a euclidean coordinate system assigns a numeric vector* $\mathbf{v}$ *to a geometric vector* $\vec{v}$, *then*

$$|\vec{v}| = \sqrt{\mathbf{v} \cdot \mathbf{v}}$$

With this proposition in mind, we next make a type of generalization that has great importance in our subject:

We export a geometric notion in the plane to the purely numeric space $\mathbf{R}^2$, *and beyond that, to* $\mathbf{R}^n$ *for any* n.

Doing so for the geometric notion of *length* in the plane, we get

DEFINITION 3.9 (Length in $\mathbf{R}^n$). We define the **length** of any *numeric* vector $\mathbf{v} = (v_1, v_2, \ldots, v_n) \in \mathbf{R}^n$ as

$$|\mathbf{v}| := \sqrt{\mathbf{v} \cdot \mathbf{v}} = \sqrt{v_1^2 + v_2^2 + \cdots + v_n^2}$$

$\square$

EXAMPLE 3.10. The numeric vectors

$$\mathbf{u} = (1, 2, 3) \quad \mathbf{v} = (1, -1, -1, 1) \quad \text{and} \quad \mathbf{w} = (1, -4, 6, -10, 6, -4, 1)$$

belong to $\mathbf{R}^3$, $\mathbf{R}^4$, and $\mathbf{R}^7$, respectively. Their lengths are

$$
\begin{aligned}
|\mathbf{u}| &= \sqrt{1^2 + 2^2 + 3^2} = \sqrt{1 + 4 + 9} = \sqrt{14} \\
|\mathbf{v}| &= \sqrt{1^2 + (-1)^2 + 1^2 + (-1)^2} = \sqrt{4} = 2 \\
|\mathbf{w}| &= \sqrt{1^2 + 4^2 + 6^2 + 10^2 + 6^2 + 4^2 + 1^2} = \sqrt{206}
\end{aligned}
$$

Similarly, note that in $\mathbf{R}^4$, the length of $(1, 1, 1, 1)$ is given by

$$\sqrt{1^2 + 1^2 + 1^2 + 1^2} = \sqrt{4} = 2$$

Indeed, for any n, the length of $(1, 1, \ldots, 1) \in \mathbf{R}^n$ is $\sqrt{n}$ by essentially the same calculation. □

We cannot "draw" nor easily imagine *geometric* vectors corresponding to the numeric vectors of $\mathbf{R}^n$ when $n > 3$. Still, our definition of *length* in $\mathbf{R}^n$ behaves exactly as it does for geometric vectors in the plane and in space. It is *positive* for all numeric vectors except $\mathbf{0} = (0, 0, \ldots, 0)$ (which has length zero), and, it multiplies "geometrically" under scalar multiplication, just as we would hope:

PROPOSITION 3.11. *Let* $\mathbf{v} \in \mathbf{R}^n$. *Then* $|\mathbf{v}| \geq 0$, *with* $|\mathbf{v}| = 0$ *only if* $\mathbf{v} = \mathbf{0}$. *Further, multiplying* $\mathbf{v}$ *by a scalar multiplies* $|\mathbf{v}|$ *by the absolute value of that scalar:*

$$|c\,\mathbf{v}| = |c|\,|\mathbf{v}|$$

PROOF. Exercise. □

3.12. Unit vectors. We call a vector $\mathbf{v} \in \mathbf{R}^n$ a **unit vector** if it has length one.

EXAMPLE 3.13. In $\mathbf{R}^2$, any vector of the form $(\cos t, \sin t)$ is a unit vector, because

$$|(\cos t, \sin t)| = \sqrt{\cos^2 t + \sin^2 t} = 1$$

□

Unit vectors are plentiful and easy to produce, because *every* non-zero vector $\mathbf{v} \in \mathbf{R}^n$ has exactly two scalar multiples with unit length, namely

$$\pm \frac{\mathbf{v}}{|\mathbf{v}|}$$

This follows from Proposition 3.11, which shows that multiplication by $\pm 1/|\mathbf{v}|$ multiplies $|\mathbf{v}|$ by its own reciprocal.

Producing a unit vector in this way (using the *plus* sign) is called *normalization*:

DEFINITION 3.14 (Normalization). If $\mathbf{0} \neq \mathbf{v} \in \mathbf{R}^n$, we call

$$\mathbf{v}/\left|\mathbf{v}\right|$$

the **normalization** of $\mathbf{v}$, and we **normalize v** when we divide it by its length. Geometrically, we imagine $\mathbf{v}/|\mathbf{v}|$ as pointing in the same direction as $\mathbf{v}$, but having unit length. $\square$

3.15. Angles. The familiar "side-side-side" theorem from plane geometry tells us that a triangle—including its angles—are determined by its side-lengths. According to the *Law of Cosines*, we can even calculate the angles of a triangle from the side-lengths:

THEOREM 3.16 (Law of Cosines). *Given a triangle with side-lengths A, B, and C, the angle θ opposite side C is determined by the formula*

$$C^2 = A^2 + B^2 - 2AB\cos\theta$$

and the requirement $0 \leq \theta \leq \pi$.

REMARK 3.17. The Law of Cosines beautifully generalizes the far more familiar **Pythagorean theorem**. For when $\theta = 90°$ (giving us a right triangle), the cosine term vanishes (since $\cos 90° = 0$), leaving

$$C^2 = A^2 + B^2$$

$\square$

Now consider two non-zero geometric vectors $\vec{a}$ and $\vec{b}$ in the plane, with numeric addresses $\mathbf{a}$ and $\mathbf{b}$ in some euclidean coordinate system. If we regard $\vec{a}$ and $\vec{b}$ as two sides of a triangle, then by the triangle rule for addition, the *third* side of this triangle is a translated copy of $\vec{c} = \vec{b} - \vec{a}$. In particular, if we write $A = |\vec{a}|$ and $B = |\vec{b}|$, then the third side of our triangle has length $C = |\vec{b} - \vec{a}|$.

Since we now know how to compute lengths algebraically in a euclidean coordinate system, however, we can compute $|\vec{a}|$, $|\vec{b}|$, and $|\vec{c}|$ from their numeric coordinates. Indeed, using the dot-product formula for length, we have

$$A^2 = \mathbf{a} \cdot \mathbf{a}, \quad B^2 = \mathbf{b} \cdot \mathbf{b}$$

and

$$\begin{aligned}
C^2 = \mathbf{c} \cdot \mathbf{c} &= (\mathbf{b} - \mathbf{a}) \cdot (\mathbf{b} - \mathbf{a}) \\
&= \mathbf{b} \cdot \mathbf{b} - \mathbf{b} \cdot \mathbf{a} - \mathbf{a} \cdot \mathbf{b} + \mathbf{a} \cdot \mathbf{a} \\
&= \mathbf{a} \cdot \mathbf{a} + \mathbf{b} \cdot \mathbf{b} - 2\,\mathbf{a} \cdot \mathbf{b} \\
&= A^2 + B^2 - 2\,\mathbf{a} \cdot \mathbf{b}
\end{aligned}$$

Comparing this with the Law of Cosines above leads immediately to the amazing connection we seek between the geometry of angles and the algebra of the dot product:

PROPOSITION 3.18. *Suppose a euclidean coordinate system in the plane (or in space) assigns numeric vectors* **a** *and* **b** *to two non-zero geometric vectors* $\vec{a}$ *and* $\vec{b}$, *respectively. Then the angle* θ *between* $\vec{a}$ *and* $\vec{b}$ *is determined by*

$$\cos\theta = \frac{\mathbf{a} \cdot \mathbf{b}}{|\mathbf{a}|\,|\mathbf{b}|}$$

The left side of this formula is purely geometric, while the right is purely algebraic. Setting them equal, we get to compute the angle between two geometric vectors using only their numeric coordinates in *any* euclidean system.

EXAMPLE 3.19. Suppose $\vec{a}$ and $\vec{b}$ are geometric vectors in the plane, with numeric addresses $\mathbf{a} = (1, \sqrt{3})$ and $\mathbf{b} = (\sqrt{3}, 1)$, respectively in a euclidean coordinate system. What is the angle θ between them?

To answer this, we use Proposition 3.18 to compute

$$\begin{aligned}
\cos\theta &= \frac{\mathbf{a} \cdot \mathbf{b}}{|\mathbf{a}|\,|\mathbf{b}|} \\
&= \frac{(1, \sqrt{3}) \cdot (\sqrt{3}, 1)}{\sqrt{1^2 + 3}\sqrt{3 + 1^2}} \\
&= \frac{2\sqrt{3}}{2 \cdot 2} = \sqrt{3}/2
\end{aligned}$$

The only angle $0 \leq \theta \leq \pi$ with cosine $\sqrt{3}/2$ is $\pi/6$, so that is the angle between $\vec{a}$ and $\vec{b}$. $\square$

By combining Proposition 3.18 with Proposition 3.8, we can now deduce the important fact we promised earlier:

PROPOSITION 3.20 (Euclidean invariance of dot products). *If* **a** *and* **b** *give the numeric addresses of geometric vectors $\vec{a}$ and $\vec{b}$ in a euclidean coordinate system, and* **a**′, **b**′ *are their numeric addresses in a different—but equivalent—euclidean system (a euclidean system with the same unit of length), then*

$$\mathbf{a} \cdot \mathbf{b} = \mathbf{a}' \cdot \mathbf{b}'$$

In short, switching between equivalent euclidean coordinate systems leaves dot products unchanged.

PROOF. First, by Proposition 3.8, we can compute the lengths of $\vec{a}$ and $\vec{b}$ equally well in both coordinate systems:

$$|\mathbf{a}| = |\vec{a}| = |\mathbf{a}'| \quad \text{and} \quad |\mathbf{b}| = |\vec{b}| = |\mathbf{b}'|$$

Using these facts, however, we can rewrite the cosine formula in Proposition 3.18 as either

$$|\vec{a}|\,|\vec{b}|\,\cos\theta = \mathbf{a} \cdot \mathbf{b}$$

in the one system, or

$$|\vec{a}|\,|\vec{b}|\,\cos\theta = \mathbf{a}' \cdot \mathbf{b}'$$

in the other. The left-hand sides agree, so the right-hand sides must also agree.   □

3.21. Angles in $\mathbf{R}^n$. We now generalize the notion of *angle* to $\mathbf{R}^n$ (for any n), just as we did for *length* in Definition 3.9.

DEFINITION 3.22 (Angles in $\mathbf{R}^n$). The **angle** θ between any two *nonzero* vectors $\mathbf{v}, \mathbf{w} \in \mathbf{R}^n$ is the unique scalar $0 \le \theta \le \pi$ satisfying

$$\cos\theta = \frac{\mathbf{v} \cdot \mathbf{w}}{|\mathbf{v}|\,|\mathbf{w}|}$$

We require $\mathbf{v}, \mathbf{w} \neq \mathbf{0}$ to keep the denominator here from vanishing. When we need to, we'll denote the angle between $\mathbf{v}$ and $\mathbf{w}$ by $\angle\,\mathbf{v}, \mathbf{w}$.

This definition has a big potential problem, and we must settle the problem before we can accept the definition. Indeed, the cosine function only assumes values between ± 1. *So unless the right side of the formula in* Definition 3.22 *stays between* -1 *and* $+1$, *it won't be the cosine of any angle* θ.

Fortunately, a fundamental result known as the *Cauchy–Schwarz inequality* guarantees the fact we need:

PROPOSITION 3.23 (Cauchy–Schwarz inequality). *For **any** two numeric vectors* $\mathbf{v}, \mathbf{w} \in \mathbf{R}^n$, *we have*

$$|\mathbf{v} \cdot \mathbf{w}| \le |\mathbf{v}|\,|\mathbf{w}|$$

Equality holds if and only if one vector is a scalar multiple of the other.

PROOF. The inequality is obvious if either vector is $\mathbf{0}$, so we may assume $\mathbf{v} \ne \mathbf{0} \ne \mathbf{w}$. It is also easy to check that when one vector is a scalar multiple of the other, the Cauchy–Schwarz inequality becomes an equality (Exercise 165). So we can also assume that $\mathbf{v}$ is *not* a multiple of $\mathbf{w}$.

The rest of the proof, while short, depends on a clever, unexpected trick. Namely, we consider the quantity

$$(11) \qquad\qquad |\mathbf{v} - t\mathbf{w}|^2 > 0$$

The key point is that, no matter what scalar we take for t, this quantity is *strictly* positive. For as the square of a length, it can only vanish if $\mathbf{v} - t\mathbf{w} = \mathbf{0}$, implying $\mathbf{v} = t\mathbf{w}$, a scalar multiple, which we already prohibited.

On the other hand, if we expand (11) using the definition of length and the distributivity of dot products, we get

$$
\begin{aligned}
|\mathbf{v} - t\mathbf{w}|^2 &= (\mathbf{v} - t\mathbf{w}) \cdot (\mathbf{v} - t\mathbf{w}) \\
&= \mathbf{v} \cdot \mathbf{v} - 2t\,\mathbf{v} \cdot \mathbf{w} + t^2 \mathbf{w} \cdot \mathbf{w}
\end{aligned}
$$

We now seek the value of t that *minimizes* this quantity using elementary Calculus:[3] we differentiate with respect to t and set the result equal to zero:

$$-2\,\mathbf{v} \cdot \mathbf{w} + 2t\,\mathbf{w} \cdot \mathbf{w} = 0 \quad \Rightarrow \quad t = (\mathbf{v} \cdot \mathbf{w})/(\mathbf{w} \cdot \mathbf{w})$$

The point is that even with this minimizing value of t, (11) is *still positive*. For by plugging it into the expansion of $|\mathbf{v} - t\mathbf{w}|^2$ above, we get

$$\mathbf{v} \cdot \mathbf{v} - 2\frac{(\mathbf{v} \cdot \mathbf{w})^2}{\mathbf{w} \cdot \mathbf{w}} + \frac{(\mathbf{v} \cdot \mathbf{w})^2}{\mathbf{w} \cdot \mathbf{w}} > 0$$

which simplifies to

$$|\mathbf{v}|^2 - \frac{(\mathbf{v} \cdot \mathbf{w})^2}{|\mathbf{w}|^2} > 0$$

---

[3]This is the only argument in our book (outside of Appendix B) that uses Calculus.

Multiply through by $|\mathbf{w}|^2$, add $(\mathbf{v} \cdot \mathbf{w})^2$ to both sides, and take square roots to see that Cauchy–Schwarz holds.   □

The Cauchy–Schwarz inequality puts Definition 3.22 on solid ground by guaranteeing that its right-hand side never exceeds 1 in absolute value.

We now apply that Definition to compute some angles in $\mathbf{R}^n$.

EXAMPLE 3.24. What is $\angle \mathbf{v}, \mathbf{w}$ when $\mathbf{v} = (1, 1, 0, 1, 1)$ and $\mathbf{w} = (1, 0, 0, 0, 1)$ in $\mathbf{R}^5$?

We have $\mathbf{v} \cdot \mathbf{w} = 2$, while $|\mathbf{v}| = 2$ and $|\mathbf{w}| = \sqrt{2}$. So by Definition 3.22,

$$\cos \theta = \frac{2}{2\sqrt{2}} = \frac{1}{\sqrt{2}} = \cos \tfrac{\pi}{4}$$

The angle between the two vectors is therefore $\pi/4$, i.e., $45°$.

Suppose we modify $\mathbf{w}$ slightly, changing the sign of its last coordinate to get $\mathbf{w}' = (1, 0, 0, 0, -1)$.

Now we find that $\mathbf{v} \cdot \mathbf{w}' = 0$, which immediately yields

$$\cos \theta = 0$$

The only angle $0 \le \theta \le \pi$ whose cosine equals zero is $\pi/2$, i.e., $90°$. In other words, $\mathbf{v}$ and $\mathbf{w}'$ form a right angle.   □

Indeed, since the cosine of an angle $0 \le \theta \le \pi$ vanishes *if and only if* $\theta = \pi/2$, we see that *non-zero vectors* $\mathbf{v}, \mathbf{w} \in \mathbf{R}^n$ *form a right angle if and only if their dot product is zero.* We therefore make the following definition:

DEFINITION 3.25. Vectors $\mathbf{v}, \mathbf{w} \ne \mathbf{0}$ in $\mathbf{R}^n$ are **perpendicular** or **orthogonal**, or **normal** (all three are synonyms) to each other if and only if $\mathbf{v} \cdot \mathbf{w} = 0$.   □

EXAMPLE 3.26. Given any vector $(x, y) \in \mathbf{R}^2$, the vector $(-y, x)$ is perpendicular to it, because $(x, y) \cdot (-y, x) = -xy + yx = 0$ $(= \cos 90°)$. Note that $(-y, x)$ also has the same *length* as (x, y). In fact, if (x, y) represents a geometric vector $\mathbf{v}$ in a euclidean coordinate system, then $(-y, x)$ represents the geometric vector we get from $\mathbf{v}$ after rotation by $90°$.   □

– **Practice** –

148. Suppose $\mathbf{v}_1 = (1, 2, 3)$ and $\mathbf{v}_2 = (4, -3, 5)$ in $\mathbf{R}^3$, and $\mathbf{w}_1 = (2, -1, 2, -1)$ and $\mathbf{w}_2 = (1, 2, -1, -2)$ in $\mathbf{R}^4$. Compute:

(a) $\mathbf{v}_1 \cdot \mathbf{v}_2$ (b) $\mathbf{v}_1 \cdot \mathbf{v}_1$ (c) $|\mathbf{v}_2|$

(d) $\mathbf{w}_1 \cdot \mathbf{w}_2$ (d) $\mathbf{w}_1 \cdot \mathbf{w}_1$ (f) $|\mathbf{w}_2|$

149. Compute the lengths of these vectors:

a) $(2, -1, 2) \in \mathbf{R}^3$ b) $(1, 2, -2, 4) \in \mathbf{R}^4$ c) $(3, 2, 1, 2, 3) \in \mathbf{R}^5$

150. Normalize each vector below:

$$\mathbf{u} = (1, 2, 3) \quad \mathbf{v} = (1, -1, -1, 1) \quad \text{and} \quad \mathbf{w} = (1, -4, 6, -10, 6, -4, 1)$$

151. The $90°$ rotation described in Example 3.26 is a linear transformation from $\mathbf{R}^2$ to itself. What 2-by-2 matrix represents it? [*Hint: Recall Theorem 5.6 of Chapter 1.*]

152. Prove Proposition 3.11

153. Compute the length, in $\mathbf{R}^n$, of

   a) The zero vector $\mathbf{0} = (0, 0, \dots, 0)$.
   b) The vector $(c, c, \dots, c)$ when c is an arbitrary constant?
   c) The vector $(1, 2, 3, \dots, n)$? (You may need to look up the formula for summing the first n squares.)

154. Show that in $\mathbf{R}^3$, every vector of the form

$$\mathbf{u} = (\sin t \cos s, \ \sin t \sin s, \ \cos t)$$

is a unit vector. (Here s and t can be any scalars.)

155. Suppose $\mathbf{v} \in \mathbf{R}^n$ is any non-zero vector. Show that for any scalar $k > 0$, we get the same result when we normalize $k\mathbf{v}$ that we get when we normalize $\mathbf{v}$. What happens if we normalize $k\mathbf{v}$ when $k < 0$?

156. Compute the angles between

    a) $(4, 0)$ and $(3, 3)$ in $\mathbf{R}^2$

    b) $(2, -2, -2, 2)$ and $(1, 3, 1, 3)$ in $\mathbf{R}^4$

    c) Any vector $\mathbf{v} \in \mathbf{R}^n$ and its *opposite,* $-\mathbf{v}$

    d) $(1, 1, 1, \ldots, 1, 1)$ and $(1, 1, 1, 1, 0, 0, 0 \ldots, 0, 0)$ in $\mathbf{R}^{16}$

157. Compute the angle between each pair of vectors, and determine whether they form the sides of an *isosceles* triangle.

    a) $(1, -1, 0)$ and $(0, 1, -1)$

    b) $(1, -1, 0)$ and $(1, 1, 1)$

    c) $(1, -1, 0$ and $(0, 1, 0)$

    d) $(1, 1, 1)$ and $(-1, -1, -1)$

158. Show that the angle between any pair of vectors in the following set is the same:

$$\{(2, 0, 1), \ (1, 2, 0), \ (0, 1, 2)\}$$

159. Find half a dozen unit vectors orthogonal to $\mathbf{w} = (1, 2, 3, 2, 1)$ in $\mathbf{R}^5$. Can you find *all* unit vectors orthogonal to $\mathbf{w}$?

160. Do the vectors $(1, 1, 1, t)$ and $(-1, -1, -1, t)$ form an angle of $\pi/3$ in $\mathbf{R}^4$ for any values of t? (Recall that $\cos \pi/3 = 1/2$.)

161. Which scalars k minimize the length of the difference between $\mathbf{v}(k) := k(1, 0, 1, 0, 1)$ and $\mathbf{u} = (1, -2, 4, -2, 1)$ in $\mathbf{R}^5$? What is the minimum length?

162. Show that $(-y, x)$ and $(y, -x)$ are the *only* vectors that are both orthogonal to (x, y), *and* have the same length as (x, y).

163. Suppose $\mathbf{x}$ and $\mathbf{y}$ are both perpendicular to a certain vector $\mathbf{u}$ in $\mathbf{R}^m$. Show that every linear combination of $\mathbf{x}$ and $\mathbf{y}$ is also perpendicular to $\mathbf{u}$.

164. Suppose $\vec{a}$ and $\vec{b}$ are geometric vectors in the plane, and a certain euclidean coordinate system assigns them numeric coordinates $\mathbf{a} = (\cos t, \ \sin t)$ and $\mathbf{b} = (\cos s, \sin s)$, respectively. Show that Proposition 3.18 correctly computes the angle between them as $\theta = t - s$.

165. In the proof of Cauchy–Schwarz (Proposition 3.23), we said it was "easy to check" that the inequality becomes an equality when $\mathbf{w}$ is a scalar multiple of $\mathbf{v}$. Explain this.

166. Sketch the gray and dashed blue axes in Figure 17, and geometric vectors of length 1 that point along the positive x- and y-axes. Call them $\vec{e}_1$ and $\vec{e}_2$, respectively.

    a) What are the numeric vectors $\mathbf{e}_1$ and $\mathbf{e}_2$ corresponding to $\vec{e}_1$ and $\vec{e}_2$ in the gray coordinate system?

    b) If we multiply $\mathbf{e}_1$ and $\mathbf{e}_2$ together entry-wise, what numeric vector do we get? What geometric vector does it represent?

    c) What are the numeric vectors $\mathbf{e}'_1$ and $\mathbf{e}'_2$ corresponding to $\vec{e}_1$ and $\vec{e}_2$ in the blue coordinate system? [*Hint: Use right-triangles and the Pythagorean theorem here.*]

    d) If we multiply $\mathbf{e}'_1$ and $\mathbf{e}'_2$ together entry-wise, what numeric vector do we get? What geometric vector does it represent? [*Hint: It should be a multiple of $\vec{e}_1$ or $\vec{e}_2$.*]

    Your answers to (b) and (d) should be different, showing again that numeric/geometric duality does not respect entry-wise multiplication.

167. Trace or copy Figure 17, and on it, sketch the geometric vector $\vec{w}$ corresponding to the numeric vector $\mathbf{w}_{\text{gray}} = (-1, 1)$ relative to the gray axes.

    (a) What numeric vector $\mathbf{w}_{\text{red}}$ represents $\vec{w}$ relative to the dashed blue axes?

    (b) Show that these dot products are the same, even though the numeric vectors are different:

$$\mathbf{v}_{\text{gray}} \cdot \mathbf{w}_{\text{gray}} = \mathbf{v}_{\text{blue}} \cdot \mathbf{w}_{\text{blue}}$$

    (c) Sketch the geometric vector $\vec{u}$ corresponding to the numeric vector $\mathbf{u}_{\text{gray}} = (0, 2)$ relative to the gray axes, and repeat (a) and (b) with the $\mathbf{w}$'s replaced by $\mathbf{u}$'s.

— ★ —

4. Lines, Planes, and Hyperplanes

4.1. Lines. If we take *all* the scalar multiples of a geometric vector $\vec{v} \neq \vec{0}$ in the plane (or in space), we get an entire line (compare Exercise 134). We call it the line **generated** (or later, **spanned**) by $\vec{v}$ (Figure 4). Note that a line generated this way always includes the origin $\vec{0}$, since $\vec{0} = 0\,\vec{v}$ is always a scalar multiple of $\vec{v}$.

Note too that *every* line through the origin can be generated this way: just take all scalar multiples of a vector $\vec{v} \neq \vec{0}$ on that line. Lines that *don't* go through the origin can be generated almost as simply, because every line in the plane (or in space) is *parallel* to some line through the origin:

OBSERVATION 4.2. *Let L be any line in the plane or in space. Let $\vec{v} \neq \vec{0}$ be a vector that generates the line L_v parallel to L through the origin, and let $\vec{p}$ be any vector whose tip lies on L. Then we get all vectors on L by adding $\vec{p}$ to all vectors on L_v :*

$$L = \quad all \quad \vec{p} + t\,\vec{v}, \quad where\ t\ varies\ over\ R.$$

PROOF. For any $t \in \mathbf{R}$, we get to the tip of $\vec{p} + t\vec{v}$ by translating $t\vec{v}$ to the tip of $\vec{p}$. As t varies, the tip of this sum clearly visits every point on the line through the tip of $\vec{p}$ and parallel to L_v (Figure 20).   □

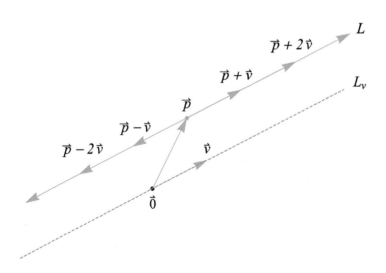

Figure 20. By adding all multiples of $\vec{v}$ to $\vec{p}$, we trace out a line (solid blue) through the tip of $\vec{p}$ and parallel to the (dashed) line generated by $\vec{v}$.

Using numeric/geometric duality, this picture gives us a *numeric* notion of line in $\mathbf{R}^2$ (or $\mathbf{R}^3$). Indeed, since a line in the plane is just the set of all $\vec{p} + t\vec{v}$, where $\vec{p}$ and $\vec{v} \neq \vec{0}$ are geometric vectors and t is any scalar, we can define the numeric analogue of a line, in $\mathbf{R}^2$, say, by replacing the *geometric* vectors $\vec{p}$ and $\vec{v} \neq \vec{0}$ in the plane, by *numeric* vectors $\mathbf{p}$ and $\mathbf{v} \neq \mathbf{0}$ in $\mathbf{R}^2$.

But why stop at $\mathbf{R}^2$? We can make the same definition in $\mathbf{R}^m$ for any m:

DEFINITION 4.3. For any positive integer m, suppose $\mathbf{p}, \mathbf{v} \in \mathbf{R}^m$, and $\mathbf{v} \neq \mathbf{0}$. We call the image of the map $L(t) := \mathbf{p} + t\mathbf{v}$ the **line through p in the v direction** or the **line traced out by** $L(t)$. $\qquad\square$

EXAMPLES 4.4. In $\mathbf{R}^4$, we can speak of "the line through $(1, 1, 0, 0)$ in the $(1, 2, -2, -1)$ direction." It is the image of the map

$$L(t) = (1, 1, 0, 0) + t\,(1, 2, -2, -1)$$

As t ranges over all scalars, $L(t)$ traces out the line.

Similarly the line through $(1, 2, 3, 4)$ in the direction of the third standard basis vector is the image of

$$L(t) = (1, 2, 3, 4) + t\,\mathbf{e}_3 = (1, 2, 3 + t, 4)$$

$\qquad\square$

REMARK 4.5. Just as we generalized the notions of length and angle from the plane to $\mathbf{R}^m$, we have now extended the geometric notion of *line*, to the purely numeric vector spaces $\mathbf{R}^m$, where m can be as big as we please. It may be difficult to accurately *visualize* a line in $\mathbf{R}^{37}$, but we can express it algebraically with no difficulty at all, using Definition 4.3.

We shall soon generalize the notion of *plane* in a similar way. $\qquad\square$

4.6. The 2-variable equation: Lines. Consider a *homogeneous* linear equation in two variables:

$$a_1 x + a_2 y = 0$$

Set $\mathbf{x} = (x, y)$ and $\mathbf{a} = (a_1, a_2)$. Assume $\mathbf{a} \neq (0, 0)$ (otherwise, the equation is trivially solved by *any* $(x, y) \in \mathbf{R}^2$). Now rewrite the left-hand side as a dot product:

$$\mathbf{a} \cdot \mathbf{x} = 0$$

Since vanishing of the dot product means perpendicularity (Proposition 3.18), we can interpret this equation to mean that $\mathbf{a}$ and $\mathbf{x}$ are orthogonal. In short,

Solving the equation $\mathbf{a} \cdot \mathbf{x} = 0$ *corresponds to finding all vectors* $\mathbf{x}$ *orthogonal to* $\mathbf{a}$ *in* R^2.

If we visualize all *geometric* vectors $\vec{x}$ orthogonal to $\vec{a}$ in the plane, we get a line (Figure 21).

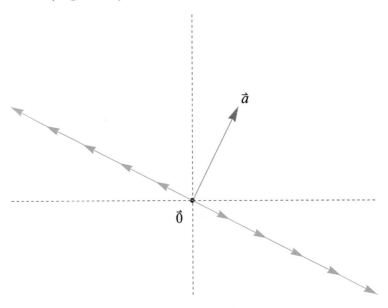

Figure 21. The vectors perpendicular to $\vec{a}$ form a *line* perpendicular to $\vec{a}$.

This is a wonderful benefit of numeric/geometric duality: if we have a homogeneous linear equation in two variables, we can visualize its solution as a line.

We may see this *algebraically* too. If we solve $a_1 x + a_2 y = 0$ by row-reducing the one-rowed matrix $[a_1\ a_2]$, we get one free column, and hence one homogeneous generator $\mathbf{h}$. Our solution mapping consequently takes the form

$$\mathbf{H}(t) = t\,\mathbf{h}$$

which is none other than the line through $\mathbf{0}$ generated by $\mathbf{h}$ (Definition 4.3). Since $\mathbf{h}$ is orthogonal to $\mathbf{a}$ (after all, it solves $\mathbf{a} \cdot \mathbf{h} = 0$), this line is orthogonal to $\mathbf{a}$.

In sum, we have

OBSERVATION 4.7. *The solution set of the homogeneous equation* $a_1 x + a_2 y = 0$ *is a line through the origin and perpendicular to* $\mathbf{a} = (a_1, a_2)$.

Now consider an **in**homogeneous two-variable equation

$$\mathbf{a} \cdot \mathbf{x} = b \qquad (b \neq 0)$$

Solve this by row-reducing the one-rowed *augmented matrix*

$$\begin{bmatrix} a_1 & a_2 & | & b \end{bmatrix}$$

Assuming $\mathbf{a} \neq (0,0)$, we *must* get a pivot when we row-reduce. That leaves *one free column*, producing one homogeneous generator $\mathbf{h}$.

The solution of the inhomogeneous equation $\mathbf{a} \cdot \mathbf{x} = b$ therefore takes the form

$$\mathbf{X}(t) = \mathbf{x}_p + t\,\mathbf{h}$$

where $\mathbf{h}$ again solves the *homogeneous* equation $\mathbf{a} \cdot \mathbf{x} = 0$.

Here, as in the homogeneous case, Definition 4.3 lets us recognize this solution set as a *line: the line through* $\mathbf{x}_p$ *in the* $\mathbf{h}$ *direction.*

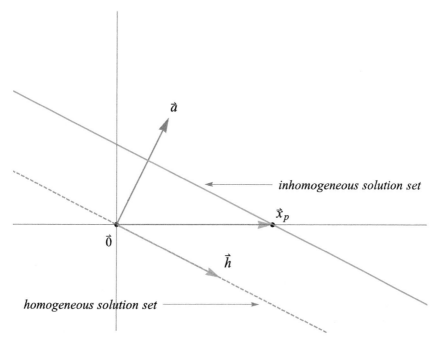

Figure 22. The solution set of an inhomogeneous equation $\mathbf{a} \cdot \mathbf{x} = b$ is a line that goes through the tip of $\vec{x}_p$, and runs parallel to the solution set of the homogeneous equation.

OBSERVATION 4.8. *The solution set of any two-variable linear equation* $a_1 x + a_2 y = b$ *is a line through the tip of (any) particular solution* $\mathbf{x}_p$ *and orthogonal to the coefficient vector* $\mathbf{a}$.

This fact is especially useful in reverse: *Given* a line known to be orthogonal to a certain vector and to go through a certain point, we can find a linear equation whose solutions are the vectors (whose tips are) on that line.

EXAMPLE 4.9. *Find a linear equation whose solution set is the line through (the tip of) $(2, 3)$ and perpendicular to $(1, -2)$.*

Since our line is perpendicular to $(1, -2)$, it runs parallel to the solution set of the *homogeneous* equation $(1, -2) \cdot (x, y) = 0$ (Figure 22), and therefore has the form

$$x - 2y = b$$

But we also know it goes through $(2, 3)$. So if we plug $(2, 3)$ into the equation, we discover what b is:

$$x - 2y = 2 - 2 \cdot 3 = -4$$

Thus, $b = -4$, and our line is the solution set of

$$x - 2y = -4$$

$\square$

4.10. Three-variable equations: Planes. Now let us analyze equations in *three*-variables using the same reasoning we applied to the two-variable case above. Their solution sets will correspond to *planes in space*.

As before, start with the homogeneous equation

$$(12) \qquad\qquad a_1 x + a_2 y + a_3 z = 0$$

Once again, we may use the dot product to write this as $\mathbf{a} \cdot \mathbf{x} = 0$, which makes an orthogonality statement. It says the solutions $\mathbf{x}$ of the equation are perpendicular to the coefficient vector $\mathbf{a} = (a_1, a_2, a_3)$. Assuming $\mathbf{a} \neq (0, 0, 0)$, and reading these numeric vectors as (euclidean) coordinates of geometric vectors in space, these solutions form a *plane through the origin.* Figure 23 illustrates.

Algebraically, we can treat equation (12) as a system of one equation in three variables, and solve it to derive a **formula** all vectors orthogonal to $\mathbf{a}$. Indeed, since $\mathbf{a} \neq \mathbf{0}$, the coefficient matrix of this "system," namely

$$\begin{bmatrix} a_1 & a_2 & a_3 \end{bmatrix}$$

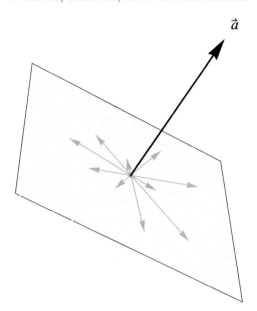

Figure 23. The vectors orthogonal to a non-zero vector $\vec{a}$ in space form a plane.

must row-reduce to a 1×3 matrix with one pivot and two free columns. The solution mapping thus takes the form

$$(13) \qquad \mathbf{x}\left(s_1, s_2\right) = s_1\,\mathbf{h}_1 + s_2\,\mathbf{h}_2$$

To visualize this, regard $\mathbf{h}_1$ and $\mathbf{h}_2$ as labeling two of the blue vectors in Figure 23. Their linear combinations $s_1\mathbf{h}_1 + s_2\mathbf{h}_2$ then fill out the entire plane orthogonal to $\vec{a}$. In particular, all the other blue vectors in Figure 23 are linear combinations of $\mathbf{h}_1$ and $\mathbf{h}_2$. This shows that as predicted, *the solution set of a homogeneous linear equation in three variables describes a plane through the origin in space.*

Moreover, in precise analogy with what we found for two-variable equations, the solution set of an *inhomogeneous* three-variable linear equation represents a plane that does *not* go through the origin.

To see this, we again recall that the solution set of the *in*homogeneous equation

$$a_1 x + a_2 y + a_3 z = b$$

takes the form

$$\mathbf{x}\left(s_1, s_2\right) = \mathbf{x}_p + s_1\,\mathbf{h}_1 + s_2\,\mathbf{h}_2$$

which differs from the solution of its homogeneous version (cf. Equation (13)) only by the addition of a particular solution $\mathbf{x}_p$. We saw that the homogeneous solution represents a plane through the origin, and orthogonal to **a**. The *in*homogeneous solution just adds the corresponding vector $\mathbf{x}_p$ to each vector in this plane, translating it away from the origin until it originates at the tip of $\mathbf{x}_p$ (Figure 24).

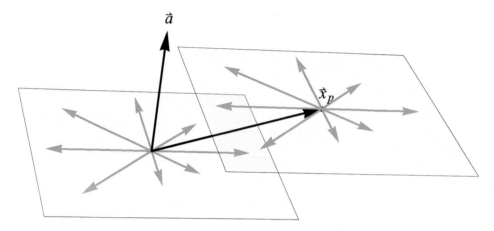

Figure 24. The solution set of $\mathbf{a} \cdot \mathbf{x} = b$ represents a plane orthogonal to the line generated by $\vec{a}$, but through $\vec{x}_p$ rather than the origin.

This is another wonderful benefit of numeric/geometric duality: we can describe any plane in space algebraically as the solution set of a single 3-variable equation:

PROPOSITION 4.11. *The solution set of any linear equation*

$$a_1 x + a_2 y + a_3 z = b$$

consists of all vectors on the plane perpendicular to the line generated by $\mathbf{a} = (a_1, a_2, a_3)$, *and containing any particular solution* $\mathbf{x}_p$ *of the equation.*

EXAMPLE 4.12. Suppose we have geometric vectors $\vec{a}$ and $\vec{p}$ in space with numeric coordinates $\mathbf{a} = (1, 1, 1)$ and $\mathbf{p} = (2, -1, -3)$, respectively. What equation describes the plane P orthogonal to the line generated by $\vec{a}$ and containing the tip of $\vec{p}$?

Proposition 4.11 lets us answer this easily. Since P is orthogonal to the line generated by $\vec{a}$, the equation we want must take the form $\mathbf{a} \cdot \mathbf{x} = b$, where $\mathbf{a} = (1, 1, 1)$. Since $\vec{p}$ lies on this plane, moreover, **p** *is a particular solution of that equation.* In other words, $\mathbf{a} \cdot \mathbf{p} = b$. Turning this around, and evaluating, we get

$$b = \mathbf{a} \cdot \mathbf{p} = (1, 1, 1) \cdot (2, -1, -3) = -2$$

In short, we may conclude that P is is the solution set of the equation $\mathbf{a} \cdot \mathbf{x} = \mathbf{a} \cdot \mathbf{p}$, that is

$$(14) \qquad\qquad x + y + z = -2$$

☐

REMARK 4.13. (Test for membership) In the Example we just presented, we found that (the tip of) a geometric vector $\vec{x}$ lies on the plane P if and only if its coordinate vector $\mathbf{x}$ satisfies (14). That equation gives us a *test for membership in P*.

For instance, suppose we ask: *Do either of the vectors with coordinates* $(1, 2, 3)$ *or* $(-3, -1, 2)$ *lie on P?* Equation (14) makes it easy to answer. The first point does *not*, since $x + y + z = 1 + 2 + 3$ does *not* equal -2. The second one *does*, however, because $-3 - 1 + 2$ does equal -2. Equation (14) makes it easy to test any point for membership in P this way.

On the other hand, if we had wanted to actually *produce* vectors on P (for instance, if we were asked to *list half a dozen vectors on P*), we would have to *solve* Equation (14). That's a different task. The coefficient matrix of (14) is $[1\ 1\ 1]$, which happens to have RRE-form, so we can solve it by inspection: its solutions are generated by the mapping

$$(15) \quad \mathbf{X}(s, t) = \begin{pmatrix} -2 \\ 0 \\ 0 \end{pmatrix} + s \begin{pmatrix} -1 \\ 1 \\ 0 \end{pmatrix} + t \begin{pmatrix} -1 \\ 0 \\ 1 \end{pmatrix} = \begin{pmatrix} -2 - s - t \\ s \\ t \end{pmatrix}$$

By plugging in values for s and t we can now produce as many vectors on P as we like.

Thus, (14) and (15) give two completely different descriptions of the same plane P. This situation typifies the "span/perp" (or "implicit/explicit") duality that we shall discuss in Chapter 5. ☐

EXAMPLE 4.14. (*Three points determine a plane.*) If three points don't all lie on a common line, we expect them to determine a plane. Our understanding now makes it easy to find an equation for that plane.

For instance, suppose we want an equation for the plane containing $(4, 1, 1)$, $(2, 2, 0)$, and $(3, 3, 1)$ (or more precisely, the tips of the geometric vectors with those coordinates.)

The equation of a plane always takes the form

(16) $$\mathbf{a} \cdot \mathbf{x} = b$$

and we want all three coordinate vectors to satisfy this equation. This gives us three inhomogeneous equations, all having b on the right-hand side:

$$\mathbf{a} \cdot (4, 1, 1) = b, \quad \mathbf{a} \cdot (2, 2, 0) = b, \quad \text{and} \quad \mathbf{a} \cdot (3, 3, 1) = b$$

Writing $\mathbf{a} = (a_1, a_2, a_3)$, these equations form a system:

$$\begin{aligned} 4a_1 + 1a_2 + 1a_3 &= b \\ 2a_1 + 2a_2 + 0a_3 &= b \\ 3a_1 + 3a_2 + 1a_3 &= b \end{aligned}$$

The a_i's are unknowns here, *but* the right-hand side b is *also* unknown. We therefore move the b's to the left side of the system (subtracting b from both sides of each equation) to get a *homogeneous* system:

$$\begin{aligned} 4a_1 + 1a_2 + 1a_3 - b &= 0 \\ 2a_1 + 2a_2 + 0a_3 - b &= 0 \\ 3a_1 + 3a_2 + 1a_3 - b &= 0 \end{aligned}$$

After a routine application of Gauss–Jordan, the coefficient matrix for this system row-reduces to RRE form as follows:

$$\begin{bmatrix} 4 & 1 & 1 & -1 \\ 2 & 2 & 0 & -1 \\ 3 & 3 & 1 & -1 \end{bmatrix} \longrightarrow \begin{bmatrix} 1 & 0 & 0 & -1/3 \\ 0 & 1 & 0 & -1/6 \\ 0 & 0 & 1 & 1/2 \end{bmatrix}$$

We have one free column, hence one homogeneous generator, and the solution of the system is thus given by the image of the mapping

$$\mathbf{H}(s) = s \begin{pmatrix} 1/3 \\ 1/6 \\ -1/2 \\ 1 \end{pmatrix}$$

Note that the problem has many solutions: we can choose any value for the scalar s. Once we choose $s \neq 0$, however, we simply set $\mathbf{a} = (a_1, a_2, a_3, b) = \mathbf{H}(s)$ and use those values in Equation (16). This gives the equation of a plane containing the given points.

For instance, if we take $s = 6$ (to clear all denominators), we get $\mathbf{a} = (a_1, a_2, a_3, b) = \mathbf{h} = (2, 1, -3, 6)$, and (16) becomes

$$2x + 1y - 3z = 6$$

This equation describes a plane containing the given points.

Note: Choosing a different value of s would work equally well; it would just give a multiple of this same equation. The plane described by a linear equation is also described by any multiple of that equation—of course—because multiplying an equation by a non-zero scalar doesn't affect its solution set.   □

4.15. Hyperplanes. By interpreting numeric vectors in $\mathbf{R}^2$ and $\mathbf{R}^3$ as coordinates of geometric vectors in the plane and space, we have seen that, once we fix a coordinate system,

- The solution set of a linear equations in 2 variables describes a line in the plane.

- The solution set of a linear equations in 3 variables describes a plane in space.

What about solution sets of linear equations in *more* variables?

The solution set of an equation in m variables lies in $\mathbf{R}^m$, and just as in $\mathbf{R}^2$ and $\mathbf{R}^3$, we can abbreviate the linear equation, namely

$$(17) \qquad a_1 x_1 + a_2 x_2 + \cdots + a_m x_m = b$$

as

$$\mathbf{a} \cdot \mathbf{x} = b \quad \text{or} \quad \mathbf{a}^*(\mathbf{x}) = b$$

Once again, (assuming $\mathbf{a} \neq \mathbf{0}$) the coefficient matrix of this one-rowed "system" must reduce to an RRE form with one pivot column, and $m - 1$ free columns. So the solution to Equation (17) has the form

$$(18) \qquad \mathbf{X}(s_1, s_2, \ldots, s_{m-1}) = \mathbf{x}_p + s_1 \mathbf{h}_1 + s_2 \mathbf{h}_2 + \cdots + s_{m-1} \mathbf{h}_{m-1}$$

where the $\mathbf{h}_i$'s here solve the *homogeneous* version of (17), namely

$$(19) \qquad \mathbf{a} \cdot \mathbf{x} = 0$$

Again, the vanishing dot product—by definition—signals orthogonality. So we can describe the solution set of (19) as *the set of all vectors orthogonal to* $\mathbf{a}$. When $m > 3$, we can't draw an m-dimensional "picture" of this set, but we can use geometric *language* (like "orthogonality") and regard Figures 22 and 24 as *useful metaphors*. And just as we exported notions of length and angle from $\mathbf{R}^2$ to $\mathbf{R}^m$, we can now give a geometric name to the sets in $\mathbf{R}^m$ that are analogous to the lines and planes that describe solution sets of linear equations in $\mathbf{R}^2$ and $\mathbf{R}^3$:

DEFINITION 4.16 (Hyperplane). A **hyperplane in $\mathbf{R}^m$** is the solution set of an m-variable linear equation $\mathbf{a} \cdot \mathbf{x} = b$ (or $\mathbf{a}^*(\mathbf{x}) = b$) with $\mathbf{a} \neq \mathbf{0}$. More specifically, this solution set is *a hyperplane orthogonal to the line generated by* $\mathbf{a}$.

When $b = 0$ (so the hyperplane solves $\mathbf{a} \cdot \mathbf{x} = 0$), this hyperplane goes through the origin, and we call it a *perp*, denoted $\mathbf{a}^\perp$.   □

Note that *this definition applies for any* m. When $m = 2$, *hyperplane* just means a *line in the plane* (Observation 4.7). Similarly, when $m = 3$, a hyperplane is a *plane in space* (Proposition 4.11). Thus, hyperplanes really are the $\mathbf{R}^m$-analogues of lines in the plane and planes in space. A hyperplane in $\mathbf{R}^m$ is a linear "copy" of $\mathbf{R}^{m-1}$ situated in $\mathbf{R}^m$.

Hyperplane provides a geometric term for a purely algebraic object: the solution set of a linear equation.

By *solving* $\mathbf{a} \cdot \mathbf{x} = b$ to get (18), we can always derive a formula for the points on a hyperplane, and this formula shows that we can express any hyperplane as $\mathbf{x}_p + \mathbf{a}^\perp$, where $\mathbf{a}^\perp$ is the *homogeneous* solution set. We call it $\mathbf{a}^\perp$ because solutions of the homogeneous equation $\mathbf{a} \cdot \mathbf{x} = 0$ are vectors $\mathbf{x}$ *perpendicular* to $\mathbf{a}$. Geometrically, $\mathbf{a}^\perp$ is suggested by the blue plane in Figure 23 (which really is a picture of $\mathbf{a}^\perp$ in the $m = 3$ case). When $m > 3$, we may still visualize $\mathbf{a}^\perp$ as a "higher-dimensional version" of the plane in that figure.

EXAMPLE 4.17. The solution set of the equation

$$(20) \qquad\qquad x_1 - x_2 + x_3 - x_4 = 2$$

form a hyperplane in $\mathbf{R}^4$. It is orthogonal to the line in $\mathbf{R}^4$ generated by the coefficient vector $\mathbf{a} = (1, -1, 1, -1)$. It does *not* go through the origin because the equation is not homogeneous (and $(0, 0, 0, 0)$ clearly fails to solve the equation). But when we solve the equation to get the solution mapping

$$(21) \quad \mathbf{X}(s_1, s_2, s_3) = \begin{pmatrix} 2 \\ 0 \\ 0 \\ 0 \end{pmatrix} + s_1 \begin{pmatrix} 1 \\ 1 \\ 0 \\ 0 \end{pmatrix} + s_2 \begin{pmatrix} -1 \\ 0 \\ 1 \\ 0 \end{pmatrix} + s_3 \begin{pmatrix} 1 \\ 0 \\ 0 \\ 1 \end{pmatrix}$$

we see that this hyperplane can be written as $(2, 0, 0, 0) + \mathbf{a}^{\perp}$, where $\mathbf{a}^{\perp}$ (the hyperplane orthogonal to $\mathbf{a}$) *does* contain the origin (just set the s_i's all equal to zero). The hyperplane defined by (20) may be visualized as "parallel" to $\mathbf{a}^{\perp}$, with "x_1-intercept" $(2, 0, 0, 0)$. $\quad\square$

– **Practice** –

168. Suppose $\mathbf{a}$ and $\mathbf{b}$ *both* lie on the line traced out by $L(t) = \mathbf{p} + t\mathbf{v}$ in $\mathbf{R}^m$. Show that $\mathbf{b} - \mathbf{a}$ must then be a multiple of $\mathbf{v}$.

169. Suppose we have two lines $L(t) = \mathbf{p} + t\mathbf{v}$, and $L'(t) = \mathbf{p}' + t\mathbf{v}$ in $\mathbf{R}^m$, with the same $\mathbf{v} \neq \mathbf{0}$, but with $\mathbf{p} \neq \mathbf{p}'$. Show that (just as for lines in the plane or space) these two lines either have *no* point in common, or *all* their points in common.

170. The line that goes through $(2, 1)$ and perpendicular to $(1, 2)$ forms the solution set of what linear equation?

171. The line that goes through $(-1, 3)$ and perpendicular to $(0, 1)$ forms the solution set of what linear equation?

172. We all learn that the graph of $y = mx + b$ is a line with slope m and y-intercept b.

   a) What vector (or vectors) $\mathbf{a} \in \mathbf{R}^2$ will make the solution set of $\mathbf{a} \cdot \mathbf{x} = b$ coincide with the graph of $mx + b$?

   b) What vector (or vectors) $\mathbf{a} \in \mathbf{R}^2$ will make the solution set of $\mathbf{a} \cdot \mathbf{x} = b$ coincide with a line parallel to the y-axis?

   (Note that the lines in (b) have "infinite slope," hence cannot be expressed as $y = mx + b$ for any scalar m. *All* lines can be described as solution sets of inhomogeneous linear equations, however.)

173. Two perpendicular lines in the plane intersect at $(-5, 2)$. One of them is parallel to $(1, 1)$. Write two linear equations whose solution sets describes each of these lines. Their intersection point represents the solution of what linear system?

174. Two perpendicular lines in the plane intersect at $(1,1)$. One of them is parallel to $(-5,2)$. Write two linear equations whose solution sets describes each of these lines. Their intersection point represents the solution of what linear system?

175. For what scalar b does the plane...

    a) ...of solutions to $3x - y + 7z = b$ contain the the (tip of the) vector with coordinates $(7, 7, -2)$?

    b) ...of solutions to $4x + 6y - 8z = b$ contain the (tip of the) vector with coordinates $(7, 7, -2)$?

176. For what scalar c does the plane...

    a) ...of solutions to $-x + cy + 3z = 5$ contain the tip of the vector with coordinates $(1, 1, 1)$?

    b) ...of solutions to $6x - 8y + cz = 11$ contain the tip of the vector with coordinates $(3, 2, -9)$?

177. Suppose $\vec{a}$ and $\vec{p}$ are geometric vectors in space with respective numeric coordinates $\mathbf{a} = (1, -2, 1)$ and $\mathbf{p} = (1, 2, 3)$ in some euclidean coordinate system. Find a linear equation that describes the plane perpendicular to $\vec{a}$ and containing (the tip of) $\vec{p}$.

178. Suppose $\vec{a}$ and $\vec{p}$ are geometric vectors in space with respective numeric coordinates $\mathbf{a} = (1, 2, 3)$ and $\mathbf{p} = (1, -2, 1)$ in some euclidean coordinate system. Find a linear equation that describes the plane perpendicular to $\vec{a}$ and containing (the tip of) $\vec{p}$.

179. Find three vectors on the plane $2x - 3y + 4z = 3$.

180. Find three vectors on the plane $-x + 7y + 2z = -1$.

181. Find:

    a) A linear equation for the plane containing the (tips of the) geometric vectors having coordinates $(3, 2, 1)$, $(4, 4, -3)$ and $(5, 3, 2)$ in $\mathbf{R}^3$.

    b) A mapping whose image consists of all points on the plane described by $3x - 2y - z = 12$.

182. Find:

    a) A linear equation describing the plane containing the (tips of the) geometric vectors whose coordinates are $(1, -2, 3)$, $(-2, 3, 1)$, and $(3, 1, -2)$ in $\mathbf{R}^3$.

    b) A mapping whose image consists of all points on the plane described by $x - y + z = 1$.

183. Suppose $\vec{a}$ and $\vec{b}$ are *different* geometric vectors in space with repective numeric coordinates $\mathbf{a} \neq \mathbf{0}$ and $\mathbf{b} \neq \mathbf{0}$ in some euclidean coordinate system. Consider two planes: The plane containing $\vec{b}$ and perpendicular to $\vec{a}$, and the plane containing $\vec{a}$ perpendicular to $\vec{b}$. Could these be the same under any circumstances? Why or why not?

184. Find an equation for the hyperplane through the origin and orthogonal to the vector

    a) $(3, -2, 1, -2, 3) \in \mathbf{R}^5$ b) $(2, 0, 2, 0, 2, 0, 2) \in \mathbf{R}^7$
    c) $(8, 4, 2, 1) \in \mathbf{R}^4$ d) $(-1, 1, -1, 1, -1, 1, -1, 1) \in \mathbf{R}^8$

185. Find a vector that generates the line perpendicular to the hyperplane of solutions to

    a) $3x - 4y = -1$ in $\mathbf{R}^2$

    b) $-2x_1 + 3x_2 - 2x_3 + x_4 = -7$ in $\mathbf{R}^4$

    c) $x_1 - x_2 + x_3 - x_4 + x_5 - x_6 = 0$ in $\mathbf{R}^6$

    d) $-3x_1 + 3x_3 - 3x_6 + 3x_9 = 5$ in $\mathbf{R}^{10}$

186. Find a **unit** vector that generates the line perpendicular to

    a) the line defined by $x + y = 0$ in the plane.

    b) the plane defined by $z = 2 + 3x - 4y$ in space.

    c) the hyperplane defined by $x_1 + x_2 + x_3 + x_4 = 15$ in $\mathbf{R}^4$.

187. Find the equation of a line perpendicular to the hyperplane of solutions to each given equation, and containing each the given point **p**.

    a) $4x - 3y = 0$, $\mathbf{p} = (3, 4)$

    b) $-3x_1 + 2x_2 - x_3 + 2x_4 = 0$, $\mathbf{p} = (1, 1, 1, 1)$

    c) $x_1 - x_2 + x_3 - x_4 + x_5 - x_6 = \pi$, $\mathbf{p} = (5, 0, -2, 0, 5)$

    d) $-3x_1 + x_3 - x_6 + 3x_9 = -8$, $\mathbf{p} = (1, -2, 3, -4, 5, -4, 3, -2, 1)$

188. Find a mapping whose image is the hyperplane of solutions to the equation

$$2x - 3y + 4z - 5w = 10$$

in $\mathbf{R}^4$.

189. Find a mapping whose image is the hyperplane of solutions to the equation

$$x_1 + 2x_2 - 3x_3 + 4x_4 = 5$$

in $\mathbf{R}^4$.

— ⋆ —

5. System Geometry and Row/Column Duality

We end this chapter by revisiting the algebraic subject of Chapter 2— linear systems—from our new geometric viewpoint.

A linear *system* is a set of one or more linear equations. The *solutions* of such a system are the vectors that simultaneously solve *each* of its equations.

Using the geometric term we have just defined, however, each equation in a system defines a *hyperplane*. So from this standpoint, *the solutions of the* system *are the vectors that simultaneously lie on* all *of these hyperplanes*. Equivalently,

*The solutions of a linear system are the vectors in the **intersection** of the hyperplanes described by the individual equations of the system.*

This is the *geometric* interpretation of the *row problem* associated with a linear system. Each *row* in the system's augmented matrix corresponds to one equation, and hence to a *hyperplane*—the hyperplane of solutions to the equation in that row. Each row corresponds to a hyperplane, and the solution of the system is the *intersection* of these hyperplanes.

EXAMPLE 5.1. Consider the linear system

$$
\begin{aligned}
x + 2y &= 4 \\
2x + y &= 5
\end{aligned}
$$

Each equation in this system defines a hyperplane in $\mathbf{R}^2$, that is, a line in the plane. The intersection of the two lines they define locates the solution of the system, as Figure 25 illustrates.

□

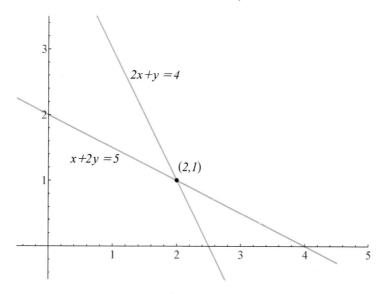

Figure 25. The solution of a linear system of two equations in two variables locates the intersection of two lines in the plane.

EXAMPLE 5.2. Consider this system of two equations in *three* variables:

$$\begin{aligned} x + 2y + z &= 4 \\ 2x + y - z &= 5 \end{aligned}$$

Again, each equation defines a hyperplane, and in $\mathbf{R}^3$, hyperplanes correspond (by numeric/geometric duality) to planes in space. These two planes intersect in a line, which therefore locates the solutions of the system. Each point on the line of intersection corresponds to one solution of the equation. Note that this gives infinitely many solutions (cf. Figure 26).

$\square$

Note that two distinct planes in space don't *always* intersect in a line. They can be parallel, failing to intersect at all. This corresponds to a 2-by-3 linear system with *no* solution.

Similarly, systems of *three* equations in three variables define *three* hyperplanes in $\mathbf{R}^3$, which we can depict as three planes in space. A configuration of three planes may intersect in a single point, or in a line, or not at all, as the pictures below illustrate. Note that unlike *pairs* of planes, three planes can fail to intersect without being parallel. Figure 27 shows some of the possibilities.

$x+2y-6z = -4$

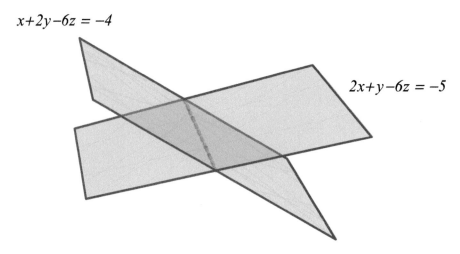

$2x+y-6z = -5$

Figure 26. A system of two equations in 3 variables corresponds to a pair of planes in space. The solutions of the system are the vectors whose tips lie on the intersection of the planes—in this case, a line. The *row* interpretation of the system asks for this intersection.

These images can help us think geometrically about linear systems in more than three variables, even if we can't actually draw pictures of them.

For instance, a system of three equations in four variables corresponds to three hyperplanes in $\mathbf{R}^4$. Solving the system means finding the intersection of these three hyperplanes. The "size" of this intersection depends on the number of free variables in the system: each free variable gives a new homogeneous generator for the system and hence "more" solutions. We develop a precise language for discussing such matters in Chapter 5.

5.3. Row/column duality. With these pictures in mind, we revisit a duality first encountered in Section 1.6, where we saw that any $n \times m$ linear system can be regarded as either a *row problem* in $\mathbf{R}^m$, or as a *column problem* in $\mathbf{R}^n$. We now interpret these problems more geometrically.

The situation is somewhat reminiscent of the "figure/ground" duality in M.C. Escher's beautiful 1938 woodcut *Day and Night* (Figure 28). We can see it as a picture of white birds flying east, or as a picture of black birds flying west. By focusing on one or the other, the image can be described in two completely different, but equally valid ways.

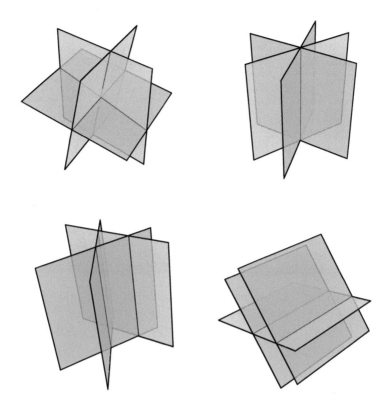

Figure 27. Typically, three planes in space intersect in a single point (upper left), corresponding to a 3-by-3 linear system with exactly one solution. But such a system may have an entire line of solutions (upper right), or none at all (bottom left and right).

In the same way, any linear system

$$
\begin{array}{rcl}
a_{11}x_1 + a_{12}x_2 + a_{13}x_3 + \cdots + a_{1m}x_m &=& b_1 \\
a_{21}x_1 + a_{22}x_2 + a_{23}x_3 + \cdots + a_{2m}x_m &=& b_2 \\
\vdots \qquad\qquad\qquad\qquad \vdots &=& \vdots \\
a_{n1}x_1 + a_{n2}x_2 + a_{n3}x_3 + \cdots + a_{nm}x_m &=& b_n
\end{array}
$$

(22)

can be seen as a row problem, or quite differently, as a column problem. Let $\mathbf{A} = [a_{ij}]$ denote the coefficient matrix of a system with right-hand side $\mathbf{b} = (b_1, b_2, \ldots, b_n)$. We may either focus on its n rows

$$\mathbf{r}_1(\mathbf{A}), \mathbf{r}_2(\mathbf{A}), \cdots, r_n(\mathbf{A})$$

or on its m columns

$$\mathbf{c}_1(\mathbf{A}), \mathbf{c}_2(\mathbf{A}), \cdots, c_m(\mathbf{A})$$

Figure 28. Which is it? White birds flying east, or black birds flying west? M.C. Escher's *Day and Night*.

If we focus on rows, the system looks like this:

$$
\begin{aligned}
r_1(\mathbf{A}) \cdot \mathbf{x} &= b_1 \\
r_2(\mathbf{A}) \cdot \mathbf{x} &= b_2 \\
&\ \vdots \\
r_n(\mathbf{A}) \cdot \mathbf{x} &= b_n
\end{aligned}
$$

(23)

Each equation here defines a hyperplane *in* $\mathbf{R}^m$. Specifically, the ith equation defines a hyperplane orthogonal to the line generated by $\mathbf{r}_i(\mathbf{A})$. *Solving the system*, from this viewpoint, means finding every $\mathbf{x} \in \mathbf{R}^m$ that lies in the *intersection* of these n hyperplanes.

If, alternatively, we focus on *columns*, the system poses a problem in $\mathbf{R}^n$—not $\mathbf{R}^m$. From the column perspective, the same system (22) looks like this:

$$(24) \qquad x_1\, \mathbf{c}_1(\mathbf{A}) + x_2\, \mathbf{c}_2(\mathbf{A}) + \cdots + x_m \mathbf{c}_m(\mathbf{A}) = \mathbf{b}$$

It asks us to express the right-hand side vector $\mathbf{b}$, *which lies in* $\mathbf{R}^n$, *not* $\mathbf{R}^m$ as a linear combination of the columns of $\mathbf{A}$ (which also lie in $\mathbf{R}^n$). Here, *solving the system* means finding all ways of making this linear combination reach its target vector $\mathbf{b}$. Geometrically, we're asking what multiples of the columns of $\mathbf{A}$ we can chain together "tail to tip" in $\mathbf{R}^n$ to reach the tip of $\mathbf{b}$.

Just as the white birds flying east and the black birds moving west in Figure 28 are actually two aspects of the same drawing, so (23) and

(24) are actually two aspects of the same system. In some underlying mathematical sense they are one problem, with one solution, and one method of solution: writing down the augmented matrix of the system, row-reducing, and then extracting particular solution and homogeneous generators.

At the same time, however, the system has two entirely different interpretations: the *row* problem in $\mathbf{R}^m$, and the *column* problem in $\mathbf{R}^n$. They are quite different even when $m = n$. We summarize them again for emphasis:

Row problem: Each row of the augmented matrix describes a hyperplane in $\mathbf{R}^m$. We seek the intersection of those hyperplanes.

Column problem: Each column of the coefficient matrix forms a vector in $\mathbf{R}^n$. We seek to express the right-hand side vector $\mathbf{b}$ (also in $\mathbf{R}^n$) as a linear combination of these columns.

EXAMPLE 5.4. Consider the row and column problems associated with the system

$$
(25) \qquad \begin{array}{rcl} x + 2y - 6z & = & -4 \\ 2x + y - 6z & = & -5 \end{array}
$$

The augmented matrix for this system row-reduces to

$$
\begin{bmatrix} 1 & 0 & -2 & -2 \\ 0 & 1 & -2 & -1 \end{bmatrix}
$$

From this, we easily read off the particular solution $\mathbf{x}_p = (-2, -1, 0)$, and the homogeneous generator $\mathbf{h} = (2, 2, 1)$. The solutions of the system are then given by the image of the mapping

$$
\mathbf{X}(s) = \begin{pmatrix} -2 \\ -1 \\ 0 \end{pmatrix} + s \begin{pmatrix} 2 \\ 2 \\ 1 \end{pmatrix}
$$

The **row problem** here takes place in $\mathbf{R}^3$, where each equation represents a plane in space. From this perspective, our solution mapping $\mathbf{X}$ traces out the *line* in $\mathbf{R}^3$ where these planes intersect: the line through $(-2, -1, 0)$ in the $(2, 2, 1)$ direction (Figure 26 again).

The **column problem** is entirely different (Figure 29). It takes place in $\mathbf{R}^2$, and asks us for all possible ways to express the right-hand side vector $(-4, -5)$ as a linear combination of the column vectors

$$
\begin{pmatrix} 1 \\ 2 \end{pmatrix}, \begin{pmatrix} 2 \\ 1 \end{pmatrix} \quad \text{and} \quad \begin{pmatrix} -6 \\ -6 \end{pmatrix}
$$

Again, our solution mapping $\mathbf{X}(s)$ gives all possible answers. For instance $\mathbf{X}(0) = (-2, -1, 0)$, so we have

$$-2 \begin{pmatrix} 1 \\ 2 \end{pmatrix} - 1 \begin{pmatrix} 2 \\ 1 \end{pmatrix} + 0 \begin{pmatrix} -6 \\ -6 \end{pmatrix} = \begin{pmatrix} -4 \\ -5 \end{pmatrix}$$

Similarly, $\mathbf{X}(-1) = (-4, -3, -1)$, so we also have

$$-4 \begin{pmatrix} 1 \\ 2 \end{pmatrix} - 3 \begin{pmatrix} 2 \\ 1 \end{pmatrix} - 1 \begin{pmatrix} -6 \\ -6 \end{pmatrix} = \begin{pmatrix} -4 \\ -5 \end{pmatrix}$$

Indeed, for any scalar s, we get a linear combination of the columns that adds up to $(-4, -5)$.

$$(2s - 2) \begin{pmatrix} 1 \\ 2 \end{pmatrix} + (2s - 1) \begin{pmatrix} 2 \\ 1 \end{pmatrix} + s \begin{pmatrix} -6 \\ -6 \end{pmatrix} = \begin{pmatrix} -4 \\ -5 \end{pmatrix}$$

$\square$

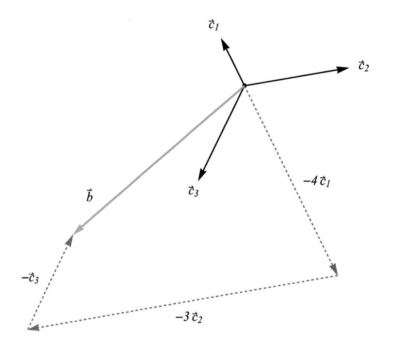

Figure 29. The column problem. This interpretation of the system (25) asks which scalar multiples of the columns c_i (black arrows) add up to $\mathbf{b}$ (blue arrow). The dashed gray arrows illustrate one of the many solutions, namely $(-4, -3, -1)$.

– Practice –

190. Row problem vs. column problem:

    a) The equations $-x+2y = 3$ and $x+2y = 1$ both describe lines in the plane (using the standard euclidean coordinate system). Sketch both lines and find their intersection.

    b) What linear combination of $(-1, 1)$ and $(2, 2)$ adds up to $(3, 1)$ in $\mathbf{R}^2$? Make a sketch like Figure 29 to illustrate your answer.

191. Row problem vs. column problem:

    a) Find the intersection of the planes described by $x - 2y + z = 1$ and $4x - 3y - z = -2$ in $\mathbf{R}^3$.

    b) What linear combinations of the vectors $(1, 4)$, $(-2, -3)$, and $(1, -1)$ in $\mathbf{R}^2$ add up to $(1, -2)$?

192. Row problem vs. column problem:

    a) The homogeneous equations $x_1 + x_3 + 2x_4 = 0$, $x_1 + x_2 - 2x_4 = 0$ and $x_2 + x_3 + 2x_4 = 0$ all describe hyperplanes in $\mathbf{R}^4$. Find all points in the intersection of these hyperplanes.

    b) What linear combinations of $(1, 1, 0)$, $(0, 1, 1)$, $(1, 0, 1)$ and $(2, -2, 2)$ add up to $\mathbf{0} \in \mathbf{R}^3$?

193. Solve this system:

$$\begin{aligned} x + y &= 5 \\ 2x - y &= 4 \end{aligned}$$

    a) Illustrate the "row problem" for this system by sketching a line in the plane for each equation. Interpret the lack of solution geometrically from this viewpoint.

    b) What is the "column problem" associated with the system above? Interpret its solution geometrically by sketching a linear combination that adds up to $(5, 4)$ (Compare Figure 29).

194. The system below has no solution:

$$3x - y = 1$$
$$-3x + y = 1$$

a) Illustrate the "row problem" for this system by sketching a line in the plane for each equation. Interpret the lack of solution geometrically from this viewpoint.

b) What is the "column problem" associated with the system above? Interpret its lack of solution geometrically in terms of linear combinations.

195. Consider three planes in space that go through the origin and are respectively perpendicular to $(3, 0, -1)$, $(2, 1, 0)$, and $(1, -1, -1)$. Determine whether these planes intersect in just the origin, an entire line, or a complete plane.

196. Consider three planes in space that go through the origin and are respectively perpendicular to $(1, 0, -1)$, $(1, 1, 1)$, and $(0, 1, -1)$. Determine whether these planes intersect in just the origin, an entire line, or a complete plane.

197. The solutions sets of $x - y = 1$, $y + z = 1$, and $z + x = 1$ each describe a plane in space. Describe the set of points that lie on all three planes.

— ★ —

CHAPTER 4

The Algebra of Matrices

Our overarching goal, recall, is to understand linear transformations, and as we saw in Chapter 1, every linear transformation is represented by a matrix. To progress further toward our goal, not surprisingly, we now need a deeper understanding of matrices. So far, we have used them in two basic ways:

- To represent linear transformations, and

- As a notational tool for solving systems with the Gauss-Jordan algorithm.

We shall now see that matrices encode additional algebraic and geometric meaning.

Algebraically, matrices can be added and scalar-multiplied, just like numeric and geometric vectors. Much more importantly, however, we can *multiply* compatible pairs of matrices to produce new matrices in a powerful extension of matrix/vector multiplication. The importance of this *matrix product* is hard to overstate. We begin to explore it in Section 1.9 below, after familiarizing ourselves with matrix addition and scalar multiplication.

1. Matrix Operations

1.1. Matrix addition. As we saw in Chapter 1, any two matrices of the same size can be added or subtracted entry-by-entry, just as we add or subtract vectors. For instance

$$\begin{bmatrix} 3 & -1 \\ 1 & 2 \end{bmatrix} + \begin{bmatrix} 4 & 4 \\ 4 & 4 \end{bmatrix} = \begin{bmatrix} 3+4 & -1+4 \\ 1+4 & 2+4 \end{bmatrix} = \begin{bmatrix} 7 & 3 \\ 5 & 6 \end{bmatrix}$$

$$\begin{bmatrix} -1 & 2 \\ 4 & 3 \\ -2 & 1 \end{bmatrix} - \begin{bmatrix} 1 & -1 \\ -1 & 1 \\ 1 & -1 \end{bmatrix} = \begin{bmatrix} -2 & 3 \\ 5 & 2 \\ -3 & 2 \end{bmatrix}$$

We may describe this operation formally as follows:

DEFINITION 1.2 (Matrix addition). The **sum** $\mathbf{C} = [c_{ij}]$ of two $n \times m$ matrices $\mathbf{A} = [a_{ij}]$ and $\mathbf{B} = [b_{ij}]$ is again an $n \times m$ matrix. We get its entries from the simple rule $c_{ij} = a_{ij} + b_{ij}$. Addition requires that both matrices have the same size (the same numbers of rows and columns).

As an entry-by-entry extension of scalar addition, which is commutative and associative, matrix addition inherits both those properties:

PROPOSITION 1.3. *Matrix addition is commutative and associative: For any matrices* $\mathbf{A}$, $\mathbf{B}$, *and* $\mathbf{C}$ *of the same size, we have*

$$\mathbf{A} + \mathbf{B} = \mathbf{B} + \mathbf{A} , \qquad \mathbf{A} + (\mathbf{B} + \mathbf{C}) = (\mathbf{A} + \mathbf{B}) + \mathbf{C}$$

PROOF. See Exercise 202.   □

1.4. Scalar multiplication.

In Chapter 1, we saw how to scalar multiply matrices entry-by-entry. Formally, given any scalar λ (the Greek letter "lambda," a common symbol for scalars when dealing with matrices), we have

$$\lambda \begin{bmatrix} 1 & 0 & 3 \\ 0 & 2 & -1 \end{bmatrix} = \begin{bmatrix} \lambda & 0 & 3\lambda \\ 0 & 2\lambda & -\lambda \end{bmatrix}$$

Here's a careful definition:

DEFINITION 1.5 (Scalar multiplication of matrices). The product of a matrix $\mathbf{A} = [a_{ij}]$ with a scalar λ is denoted by $\lambda \mathbf{A}$. The entry in row i, column j of $\lambda \mathbf{A}$ is simply $\lambda \, a_{ij}$, so that $\lambda \mathbf{A} = [\lambda \, a_{ij}]$.   □

Scalar multiplication inherits certain properties immediately from elementary arithmetic just like matrix addition does.

PROPOSITION 1.6. *Scalar multiplication distributes over matrix addition, so that for any two matrices* $\mathbf{A}$ *and* $\mathbf{B}$ *of the same size, and for any scalar* λ, *we have*

$$\lambda \, (\mathbf{A} + \mathbf{B}) = \lambda \, \mathbf{A} + \lambda \, \mathbf{B}$$

Moreover, $(\lambda \mu) \, \mathbf{A} = \lambda \, (\mu \mathbf{A})$ *for any scalar* μ.

("μ" is the Greek letter "mu.")

As with the commutativity and associativity of addition, this matrix statement follows directly from the corresponding properties of scalars, since matrix addition and scalar multiplication both operate entry-by-entry. We leave the proof as an exercise.

1.7. Linear combination. Since we can add and scalar multiply matrices, we can also linearly combine them. For instance:

EXAMPLE 1.8. Consider these three 3×2 matrices:

$$\mathbf{A} = \begin{bmatrix} 0 & 1 \\ 1 & 0 \\ 0 & 1 \end{bmatrix} \qquad \mathbf{B} = \begin{bmatrix} 3 & -2 \\ 2 & -3 \\ 1 & 1 \end{bmatrix} \qquad \mathbf{C} = \begin{bmatrix} -1 & 2 \\ -3 & 4 \\ -1 & 2 \end{bmatrix}$$

Let us compute a typical linear combination; say $5\mathbf{A} + 2\mathbf{B} - \mathbf{C}$:

$$5\mathbf{A} + 2\mathbf{B} - \mathbf{C}$$

$$= 5 \begin{bmatrix} 0 & 1 \\ 1 & 0 \\ 0 & 1 \end{bmatrix} + 2 \begin{bmatrix} 3 & -2 \\ 2 & -3 \\ 1 & 1 \end{bmatrix} - \begin{bmatrix} -1 & 2 \\ -3 & 4 \\ -1 & 2 \end{bmatrix}$$

$$= \begin{bmatrix} 0 & 5 \\ 5 & 0 \\ 0 & 5 \end{bmatrix} + \begin{bmatrix} 6 & -4 \\ 4 & -6 \\ 2 & 2 \end{bmatrix} - \begin{bmatrix} -1 & 2 \\ -3 & 4 \\ -1 & 2 \end{bmatrix}$$

$$= \begin{bmatrix} 0+6+1 & 5-4-2 \\ 5+4+3 & 0-6-4 \\ 0+2+1 & 5+2-2 \end{bmatrix} = \begin{bmatrix} 7 & -1 \\ 12 & -10 \\ 3 & 5 \end{bmatrix}$$

$\square$

1.9. Matrix multiplication. Addition and scalar multiplication are too simple to have consequences of real interest. Indeed, for purposes of addition and scalar multiplication, $n \times m$ matrices are really no different than numeric vectors in $\mathbf{R}^{nm}$ (i.e., vectors with $n \cdot m$ entries) whose entries are arranged in several rows and columns instead of just one.

True *matrix multiplication*, on the other hand, will have profound new implications. We can define it easily though, as a column-by-column extension of matrix/vector multiplication.

Indeed, given two matrices $\mathbf{A}$ and $\mathbf{B}$, we compute the product $\mathbf{AB}$ column-by-column as follows:

To get column j of $\mathbf{AB}$, *we multiply* $\mathbf{A}$ *times column j of* $\mathbf{B}$:

$$(26) \qquad\qquad \mathbf{c}_j(\mathbf{AB}) = \mathbf{A}\,\mathbf{c}_j(\mathbf{B})$$

We shall see other descriptions of matrix multiplication, but this is the simplest and most illuminating. It is the **column-wise** definition of matrix multiplication.

Note that the right-hand side of (26) makes no sense unless the number of *columns* in $\mathbf{A}$ matches the number of *entries* in $\mathbf{c}_j(\mathbf{B})$ (which equals the number of *rows* of $\mathbf{B}$). If those two numbers match, the formula above makes sense, and shows how to compute every column of AB — and thereby $\mathbf{AB}$ in full—using matrix/vector multiplication. If they don't match, the product simply isn't defined.

Here is a more precise statement that gives a *row-wise* definition too:

DEFINITION 1.10 (Matrix multiplication). The product of an $n \times k$ matrix $\mathbf{A}_{n\times k}$ and a $k \times m$ matrix $\mathbf{B}_{k\times m}$ will have n rows and m columns:

$$\mathbf{A}_{n\times k} \times \mathbf{B}_{k\times m} = (\mathbf{AB})_{n\times m}$$

Note that the k's disappear. Column-wise, $\mathbf{AB}$ is the matrix whose jth column is $\mathbf{A}\,\mathbf{c}_j(\mathbf{B})$:

$$\mathbf{AB} = \left[\begin{array}{cccc} \mathbf{A}\,\mathbf{c}_1(\mathbf{B}) & \mathbf{A}\,\mathbf{c}_2(\mathbf{B}) & \cdots & \mathbf{A}\,\mathbf{c}_m(\mathbf{B}) \\ | & | & \cdots & | \end{array} \right]$$

Alternatively, we can define $\mathbf{AB}$ using dot products and "fast" matrix/vector multiplication as described in Section 1.31 of Chapter 1. . In this formulation, the (i, j) entry of $\mathbf{AB}$ is $\mathbf{r}_i(\mathbf{A}) \cdot \mathbf{c}_j(\mathbf{B})$:

$$\mathbf{AB} = \left[\begin{array}{cccc} \mathbf{r}_1(\mathbf{A}) \cdot \mathbf{c}_1(\mathbf{B}) & \mathbf{r}_1(\mathbf{A}) \cdot \mathbf{c}_2(\mathbf{B}) & \cdots & \mathbf{r}_1(\mathbf{A}) \cdot \mathbf{c}_m(\mathbf{B}) \\ \mathbf{r}_2(\mathbf{A}) \cdot \mathbf{c}_1(\mathbf{B}) & \mathbf{r}_2(\mathbf{A}) \cdot \mathbf{c}_2(\mathbf{B}) & \cdots & \mathbf{r}_2(\mathbf{A}) \cdot \mathbf{c}_m(\mathbf{B}) \\ \vdots & \vdots & & \vdots \\ \mathbf{r}_n(\mathbf{A}) \cdot \mathbf{c}_1(\mathbf{B}) & \mathbf{r}_n(\mathbf{A}) \cdot \mathbf{c}_2(\mathbf{B}) & \cdots & \mathbf{r}_n(\mathbf{A}) \cdot \mathbf{c}_m(\mathbf{B}) \end{array} \right]$$

$\square$

The column-wise description above is usually the best way to *understand* matrix multiplication, but the dot-product formula gives a convenient way to *compute* matrix products. Either method gives the same result.

EXAMPLE 1.11.

$$\mathbf{A} = \begin{bmatrix} 1 & 2 \\ 3 & 4 \\ 5 & 6 \end{bmatrix} \qquad \mathbf{B} = \begin{bmatrix} 0 & 3 \\ -3 & 0 \end{bmatrix}$$

Here $\mathbf{A}$ has two columns, and $\mathbf{B}$ has two rows. The product $\mathbf{AB}$ therefore exists. By the column-wise definition, its first column is

$$\mathbf{A}\begin{pmatrix} 0 \\ -3 \end{pmatrix} = 0\begin{pmatrix} 1 \\ 3 \\ 5 \end{pmatrix} - 3\begin{pmatrix} 2 \\ 4 \\ 6 \end{pmatrix} = \begin{pmatrix} -6 \\ -12 \\ -18 \end{pmatrix}$$

while the second column is

$$\mathbf{A}\begin{pmatrix} 3 \\ 0 \end{pmatrix} = 3\begin{pmatrix} 1 \\ 3 \\ 5 \end{pmatrix} + 0\begin{pmatrix} 2 \\ 4 \\ 6 \end{pmatrix} = \begin{pmatrix} 3 \\ 9 \\ 15 \end{pmatrix}$$

So

$$\mathbf{AB} = \begin{bmatrix} -6 & 3 \\ -12 & 9 \\ -18 & 15 \end{bmatrix}$$

The dot-product formula from our definition gives the same result, though perhaps more cleanly:

$$\mathbf{AB} = \begin{bmatrix} (1,2)\cdot(0,-3) & (1,2)\cdot(3,0) \\ (3,4)\cdot(0,-3) & (3,4)\cdot(3,0) \\ (5,6)\cdot(0,-3) & (5,6)\cdot(3,0) \end{bmatrix} = \begin{bmatrix} -6 & 3 \\ -12 & 9 \\ -18 & 15 \end{bmatrix}$$

Note that the reverse product $\mathbf{BA}$ is **not** defined here, since $\mathbf{B}$ has two columns, which does *not* match the number of rows (three) in $\mathbf{A}$. On the other hand, if we replace $\mathbf{A}$ by a 2×2 matrix, say

$$\mathbf{A} = \begin{bmatrix} 1 & 2 \\ 3 & 4 \end{bmatrix}$$

then $\mathbf{A}$ and $\mathbf{B}$ each have two rows and two columns, so *both* $\mathbf{AB}$ and $\mathbf{BA}$ are defined. On the one hand, we have

$$\mathbf{AB} \;=\; \begin{bmatrix} 1 & 2 \\ 3 & 4 \end{bmatrix} \begin{bmatrix} 0 & 3 \\ -3 & 0 \end{bmatrix}$$

$$=\; \begin{bmatrix} (1,2)\cdot(0,-3) & (1,2)\cdot(3,0) \\[2ex] (3,4)\cdot(0,-3) & (3,4)\cdot(3,0) \end{bmatrix}$$

$$=\; \begin{bmatrix} -6 & 3 \\ -12 & 9 \end{bmatrix}$$

while on the other, we have

$$\mathbf{BA} \;=\; \begin{bmatrix} 0 & 3 \\ -3 & 0 \end{bmatrix} \begin{bmatrix} 1 & 2 \\ 3 & 4 \end{bmatrix}$$

$$=\; \begin{bmatrix} (0,3)\cdot(1,3) & (0,3)\cdot(2,4) \\[2ex] (-3,0)\cdot(1,3) & (-3,0)\cdot(2,4) \end{bmatrix}$$

$$=\; \begin{bmatrix} 9 & 12 \\ -3 & -6 \end{bmatrix}$$

$\square$

Here $\mathbf{AB}$ and $\mathbf{BA}$ are both *defined* in the example above, but note: *they are not equal.*

Indeed, a little experimentation with 2×2 matrices quickly shows that $\mathbf{AB}$ and $\mathbf{BA}$ *usually* don't agree. Thus, *even when defined, matrix multiplication is generally not commutative.*

This is so important that we state it formally:

OBSERVATION 1.12. *Matrix multiplication is **not** commutative. Even when $\mathbf{AB}$ and $\mathbf{BA}$ are both defined, they do not typically agree.*

Some pairs of matrices do commute, but by far, most do *not*, and this means that when doing algebra with matrices, *one must take care not to change the order of multiplication.* For students accustomed only to the algebra of scalars, this is something new and different, and hence a common source of mistakes. Be careful about it.

For an extreme example of non-commutativity, take a $1 \times n$ matrix $\mathbf{V}$, and an $n \times 1$ matrix $\mathbf{W}$. Then $\mathbf{VW}$ and $\mathbf{WV}$ are both defined, with

$$\mathbf{V}_{1\times n}\mathbf{W}_{n\times 1} \;=\; (\mathbf{VW})_{1\times 1}$$
$$\mathbf{W}_{n\times 1}\mathbf{V}_{1\times n} \;=\; (\mathbf{VW})_{n\times n}$$

Here, $\mathbf{VW}$ is a 1×1 matrix—basically, just a scalar—while $\mathbf{WV}$ is much bigger: $n \times n$.

EXAMPLE 1.13. For instance, let $n = 3$ and take

$$\mathbf{V}_{1\times 3} = [\, 3 \;\; -2 \;\; 5 \,] \;, \quad \mathbf{W}_{3\times 1} = \begin{bmatrix} -1 \\ 1 \\ -1 \end{bmatrix}$$

The columns of $\mathbf{V}$ and the rows of $\mathbf{W}$ have only one entry each—they are individual scalars. So the dot-product formula for matrix multiplication yields

$$\mathbf{VW} = [\, 3(-1) - 2(1) + 5(-1) \,] = [-10]$$

Indeed, when we multiply an $n \times 1$ matrix by a $1 \times n$ matrix, we're essentially just treating both as numeric vectors, and computing their dot product.

Multiplying them in the other order produces a very different result, however:

$$\mathbf{WV} \;=\; \begin{bmatrix} -1\,(3) & -1\,(-2) & -1\,(5) \\ 1\,(3) & 1\,(-2) & 1\,(5) \\ -1\,(3) & -1\,(-2) & -1\,(5) \end{bmatrix} = \begin{bmatrix} -3 & 2 & -5 \\ 3 & -2 & 5 \\ -3 & 2 & -5 \end{bmatrix}$$

$\square$

REMARK 1.14. While matrix products don't *typically* commute, they sometimes do, and the question naturally arises: *Is there some property that commuting matrices share with each other?*

Indeed there is such a property, and it has important ramifications, especially for square matrices. We won't be in a position to understand it well until Chapter 7, but we can state a rough version of it now, using the notion of *eigenvector* that we introduced in Section 1.35 of Chapter 1:

When $\mathbf{A}$ and $\mathbf{B}$ are $n \times n$ matrices, we have $\mathbf{AB} = \mathbf{BA}$ if and only if $\mathbf{A}$ and $\mathbf{B}$ share a suitably large set of eigenvectors.

We can't be more precise yet about what "suitably large" means here because it requires notions we won't introduce for some time (see Exercise 500). $\square$

1.15. Distributivity and associativity. While matrix multiplication is not commutative, it *does* have other familiar properties:

PROPOSITION 1.16. *Matrix multiplication* distributes *over matrix addition from both sides, so that (when sizes match appropriately) we always have*

$$\mathbf{A}(\mathbf{B} + \mathbf{C}) = \mathbf{A}\mathbf{B} + \mathbf{A}\mathbf{C}$$
$$(\mathbf{B} + \mathbf{C})\mathbf{A} = \mathbf{B}\mathbf{A} + \mathbf{C}\mathbf{A}$$

PROOF. Both identities follow easily from the dot-product description of matrix multiplication in Definition 1.10. The first identity can also be proven very easily from the column-by-column description there. The second is easier using the row-by-row description of Exercise 210. We leave the details to the reader. □

PROPOSITION 1.17. *Matrix multiplication is* associative*: Any three compatibly-sized matrices* $\mathbf{A}$, $\mathbf{B}$, *and* $\mathbf{C}$ *obey the rule*

$$\mathbf{A}(\mathbf{B}\mathbf{C}) = (\mathbf{A}\mathbf{B})\mathbf{C}$$

PROOF. It suffices to prove this entry-by-entry, showing that the entries of $\mathbf{A}(\mathbf{B}\mathbf{C})$ and $(\mathbf{A}\mathbf{B})\mathbf{C}$ agree in each position.

According to the dot-product description of matrix multiplication, the entry of $\mathbf{A}(\mathbf{B}\mathbf{C})$ in row i, column j, is given by

$$\mathbf{r}_i(\mathbf{A}) \cdot \mathbf{c}_j(\mathbf{B}\mathbf{C})$$

The column-by-column description of $\mathbf{B}\mathbf{C}$, however, tells us that

$$\mathbf{c}_j(\mathbf{B}\mathbf{C}) = \mathbf{B}\,\mathbf{c}_j(\mathbf{C})$$

Putting these two facts together, we see that the entry of $\mathbf{A}(\mathbf{B}\mathbf{C})$ in row i, column j is

(27) $$\mathbf{r}_i(\mathbf{A}) \cdot (\mathbf{B}\,\mathbf{c}_j(\mathbf{C}))$$

Now the jth column of $\mathbf{C}$, is given by $\mathbf{c}_j(\mathbf{C}) = (c_{1j}, c_{2j}, \ldots, c_{kj})$. So when we expand out $\mathbf{B}\,\mathbf{c}_j(\mathbf{C})$ using the column description of matrix/vector multiplication, we get

$$\mathbf{B}\,\mathbf{c}_j(\mathbf{C}) = c_{1j}\,\mathbf{c}_1(\mathbf{B}) + c_{2j}\,\mathbf{c}_2(\mathbf{B}) + \cdots + c_{kj}\,\mathbf{c}_k(\mathbf{B})$$

Putting this into Equation (27) and using the distributive law, the i, j entry of $\mathbf{A}(\mathbf{B}\mathbf{C})$ now expands out to

$$c_{1j}\,\mathbf{r}_i(\mathbf{A}) \cdot \mathbf{c}_1(\mathbf{B}) + c_{2j}\,\mathbf{r}_i(\mathbf{A}) \cdot \mathbf{c}_2(\mathbf{B}) + \cdots + c_{kj}\,\mathbf{r}_1(\mathbf{A}) \cdot \mathbf{c}_k(\mathbf{B})$$

Here the $\mathbf{r}_i(\mathbf{A}) \cdot \mathbf{c}_j(\mathbf{B})$ terms are exactly—according to our entry-by-entry rule for multiplication—the entries in row i of $\mathbf{AB}$. So the formula above can be seen as the dot product

$$\mathbf{c}_j(\mathbf{C}) \cdot \mathbf{r}_i(\mathbf{AB}) = \mathbf{r}_i(\mathbf{AB}) \cdot \mathbf{c}_j(\mathbf{C})$$

The right side above now gives precisely the entry-by-entry formula for the i, j entry of $(\mathbf{AB})\mathbf{C}$. Since we started with the i, j entry $\mathbf{A}(\mathbf{BC})$, this is precisely what we wanted to know.   □

1.18. Composition of linear maps. One of the first things we learned about matrices is that *they represent linear mappings*. By Definition 5.1 of Chapter 1, an $n \times m$ matrix $\mathbf{A}$ represents a linear transformation $T : \mathbf{R}^m \to \mathbf{R}^m$ via the simple formula

$$T(\mathbf{x}) := \mathbf{A}\mathbf{x}$$

In other words, $\mathbf{A}$ represents T when we can compute $T(\mathbf{x})$ for any input $\mathbf{x}$ just by multiplying $\mathbf{x}$ by $\mathbf{A}$.

It turns out that matrix/matrix multiplication *also* has a simple meaning for the linear maps the matrices represent.

OBSERVATION 1.19. *If $\mathbf{A}$ and $\mathbf{B}$ represent a linear maps $T : \mathbf{R}^k \to \mathbf{R}^n$, and $S : \mathbf{R}^m \to \mathbf{R}^k$ respectively, then the product $\mathbf{AB}$ represents the composed map $T \circ S : \mathbf{R}^m \to \mathbf{R}^n$.*

PROOF. This is an easy calculation. Let $\mathbf{x} \in \mathbf{R}^m$ be any input. Then since $\mathbf{A}$ and $\mathbf{B}$ represent T and S respectively, we have

$$(T \circ S)(\mathbf{x}) := T(S(\mathbf{x})) = \mathbf{A}S(\mathbf{x}) = \mathbf{A}(\mathbf{Bx}) = (\mathbf{AB})\mathbf{x}$$

In short, $(T \circ S)(\mathbf{x}) = (\mathbf{AB})\mathbf{x}$ for any input $\mathbf{x}$, which is the same as saying that $\mathbf{AB}$ represents $T \circ S$.   □

– **Practice** –

198. Compute the matrix sum $\mathbf{A} + \mathbf{B}$, where:

a) $\mathbf{A} = \begin{bmatrix} 0 & -3 \\ -2 & 5 \\ -3 & 0 \end{bmatrix}$, $\mathbf{B} = \begin{bmatrix} 1 & -1 \\ 2 & -2 \\ 3 & -3 \end{bmatrix}$

b) $\mathbf{A} = \begin{bmatrix} 1 & -2 & 1 \\ 1 & 0 & -1 \end{bmatrix}$, $\mathbf{B} = \begin{bmatrix} 3 & 3 & 3 \\ 2 & 3 & -2 \end{bmatrix}$

c) $\quad \mathbf{A} = \begin{bmatrix} 1 & 0 & 0 \\ 0 & 2 & 0 \\ 0 & 0 & 3 \end{bmatrix}, \quad \mathbf{B} = \begin{bmatrix} 0 & 0 & -1 \\ 0 & -2 & 0 \\ -3 & 0 & 0 \end{bmatrix}$

199. Compute the scalar multiple $\lambda\mathbf{A}$, where:

a) $\quad \mathbf{A} = \begin{bmatrix} 0 & -3 \\ -2 & 5 \\ -3 & 0 \end{bmatrix}, \quad \lambda = -2$

b) $\quad \mathbf{A} = \begin{bmatrix} 1 & -2 & 1 \\ 1 & 0 & -1 \end{bmatrix}, \quad \lambda = \pi$

c) $\quad \mathbf{A} = \begin{bmatrix} 0 & 0 & -1 \\ 0 & -2 & 0 \\ -3 & 0 & 0 \end{bmatrix}, \quad \lambda = 111$

200. Letting

$$\mathbf{A} = \begin{bmatrix} 1 & 2 \\ 2 & 3 \end{bmatrix}, \quad \mathbf{B} = \begin{bmatrix} 0 & 1 \\ -1 & 0 \end{bmatrix}, \quad \mathbf{C} = \begin{bmatrix} 1 & 1 \\ 1 & 1 \end{bmatrix}$$

Compute these linear combinations:

a) $\mathbf{A} - \mathbf{B} + \mathbf{C}$　　b) $3\mathbf{A} - 4\mathbf{C}$　　c) $-2\mathbf{A} + 3\mathbf{B} + \pi\mathbf{C}$

201. Let

$$\mathbf{A} = \begin{bmatrix} 2 & 1 & 0 \\ -1 & 2 & 1 \\ 0 & -1 & 2 \\ 1 & 0 & -1 \end{bmatrix} \quad \mathbf{B} = \begin{bmatrix} 1 & 2 & 3 \\ 2 & 2 & 3 \\ 3 & 3 & 3 \\ -1 & -1 & -1 \end{bmatrix}$$

Compute the indicated linear combinations. (In (d), λ and μ represent arbitrary scalars.)

a) $\mathbf{A} + \mathbf{B}$　　b) $\mathbf{A} - \mathbf{B}$　　c) $3\mathbf{A} - 2\mathbf{B}$　　d) $\lambda\mathbf{A} + \mu\mathbf{B}$

202. Prove:

a) The first assertion of Proposition 1.3 by showing that for any row i and any column j, the i,j entry of $\mathbf{A} + \mathbf{B}$ equals that of $\mathbf{B} + \mathbf{A}$. (If equality holds for every entry, it holds for the matrices themselves.)

b) The second assertion using the same "entry-by-entry" strategy.

203. Prove the first conclusion of Proposition 1.6 by comparing the general i,j entry of $\lambda(\mathbf{A} + \mathbf{B})$ with that of $\lambda\mathbf{A} + \lambda\mathbf{B}$. Prove the second assertion similarly by comparing the general i,j entry of $(\lambda\mu)\mathbf{A}$ with that of $\lambda(\mu\mathbf{A})$.

204. Compute $\mathbf{AB}$ and $\mathbf{BA}$ given

a) $\quad \mathbf{A} = \begin{bmatrix} 1 & 2 & 3 \end{bmatrix}, \quad \mathbf{B} = \begin{bmatrix} -1 \\ 2 \\ -1 \end{bmatrix}$

b) $\quad \mathbf{A} = \begin{bmatrix} 1 & 0 \\ 0 & -1 \end{bmatrix}, \quad \mathbf{B} = \begin{bmatrix} 0 & 1 \\ 1 & 0 \end{bmatrix}$

c) $\quad \mathbf{A} = \begin{bmatrix} 2 & -1 & 3 \\ -1 & 2 & -1 \end{bmatrix}, \quad \mathbf{B} = \begin{bmatrix} 2 & -1 \\ -1 & 2 \\ 3 & -1 \end{bmatrix}$

205. Let

$$\mathbf{A} = \begin{pmatrix} 1 & 1/2 \\ 1/2 & 1/3 \end{pmatrix} \quad \text{and} \quad \mathbf{B} = \begin{pmatrix} 4 & -6 \\ -6 & 12 \end{pmatrix}.$$

Compute $\mathbf{AB}$ and $\mathbf{BA}$. Are they equal?

206. Let

$$\mathbf{A} = \begin{bmatrix} 1 & 0 \\ 0 & -1 \end{bmatrix} \quad \text{and} \quad \mathbf{B} = \begin{bmatrix} 0 & -1 \\ 1 & 0 \end{bmatrix}$$

Compute $\mathbf{AB}$ and $\mathbf{BA}$. Are they equal?

207. Let

$$A = \begin{bmatrix} 1 & -2 & 3 \\ -2 & 3 & 1 \\ 3 & 1 & -2 \end{bmatrix}$$

$$\mathbf{x} = (1, 1, 1), \quad \mathbf{y} = (1, -1, 0), \quad \mathbf{z} = (0, 1, -1)$$

Let **B** denote the matrix with *columns* **x**, **y**, and **z**:

$$\mathbf{B} = \begin{bmatrix} \mathbf{x} & \mathbf{y} & \mathbf{z} \\ | & | & | \\ | & | & | \end{bmatrix}$$

Compute

a) **Ax**   b) **Ay**   c) **Az**   d) **AB**

Using the column description of matrix multiplication and the answers to (a), (b), and (c), you should be able to answer (d) without doing any further calculations.

208. Suppose **A** has n rows and m columns, and that **AB** and **BA** are both defined. What is the size of **B**?

209. Any square matrix can be multiplied by itself any (whole) number of times. So when **A** is square, we can evaluate its *powers* $\mathbf{A}^2, \mathbf{A}^3$, etc.

a) Compute $\mathbf{A}^k$ for $k = 2, 3, 4$ with

$$A = \begin{bmatrix} 0 & -1 \\ 1 & 0 \end{bmatrix}$$

What will higher powers give?

b) Compute $\mathbf{A}^k$ for $k = 2, 3, 4$ with

$$A = \begin{bmatrix} -1 & 0 \\ 0 & 2 \end{bmatrix}$$

What will higher powers give?

c) Compute $\mathbf{A}^k$ for $k = 2, 3$ with

$$A = \frac{1}{2} \begin{bmatrix} -1 & -\sqrt{3} \\ \sqrt{3} & -1 \end{bmatrix}$$

What will higher powers give?

210. Suppose $\mathbf{B}$ is a $k \times n$ matrix and

$$\mathbf{X}_{1 \times k} = \begin{bmatrix} x_1 & x_2 & \cdots & x_k \end{bmatrix}$$

a) Show that we have

$$\mathbf{XB} = x_1 \mathbf{r}_1(\mathbf{B}) + x_2 \mathbf{r}_2(\mathbf{B}) + \cdots + x_k \mathbf{r}_k(\mathbf{B})$$

This "vector/matrix" product is the row-analogue of the matrix/vector multiplication we defined in Chapter 1 (Definition 1.20).

b) The vector/matrix product above gives yet another "row-by-row" way of describing matrix multiplication, in which the ith row of $\mathbf{AB}$ is given by the vector/matrix product $\mathbf{r}_i(\mathbf{A})\mathbf{B}$:

$$\mathbf{A}_{n \times k}\mathbf{B}_{k \times m} = \begin{bmatrix} \mathbf{r}_1(\mathbf{A})\mathbf{B} \\ \mathbf{r}_2(\mathbf{A})\mathbf{B} \\ \vdots \\ \mathbf{r}_n(\mathbf{A})\mathbf{B} \end{bmatrix}$$

Does this give the same result as the dot-product description in Definition 1.10? Explain.

211. Write a proof of Proposition 1.16. [*Hint: Use the dot-product description of matrix multiplication, along with distributivity of the dot-product over vector addition (Corollary 1.30).*]

2. Special Matrices

2.1. The transpose. The rows of an $n \times m$ matrix $\mathbf{A}$ lie in $\mathbf{R}^m$. When we take these same rows and write them vertically as *columns*, we get a new matrix called the *transpose* of $\mathbf{A}$. The transpose of an $n \times m$ matrix is $m \times n$.

DEFINITION 2.2. The **transpose** of an $n \times m$ matrix $\mathbf{A}$, denoted $\mathbf{A}^T$, is the matrix whose ith row is the ith *column* of $\mathbf{A}$. It follows that the jth *column* of $\mathbf{A}^T$ is also the jth *row* of $\mathbf{A}$. In symbols,

$$\mathbf{r}_i\left(\mathbf{A}^T\right) = \mathbf{c}_i\left(\mathbf{A}\right) \quad \text{and} \quad \mathbf{c}_j\left(\mathbf{A}^T\right) = \mathbf{r}_j\left(\mathbf{A}\right)$$

If we write a_{ij} and a'_{ij} for the entries in row i, column j of $\mathbf{A}$ and $\mathbf{A}^T$ respectively, the definition of $\mathbf{A}^T$ says that the entry in row i, column j of $\mathbf{A}$ turns up in row j column i of $\mathbf{A}^T$.

Symbolically,

(28) $a'_{ij} = a_{ji}$

for all $i = 1, 2, \ldots, n$ and $j = 1, 2, \ldots, m$.   □

Visually, we get $\mathbf{A}^T$ by "reflecting" $\mathbf{A}$ across its diagonal. Here are some examples.

EXAMPLE 2.3. The rows of any matrix form the columns of its transpose, which can also be constructed by reflecting the original matrix across the diagonal that heads southeast from the upper-left corner. Verify that by inspection for each pair below:

$$\mathbf{A} = \begin{bmatrix} \mathbf{1} & 2 & 3 \\ 4 & \mathbf{5} & 6 \end{bmatrix} \qquad \mathbf{A}^T = \begin{bmatrix} \mathbf{1} & 4 \\ 2 & \mathbf{5} \\ 3 & 6 \end{bmatrix}$$

$$\mathbf{B} = \begin{bmatrix} \mathbf{1} & 2 & 3 \\ 0 & \mathbf{5} & 6 \\ 0 & 0 & \mathbf{7} \end{bmatrix} \qquad \mathbf{B}^T = \begin{bmatrix} \mathbf{1} & 0 & 0 \\ 2 & \mathbf{5} & 0 \\ 3 & 6 & \mathbf{7} \end{bmatrix}$$

$$\mathbf{V} = \begin{bmatrix} 1 \\ 0 \\ -2 \\ 0 \\ 1 \end{bmatrix} \qquad \mathbf{V}^T = \begin{bmatrix} 1 & 0 & -2 & 0 & 1 \end{bmatrix}$$

□

Note that the *diagonal* ($i = j$) entries of a matrix—the a_{ii}'s (marked with bold in the examples above)—never change when we transpose. But certain matrices don't change at all—or change only the signs of entries—when transposed:

DEFINITION 2.4. A matrix $\mathbf{A}$ is **symmetric** if it *equals* its transpose: $\mathbf{A}^T = \mathbf{A}$. If, contrastingly, transposition changes the *signs* (but not the absolute values) of all entries, so that $\mathbf{A}^T = -\mathbf{A}$, we say that $\mathbf{A}$ is **skew-symmetric**.   □

Note that both symmetric and skew-symmetric matrices are necessarily always square.

EXAMPLE 2.5. Here are two *symmetric* matrices. Verify by inspection that transposition does not change them: $\mathbf{S}^T = \mathbf{S}$ and $\mathbf{Q}^T = \mathbf{Q}$.

$$\mathbf{S} = \begin{bmatrix} 1 & 2 \\ 2 & 3 \end{bmatrix} \qquad \mathbf{Q} = \begin{bmatrix} 1 & 2 & -3 \\ 2 & 0 & -2 \\ -3 & -2 & 1 \end{bmatrix}$$

Below we exhibit two *skew-symmetric* matrices. The reader should verify that transposition changes the sign of every entry, so that $\mathbf{X}^T = -\mathbf{X}$ and $\mathbf{Y}^T = -\mathbf{Y}$.

$$\mathbf{X} = \begin{bmatrix} 0 & 2 \\ -2 & 0 \end{bmatrix} \qquad \mathbf{Y} = \begin{bmatrix} 0 & 2 & -3 \\ -2 & 0 & -2 \\ 3 & 2 & 0 \end{bmatrix}$$

$\square$

Two key properties of the transpose are easy to prove:

PROPOSITION 2.6. *Transposition distributes over addition, and done twice, returns any matrix to itself. In symbols, for any square matrices* $\mathbf{A}$ *and* $\mathbf{B}$, *we have*

$$(\mathbf{A} + \mathbf{B})^T = \mathbf{A}^T + \mathbf{B}^T \quad and \quad \left(\mathbf{A}^T\right)^T = \mathbf{A}$$

PROOF. Compute the general i, j entry on both sides of each identity and check that they agree. We leave this as an exercise. $\square$

Transposition does *not* "distribute" over matrix multiplication like it does over addition, however—at least, not quite. There's a twist. To see this, consider the matrices $\mathbf{A}_{2\times 3}$ and $\mathbf{B}_{3\times 3}$ of Example 2.3 above. We can form the product $\mathbf{AB}$ in the usual way:

$$\mathbf{AB} = \begin{bmatrix} 1 & 2 & 3 \\ 4 & 5 & 6 \end{bmatrix} \begin{bmatrix} 1 & 2 & 3 \\ 0 & 5 & 6 \\ 0 & 0 & 7 \end{bmatrix} = \begin{bmatrix} 1 & 12 & 36 \\ 4 & 33 & 84 \end{bmatrix}$$

The product of *transposes*, namely $\mathbf{A}^T\mathbf{B}^T$, is not even defined, however, since $\mathbf{A}^T$ has 2 columns, while $\mathbf{B}^T$ has 3 rows. On the other hand, if we reverse the order, we *can* multiply:

$$\mathbf{B}^T\mathbf{A}^T = \begin{bmatrix} 1 & 0 & 0 \\ 2 & 5 & 0 \\ 3 & 6 & 7 \end{bmatrix} \begin{bmatrix} 1 & 4 \\ 2 & 5 \\ 3 & 6 \end{bmatrix} = \begin{bmatrix} 1 & 4 \\ 12 & 33 \\ 36 & 84 \end{bmatrix}$$

Notice that $\mathbf{AB}$ and $\mathbf{B}^T\mathbf{A}^T$ above turn out to be transposes of each other. This is no accident:

PROPOSITION 2.7. *Whenever* $\mathbf{AB}$ *is defined, we have*

$$(\mathbf{AB})^T = \mathbf{B}^T\mathbf{A}^T$$

*In words, the transpose of a product is the product of the transposes **in reverse order**.*

PROOF. This follows easily from the dot-product formula for matrix multiplication. Indeed, let c_{ij} and $\tilde{c}_{ij}$ denote the entries in row i and column j of $(\mathbf{AB})^T$ and $\mathbf{B}^T\mathbf{A}^T$, respectively. We want to show that $c_{ij} = \tilde{c}_{ij}$.

To do so, recall that by definition of transpose, c_{ij} is the j, i entry of $\mathbf{AB}$, which we can compute by dotting row j of $\mathbf{A}$ with column i of $\mathbf{B}$:

$$c_{ij} = \mathbf{r}_j(\mathbf{A}) \cdot \mathbf{c}_i(\mathbf{B})$$

Compare this with $\tilde{c}_{ij}$. The dot-product formula for entries of $\mathbf{B}^T\mathbf{A}^T$ computes it as

$$\tilde{c}_{ij} = \mathbf{r}_i\left(\mathbf{B}^T\right) \cdot \mathbf{c}_j\left(\mathbf{A}^T\right)$$

But $\mathbf{r}_i(\mathbf{B}^T) = \mathbf{c}_i(\mathbf{B})$, by definition of the transpose, and likewise, $\mathbf{c}_j(\mathbf{A}^T) = \mathbf{r}_j(\mathbf{A})$. Substitute these into the formula above for $\tilde{c}_{ij}$ and we immediately see (since dot products commute) that $\tilde{c}_{ij} = c_{ij}$ as hoped. $\qquad\square$

As a special case of this Proposition, we get a fundamental fact about the relationship between matrix/vector multiplication (which we know represents linear transformation) and dot products:

COROLLARY 2.8. *Suppose* $\mathbf{A}$ *is an* $n \times m$ *matrix. Then for any* $\mathbf{x} \in \mathbf{R}^m$ *and* $\mathbf{y} \in \mathbf{R}^n$, *we have*

$$\mathbf{Ax} \cdot \mathbf{y} = \mathbf{x} \cdot \mathbf{A}^T\mathbf{y}$$

In other words, when we have a matrix/vector product on one side of a dot product, we can always move the matrix onto the other factor, provided *we first transpose it.*

Note too that the dot product on the left takes place in $\mathbf{R}^n$, while the one on the right happens in $\mathbf{R}^m$.

PROOF. To prove the Corollary, we interpret $\mathbf{x}$ and $\mathbf{y}$ as $m \times 1$ and $n \times 1$ matrices $[\mathbf{x}]$ and $[\mathbf{y}]$. The dot products can then be interpreted as matrix products. For example, we have

$$\mathbf{Ax} \cdot \mathbf{y} = [\mathbf{y}]^T\mathbf{A}[\mathbf{x}]$$

The desired formula then follows immediately when we transpose both sides of this identity and apply the Proposition. We leave the details as an exercise. □

2.9. Triangular and diagonal matrices. Consider an arbitrary *square* matrix $\mathbf{A} = [a_{ij}]_{n \times n}$. The entries of $\mathbf{A}$ whose row and column indices are *equal*, like a_{11}, a_{22}, etc., are called **diagonal** entries, because they lie along the main **diagonal** joining the upper-leftmost entry a_{11} to the lower-rightmost entry a_{nn}. For instance, the diagonal entries of the following matrix, a_{11}, a_{22}, and a_{33}, are in bold:

$$\begin{bmatrix} \mathbf{1} & 2 & 3 \\ 4 & \mathbf{5} & 6 \\ 7 & 8 & \mathbf{9} \end{bmatrix}$$

Further,

- When $i > j$ (row index > column index), a_{ij} lies *below* the diagonal.

- When $i < j$ (row index < column index), a_{ij} lies *above* the diagonal.

DEFINITION 2.10. A square matrix is **upper-triangular** when all entries *below* its diagonal *vanish*: $a_{ij} = 0$ whenever $i > j$. For instance, the following matrices are all upper triangular:

$$\begin{bmatrix} 2 & -3 & 4 \\ 0 & 5 & -3 \\ 0 & 0 & 2 \end{bmatrix}, \quad \begin{bmatrix} 1 & 0 & 1 & 0 \\ 0 & 1 & 0 & 1 \\ 0 & 0 & 1 & 0 \\ 0 & 0 & 0 & 1 \end{bmatrix}, \quad \begin{bmatrix} 0 & 0 \\ 0 & 0 \end{bmatrix}$$

Note that the zero matrix on the right *is* upper-triangular, since its entries below the diagonal vanish. The entries on the diagonal and above the diagonal vanish too, but that's irrelevant. All that's required for upper-triangularity is for every entry *below* the diagonal to vanish.

Similarly, a square matrix is **lower triangular** if all entries *above* its diagonal *vanish*. That is, $a_{ij} = 0$ whenever $i < j$, as in these examples:

$$\begin{bmatrix} 2 & 0 & 0 \\ -3 & 5 & 0 \\ 4 & -3 & 2 \end{bmatrix}, \quad \begin{bmatrix} 1 & 0 & 0 & 0 \\ -2 & 1 & 0 & 0 \\ 1 & -2 & 1 & 0 \\ 3 & 1 & -2 & 1 \end{bmatrix}, \quad \begin{bmatrix} 2 & 0 \\ 0 & 3 \end{bmatrix}$$

The matrix on the right is lower triangular because its sole entry *above* the diagonal vanishes. Note that this matrix is also *upper* triangular, since the entry below the diagonal vanishes too.

We say that $\mathbf{A}_{n \times n}$ is **triangular** if it is *either* upper *or* lower triangular. If $A_{n \times n}$ is *both* upper *and* lower triangular, we call it a **diagonal** matrix. In a diagonal matrix, all entries above *and* below the diagonal vanish. The diagonal entries themselves can be anything—even zeros. Each of the following matrices, for instance, is diagonal:

$$\begin{bmatrix} 2 & 0 \\ 0 & -3 \end{bmatrix}, \quad \begin{bmatrix} 9 & 0 & 0 \\ 0 & 1 & 0 \\ 0 & 0 & 2 \end{bmatrix}, \quad \begin{bmatrix} 4 & 0 & 0 & 0 \\ 0 & 0 & 0 & 0 \\ 0 & 0 & 4 & 0 \\ 0 & 0 & 0 & 2 \end{bmatrix}$$

$\square$

The Venn diagram in Figure 1 depicts the relationships between upper-triangular, lower triangular, and diagonal matrices.

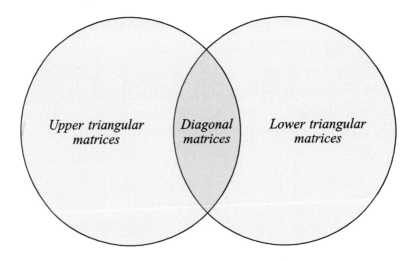

Figure 1. Triangular matrices. The overlapping sets of upper and lower triangular matrices together form the universe of triangular matrices. Matrices in the intersection of these two sets are diagonal.

Triangular and diagonal matrices play key roles in the algebra of matrices because they form "subalgebras." By this we mean that sums and products of upper-triangular matrices remain upper-triangular, and similarly for lower triangular and diagonal matrices. The subalgebra of diagonal matrices is even *commutative*: $\mathbf{AB} = \mathbf{BA}$ when

A and **B** are diagonal. We only scratch the surface of these concepts here, mainly through the exercises at the end of this section.

2.11. Identity matrix. For any positive integer n, the $n \times n$ diagonal matrix whose diagonal entries are all 1's, namely

$$\mathbf{I}_n := \begin{bmatrix} 1 & 0 & 0 & \cdots & 0 \\ 0 & 1 & 0 & \cdots & 0 \\ 0 & 0 & 1 & \cdots & 0 \\ & & \vdots & & \vdots \\ 0 & 0 & 0 & \cdots & 1 \end{bmatrix}$$

is called the $n \times n$ **identity matrix**. It is the *multiplicative identity* for matrix algebra, meaning that *multiplication by* $\mathbf{I}_n$ *from either side leaves a matrix unchanged.* In other words, for any matrix $\mathbf{A} = \mathbf{A}_{n \times m}$, we have

$$\mathbf{I}_n \, \mathbf{A} = \mathbf{A} \, \mathbf{I}_m = \mathbf{A}$$

This follows easily from the column-by-column description of matrix multiplication, and it shows that the identity matrix $\mathbf{I}_n$ plays the same multiplicative role for matrices that 1 plays for scalars.

There is also an *additive* identity matrix: the **zero matrix** $\mathbf{0}_{n \times m}$. It is populated entirely by zeros, and clearly,

$$\mathbf{0} + \mathbf{A} = \mathbf{A} + \mathbf{0} = \mathbf{A}$$

for any $n \times m$ matrix **A**.

– **Practice** –

212. Compute the transposes of each matrix below. Which are symmetric? Skew-symmetric?

$$\mathbf{A} = \begin{bmatrix} 1 & 1 \\ 2 & 5 \end{bmatrix} \qquad \mathbf{B} = \begin{bmatrix} 1 & 2 \\ 2 & 5 \end{bmatrix}$$

$$\mathbf{C} = \begin{bmatrix} 1 & 2 \\ 3 & 4 \\ 5 & 6 \end{bmatrix} \qquad \mathbf{D} = \begin{bmatrix} 1 & 2 & 3 \\ 4 & 5 & 6 \end{bmatrix}$$

$$\mathbf{E} = \begin{bmatrix} 0 & 1 & 2 \\ 1 & 3 & 1 \\ 2 & 1 & 0 \end{bmatrix} \qquad \mathbf{F} = \begin{bmatrix} 0 & 1 & -2 \\ -1 & 0 & 1 \\ 2 & -1 & 0 \end{bmatrix}$$

213. What matrix $\mathbf{A}$ represents the linear map $\mathbf{R}^4 \to \mathbf{R}^7$ given by

$$P(x_1, x_2, x_3, x_4) = (x_1, x_2, x_3, x_4, 0, 0, 0)$$

Give a similar formula for the linear map represented by $\mathbf{A}^T$. Which of these maps (if either) is onto? Is either one-to-one?

214. Explain why

    a) Symmetric and skew-symmetric matrices are always square.

    b) The *diagonal* entries of a skew-symmetric must always vanish.

215. With reference to the matrices in Exercise 212 above,

    a) Compute $(\mathbf{AD})^T$, $\mathbf{A}^T\mathbf{D}^T$, $(\mathbf{DA})^T$, and $\mathbf{D}^T\mathbf{A}^T$. Which pairs are the same? Compare Proposition 2.7.

    b) Compute $\mathbf{A} + \mathbf{A}^T$ and $\mathbf{A} - \mathbf{A}^T$. Is either symmetric or skew-symmetric? Compare statement (a) in Exercise 214 below.

    c) Compute $\mathbf{C}^T\mathbf{C}$ and $\mathbf{DD}^T$. Is either one symmetric or skew-symmetric? Compare the statement of Exercise 216 below.

216. Show that

    a) If $\mathbf{A}$ is any square matrix, then $\mathbf{A} + \mathbf{A}^T$ is symmetric, while $\mathbf{A} - \mathbf{A}^T$ is skew-symmetric.

    b) That *every* square matrix can be written as the sum of a symmetric matrix and a skew-symmetric matrix.

217. Prove: *If $\mathbf{A}$ is an $n \times m$ matrix, then $\mathbf{A}^T\mathbf{A}$ and $\mathbf{A}\mathbf{A}^T$ are always defined, and are both **symmetric**.*

218. Arguing that corresponding entries are equal, write a proof of Proposition 2.6.

219. Verify that Corollary 2.8 holds with

    a) $\mathbf{x} = (2, -7, 1)$, $\mathbf{y} = (3, -2)$, and

$$\mathbf{A} = \begin{bmatrix} 2 & -3 & 5 \\ 5 & 0 & -2 \end{bmatrix}$$

    b) $\mathbf{x} = (1, -1)$, $\mathbf{y} = (2, -1, 3, -2)$, and

$$\mathbf{A} = \begin{bmatrix} 1 & 3 \\ 2 & -4 \\ 4 & -1 \\ 0 & 5 \end{bmatrix}$$

220. [*Hint: Review Observation 1.12 in Chapter 1 before doing this exercise.*]

    a) Verify that for any $\mathbf{x} = (x, y) \in \mathbf{R}^2$, we have $\mathbf{I}_2\mathbf{x} = \mathbf{x}$, and that for any $\mathbf{x} = (x, y, z) \in \mathbf{R}^3$, we have $\mathbf{I}_3\mathbf{x} = \mathbf{x}$.

    b) Use the column description of matrix/vector multiplication to show that $\mathbf{I}_n\mathbf{x} = \mathbf{x}$ for any $\mathbf{x} = (x_1, x_2, \ldots, x_n) \in \mathbf{R}^n$.

221. A familiar and important algebraic property of scalars says this:

$$\text{If } ab = 0, \text{ then either } a = 0 \text{ or } b = 0$$

The corresponding fact is **false** for matrices!

   a) Find a pair of 2-by-2 matrices $\mathbf{A}$ and $\mathbf{B}$ such that $\mathbf{AB} = \mathbf{0}$, but $\mathbf{A} \neq \mathbf{0}$ and $\mathbf{B} \neq \mathbf{0}$. (Your matrices may have some zero entries—but each must have at least *one* non-zero entry.)

   b) Do this without putting *any* zeros in either matrix! [*Hint: Can you make the rows of $\mathbf{A}$ orthogonal to the columns of $\mathbf{B}$?*]

222. What form do the powers $\mathbf{A}^2, \mathbf{A}^3$, etc. take when $\mathbf{A}$ is a diagonal matrix? (Experiment with the 2-by-2 and 3-by-3 cases first to get a sense of what to expect.)

223. Suppose $\mathbf{A}$ and $\mathbf{B}$ are both diagonal matrices.

   a) Show that the product $\mathbf{AB}$ will again be diagonal. Is it the same as $\mathbf{BA}$?

   b) Matrices don't *have* to be diagonal to commute. Find a non-diagonal 2×2 matrix that commutes with

$$\begin{bmatrix} 1 & 2 \\ 2 & 1 \end{bmatrix}$$

224. Suppose $\mathbf{A}$ and $\mathbf{B}$ are both triangular matrices.

   a) Show that in the 2×2 case, $\mathbf{AB}$ will be upper-triangular if both $\mathbf{A}$ and $\mathbf{B}$ are. Do this directly by computing the product of

$$\mathbf{A} = \begin{bmatrix} a_{11} & a_{12} \\ 0 & a_{22} \end{bmatrix} \quad \text{and} \quad \mathbf{B} = \begin{bmatrix} b_{11} & b_{12} \\ 0 & b_{22} \end{bmatrix}$$

   b) Show that $\mathbf{AB}$ will be upper-triangular if both $\mathbf{A}_{n \times n}$ and $\mathbf{B}_{n \times n}$ are upper-triangular for any n. (Since the i, j entry of $\mathbf{AB}$ is $\mathbf{r}_i(\mathbf{A}) \cdot \mathbf{c}_j(\mathbf{B}) = 0$, you need to argue that $\mathbf{r}_i(\mathbf{A}) \cdot \mathbf{c}_j(\mathbf{B}) = 0$ when $i > j$.)

   c) Is the product of 2×2 *lower* triangular matrices again *lower* triangular? How about in the $n \times n$ case? (You can deduce these answers easily from (b) above by taking transposes.)

225. Compute all powers $\mathbf{A}$, $\mathbf{A}^2$, $\mathbf{A}^3$, etc., for the matrices

$$\text{a) } \mathbf{A} = \begin{bmatrix} 0 & 1 & 0 \\ 0 & 0 & 1 \\ 0 & 0 & 0 \end{bmatrix} \qquad\qquad \text{b) } \mathbf{A} = \begin{bmatrix} 1 & 1 & 0 \\ 0 & 1 & 1 \\ 0 & 0 & 1 \end{bmatrix}$$

$$-\star-$$

3. Matrix Inversion

Recall that scalars a and b with $ab = 1$ are called *reciprocals*. In that case we also call them *multiplicative inverses*. Some *matrices* have multiplicative inverses too:

DEFINITION 3.1. (Matrix Inverse) When $\mathbf{A}$ and $\mathbf{B}$ are matrices with $\mathbf{AB} = \mathbf{BA} = \mathbf{I}_n$ for some n, we say that $\mathbf{B}$ is the **inverse** of $\mathbf{A}$, written $\mathbf{B} = \mathbf{A}^{-1}$. Likewise, $\mathbf{A}$ is the inverse of $\mathbf{B}$ (i.e., $\mathbf{A} = \mathbf{B}^{-1}$).

By an easy exercise, the inverse condition $\mathbf{AB} = \mathbf{BA} = \mathbf{I}_n$ forces both $\mathbf{A}$ and $\mathbf{B}$ to be $n \times n$. In particular, **only square matrices have inverses**.   □

This definition actually assumes a bit more than it needs to: it would have sufficed to require just $\mathbf{AB} = \mathbf{I}_n$ (or just $\mathbf{BA} = \mathbf{I}_n$), since each of these "one-sided" facts implies the other, thanks to the following:

PROPOSITION 3.2. *For* **square** *matrices,* $\mathbf{AB} = \mathbf{I}_n$ *forces* $\mathbf{BA} = \mathbf{I}_n$ *and vice-versa.*

Rather than interrupt the flow of ideas here, we leave the proof of this fact to the end of this section.

EXAMPLE 3.3. Consider the 2×2 matrices

$$\mathbf{A} = \begin{bmatrix} 3 & 4 \\ 5 & 7 \end{bmatrix} \quad \text{and} \quad \mathbf{B} = \begin{bmatrix} 7 & -4 \\ -5 & 3 \end{bmatrix}$$

These are inverse to each other, since $\mathbf{AB} = \mathbf{BA} = \mathbf{I}_2$. Indeed, we may compute

$$\mathbf{BA} \;=\; \begin{bmatrix} 7 & -4 \\ -5 & 3 \end{bmatrix} \begin{bmatrix} 3 & 4 \\ 5 & 7 \end{bmatrix}$$

$$=\; \begin{bmatrix} 7 \cdot 3 - 4 \cdot 5 & 7 \cdot 4 - 4 \cdot 7 \\ -3 \cdot 5 + 3 \cdot 5 & -5 \cdot 4 + 3 \cdot 7 \end{bmatrix}$$

$$=\; \begin{bmatrix} 21 - 20 & 28 - 28 \\ -15 + 15 & -20 + 21 \end{bmatrix} \;=\; \begin{bmatrix} 1 & 0 \\ 0 & 1 \end{bmatrix}$$

The reader should check that $\mathbf{AB} = \mathbf{I}_2$ as well. So we can express the relationship between these two matrices by writing

$$\begin{bmatrix} 3 & 4 \\ 5 & 7 \end{bmatrix}^{-1} = \begin{bmatrix} 7 & -4 \\ -5 & 3 \end{bmatrix}$$

or

$$\begin{bmatrix} 3 & 4 \\ 5 & 7 \end{bmatrix} = \begin{bmatrix} 7 & -4 \\ -5 & 3 \end{bmatrix}^{-1}$$

□

All scalars except zero have reciprocals. That's *not* true for matrices: *many square matrices have no inverse.* For instance,

$$\mathbf{A} = \begin{bmatrix} 1 & 2 \\ 2 & 4 \end{bmatrix}$$

has no inverse. Indeed, the second column of $\mathbf{BA}$ will always be twice the first column, no matter what $\mathbf{B}$ is. Can you see why? [*Hint: Matrix multiplication operates column-wise.*] Since the second column of $\mathbf{I}_2$ is **not** twice its first, this makes it impossible to arrange $\mathbf{BA} = \mathbf{I}_2$.

DEFINITION 3.4. A square matrix that has no inverse is called **non-invertible** or **singular**.   □

The question as to which square matrices are invertible and which are not is crucial, and we shall learn several different ways to answer it. Like non-commutativity, the non-invertibility of many square matrices makes matrix algebra much subtler than that of scalars. Still, matrices are like scalars in some respects. For instance:

PROPOSITION 3.5. *A square matrix has at most one inverse.*

PROOF. Suppose $\mathbf{A}_{n \times n}$ had two inverses, $\mathbf{B}$ and $\mathbf{B}'$. Then $\mathbf{AB} - \mathbf{AB}' = \mathbf{I}_n - \mathbf{I}_n = \mathbf{0}_{n \times n}$. The distributive law then yields $\mathbf{A}(\mathbf{B} - \mathbf{B}') = \mathbf{0}_{n \times n}$. Multiply both sides of this equation—on the left—by $\mathbf{B}$ to get

$$\mathbf{BA}(\mathbf{B} - \mathbf{B}') = \mathbf{B}\,\mathbf{0}$$

Since $\mathbf{BA} = \mathbf{I}_n$, and $\mathbf{B}\,\mathbf{0} = \mathbf{0}$, the equation now reduces to

$$\mathbf{B} - \mathbf{B}' = \mathbf{0}$$

which forces $\mathbf{B} = \mathbf{B}'$. The two supposed inverses are thus equal. □

3.6. Three questions. Three questions should now be forming in the reader's mind:

1) What are inverses good for?

2) How can we tell if a matrix has an inverse?

3) If a matrix has an inverse, how can we find it?

As mentioned above, there are several ways to answer Question (2)— that is, to test a matrix for invertibility. In the end, most of them rely on the Gauss-Jordan algorithm. In fact, we'll show below how to settle both (2) and (3) very nicely using Gauss-Jordan.

Question (1) is broader. What are inverses good for? We should start by saying that invertibility (or the lack of it) will play a big role in several of our future explorations. Just knowing whether a matrix is invertible or not (i.e., answering Question (2) above) is often as useful as knowing what the inverse actually is, in answer to Question (3)).

Nevertheless, we can start to make the case for the importance of inverses by considering the simple situation of a linear system with square coefficient matrix $\mathbf{A}$. Suppose we want to solve

$$\mathbf{Ax} = \mathbf{b}$$

If we know an inverse for $\mathbf{A}$, we can solve the system with a single matrix/vector multiplication. Indeed, when $\mathbf{A}$ is invertible, the system above has exactly one solution, given by $\mathbf{x} = \mathbf{A}^{-1}\mathbf{b}$.

To see this, just start with $\mathbf{Ax} = \mathbf{b}$, multiply on the left by $\mathbf{A}^{-1}$, and simplify:

$$
\begin{aligned}
\mathbf{Ax} &= \mathbf{b} \\
\mathbf{A}^{-1}\mathbf{Ax} &= \mathbf{A}^{-1}\mathbf{b} \\
\mathbf{I}_n\mathbf{x} &= \mathbf{A}^{-1}\mathbf{b} \\
\mathbf{x} &= \mathbf{A}^{-1}\mathbf{b}
\end{aligned}
$$

This calculation proves some simple but important facts:

PROPOSITION 3.7. *If an $n \times n$ matrix $\mathbf{A}$ is invertible, then*

i) *The linear system $\mathbf{Ax} = \mathbf{b}$ has exactly one solution for each $\mathbf{b} \in \mathbf{R}^n$.*

ii) *The linear transformation T represented by $\mathbf{A}$ is both one-to-one and onto.*

PROOF. Conclusion (ii) means that every $\mathbf{b}$ in $\mathbf{R}^n$ has at most one pre-image (T is one-to-one) and at *least* one (T is onto). So Conclusion (ii) is really the same as Conclusion (i), which says that each $\mathbf{b}$ has *exactly* one pre-image. It therefore suffices to prove Conclusion (i).
For that, simply substitute $\mathbf{A}^{-1}\mathbf{b}$ for $\mathbf{x}$, and see that $\mathbf{A}^{-1}\mathbf{b}$ does solve the system. At the same time, by the calculation preceding the Proposition, any $\mathbf{x}$ that solves the system has to equal $\mathbf{A}^{-1}\mathbf{b}$. Thus, $\mathbf{A}^{-1}\mathbf{b}$ is the one and only solution of $\mathbf{Ax} = \mathbf{b}$ when $\mathbf{A}$ is invertible. $\square$

EXAMPLE 3.8. Suppose we want to solve

$$3x + 4y = b_1$$
$$5x + 7y = b_2$$

This system can be abbreviated as $\mathbf{Ax} = \mathbf{b}$, where

$$\mathbf{A} = \begin{bmatrix} 3 & 4 \\ 5 & 7 \end{bmatrix}, \quad \text{and} \quad \mathbf{b} = \begin{bmatrix} b_1 \\ b_2 \end{bmatrix}$$

We saw in Example 3.3 that the inverse of this matrix $\mathbf{A}$ is

$$\mathbf{A}^{-1} = \begin{bmatrix} 7 & -4 \\ -5 & 3 \end{bmatrix}$$

So for **any** $(b_1, b_2) \in \mathbf{R}^2$, the unique solution to $\mathbf{Ax} = \mathbf{b}$ is

$$\mathbf{A}^{-1}\mathbf{b} = \begin{bmatrix} 7 & -4 \\ -5 & 3 \end{bmatrix} \begin{bmatrix} b_1 \\ b_2 \end{bmatrix}$$
$$= \begin{bmatrix} 7b_1 - 4b_2 \\ -5b_1 + 3b_2 \end{bmatrix}$$

The inverse here gives a **general formula** for the solution of the system, no matter what right-hand side $\mathbf{b}$ we target. $\square$

3.9. The inverse of a product. When $n \times n$ matrices $\mathbf{A}$ and $\mathbf{B}$ are *both* invertible, so is the product $\mathbf{AB}$, and as we might hope, the inverse of $\mathbf{AB}$ is the product of the inverses—*but in reverse order.* The *inverse* of the product thus exhibits the same order-reversal that we encountered for the *transpose* of a product:

PROPOSITION 3.10. *If $n \times n$ matrices $\mathbf{A}$ and $\mathbf{B}$ are invertible, then $\mathbf{AB}$ is invertible too, with*

$$(\mathbf{AB})^{-1} = \mathbf{B}^{-1}\mathbf{A}^{-1}$$

PROOF. If $\mathbf{A}$ and $\mathbf{B}$ are both invertible, and we multiply $\mathbf{AB}$ on the right by $\mathbf{B}^{-1}\mathbf{A}^{-1}$, the associativity of matrix multiplication lets us deduce

$$(\mathbf{AB})(\mathbf{B}^{-1}\mathbf{A}^{-1}) = \mathbf{A}(\mathbf{BB}^{-1})\mathbf{A}^{-1} = \mathbf{A}\mathbf{I}_n\mathbf{A}^{-1} = \mathbf{AA}^{-1} = \mathbf{I}_n$$

In exactly the same way (or by Proposition 3.2) we likewise get

$$(\mathbf{B}^{-1}\mathbf{A}^{-1})(\mathbf{AB}) = \mathbf{I}_n$$

So $\mathbf{AB}$ is invertible, with inverse given by $\mathbf{B}^{-1}\mathbf{A}^{-1}$. $\quad\square$

We conclude this section by fulfilling our promise (stated as Proposition 3.2) to prove that a matrix that inverts $\mathbf{A}$ from the right automatically inverts it from the left as well.

PROOF OF PROPOSITION 3.2. Suppose $\mathbf{A}$ and $\mathbf{B}$ are $n \times n$ matrices with $\mathbf{AB} = \mathbf{I}$. Then for any $\mathbf{b} \in \mathbf{R}^n$, we have

$$(29) \qquad\qquad \mathbf{ABb} = \mathbf{Ib} = \mathbf{b}$$

This shows we can solve $\mathbf{Ax} = \mathbf{b}$ for any target $\mathbf{b} \in \mathbf{R}^n$—just take $\mathbf{x} = \mathbf{Bb}$. The linear transformation T represented by $\mathbf{A}$ is therefore *onto*, and by Corollary 6.12 of Chapter 2, that makes it one-to-one as well. In particular, T never sends different inputs to the same output. With this in mind, take any $\mathbf{x} \in \mathbf{R}^n$, and replace $\mathbf{b}$ by $\mathbf{Ax}$ in (29), to get

$$\mathbf{ABAx} = \mathbf{Ax}$$

Since $\mathbf{A}$ represents T, we can see this as $T(\mathbf{BAx}) = T(\mathbf{x})$. In other words, T maps $\mathbf{BAx}$ and $\mathbf{x}$ to the same output. But we just deduced that T is one-to-one; it *never* sends different outputs to the same input. So these inputs must be the same: we must have $\mathbf{BAx} = \mathbf{x}$ for every $\mathbf{x}$, which forces $\mathbf{BA} = \mathbf{I}$.

In short, $\mathbf{AB} = \mathbf{I}$ implies $\mathbf{BA} = \mathbf{I}$, as hoped. $\quad\square$

3.11. The Inversion Algorithm. We now tackle questions (2) and (3) from Section 3.6 above: How do we tell if a square matrix has an inverse? When it does, how do we find it?

One algorithm answers both these questions—and it amounts to little more than Gauss-Jordan, applied to a "super-augmented" matrix. We first present the algorithm. After we practice with it, we'll figure out why it works.

Inversion Algorithm. To determine whether $\mathbf{A}_{n \times n}$ is invertible, and to find the inverse if it is, take these steps:

1) Set up the $n \times 2n$ super-augmented matrix $\left[\mathbf{A} \,\middle|\, \mathbf{I}_n\right]$.

2) Row-reduce using Gauss-Jordan until the left half reaches RRE form. The resulting matrix will then look like this:

$$\left[\text{RRE}(\mathbf{A}) \,\middle|\, \mathbf{X}\right]$$

where $\mathbf{X}$ is some square matrix.

3) If $\text{RRE}(\mathbf{A}) = \mathbf{I}_n$, then $\mathbf{A}$ **is** invertible, with $\mathbf{A}^{-1} = \mathbf{X}$ (!)

4) If $\text{RRE}(\mathbf{A}) \neq \mathbf{I}_n$ then $\mathbf{A}$ is **not** invertible. For when we see (at any point during the row-reduction) that $\text{RRE}(\mathbf{A}) \neq \mathbf{I}_n$ (e.g., if the matrix develops a row or column of zeros) we can immediately conclude that $\mathbf{A}$ has no inverse.

EXAMPLE 3.12. Let us find the inverse of

$$\mathbf{A} = \begin{bmatrix} 2 & 5 \\ 3 & 7 \end{bmatrix}$$

According to the instructions above, we should proceed by row-reducing the super-augmented matrix $[\mathbf{A} \,|\, \mathbf{I}_2]$. This goes as follows:

$$\begin{bmatrix} 2 & 5 & | & 1 & 0 \\ 3 & 7 & | & 0 & 1 \end{bmatrix} \xrightarrow{\text{r1/2}} \begin{bmatrix} 1 & 5/2 & | & 1/2 & 0 \\ 3 & 7 & | & 0 & 1 \end{bmatrix}$$

$$\xrightarrow{\text{r2-3r1}} \begin{bmatrix} 1 & 5/2 & | & 1/2 & 0 \\ 0 & -1/2 & | & -3/2 & 1 \end{bmatrix} \xrightarrow{(-2)\cdot\text{r2}} \begin{bmatrix} 1 & 5/2 & | & 1/2 & 0 \\ 0 & 1 & | & 3 & -2 \end{bmatrix}$$

$$\xrightarrow{\text{r1-(5/2)r2}} \begin{bmatrix} 1 & 0 & | & -7 & 5 \\ 0 & 1 & | & 3 & -2 \end{bmatrix}$$

At this point, the left side has reached RRE form—and equals $\mathbf{I}_2$. The original matrix $\mathbf{A}$ is therefore invertible, with its inverse given by the right-hand block:

$$\begin{bmatrix} 2 & 5 \\ 3 & 7 \end{bmatrix}^{-1} = \begin{bmatrix} -7 & 5 \\ 3 & -2 \end{bmatrix}$$

Multiplying this result by $\mathbf{A}$ does gives $\mathbf{I}_2$—our calculation checks out. $\square$

EXAMPLE 3.13. Similarly, we find the inverse of

$$\begin{bmatrix} 1 & 2 & 0 \\ 0 & 1 & 0 \\ 0 & 3 & 1 \end{bmatrix}$$

by augmenting it with $\mathbf{I}_3$ and row-reducing:

$$\left[\begin{array}{ccc|ccc} 1 & 2 & 0 & 1 & 0 & 0 \\ 0 & 1 & 0 & 0 & 1 & 0 \\ 0 & 3 & 1 & 0 & 0 & 1 \end{array}\right] \xrightarrow{\text{r3-3r2}} \left[\begin{array}{ccc|ccc} 1 & 2 & 0 & 1 & 0 & 0 \\ 0 & 1 & 0 & 0 & 1 & 0 \\ 0 & 0 & 1 & 0 & -3 & 1 \end{array}\right]$$

$$\xrightarrow{\text{r1-2r2}} \left[\begin{array}{ccc|ccc} 1 & 0 & 0 & 1 & -2 & 0 \\ 0 & 1 & 0 & 0 & 1 & 0 \\ 0 & 0 & 1 & 0 & -3 & 1 \end{array}\right]$$

Since that left-hand block has again reached the identity matrix, we conclude that the original matrix was indeed invertible, with

$$\begin{bmatrix} 1 & 2 & 0 \\ 0 & 1 & 0 \\ 0 & 3 & 1 \end{bmatrix}^{-1} = \begin{bmatrix} 1 & -2 & 0 \\ 0 & 1 & 0 \\ 0 & -3 & 1 \end{bmatrix}$$

Again, it's easy to check our result: Multiplying the original matrix by the supposed inverse gives $\mathbf{I}_3$, as it should. $\square$

EXAMPLE 3.14. As a final example, we show that the matrix

$$\begin{bmatrix} 1 & -1 & 0 \\ 0 & 1 & -1 \\ -1 & 0 & 1 \end{bmatrix}$$

is *singular*—it has *no* inverse. To do so, we use the same algorithm:

$$\begin{bmatrix} 1 & -1 & 0 & | & 1 & 0 & 0 \\ 0 & 1 & -1 & | & 0 & 1 & 0 \\ -1 & 0 & 1 & | & 0 & 0 & 1 \end{bmatrix} \xrightarrow{\text{r3+r1}} \begin{bmatrix} 1 & -1 & 0 & | & 1 & 0 & 0 \\ 0 & 1 & -1 & | & 0 & 1 & 0 \\ 0 & -1 & 1 & | & 1 & 0 & 1 \end{bmatrix}$$

$$\xrightarrow{\text{r3+r2}} \begin{bmatrix} 1 & -1 & 0 & | & 1 & 0 & 0 \\ 0 & 1 & -1 & | & 0 & 1 & 0 \\ 0 & 0 & 0 & | & 1 & 1 & 1 \end{bmatrix}$$

At this point, the row of zeros at the bottom of the left-hand block clearly tells us that $\mathbf{A}$ will **not** row-reduce to $\mathbf{I}_3$. According to our algorithm, that means **A has no inverse.** $\square$

– Practice –

226. We found a pair of inverse matrices in Example 3.3, namely

$$\mathbf{A} = \begin{bmatrix} 3 & 4 \\ 5 & 7 \end{bmatrix} \quad \text{and} \quad \mathbf{A}^{-1} = \begin{bmatrix} 7 & -4 \\ -5 & 3 \end{bmatrix}$$

a) Show that the following also invert each other:

$$\mathbf{B} = \begin{bmatrix} 1 & 2 \\ 2 & 3 \end{bmatrix} \quad \text{and} \quad \mathbf{B}^{-1} = \begin{bmatrix} -3 & 2 \\ 2 & -1 \end{bmatrix}$$

b) Compute the matrix products $\mathbf{AB}$, $\mathbf{A}^{-1}\mathbf{B}^{-1}$, and $\mathbf{B}^{-1}\mathbf{A}^{-1}$.

c) Show that $(\mathbf{AB})(\mathbf{B}^{-1}\mathbf{A}^{-1}) = \mathbf{I}_2$, while $(\mathbf{AB})(\mathbf{A}^{-1}\mathbf{B}^{-1})$ is very different from $\mathbf{I}_2$.

227. Verify that these pairs of matrices are inverses.

a)
$$\begin{bmatrix} 2 & 1 & 0 \\ 9 & 4 & 0 \\ 0 & 1 & 1 \end{bmatrix} \quad \text{and} \quad \begin{bmatrix} -4 & 1 & 0 \\ 9 & -2 & 0 \\ -9 & 2 & 1 \end{bmatrix}$$

b)
$$\begin{bmatrix} 1 & 0 & 0 \\ 2 & 1 & 0 \\ 3 & 2 & 1 \end{bmatrix} \quad \text{and} \quad \begin{bmatrix} 1 & 0 & 0 \\ -2 & 1 & 0 \\ 1 & -2 & 1 \end{bmatrix}$$

c)
$$\begin{bmatrix} 0 & 0 & 1 & 0 \\ 0 & 0 & 0 & 1 \\ 1 & 0 & 0 & 0 \\ 0 & 1 & 0 & 0 \end{bmatrix} \quad \text{and} \quad \begin{bmatrix} 0 & 0 & 1 & 0 \\ 0 & 0 & 0 & 1 \\ 1 & 0 & 0 & 0 \\ 0 & 1 & 0 & 0 \end{bmatrix}$$

228. Imitate Example 3.8 to find a general formula for the solution of

$$\begin{aligned} x + 2y &= b_1 \\ 2x + 3y &= b_2 \end{aligned}$$

Use the invertibility of the coefficient matrix $\mathbf{B}$, which (along with its inverse) appears in part (a) of Exercise 226 above.

229. If $a, b \neq 0$, what values must c and d have if

$$\begin{bmatrix} c & 0 \\ 0 & d \end{bmatrix} = \begin{bmatrix} a & 0 \\ 0 & b \end{bmatrix}^{-1} ?$$

230. Generalizing from Exercise 229, what diagonal matrix will invert $\mathbf{D}$ below, assuming $a, b,$ and c are all non-zero?

$$\mathbf{D} = \begin{bmatrix} a & 0 & 0 \\ 0 & b & 0 \\ 0 & 0 & c \end{bmatrix}$$

State a reason why $\mathbf{D}$ *cannot* have an inverse if $a = 0$.

231. Suppose $\mathbf{A}$ is invertible. Show that the transpose of $\mathbf{A}^{-1}$ must equal the inverse $\mathbf{A}^T$ (show that $(\mathbf{A}^{-1})^T$ inverts $\mathbf{A}^T$), thus proving that transposes of invertible matrices are invertible. [*Hint: Use Proposition 2.7.*]

232. Determine whether each 2×2 matrix below is invertible. If it is, find its inverse.

a) $\begin{bmatrix} 4 & 3 \\ 2 & 1 \end{bmatrix}$ b) $\begin{bmatrix} 0 & -3 \\ 3 & 0 \end{bmatrix}$ c) $\begin{bmatrix} 1 & \frac{1}{2} \\ \frac{1}{2} & \frac{1}{3} \end{bmatrix}$ d) $\begin{bmatrix} 0 & 1 \\ 1 & 0 \end{bmatrix}$

233. Determine whether each 3×3 matrix below is invertible. If it is, find its inverse.

a) $\begin{bmatrix} 1 & 2 & 0 \\ 1 & 3 & 2 \\ 0 & 1 & 1 \end{bmatrix}$ b) $\begin{bmatrix} 1 & 2 & 0 \\ 1 & 3 & 1 \\ 0 & 1 & 1 \end{bmatrix}$

c) $\begin{bmatrix} 1 & 1 & 1 \\ 1 & 0 & 1 \\ 1 & 1 & 1 \end{bmatrix}$ d) $\begin{bmatrix} 1 & 0 & 1 \\ 1 & 1 & 0 \\ 1 & 1 & 1 \end{bmatrix}$

234. Determine whether each 4×4 matrix below is invertible. If it is, find its inverse.

a) $\begin{bmatrix} 1 & 1 & 1 & 1 \\ 1 & 1 & -1 & -1 \\ 1 & -1 & -1 & 1 \\ 1 & -1 & 1 & -1 \end{bmatrix}$ b) $\begin{bmatrix} 2 & 1 & 1 & 1 \\ 1 & 0 & -1 & -1 \\ 1 & -1 & 0 & 1 \\ 1 & -1 & 1 & 2 \end{bmatrix}$

235. Show that the general 2×2 matrix

$$\mathbf{A} = \begin{bmatrix} a & b \\ c & d \end{bmatrix}$$

is invertible if and only if $ad - bc \neq 0$, in which case $\mathbf{A}^{-1}$ is given by

$$\mathbf{A}^{-1} = \frac{1}{ad - bc} \begin{bmatrix} d & -b \\ -c & a \end{bmatrix}$$

This general formula for the inverse of a 2×2 matrix is worth memorizing.

236. For what values of λ (if any) are the matrices below singular, i.e., **non**-invertible?

a) $\begin{bmatrix} \lambda & 3 \\ 3 & \lambda \end{bmatrix}$ b) $\begin{bmatrix} 1-\lambda & 5 \\ 5 & 1-\lambda \end{bmatrix}$ c) $\begin{bmatrix} 1-\lambda & 5 \\ 0 & 2-\lambda \end{bmatrix}$

d) $\begin{bmatrix} \lambda & 1 \\ -1 & \lambda \end{bmatrix}$

237. Prove these easy properties of the matrix inverse.

a) $(\mathbf{A}^{-1})^{-1} = \mathbf{A}$

b) If $\mathbf{A}$ is invertible, then so is $k\mathbf{A}$ for any scalar $k \neq 0$, with
$$(k\mathbf{A})^{-1} = \frac{1}{k}\mathbf{A}^{-1}$$

c) If $\mathbf{A}$ is invertible, so is $\mathbf{A}^p$ for any positive integer p, with
$$(\mathbf{A}^p)^{-1} = (\mathbf{A}^{-1})^p$$

238. Proposition 3.10 tells us that the product of invertible $n \times n$ matrices is again invertible. Is the same true for sums? That is, does invertibility of $\mathbf{A}$ and $\mathbf{B}$ imply that of $\mathbf{A}+\mathbf{B}$? Prove it does, or show (by counterexample) that it does not.

239. If an upper-triangular matrix is invertible, its inverse will also be upper-triangular. Can you explain this as a consequence of the inversion algorithm?

— ★ —

4. A Logical Digression

We now want to *understand* the inversion algorithm—to see why it works. For this, and increasingly as we go forward, we will use mathematical *reasoning* at least as much as *calculation*. To prepare for that, we digress to briefly discuss some basic facts about logic.

Specifically, we present the four fundamental templates for the relationship between a pair of mathematical claims, or *assertions* P and Q. These four templates are known as **statement**, **converse**, **obverse**, and **contrapositive**.

4.1. Statements. Every mathematical theorem makes a statement about some set of objects, where by *statement*, we specifically mean an assertion of the form

If P then Q

This is the same as saying P *implies* Q. Either way, we express it symbolically like this:

$$P \Rightarrow Q$$

We call P the **hypothesis**, and Q the **conclusion** of the statement. Here is a simple example:

Let X be a quadrilateral in the plane. If X is a square, then X is a rectangle.

The opening phrase here, *Let X be a plane quadrilateral*, just specifies the set of objects under discussion: quadrilaterals in the plane. Then comes the statement: *If P then Q*, where P and Q are these assertions:

P : *X is a square*

Q : *X is a rectangle*

In this simple example, the classic *"Let... If... then"* structure is clear. Every theorem has this same structure, even when it isn't so obvious. For example, we could have stated the very same theorem about quadrilaterals more succinctly as follows:

All squares are rectangles.

This version doesn't bother to say that it concerns quadrilaterals—it assumes the reader will realize that. It doesn't use the words *if* or *then* either, but it still means the same thing: *if X is a square, then X is a rectangle.*

Every mathematical theorem, proposition, corollary, and lemma can be given this same *"Let... If... then"* structure.

We can illustrate this structure with a diagram that depicts the set we're considering as a rectangular "universe" containing two circular subsets: the subset of objects for which P is true, and the subset for which Q holds (Figure 2).

The statement above about squares and rectangles happens to be true, so we can call it a theorem. Of course, we may easily construct *false* statements too. Consider this statement about the set of polygons:

If a polygon X has equal sides, then X is a rectangle.

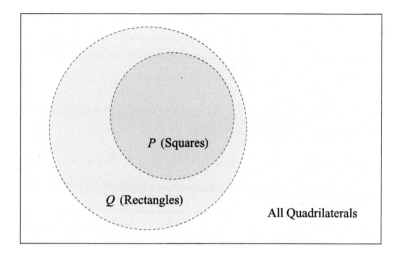

Figure 2. Typical true statement (theorem). The set
for which Q is true *contains* the set for which P is true,
which makes $P \Rightarrow Q$ true: *If* a quadrilateral lies in P, *then*
it certainly lies in the larger set Q.

This is false—but it's still a perfectly well-formed statement. It has a
perfectly good hypothesis P—that X has equal sides—and a perfectly
good conclusion Q: that X is a rectangle. It just happens to be false.
Figure 3 illustrates.

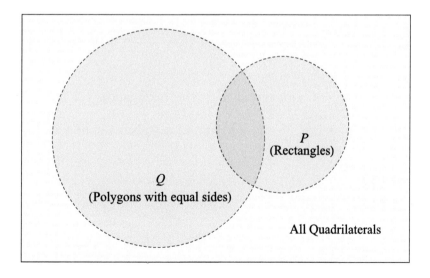

Figure 3. False statement. The set Q overlaps, but does
not *contain* P, so $P \not\Rightarrow Q$. Some polygons with equal sides
are rectangles, but some are not.

4.2. Converse. When we swap the order of hypothesis and conclusion to form the reverse implication

$$\textit{If } Q \textit{ then } P$$

we get a new statement called the **converse** of the original. The crucial thing to realize about the converse is this:

The converse is independent of the statement.

That is: *The converse of a true statement can be true or false—neither is automatically the case.*

EXAMPLE 4.3. The square/rectangle statement above is a great example to remember, because it shows that *the converse can be false even though the statement is true*:

    Statement: *If X is a square, then X is a rectangle.* (True)
    Converse: *If X is a rectangle, then X is a square.* (False)

                                                  □

The converse of a true statement doesn't *have* to be false, however. For instance, consider the following variant of the example above:

EXAMPLE 4.4. Suppose we have assertions P and Q as follows:

    $P:$ *X is a square.*
    $Q:$ *X is a rectangle with four equal sides.*

With these values for P and Q, *both* the statement and its converse are true, as follows.

Statement:
    If X is a square, then X is a rectangle with four equal sides.

Converse:
    If X is a rectangle with four equal sides, then X is a square.

Both the statement and converse here are clearly true.   □

4.5. Equivalence. The converse of a true statement isn't *necessarily* true, as in Example 4.3, but a statement and its converse *may* both hold, as in Example 4.4. In that situation, we have both $P \Rightarrow Q$ *and* $Q \Rightarrow P$, and we say that the P and Q are **equivalent** assertions, written

$$P \iff Q$$

When we have equivalent assertions, we usually economize. Rather than assert both the statement and its converse separately, we usually adopt one of two strategies. The simplest of these is just to list the assertions and say they are equivalent. For example,

PROPOSITION 4.6. *Let X be a quadrilateral. Then the following are equivalent:*

- X *is a square.*

- X *is a rectangle with four equal sides*

This is logically the same as saying that each assertion implies the other, and we will employ this formula whenever convenient—especially when we have more than just two equivalent assertions. The first important example of this occurs in the next section (Theorem 5.7).

Typically, however, when we just have two equivalent assertions, we don't construct a "TFAE" (*The Following Are Equivalent*) statement like the one above. Instead, we use *"if and only if"* like this:

$$P \text{ if and only if } Q.$$

Here, *only if* means P can't hold without Q. As we shall shortly see ("contrapositive" below), this is just another way of saying $P \Rightarrow Q$. The *if* clause, on the other hand, reverses the implication, claiming that $Q \Rightarrow P$. In other words, the *if* clause states the converse. So *if and only if* packs both the statement and its converse into one short formula.

For example, we could state Proposition 4.6 above using *if and only if* like this:

PROPOSITION 4.7. *X is a square if and only if X is a rectangle with four equal sides.*

Read Proposition 4.7 carefully to convince yourself that it means both

> *Every square is a rectangle with four equal sides.*

and

> *Every rectangle with four equal sides is a square.*

The matrix inversion algorithm provides another example of *if and only if.* For it asserts (among other things) that

PROPOSITION 4.8. *An $n \times n$ matrix* **A** *is invertible if and only if* $\text{RRE}(\mathbf{A}) = \mathbf{I}_n.$

Again, *if and only if* asserts that *both the statement and its converse hold.* Diagrammatically, it means that in the set of objects under consideration, those for which P holds are *the same* as those for which Q holds (Figure 4). Theorems of this kind are especially interesting because they pack two truths into one result: they tell us that a certain statement and its converse are *both* true.

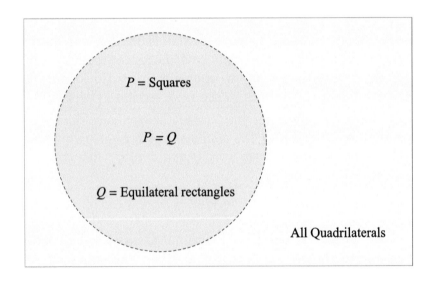

Figure 4. If and only if. When the set for which Q holds *is the same* as the set for which P holds, we have P *if and only if* Q, written $P \Leftrightarrow Q.$

4.9. Contrapositive and obverse. We get the **obverse** of $P \Rightarrow Q$ by *negating* P and Q both. If we *negate and swap* P and Q, we get the **contrapositive.** Statement, converse, obverse, and contrapositive thus take the following forms:

Statement	*If P, then Q.*
Converse	*If Q, then P.*
Obverse	*If P is false, then Q is false.*
Contrapositive	*If Q is false, then P is false.*

EXAMPLE 4.10. Again consider the simple true statement

If X is a square, then X is a rectangle.

The converse of this statement (*If X is a rectangle, then X is a square*) is *false*, as we have seen. But the *contrapositive*,

If X is not a rectangle, then X is not a square.

is certainly true.   □

The contrapositive (as the Example above suggests) *is always equivalent to the original statement*, and this equivalence of statement and contrapositive lies at the very heart of mathematical reasoning. It is easy to understand diagrammatically: we depict it in Figure 5.

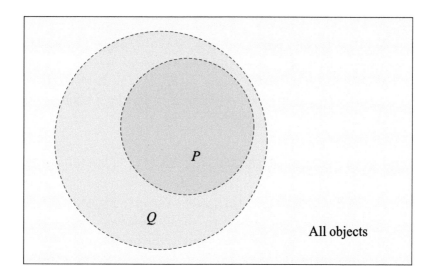

Figure 5. Equivalence of statement and contrapositive. The statement $P \Rightarrow Q$ is true here: If a point belongs to P ("P is true"), then certainly it belongs to Q ("Q is true."). The contrapositive says that when a point lies *outside* Q (gray: "Q is false"), then it certainly lies outside of P ("P is false."). Both statement and converse amount to saying "Q contains P," so they are equivalent.

To anyone serious about learning mathematics, the equivalence of statement and contrapositive must become second nature, for it is one of the most useful tools in the mathematical toolbox. Here's why:

Though logically equivalent to the statement, the contrapositive is often easier to prove.

The next Example illustrates this within linear algebra.

EXAMPLE 4.11. Consider these claims about a square matrix $\mathbf{A}_{n \times n}$:

$$P: \quad \mathbf{A}\mathbf{x} = \mathbf{0} \ \text{ has a non-trivial solution.}$$
$$Q: \quad \mathbf{A} \ \text{ is singular.}$$

With these meanings for P and Q, the resulting **statement** asserts

OBSERVATION 4.12. *If* $\mathbf{A}$ *is an* $n \times n$ *matrix, and* $\mathbf{A}\mathbf{x} = \mathbf{0}$ *has a non-trivial solution, then* $\mathbf{A}$ *is singular.*

While this is true, it is actually easier to prove the *contrapositive*, namely

OBSERVATION 4.13. *If* $\mathbf{A}_{n \times n}$ *is invertible, then* $\mathbf{A}\mathbf{x} = \mathbf{0}$ *has only the trivial solution.*

Logically, the two Observations mean exactly the same thing. But the latter is easier to prove. Indeed, it's a special case of Proposition 3.7, which says that when $\mathbf{A}$ is invertible, the general system $\mathbf{A}\mathbf{x} = \mathbf{b}$ has exactly one solution no matter what $\mathbf{b}$ is. When $\mathbf{b} = \mathbf{0}$, we already know one solution—the trivial solution $\mathbf{x} = \mathbf{0}$. So if there's only one solution, that must be it. □

So much for the virtues of the contrapositive. What about the **obverse**?

A glance at the definitions shows that the obverse is the contrapositive of—and hence logically equivalent to—the converse. For that reason one seldom encounters the obverse directly. If the converse is true, then the obverse automatically holds, and one need not state it separately. If the converse is false, so is the obverse, which then holds little interest.

4.14. About negation. The negation (or falsification) of a simple assertion like X *is a rectangle* is pretty obvious: it is X *is not a rectangle*. More care is required for qualified assertions involving expressions like *some*, *all*, *every*, and *there exists*.

Suppose, for instance, we make an assertion like

All math professors are boring.

To negate this *all* assertion, do **not** negate the word *all* to get *none*. Instead, change *all* to *some* and negate the property (here *boring*). So the correct way to say *it is false that all math professors are boring* is

> *Some math professor is not boring.* (Correct negation)

not

> *No math professors are boring.* (Incorrect negation)

The second formulation is wrong because it is *too strong*. The statement *all math professors are boring* is clearly false if *even one* math professor is not boring.

In general, for a statement of the form

> *All X have property Y.*

the correct negation is

> *Some X does* not *have property Y.*

Conversely, the negation of the *some* statement above is the *all* statement before it.

Since *every* is closely related to *all*, similar considerations hold for it. For instance, to negate a statement of the form

> *For every X, there exists a Y with property Z.*

we assert

> *For* some *X, there exists* no *Y with property Z.*

Considered cautiously, these facts are common sense, but they are easy to get wrong if one is hasty or careless.

EXAMPLE 4.15. A correct negation of

> *Every pot has a cover.*

is

> Some *pot has no cover.*

$\square$

Here are some mathematical examples:

EXAMPLE 4.16. One of the most famous unsolved problems in mathematics is the *Goldbach* conjecture, which states:

> *Every even number greater than 2 is the sum of two primes.*

The negation of this statement (the assertion that Goldbach's conjecture is false) would be stated like this:

> Some *even number greater than 2 is* not *the sum of two primes.*

In a more linear algebraic spirit, let $\mathbf{A}$ be a fixed $n \times m$ matrix, and consider this statement:

> *For some* $\mathbf{b} \in \mathbf{R}^n$, *the system* $\mathbf{A}\mathbf{x} = \mathbf{b}$ *has no solution.*

The correct negation of this statement is

> *For* all $\mathbf{b} \in \mathbf{R}^n$, *the system* $\mathbf{A}\mathbf{x} = \mathbf{b}$ *has at least one solution.*

Similarly, the negation of

> *There exists a* $\mathbf{b} \in \mathbf{R}^n$ *for which* $\mathbf{A}\mathbf{x} = \mathbf{b}$ *has many solutions.*

would be

> *For* every $\mathbf{b} \in \mathbf{R}^n$, *the system* $\mathbf{A}\mathbf{x} = \mathbf{b}$ *has at most* one *solution.*

□

Finally, we mention the negation of compound statements—statements with *and* or *or*, which get *swapped* by negation. Consider, for instance, the statement

(30)   *All students are both ambitious* and *smart.*

To negate this, we change the *all* to *some* as discussed above. *But we must also change both to either and the* and *to an* or:

> *Some student is either not ambitious* or *not smart.*

For the existence of even one student who is smart, but not ambitious— *or* one student who is ambitious but not smart—clearly falsifies statement (30).

Similarly, to negate an *or* statement like

> *That child is either hungry* or *tired.*

we must change the *or* to an *and*:

That child is not hungry, and *she's not tired.*

Figure 6 depicts the logic of *and* and *or* using a Venn diagram.

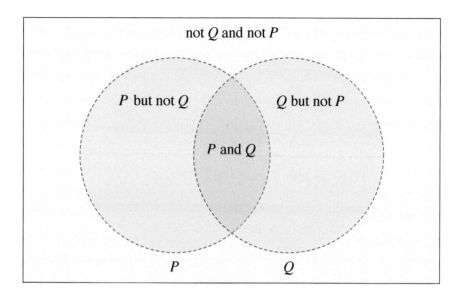

Figure 6. And/Or. Blue discs P and Q are the points for which P and Q respectively are true. The dark blue *intersection* depicts "P and Q." By "P or Q," we mean the *union* of the light or dark regions, so the negation of "P or Q" shows up here as the gray region: "not P *and* not Q." The negation of the dark blue region is the union of the light blue and gray regions, corresponding to "not P *or* not Q."

4.17. Summary. The four named implications between assertions P, Q, and their negations (falsifications) come in two equivalent pairs:

$$\begin{cases} \text{Statement:} & \textit{If } P \textit{ then } Q \\ \text{Contrapositive:} & \textit{If } Q \textit{ is false, then } P \textit{ is false} \end{cases}$$

$$\begin{cases} \text{Converse:} & \textit{If } Q \textit{ then } P \\ \text{Obverse:} & \textit{If } P \textit{ is false then } Q \textit{ is false} \end{cases}$$

Within each pair, both statements hold or fail together—both are true, or both are false. But the truth of each *pair* is completely independent of the other pair. Either pair can be false when the other is true.

When the statement and converse (and hence all four assertions) are simultaneously true, we say that P and Q are *equivalent*, and often use an *if and only if* formulation to express that. Finally, when we want to claim that several assertions are all equivalent to each other, we usually say *the following are equivalent*, and then list the assertions in question.

With regard to negation, *all* and *for every* statements are typically negated by a *some* statement, and vice-versa. The conjunctions *and* and *or* replace each other when we negate a compound statement that one of them joins.

− **Practice** −

240. Consider these assertions

$$P : \quad \textbf{A} \text{ is an invertible matrix}$$

$$Q : \quad \textbf{A} \text{ is a square matrix}$$

a) How would you phrase the *statement* associated with these assertions? Is it true? Why or why not?

b) How would you phrase the *converse* associated with these assertions? Is it true? Why or why not?

241. Suppose $\textbf{u}$, $\textbf{v}$, and $\textbf{w} \in \textbf{R}^n$. Consider the assertions

$$P : \quad \textbf{u} \cdot \textbf{w} = 0 \text{ and } \textbf{v} \cdot \textbf{w} = 0$$

$$Q : \quad (\textbf{u} + \textbf{v}) \cdot \textbf{w} = 0$$

a) How would you phrase the *statement* associated with these assertions? Is it true? Why or why not?

b) How would you phrase the *converse* associated with these assertions? Is it true? Why or why not?

242. Suppose $T : \textbf{R}^n \to \textbf{R}^n$ is a linear transformation. (Note that domain and range are the same here.) Consider the assertions

$$P : \quad T \text{ is one-to-one}$$

$$Q : \quad T \text{ is onto}$$

a) How would you phrase the *statement* associated with these assertions? Is it true? Why or why not?

b) How would you phrase the *converse* associated with these assertions? Is it true? Why or why not?

243. Consider again the assertions

$$P: \quad \mathbf{A} \text{ is an invertible matrix}$$
$$Q: \quad \mathbf{A} \text{ is a square matrix}$$

a) How would you phrase the *contrapositive* associated with these assertions? Is it true? Why or why not?

b) How would you phrase the *obverse* associated with these assertions? Is it true? Why or why not?

c) What *if and only if* statement would assert the equivalence of statement and converse for these two assertions? Would it be true or false? Justify your answer.

244. Suppose $\mathbf{u}$, $\mathbf{v}$, and $\mathbf{w} \in \mathbf{R}^n$. Consider the assertions

$$P: \quad \mathbf{u} \cdot \mathbf{w} = 0 \text{ and } \mathbf{v} \cdot \mathbf{w} = 0$$
$$Q: \quad (\mathbf{u} + \mathbf{v}) \cdot \mathbf{w} = 0$$

a) How would you phrase the *contrapositive* associated with these assertions? Is it true? Why or why not?

b) How would you phrase the *obverse* associated with these assertions? Is it true? Why or why not?

c) What *if and only if* statement would assert the equivalence of statement and converse for these two assertions? Would it be true or false? Justify your answer.

245. Let P be the assertion that X *is a right triangle with short sides* A, B *and hypotenuse* C.

a) What assertion Q now makes the statement *If P then Q* into the Pythgorean Theorem?

b) State the converse. Is it true?

246. Identify the hypothesis and conclusion (P and Q) implicit in these statements:

a) *A transformation represented by an invertible matrix is onto.*

b) *A free column in* RRE($\mathbf{A}$) *means the transformation represented by* $\mathbf{A}$ *isn't one-to-one.*

c) *Products of* $n \times n$ *diagonal matrices are again diagonal.*

247. Consider this statement (from Theorem 5.6): *Every linear transformation is represented by a matrix.*

    a) Rephrase it in *"If P then Q"* form.

    b) Formulate the contrapositive. Is it true?

    c) Formulate the negation of the statement. Is it true?

248. Suppose $\mathbf{A}_{n \times m}$ is a matrix and $\mathbf{b} \in \mathbf{R}^n$. Consider the statement (compare Theorem 4.1): *If $\mathbf{x}'$ and $\mathbf{x}''$ in $\mathbf{R}^m$ both solve $\mathbf{Ax} = \mathbf{b}$, then $\mathbf{x}' - \mathbf{x}''$ solves $\mathbf{Ax} = \mathbf{0}$.*

    a) Formulate the contrapositive.

    b) Formulate the converse.

    c) Show that the converse is false by constructing a counterexample using

$$\mathbf{A} = \begin{bmatrix} 1 & 1 \\ 0 & 0 \end{bmatrix} \quad \text{and} \quad \mathbf{b} = \begin{pmatrix} 2 \\ 0 \end{pmatrix}$$

249. Rewrite each statement using *if and only if*:

    a) *A square matrix and its transpose are either both invertible, or both singular.*

    b) *A linear transformation $T : \mathbf{R}^n \to \mathbf{R}^n$ that is onto is also one-to-one, and vice-versa.*

    c) *Invertibility of $\mathbf{A}$ is equivalent to $\mathbf{Ax} = \mathbf{0}$ having a non-trivial solution.*

250. Write the negation of each statement, and say which, if either, is true: the statement or its negation.

    a) *Every linear transformation from $\mathbf{R}^m$ to $\mathbf{R}^n$ is onto.*

    b) *There exists a 2×2 matrix $\mathbf{A}$ which does not row-reduce to the identity matrix.*

    c) *Every vector in $\mathbf{R}^3$ except the origin has a non-zero dot product with $(1, 1, 1)$.*

    d) *All the diagonal entries in any skew-symmetric matrix are zeros.*

    e) *Every symmetric 3×3 matrix has a pair of equal entries.*

251. Write the negation of each statement.

   a) *The mapping is both one-to-one and onto.*

   b) *The mapping commutes with both vector addition and scalar multiplication.*

   c) *The matrix is either upper- or lower-triangular.*

   d) *The vector is neither orthogonal to $(1, 1, 0)$, nor to $(0, 1, 1)$.* [*Hint: "Neither P nor Q" means "Not P **and** not Q."*]

252. Write the negation of each statement, and say which, if either, is true: the statement or its negation.

   a) *Every linear transformation from $\mathbf{R}^n$ to $\mathbf{R}^n$ is either one-to-one or onto.*

   b) *Every linear system either has a solution or has more unknowns than equations.*

   c) *Every diagonal matrix is both square and symmetric.*

   d) *No triangular matrix is both symmetric and invertible.*

253. Consider an $n \times m$ matrix $\mathbf{A}$, a vector $\mathbf{b} \in \mathbf{R}^n$, and the assertions

   $P:$ $m \geq n$

   $Q:$ For every $\mathbf{b} \in \mathbf{R}^n$, $\mathbf{Ax} = \mathbf{b}$ has one or more solutions.

   a) How would you phrase the *statement* associated with these assertions? Is it true? Why or why not?

   b) How would you phrase the *converse* associated with these assertions? Is it true? Why or why not?

   c) How would you phrase the *contrapositive* associated with these assertions? Is it true? Why or why not?

   d) How would you phrase the *obverse* associated with these assertions? Is it true? Why or why not?

   e) What *if and only if* statement would assert the equivalence of statement and converse for these two assertions? Would it be true or false? Justify your answer.

— ★ —

5. The Logic of the Inversion Algorithm

The inversion algorithm is powerful and efficient. Simple as it is, when implemented on a computer it can invert truly enormous square matrices. It isn't magic, however—we can understand it completely.

The inversion algorithm accomplishes two things. It tells us that

- **A** is **not** invertible when $\text{RRE}(\mathbf{A}) \neq \mathbf{I}_n$, and

- **A is** invertible when $\text{RRE}(\mathbf{A}) = \mathbf{I}_n$, in which case it actually constructs $\mathbf{A}^{-1}$.

As a first step toward understanding why it works, we record an Observation relevant to both points above.

OBSERVATION 5.1. *Let* **A** *be any* $n \times n$ *matrix. Then* $\text{RRE}(\mathbf{A})$ *has a free column if and only if* $\text{RRE}(\mathbf{A}) \neq \mathbf{I}_n$.

PROOF. When we apply Gauss-Jordan, either we get a free column, or we don't. If we get a free column, $\text{RRE}(\mathbf{A})$ *cannot* be $\mathbf{I}_n$, because $\mathbf{I}_n$ has a pivot in every column.

On the other hand, when we *don't* get a free column, all n columns of $\text{RRE}(\mathbf{A})$ have a pivot. Since there are exactly n rows too, we also get a pivot in every row. The pivot in row 1 must lie in column 1, since all other pivots are below (and hence to the right) of it, and the remaining $n-1$ pivots must fit into the remaining $n-1$ columns. Similarly, the pivot in row 2 must lie in column 2, since all $n-2$ remaining pivots are below, and hence to the right of *it*, and they must fit into the remaining $n-2$ columns. In this same way, it follows easily that for every $k = 3, 4, \ldots, n$, the kth pivot occupies the (k, k) position. The pivots hence fill the main diagonal of $\text{RRE}(\mathbf{A})$, which must therefore equal $\mathbf{I}_n$.   □

5.2. The non-invertible case. Observation 5.1 quickly yields a proof of the first bullet point above.

Recall that we get a homogeneous generator **h** for the system $\mathbf{Ax} = \mathbf{0}$ if and only if $\text{RRE}(\mathbf{A})$ has a free column. So Observation 5.1 implies

OBSERVATION 5.3. *Let* **A** *be an* $n \times n$ *matrix. Then* $\text{RRE}(\mathbf{A}) \neq \mathbf{I}_n$ *if and only if the homogeneous system* $\mathbf{Ax} = \mathbf{0}$ *has a non-trivial solution.*

Just one more fact will let us complete the proof that $\text{RRE}(\mathbf{A}) \neq \mathbf{I}_n$ implies non-invertibility.

OBSERVATION 5.4. *If* **A** *is an* $n \times n$ *matrix, and the homogeneous system* $\mathbf{Ax} = \mathbf{0}$ *has a non-trivial solution, then* **A** *is* not *invertible.*

PROOF. The homogeneous system *always* has the trivial solution $\mathbf{x} = \mathbf{0}$. So the contrapositive to our statement can be phrased like this:

*If $\mathbf{A}$ is invertible, then $\mathbf{Ax} = \mathbf{0}$ has **only** the trivial solution.*

This we can easily prove. For when $\mathbf{A}$ is invertible, and $\mathbf{Ax} = \mathbf{0}$ we must also have $\mathbf{A}^{-1}\mathbf{Ax} = \mathbf{A}^{-1}\mathbf{0}$, which simplifies to $\mathbf{x} = \mathbf{0}$. In short, when $\mathbf{A}$ has a inverse, it makes every solution of $\mathbf{Ax} = \mathbf{0}$ necessarily trivial. This proves the contrapositive, and hence the statement. $\square$

By chaining together Observations 5.1, 5.3, and 5.4, we now get the first property of the inversion algorithm: *When $\mathbf{A}$ is square and $\mathrm{RRE}(\mathbf{A}) \neq \mathbf{I}_n$, $\mathbf{A}$ is not invertible.* The reasoning goes like this:

- (Observation 5.1) If $\mathbf{A}$ is square, and $\mathrm{RRE}(\mathbf{A}) \neq \mathbf{I}_n$, then $\mathbf{A}$ has a free column.

- (Observation 5.3) If $\mathrm{RRE}(\mathbf{A})$ has a free column, then $\mathbf{Ax} = \mathbf{0}$ has a non-trivial solution.

- (Observation 5.4) If $\mathbf{Ax} = \mathbf{0}$ has a non-trivial solution, $\mathbf{A}$ cannot be invertible.

This proves the inversion algorithm's first claim: *If $\mathbf{A}$ is square, and $\mathrm{RRE}(\mathbf{A}) \neq \mathbf{I}_n$, then $\mathbf{A}$ is not invertible.*

5.5. The invertible case. We now want to understand the inversion algorithm's other, more ambitious claim: *When $\mathbf{A}$ does row-reduce to the identity, $\mathbf{A}$ has an inverse, and the algorithm constructs $\mathbf{A}^{-1}$.*

To start, we note that when $\mathrm{RRE}(\mathbf{A}) = \mathbf{I}_n$, we always get exactly one solution to $\mathbf{Ax} = \mathbf{b}$ no matter what $\mathbf{b} \in \mathbf{R}^n$ we target. For when we row-reduce the augmented matrix $[\mathbf{A} \,|\, \mathbf{b}]$ in that situation, we get $[\mathbf{I}_n \,|\, \mathbf{x}_p]$, where $\mathbf{x}_p$ solves the system, and further, $\mathbf{x}_p$ will be the *only* solution, since $\mathrm{RRE}(\mathbf{A}) = \mathbf{I}_n$ has no free columns.

With this unique solvability in mind, let us solve **each** of the following systems:

$$\mathbf{Ax} = \mathbf{e}_j, \qquad j = 1, 2, \ldots n$$

Here the $\mathbf{e}_j$'s are the **standard basis vectors** for $\mathbf{R}^n$ that we introduced in Chapter 1 (Definition 1.11). As noted above, $\mathbf{Ax} = \mathbf{e}_j$ will have exactly one solution $\mathbf{x}_p^{(j)}$ for each j, and row-reduction will place that solution in the last column of the reduced augmented matrix:

$$\left[\,\mathbf{A} \,\middle|\, \mathbf{e}_j\,\right] \;\longrightarrow\; \left[\,\mathbf{I}_n \,\middle|\, \mathbf{x}_p^{(j)}\,\right]$$

Now consider what happens if we arrange these $\mathbf{x}_p^{(j)}$'s in columns to form a matrix $\mathbf{X}$ like this:

$$\mathbf{X} = \left[\begin{array}{cccc} \mathbf{x}_p^{(1)} & \mathbf{x}_p^{(2)} & \cdots & \mathbf{x}_p^{(n)} \\ | & | & \cdots & | \end{array}\right]$$

Since matrix multiplication operates column-by-column, and $\mathbf{A}\mathbf{x}_p^{(j)} = \mathbf{e}_j$, for each j, we now easily deduce that

$$\mathbf{A}\mathbf{X} = \left[\begin{array}{cccc} \mathbf{A}\mathbf{x}_p^{(1)} & \mathbf{A}\mathbf{x}_p^{(2)} & \cdots & \mathbf{A}\mathbf{x}_p^{(n)} \\ | & | & \cdots & | \end{array}\right] = \left[\begin{array}{cccc} \mathbf{e}_1 & \mathbf{e}_2 & \cdots & \mathbf{e}_n \\ | & | & \cdots & | \end{array}\right]$$

Crucially, the matrix on the right above is none other than the identity $\mathbf{I}_n$! So we have $\mathbf{A}\mathbf{X} = \mathbf{I}_n$, which means $\mathbf{X} = \mathbf{A}^{-1}$, as claimed.

This confirms that if $\text{RRE}(\mathbf{A}) = \mathbf{I}_n$, then $\mathbf{A}$ is invertible, and more— we can construct the inverse. *It is the matrix* $\mathbf{X}$ *whose columns are the individual solutions of* $\mathbf{A}\mathbf{x} = \mathbf{e}_j$ *for* $j = 1, 2, \ldots, n$.

That leaves only one thing to check: *This* $\mathbf{X}$ *is the same matrix that the inversion algorithm constructs.*

To verify this, simply note that the only difference between the inversion algorithm and the process we just described is this: We solved each of the problems $\mathbf{A}\mathbf{x} = \mathbf{e}_j$ separately, while *the inversion algorithm solves them simultaneously.* We solved them serially, whereas the inversion algorithm solves them in parallel. It does so by augmenting $\mathbf{A}$ with the entire identity matrix $\mathbf{I}_n$, whose n columns are precisely the vectors $\mathbf{e}_j$ we used as individual right-hand sides. Row-reducing the super-augmented matrix $[\,\mathbf{A}\,|\,\mathbf{I}_n\,]$ thus performs all of the row-reductions in one ambitious stroke:

$$\left[\,\mathbf{A}\,|\,\mathbf{I}_n\,\right] = \left[\,\mathbf{A}\,|\,\mathbf{e}_1\;\mathbf{e}_2\;\cdots\mathbf{e}_n\,\right] \longrightarrow \left[\,\mathbf{I}_n\,|\,\mathbf{x}_p^{(1)}\;\mathbf{x}_p^{(2)}\;\cdots\;\mathbf{x}_p^{(n)}\,\right]$$

The solutions of these n problems then constitute the last n columns of the reduced super-augmented matrix to form $\mathbf{X} = \mathbf{A}^{-1}$.

This is why the inversion algorithm works when $\text{RRE}(\mathbf{A}) = \mathbf{I}_n$. Augmenting $\mathbf{A}$ with the identity matrix and then row-reducing solves the n systems $\mathbf{A}\mathbf{x} = \mathbf{e}_j$ all at once. For the reasons given above, these solutions form the columns of $\mathbf{X} = \mathbf{A}^{-1}$.

5.6. Final notes on inversion. As a final reflection on the inversion algorithm, we observe that beyond explaining why the algorithm works, our reasoning reveals several properties *equivalent* to invertibility. We collect these together in classic TFAE (*The Following Are Equivalent*) format.

THEOREM 5.7. *For any $n \times n$ matrix $\mathbf{A}$, the following are equivalent:*

a) $\mathbf{A}$ *is invertible*

b) *The homogeneous system* $\mathbf{Ax} = \mathbf{0}$ *has only the trivial solution*

c) $\mathrm{RRE}(\mathbf{A}) = \mathbf{I}_n$

d) *Given any* $\mathbf{b} \in \mathbf{R}^n$, *we can solve* $\mathbf{Ax} = \mathbf{b}$ *for* $\mathbf{x}$.

PROOF. We use the standard procedure for proving a TFAE statement: we show that each assertion implies the next, and that the last implies the first:

$$(a) \;\Rightarrow\; (b) \;\Rightarrow\; (c) \;\Rightarrow\; (d) \;\Rightarrow\; (a)$$

By doing so, we show that each of the statement implies all the others, thereby demonstrating their equivalence.

(a) $\Rightarrow$ **(b)**: This is just the contrapositive of Observation 5.4, as we noted in proving that Observation.

(b) $\Rightarrow$ **(c)**: This follows immediately from Observation 5.3. (It's the contrapositive of the "only if" clause there.)

(c) $\Rightarrow$ **(d)**: The only obstacle to solving $\mathbf{Ax} = \mathbf{b}$ is a row of zeros in $\mathrm{RRE}(\mathbf{A})$. (This is a version of Proposition 6.7 in Chapter 2). Since $\mathbf{I}_n$ has a pivot in every row, $\mathrm{RRE}(\mathbf{A}) = \mathbf{I}_n$ implies solvability of $\mathbf{Ax} = \mathbf{b}$ for every $\mathbf{b}$. (In fact, it implies *unique* solvability, since $[\mathbf{A}|\mathbf{b}]$ then reduces to $[\mathbf{I}_n|\mathbf{x}_p]$, with $\mathbf{x}_p$ as the unique solution.)

(d) $\Rightarrow$ **(a)**: If $\mathbf{Ax} = \mathbf{b}$ has a solution for each $\mathbf{b} \in \mathbf{R}^n$, then we can certainly solve $\mathbf{Ax} = \mathbf{e}_j$ for each of the standard basis vectors (columns of $\mathbf{I}_n$) $\mathbf{e}_j$, with $j = 1, 2, \ldots, n$. We saw in analyzing the inversion algorithm that the solutions of these n problems form the columns of $\mathbf{A}^{-1}$. So (d) implies the invertibility of $\mathbf{A}$ claimed by (a).

Our proof is complete.   □

– Practice –

254. The inversion algorithm works (partly) because we can solve several inhomogeneous systems of the form $\mathbf{Ax} = \mathbf{b}_j$ in one stroke by reducing the super-augmented matrix $[\mathbf{A} \,|\, \mathbf{B}]$, where the columns of $\mathbf{B}$ are the $\mathbf{b}_j$'s.

a) Use this method to simultaneously solve the three systems $\mathbf{Ax} = \mathbf{b}_j$ with

$$\mathbf{A} = \begin{bmatrix} 3 & -1 \\ 5 & -2 \end{bmatrix}, \quad \mathbf{b}_1 = \begin{pmatrix} 5 \\ 8 \end{pmatrix}, \quad \mathbf{b}_2 = \begin{pmatrix} 1 \\ 1 \end{pmatrix}, \quad \mathbf{b}_3 = \begin{pmatrix} 4 \\ 7 \end{pmatrix}$$

b) Likewise, simultaneously solve the three systems $\mathbf{Ax} = \mathbf{b}_j$ with

$$\mathbf{A} = \begin{bmatrix} 1 & -2 & 0 \\ -2 & 1 & -2 \\ 0 & -2 & 1 \end{bmatrix}$$

and

$$\mathbf{b}_1 = \begin{pmatrix} 3 \\ 6 \\ 1 \end{pmatrix}, \quad \mathbf{b}_2 = \begin{pmatrix} -4 \\ -3 \\ -5 \end{pmatrix}, \quad \mathbf{b}_3 = \begin{pmatrix} 1 \\ -5 \\ 0 \end{pmatrix}$$

255. The technique in Exercise 254 can even be used when the coefficient matrix $\mathbf{A}$ isn't invertible. Try it on these examples.

a) Simultaneously solve the three systems $\mathbf{Ax} = \mathbf{b}_j$ with

$$\mathbf{A} = \begin{bmatrix} 1 & 2 \\ 3 & 6 \end{bmatrix}, \quad \mathbf{b}_1 = \begin{pmatrix} 2 \\ 6 \end{pmatrix}, \quad \mathbf{b}_2 = \begin{pmatrix} -5 \\ -15 \end{pmatrix}, \quad \mathbf{b}_3 = \begin{pmatrix} 4 \\ 2 \end{pmatrix}$$

b) Similarly, solve the three systems $\mathbf{Ax} = \mathbf{b}_j$ with

$$\mathbf{A} = \begin{bmatrix} 1 & 0 & -1 \\ 1 & 1 & 0 \\ 0 & 1 & 1 \end{bmatrix}$$

and

$$\mathbf{b}_1 = \begin{pmatrix} -3 \\ 0 \\ 3 \end{pmatrix}, \quad \mathbf{b}_2 = \begin{pmatrix} 4 \\ 5 \\ 1 \end{pmatrix}, \quad \mathbf{b}_3 = \begin{pmatrix} 3 \\ 0 \\ -1 \end{pmatrix}$$

256. Answer and explain:

    a) What condition determines whether a diagonal matrix is invertible?

    b) What condition determines whether a triangular matrix is invertible?

257. Suppose $\mathbf{A}$ is a (not necessarily square) matrix with n rows, and we can solve $\mathbf{Ax} = \mathbf{e}_j$ (the jth standard basis vector) for each $j = 1, 2, \ldots n$. Show that $\mathbf{A}$ represents a transformation that is *onto* $\mathbf{R}^n$. [*Hint: Use Observation 1.12.*]

258. As in Exercise 257, suppose $\mathbf{A}$ is a (not necessarily square) $n \times m$ matrix for which we can solve $\mathbf{Ax} = \mathbf{e}_j$ for each $j = 1, 2, \ldots n$.

    a) Show that $\mathbf{A}$ has a "right inverse" $\mathbf{B}$—a matrix $\mathbf{B}$ with $\mathbf{AB} = \mathbf{I}_n$.

    b) What size will $\mathbf{B}$ have? Which inequality must hold: $n \geq m$, or $n \leq m$?

    c) Why doesn't the existence of this right-inverse make $\mathbf{A}$ invertible by Proposition 3.2?

259. Prove that we could add this fifth equivalent assertion to Theorem 5.7: *The transformation represented by $\mathbf{A}$ is both one-to-one and onto.*

260. (Compare Exercises 254 and 255.) Let $\mathbf{A}$ be an invertible $n \times n$ matrix, and suppose $\mathbf{B}_{n \times m}$ is any matrix with n rows. Show that by row-reducing the super-augmented matrix $[\mathbf{A} \,|\, \mathbf{B}]$ we always get $[\mathbf{I}_n \,|\, \mathbf{A}^{-1}\mathbf{B}]$. (This procedure lets us compute $\mathbf{A}^{-1}\mathbf{B}$ without first computing $\mathbf{A}^{-1}$.)

261. Proposition 3.10 says that when $\mathbf{A}_{n \times n}$ and $\mathbf{B}_{n \times n}$ are both invertible, so is $\mathbf{AB}$. If $\mathbf{B}$ is *not* invertible, then by Theorem 5.7, the homogeneous system $\mathbf{Bx} = \mathbf{0}$ has a non-trivial solution.

    a) Use this fact to argue that when $\mathbf{B}$ is not invertible, the equation $(\mathbf{AB})\mathbf{x} = \mathbf{0}$ *also* has a non-trivial solution, so that $\mathbf{AB}$ is not invertible either. Deduce that invertibility of $\mathbf{AB}$ implies invertibility of $\mathbf{B}$.

b) Does invertibility of $\mathbf{AB}$ imply that of $\mathbf{A}$ too? [*Hint: Assume* $\mathbf{A}$ *is not invertible and prove the contrapositive.*]

c) Together, what result do (a) and (b) yield? State it as a proposition.

262. Suppose $\mathbf{A}$ represents a linear map $T : \mathbf{R}^n \to \mathbf{R}^n$ that is an *isomorphism* (i.e., one-to-one and onto).

a) Show that $\mathbf{A}$ is then invertible.

b) Deduce that $\mathbf{A}^{-1}$ represents the *inverse mapping* $T^{-1} : \mathbf{R}^n \to \mathbf{R}^n$ (see Definition 2.18).

c) Conclude that the inverse of a linear isomorphism is always another linear isomorphism.

6. Determinants

By row-reducing the super-augmented matrix

$$[\mathbf{A}\,|\,\mathbf{I}] = \left[\begin{array}{cc|cc} a & b & 1 & 0 \\ c & d & 0 & 1 \end{array}\right]$$

one easily derives this general inversion formula for 2×2 matrices (compare Exercise 235):

$$\begin{bmatrix} a & b \\ c & d \end{bmatrix}^{-1} = \frac{1}{ad - bc} \begin{bmatrix} d & -b \\ -c & a \end{bmatrix}$$

This formula makes no sense when the denominator $ad - bc$ equals zero, but that is actually good. For when $ad - bc = 0$, we have

$$\begin{bmatrix} a & b \\ c & d \end{bmatrix} \begin{pmatrix} d \\ -c \end{pmatrix} = \begin{pmatrix} ad - bc \\ dc - dc \end{pmatrix} = \mathbf{0}$$

which means that either

i) We have $d = c = 0$, in which case the second row of $\mathbf{A}$ contains only zeros, so that $\mathbf{A}$ can't reduce to the identity, and thus has no inverse, or

ii) The vector $(d, -c)$ is a non-trivial solution to the homogeneous system. By Theorem 5.7, we again conclude that $\mathbf{A}$ has no inverse.

Either way, we see that when $ad - bc = 0$, then $\mathbf{A}$ has no inverse.

The quantity $ad - bc$ thus packs a crucial piece of information into a single scalar. Namely, *the* 2×2 *matrix*

$$\mathbf{A} = \begin{bmatrix} a & b \\ c & d \end{bmatrix}$$

is invertible if and only if $ad - bc \neq 0$.

This single scalar $ad - bc$ *determines* whether $\mathbf{A}$ is invertible or singular, and for this reason, it is known as the **determinant** of $\mathbf{A}$, denoted by $\det \mathbf{A}$ or $|\mathbf{A}|$.

EXAMPLE 6.1. These two matrices differ only slightly:

$$\mathbf{A} = \begin{bmatrix} 1 & 2 \\ 2 & 3 \end{bmatrix} \quad \text{and} \quad \mathbf{B} = \begin{bmatrix} 1 & 2 \\ 2 & 4 \end{bmatrix}$$

Yet one is invertible, the other isn't, as the determinant quickly shows:

$$\det \mathbf{A} = 1 \cdot 3 - 2 \cdot 2 = 3 - 4 = -1 \neq 0$$

so $\mathbf{A}$ has an inverse, while

$$\det \mathbf{B} = 1 \cdot 4 - 2 \cdot 2 = 4 - 4 = 0$$

so $\mathbf{B}$ does not. $\qquad\qquad\qquad\qquad\qquad\qquad\qquad\qquad\qquad\qquad$ □

For 2×2 matrices, the determinant thus gives an easy test for invertibility. What about larger matrices? Do 3×3 matrices have a determinant formula? How about 4×4 or general $n \times n$ matrices?

The answer is *yes*. There *is* a determinant formula for square matrices of any size. Unfortunately, the formula gets horribly complicated as n grows. For a 3×3 matrix $\mathbf{A} = [a_{ij}]$, it goes like this:

$$(31) \qquad \begin{aligned} \det \mathbf{A} \;=\; & a_{11}a_{22}a_{33} + a_{12}a_{23}a_{31} + a_{13}a_{21}a_{32} \\ & -a_{13}a_{22}a_{31} - a_{12}a_{21}a_{33} - a_{11}a_{23}a_{32} \end{aligned}$$

Where the 2×2 formula had just two quadratic terms, the 3×3 formula has *six cubic* terms. Things have definitely gotten uglier, though perhaps not completely out of hand. They go from bad to worse very quickly as n increases, however. The 4×4 determinant involves 24 fourfold products:

$$
\begin{aligned}
\det \mathbf{A}_{4\times 4} \;=\;& a_{14}a_{23}a_{32}a_{41} - a_{13}a_{24}a_{32}a_{41} - a_{14}a_{22}a_{33}a_{41} \\
&+ a_{12}a_{24}a_{33}a_{41} + a_{13}a_{22}a_{34}a_{41} - a_{12}a_{23}a_{34}a_{41} \\
&- a_{14}a_{23}a_{31}a_{42} + a_{13}a_{24}a_{31}a_{42} + a_{14}a_{21}a_{33}a_{42} \\
&- a_{11}a_{24}a_{33}a_{42} - a_{13}a_{21}a_{34}a_{42} + a_{11}a_{23}a_{34}a_{42} \\
&+ a_{14}a_{22}a_{31}a_{43} - a_{12}a_{24}a_{31}a_{43} - a_{14}a_{21}a_{32}a_{43} \\
&+ a_{11}a_{24}a_{32}a_{43} + a_{12}a_{21}a_{34}a_{43} - a_{11}a_{22}a_{34}a_{43} \\
&- a_{13}a_{22}a_{31}a_{44} + a_{12}a_{23}a_{31}a_{44} + a_{13}a_{21}a_{32}a_{44} \\
&- a_{11}a_{23}a_{32}a_{44} - a_{12}a_{21}a_{33}a_{44} + a_{11}a_{22}a_{33}a_{44}
\end{aligned}
$$

(32)

This pushes the limits of practical hand calculation, and the number of terms grows factorially with n, rapidly getting too big even for computers! Nevertheless, there's a pattern to these formulae, and by using summation notation we can write the general determinant quite compactly. For an $n \times n$ matrix $\mathbf{A} = [a_{ij}]$, it takes this form:

$$
(33) \qquad \det \mathbf{A} = \sum_{P \in S_n} \operatorname{sign}(P)\, a_{1P(1)} a_{2P(2)} a_{3P(3)} \cdots a_{nP(n)}
$$

The sum here runs over the set S_n of all permutations of the first n natural numbers. There are $n!$ such permutations P, and there is a rule for attaching a *sign* ± 1 to each, denoted by $\operatorname{sign}(P)$ in the formula.

In Appendix A we give a full explanation for this formula, its notation and properties. For now, we content ourselves with merely knowing that such a formula exists, and that by using it, one can easily prove that determinants obey four basic rules:

PROPOSITION 6.2 (Determinant rules). *The determinant function assigns a scalar $|\mathbf{A}|$ to every $n \times n$ matrix $\mathbf{A} = [a_{ij}]$. Among its properties are these:*

    a) *When $\mathbf{A}$ is triangular, $|\mathbf{A}| = a_{11}a_{22}\ldots a_{nn}$ (the diagonal product).*

    b) *Swapping two rows of $\mathbf{A}$ changes only the sign of $|\mathbf{A}|$.*

    c) *Multiplying any single row of $\mathbf{A}$ by a scalar λ multiplies $|\mathbf{A}|$ by λ as well.*

    d) *Adding a multiple of one row of $\mathbf{A}$ to another row of $\mathbf{A}$ does not change $|\mathbf{A}|$.*

PROOF. It is easy to verify all these properties in the 2×2 (or even 3×3) case, and we leave the reader do so in Exercises 271 and 272. Using the general determinant formula (33), the $n \times n$ case requires only slightly more effort. We give those details in Appendix A.   □

Proposition 6.2 shows how *the elementary row operations* affect the determinant—and how to compute the determinant of any triangular matrix. The beauty of this is that it gives us precisely what we need to compute *any* determinant efficiently, even for large matrices. All we need is (what else?) row-reduction.

For when we row-reduce a square matrix $\mathbf{A}$ to a *triangular* matrix $\mathbf{T}$ (full RRE form isn't needed for determinants), the Proposition tells us

- The determinant of the *final* matrix $\mathbf{T}$, and
- How the determinant of the *original* matrix changed with each row-operation along the way.

Combining these facts, we can easily recover the determinant of the *original* matrix $|\mathbf{A}|$. For any determinant bigger than 2×2, this is usually so much faster than the general formula (33) that we can largely forget about the latter from now on.

A couple of examples will illustrate the technique.

EXAMPLE 6.3. Let us compute the determinant of

$$\mathbf{A} = \begin{bmatrix} 1 & 2 & 3 \\ -1 & 0 & -3 \\ 0 & -2 & 3 \end{bmatrix}$$

To do so, we row-reduce $\mathbf{A}$ to triangular form while tracking the change each step effects on $|\mathbf{A}|$, using Proposition 6.2.

To start, we zero out the first entry in row 2 by adding row 1 to it. According to Determinant Rule (d), this doesn't change $|\mathbf{A}|$. So

$$|\mathbf{A}| = \begin{vmatrix} 1 & 2 & 3 \\ 0 & 2 & 0 \\ 0 & -2 & 3 \end{vmatrix}$$

Next we add row 2 to row 3. This gets us another zero below the diagonal, bringing the matrix to triangular form, but it still doesn't affect $|\mathbf{A}|$:

$$|\mathbf{A}| = \begin{vmatrix} 1 & 2 & 3 \\ 0 & 2 & 0 \\ 0 & 0 & 3 \end{vmatrix}$$

So far, we have used only property (d) of the Determinant Rules. But now we have a triangular matrix, so we can use property (a), which tells us

$$|\mathbf{A}| = 1 \cdot 2 \cdot 3 = 6$$

This example would have been slightly more complicated if rows 1 and 3 of the original matrix had been swapped. In that case, we would have started with

$$\mathbf{B} = \begin{bmatrix} 0 & -2 & 3 \\ -1 & 0 & -3 \\ 1 & 2 & 3 \end{bmatrix}$$

Now we cannot make $\mathbf{B}$ triangular without swapping a pair of rows. According to Determinant Rule (b), this changes the sign of the determinant. For instance, if we swap rows 1 and 3, we get

$$|\mathbf{B}| = - \begin{vmatrix} 1 & 2 & 3 \\ -1 & 0 & -3 \\ 0 & -2 & 3 \end{vmatrix} = -|\mathbf{A}|$$

Having already calculated $|\mathbf{A}| = 6$, we conclude that $|\mathbf{B}| = -6$.   □

EXAMPLE 6.4. Let us compute

$$\mathbf{X} = \begin{bmatrix} 1 & 1 & 1 & 1 \\ 1 & -1 & 1 & -1 \\ 1 & -1 & -1 & 1 \\ 1 & 1 & -1 & -1 \end{bmatrix}$$

Again, the strategy is to reduce $\mathbf{X}$ to an upper-triangular matrix. To start, we zero out the last three entries of column one by subtracting the first row from each of the other rows. Determinant Rule (d) guarantees that none of these moves changes the determinant, so we have

$$|\mathbf{X}| = \begin{vmatrix} 1 & 1 & 1 & 1 \\ 0 & -2 & 0 & -2 \\ 0 & -2 & -2 & 0 \\ 0 & 0 & -2 & -2 \end{vmatrix}$$

We get another zero below the diagonal by subtracting row 2 from row 3, again leaving the determinant unchanged:

$$|\mathbf{X}| = \begin{vmatrix} 1 & 1 & 1 & 1 \\ 0 & -2 & 0 & -2 \\ 0 & 0 & -2 & 2 \\ 0 & 0 & -2 & -2 \end{vmatrix}$$

Finally, by subtracting row 3 from row 4, we reach upper-triangular form:

$$|\mathbf{X}| = \begin{vmatrix} 1 & 1 & 1 & 1 \\ 0 & -2 & 0 & -2 \\ 0 & 0 & -2 & 2 \\ 0 & 0 & 0 & -4 \end{vmatrix}$$

Since the determinant of a triangular matrix is given by its diagonal product, we conclude that

$$|\mathbf{X}| = 1 \cdot (-2) \cdot (-2) \cdot (-4) = -16$$

We have thus computed a 4×4 determinant with *much* less effort than would have been required by the explicit formula (32).   □

6.5. A trick for 3-by-3's. Row reduction is typically the most efficient method for computing determinants of matrices larger than 3-by-3.

For 2-by-2's, we just apply the $ad - bc$ formula, as in Example 6.1, and there's a similar shortcut for 3-by-3 matrices. It's really just a mnemonic for computing the formula in Equation (31) (compare Exercise 517) but please be forewarned: *This trick only works in the 3-by-3 case; it does **not** generalize to larger matrices.*

The shortcut we have in mind

$$\mathbf{A} = \begin{bmatrix} a & b & c \\ d & e & f \\ g & h & i \end{bmatrix}$$

is depicted in Figure 7 and described below.

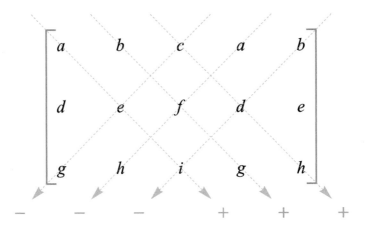

Figure 7. The trick for computing 3-by-3 determinants.

3-by-3 determinant shortcut

- Augment **A** with its own first two columns to form the 3-by-5 matrix in Figure 7.

- Locate the three complete southeast diagonals, marked by "+" signs in the figure. Compute the product along each of these: aei, bfg, and cdh.

- Locate the three complete *southwest* diagonals, marked by "−" signs in the figure. Compute the products along these too: ceg, afh, and bdi.

- Subtract the "−" products from the sum of the "+" products, and behold! We have the determinant:

(34) $$\det \mathbf{A} = aei + bfg + cdh - ceg - afh - bdi$$

EXAMPLE 6.6. We demonstrate the trick above by quickly finding the determinant of

$$\mathbf{A} = \begin{bmatrix} 2 & 3 & 4 \\ -3 & 1 & -3 \\ 4 & 3 & 2 \end{bmatrix}$$

To do so, we augment **A** with the first two columns to construct the 3×5 matrix

$$\begin{bmatrix} 2 & 3 & 4 & 2 & 3 \\ -3 & 1 & -3 & -3 & 1 \\ 4 & 3 & 2 & 4 & 3 \end{bmatrix}$$

Subtracting the three southwest diagonal products from the southeast products now gives the determinant:

$$(2 \cdot 1 \cdot 2) + (3 \cdot -3 \cdot 4) + (4 \cdot -3 \cdot 3)$$
$$-(4 \cdot 1 \cdot 4) - (2 \cdot -3 \cdot 3) - (3 \cdot -3 \cdot 2)$$
$$= \quad 4 - 36 - 36 - 16 + 18 + 18$$
$$= \quad -48$$

So $|\mathbf{A}| = -48$. $\square$

6.7. More determinant properties. The Determinant Rules in Proposition 6.2 don't just give a practical way to compute large determinants. They also have key theoretical consequences. In particular, they imply that invertibility of a square matrix is equivalent to having non-zero determinant:

COROLLARY 6.8. *These additional determinant properties hold for any square matrix* $\mathbf{A}$.

   a) *If* $\mathbf{A}$ *has two equal rows, then* $|\mathbf{A}| = 0$.

   b) *If* $\mathbf{A}$ *has a row of zeros, then* $|\mathbf{A}| = 0$.

   c) $\mathbf{A}$ *is invertible if and only if* $|\mathbf{A}| \neq 0$.

PROOF. **(a)**: If $\mathbf{A}$ has two equal rows, and we swap them, then $|\mathbf{A}|$ changes sign, by Determinant Rule (b). But when we swap two *equal* rows, we don't actually change $\mathbf{A}$—so its determinant can't change either! It follows that $-|\mathbf{A}| = |\mathbf{A}|$, which forces $|\mathbf{A}| = 0$.

(b): If $\mathbf{A}$ has a row of zeros, and we multiply that row by -1 then Determinant Rule (c) says that we multiply $|\mathbf{A}|$ by -1 too. But we're multiplying a row of *zeros* by -1, so again we don't actually change $\mathbf{A}$, nor its determinant. As in (a), this means $-|\mathbf{A}| = |\mathbf{A}|$, forcing $|\mathbf{A}| = 0$.

(c): Conclusion (c) is an *if and only if* statement. We must prove that invertibility of $\mathbf{A}$ is *equivalent* to non-zero determinant. We will do so by showing

$$|\mathbf{A}| \neq 0 \iff \text{RRE}(\mathbf{A}) = \mathbf{I}_n \iff \mathbf{A} \text{ is invertible}$$

Recall that "$\iff$" indicates *equivalence* (Section 4.5). A chain of equivalences shows equivalence of the first and last conditions, and this is what (c) asks us to prove.

Luckily, we already know the second equivalence in the chain thanks to Theorem 5.7. So we only have to show the first one, namely

(35) $$|\mathbf{A}| \neq 0 \iff \text{RRE}(\mathbf{A}) = \mathbf{I}_n$$

For this, we start by noting that we can always get from $\mathbf{A}$ to $\text{RRE}(\mathbf{A})$ using elementary row operations, and our Determinant rules tell us that each of these operations multiplies $|\mathbf{A}|$ by a *non-zero* scalar: If we swap rows, we multiply by -1. If we multiply a row by some $\lambda \neq 0$ (we never multiply by zero when row-reducing), we multiply $|\mathbf{A}|$ by λ too. Finally, if we add a multiple of one row to another, we leave $|\mathbf{A}|$ unchanged, which is the same as multiplying by $+1$. Products of non-zero scalars remain non-zero, so we *always* have

(36) $$|\mathbf{A}| = c \cdot |\text{RRE}(\mathbf{A})| \quad \text{with} \quad c \neq 0$$

But $\text{RRE}(\mathbf{A})$ will *always* be upper-triangular, because its pivots run southeast, and all rows of zeros sink to the bottom. Since $\mathbf{A}$ is square—say $n \times n$—this allows just two possibilities:

  i) *Either* we get a full set of n pivots, so that $\text{RRE}(\mathbf{A}) = \mathbf{I}_n$

  ii) *Or* we get fewer than n pivots, and the last row of $\text{RRE}(\mathbf{A})$ vanishes entirely.

In case (i), we have $|\mathbf{A}| \neq 0$ by combining Equation (36) with Determinant rule (d), which tells us that $|\mathbf{I}_n| = 1 \neq 0$.

In case (ii) $\text{RRE}(\mathbf{A})$ has a row of zeros, and property (a) of the present Corollary then says that $|\text{RRE}(\mathbf{A})| = 0$. In this case, Equation (36) yields $|\mathbf{A}| = 0$.

As (i) and (ii) are the only possibilities, we have $|\mathbf{A}| \neq 0$ *if and only if* $\text{RRE}(\mathbf{A}) = \mathbf{I}_n$. In other words, non-zero determinant is *equivalent* to $\text{RRE}(\mathbf{A}) = \mathbf{I}_n$. This proves (35), and we are done. $\square$

6.9. Determinants of products. We now present a truly amazing fact. To appreciate it, consider that the product of two matrices mixes the entries of its two factors in a rather complicated way. Even in the 3×3 case, the product of two simple matrices $\mathbf{A} = [a_{ij}]$ and $\mathbf{B} = [b_{ij}]$ is pretty monstrous:

$$\mathbf{AB} = \begin{bmatrix} a_{11}b_{11} + a_{12}b_{21} + a_{13}b_{31} & a_{11}b_{12} + a_{12}b_{22} + a_{13}b_{32} & a_{11}b_{13} + a_{12}b_{22} + a_{13}b_{33} \\ a_{21}b_{11} + a_{22}b_{21} + a_{22}b_{31} & a_{21}b_{12} + a_{22}b_{22} + a_{22}b_{32} & a_{21}b_{13} + a_{22}b_{22} + a_{22}b_{33} \\ a_{31}b_{11} + a_{32}b_{21} + a_{33}b_{31} & a_{31}b_{12} + a_{32}b_{22} + a_{33}b_{32} & a_{31}b_{13} + a_{32}b_{22} + a_{33}b_{33} \end{bmatrix}$$

Computing the determinant of this elaborate mess looks unpleasant at best, but happily, there's an amazingly simple shortcut:

THEOREM 6.10. *The determinant of a product is the product of the determinants:* $\det(\mathbf{AB}) = \det(\mathbf{A})\det(\mathbf{B})$.

PROOF. (See Section 3 of Appendix A.) □

EXAMPLE 6.11. Let us verify the Theorem for the matrices

$$\mathbf{A} = \begin{bmatrix} 2 & -3 \\ -4 & 5 \end{bmatrix} \quad \text{and} \quad \mathbf{B} = \begin{bmatrix} 3 & -1 \\ 1 & 3 \end{bmatrix}$$

It is easy to see that $|\mathbf{A}| = -2$, $|\mathbf{B}| = 10$, and hence $|\mathbf{A}||\mathbf{B}| = -20$. On the other hand, to check the Theorem, we compute

$$\mathbf{AB} = \begin{bmatrix} 2 & -3 \\ -4 & 5 \end{bmatrix}\begin{bmatrix} 3 & -1 \\ 1 & 3 \end{bmatrix} = \begin{bmatrix} 6-3 & -2-9 \\ -12+5 & 4+15 \end{bmatrix} = \begin{bmatrix} 3 & -11 \\ -7 & 19 \end{bmatrix}$$

Theorem 6.10 predicts that this matrix too will have determinant -20, and indeed, we find that $3 \cdot 19 - 7 \cdot 11 = 57 - 77 = -20$.

Note that even in this 2-by-2 example, the first calculation was much easier. □

Theorem 6.10 also yields a key fact about determinants if we apply it to compute $|\mathbf{AB}|$ when $\mathbf{B} = \mathbf{A}^{-1}$. For then by reading Theorem 6.10 backwards, we get

$$|\mathbf{A}|\,|\mathbf{A}^{-1}| = |\mathbf{AA}^{-1}| = |\mathbf{I}_n| = 1$$

This shows that the determinant of $\mathbf{A}^{-1}$, is *reciprocal* to that of $\mathbf{A}$. We state this formally for future reference:

COROLLARY 6.12. *If* $\mathbf{A}$ *is invertible, then*

$$|\mathbf{A}^{-1}| = \frac{1}{|\mathbf{A}|}$$

EXAMPLE 6.13. Let

$$\mathbf{A} = \begin{bmatrix} 9 & 2 & -5 \\ 0 & -4 & 7 \\ 0 & 0 & 3 \end{bmatrix}$$

Since $\mathbf{A}$ is triangular we can quickly compute its determinant as $9 \cdot (-4) \cdot 3 = -108$.

We can then instantly deduce that $\det \mathbf{A}^{-1} = 1/108$, even though we have not computed $\mathbf{A}^{-1}$.   □

6.14. Determinant of the transpose. Consider the typical 2×2 matrix and its transpose:

$$\mathbf{A} = \begin{bmatrix} a & b \\ c & d \end{bmatrix}, \qquad \mathbf{A}^T = \begin{bmatrix} a & c \\ b & d \end{bmatrix}$$

It's easy to see they have exactly the same determinant. For larger matrices, the same fact holds, though it's harder to see.

PROPOSITION 6.15. *For any square matrix* $\mathbf{A}$, *we have*

$$\left| \mathbf{A}^T \right| = |\mathbf{A}|$$

PROOF. See Section 2 of Appendix A, where we give a short proof based on the permutation-based determinant formula (33).   □

Proposition 6.15 is useful because the rows of $\mathbf{A}^T$ are the columns of $\mathbf{A}$. Since Proposition 6.15 says that determinants "ignore" transposition, the behavior of $|\mathbf{A}^T|$ with respect to *rows* determines that of $|\mathbf{A}|$ itself with respect to *columns*. As a result, all row properties of determinants hold for columns too. We summarize as follows:

COROLLARY 6.16. *Let* $\mathbf{A}$ *be any square matrix.*

a) *If we swap two columns of* $\mathbf{A}$ *we change the sign of* $|\mathbf{A}|$.

b) *If we multiply a column of* $\mathbf{A}$ *by a scalar, we multiply* $|\mathbf{A}|$ *by that same scalar.*

c) *If we add a multiple of one column of* $\mathbf{A}$ *to another, we leave* $|\mathbf{A}|$ *unchanged.*

d) *If* $\mathbf{A}$ *has a column of zeros, then* $|\mathbf{A}| = 0$.

e) *If* $\mathbf{A}$ *has two equal columns, then* $|\mathbf{A}| = 0$.

PROOF. We will prove (a) only. Conclusions (b)–(e) can be proven similarly, and we leave them as exercises.

(a): Let $\mathbf{A}'$ denote the matrix we get by swapping two *columns* of $\mathbf{A}$ (columns j and k, say). Then we clearly get its transpose $(\mathbf{A}')^T$ by swapping the corresponding *rows* of $\mathbf{A}^T$. Swapping of *rows* is covered by Determinant rule (b), however: it changes the sign of the determinant. So we have

$$\begin{aligned} |\mathbf{A}'| &= \left|(\mathbf{A}')^T\right| && \textit{(Proposition 6.15)} \\ &= -\left|\mathbf{A}^T\right| && \textit{(Determinant rule (b))} \\ &= -|\mathbf{A}| && \textit{(Proposition 6.15 again)} \end{aligned}$$

Thus, column-swapping changes the sign of $|\mathbf{A}|$, as claimed. □

EXAMPLE 6.17. The following matrix does not have any equal rows, nor a row of zeros. If we examine rows only, we don't easily see that $|\mathbf{A}| = 0$.

$$\mathbf{A} = \begin{bmatrix} 1 & 4 & 1 & -4 \\ 2 & 3 & 2 & -1 \\ 3 & 2 & 3 & -3 \\ 4 & 1 & 4 & -2 \end{bmatrix}$$

If we examine *columns*, however, we *do* easily see that $|\mathbf{A}| = 0$, because columns 1 and 3 are the same. Noticing column facts like this can thus save considerable effort. □

– Practice –

263. Compute the determinant of each matrix. Which are invertible?

a) $\begin{bmatrix} 0 & 1 \\ 1 & 0 \end{bmatrix}$ b) $\begin{bmatrix} 0 & -1 & -1 \\ 1 & 0 & 1 \\ -1 & 1 & 0 \end{bmatrix}$ c) $\begin{bmatrix} 1 & 0 & 1 & 1 \\ 0 & 1 & 1 & 1 \\ 1 & 1 & 1 & 0 \\ 1 & 1 & 0 & 1 \end{bmatrix}$

264. Compute the determinant of each matrix. Is either invertible?

a) $\begin{bmatrix} 1 & 1/2 \\ 1/2 & 1/3 \end{bmatrix}$ b) $\begin{bmatrix} 1 & 1/2 & 1/3 \\ 1/2 & 1/3 & 1/4 \\ 1/3 & 1/4 & 1/5 \end{bmatrix}$

265. Use row-reduction to compute the determinant of each 4×4:

a)
$$\begin{bmatrix} 1 & -3 & 0 & 0 \\ 2 & 1 & 0 & 0 \\ 0 & 0 & 1 & 5 \\ 0 & 0 & -2 & -1 \end{bmatrix}$$
b)
$$\begin{bmatrix} 1 & -2 & 0 & 0 \\ -2 & 1 & -2 & 0 \\ 0 & -2 & 1 & -2 \\ 0 & 0 & -2 & 1 \end{bmatrix}$$

c)
$$\begin{bmatrix} 0 & 0 & -1 & 2 \\ 0 & 1 & 3 & -4 \\ -5 & 6 & 1 & 0 \\ 7 & -8 & 0 & 1 \end{bmatrix}$$
d)
$$\begin{bmatrix} 1 & 0 & -1 & 2 \\ 0 & 1 & 3 & -4 \\ -5 & 6 & 1 & 0 \\ 7 & -8 & 0 & 1 \end{bmatrix}$$

266. For each pair of matrices $\mathbf{A}$ and $\mathbf{B}$, use Theorem 6.10 to compute $|\mathbf{AB}|$ without first computing $\mathbf{AB}$:

a) $\qquad \mathbf{A} = \begin{bmatrix} 3 & -9 \\ -2 & 4 \end{bmatrix} \qquad \mathbf{B} = \begin{bmatrix} 7 & 1 \\ 5 & 2 \end{bmatrix}$

b) $\qquad \mathbf{A} = \begin{bmatrix} 4 & -3 \\ 2 & -1 \end{bmatrix} \qquad \mathbf{B} = \begin{bmatrix} 10 & 0 \\ 0 & -3 \end{bmatrix}$

c) $\qquad \mathbf{A} = \begin{bmatrix} 5 & -3 \\ 0 & 5 \end{bmatrix} \qquad \mathbf{B} = \begin{bmatrix} 5 & 0 \\ -3 & 5 \end{bmatrix}$

267. Use the 3-by-3 trick from Section 6.5 to find the determinants of the four matrices in Exercise 268 below.

268. For each pair of matrices $\mathbf{A}$ and $\mathbf{B}$, use Theorem 6.10 to compute $|\mathbf{AB}|$ without first computing $\mathbf{AB}$:

a) $\qquad \mathbf{A} = \begin{bmatrix} 2 & -1 & 0 \\ -1 & 2 & -1 \\ 0 & 2 & -1 \end{bmatrix} \qquad \mathbf{B} = \begin{bmatrix} 1 & -1 & 3 \\ 0 & 1 & -1 \\ -1 & 0 & 1 \end{bmatrix}$

b) $\qquad \mathbf{A} = \begin{bmatrix} 1 & -1 & 0 \\ 0 & 1 & -1 \\ -1 & 0 & 1 \end{bmatrix} \qquad \mathbf{B} = \begin{bmatrix} 3 & -4 & 7 \\ -9 & 3 & -2 \\ 5 & -1 & 8 \end{bmatrix}$

269. Find the determinant of each matrix. What values of a (if any) make each matrix singular (i.e., *non*-invertible)?

a) $\begin{bmatrix} 1 & 1 \\ a & a^2 \end{bmatrix}$ b) $\begin{bmatrix} 1 & 1 & 1 \\ 1 & 2 & 4 \\ 1 & a & a^2 \end{bmatrix}$ c) $\begin{bmatrix} a & \sqrt{8} & 0 \\ \sqrt{8} & a & \sqrt{8} \\ 0 & \sqrt{8} & a \end{bmatrix}$

270. Verify that the trick for computing 3×3 determinants in Section 6.5 is correct by checking that Equation (34) and Equation (31) say exactly the same thing.

271. Prove that the determinant of a general 2×2 matrix obeys all three properties listed in Proposition 6.2.

272. Using Equation (31) prove that the determinant of a general 3×3 matrix obeys properties (a), (c), and (d) of Proposition 6.2. (*Optional: Check (b) too.*)

273. Using our proof of Corollary 6.16 (a) as a model, prove Conclusions (b)–(e) of that same Corollary.

274. Suppose we multiply an $n \times 1$ column matrix $\mathbf{C}_{n \times 1}$ by an $1 \times n$ row matrix $\mathbf{R}_{1 \times n}$. Is the resulting $n \times n$ matrix $\mathbf{CR}$ ever invertible? Why or why not?

275. For each statement below, give either a proof or a counterexample.

   a) For any $n \times n$ matrices $\mathbf{A}$ and $\mathbf{B}$, we have
   $$\det(\mathbf{A} + \mathbf{B}) = \det \mathbf{A} + \det \mathbf{B}$$

   b) If $\mathbf{A}$ and $\mathbf{B}$ are $n \times n$ matrices then $\det(\mathbf{AB}) = \det(\mathbf{BA})$.

276. An *anti-diagonal* matrix is an $n \times n$ matrix with zeros everywhere except for the anti-diagonal that runs from the upper-right entry a_{1n} to the lower-left entry a_{n1}. For instance these are anti-diagonal:

$$\begin{bmatrix} 0 & 2 \\ 3 & 0 \end{bmatrix} \quad \text{and} \quad \begin{bmatrix} 0 & 0 & 0 & 1 \\ 0 & 0 & 2 & 0 \\ 0 & 3 & 0 & 0 \\ 4 & 0 & 0 & 0 \end{bmatrix}$$

a) Show that the determinants of both matrices above are plus-or-minus the products of their anti-diagonal entries.

b) State and prove a proposition governing the determinant of a general $n \times n$ anti-diagonal matrix.

277. *(Difficult)* A matrix is *binary* if every entry is either a 0 or a 1.

a) How big can the determinant of a binary 2×2 matrix be?

b) How big can the determinant of a binary 3×3 matrix be?

c) How big can the determinant of a binary 4×4 matrix be?

The maximal determinant for binary $n \times n$ matrices is known for $n \leq 17$, but not for $n = 18$. For most $n > 17$, in fact, the answer is still unknown.

− ★ −

CHAPTER 5

Subspaces

We have given the name *hyperplane* to the solution set of any single non-trivial linear equation in $\mathbf{R}^n$. Geometrically, hyperplanes in $\mathbf{R}^2$ and $\mathbf{R}^3$ correspond to lines in the plane and planes in space, respectively. We visualize hyperplanes as "copies" of $\mathbf{R}^{n-1}$ situated in $\mathbf{R}^n$.

More generally, we can "linearly situate" copies of $\mathbf{R}^k$ into $\mathbf{R}^n$ for any $k < n$, not just for $k = n - 1$. We will call these copies k-*dimensional* affine subspaces. We cannot directly visualize them when $n > 3$, but we can easily understand and even axiomatize them. Indeed, Definition 1.9 below captures their essence in just two conditions.

Just as hyperplanes arise as solution sets to single linear equations, k-dimensional affine subspaces arise as solution sets to linear *systems*. As noted in Chapter 3, solutions of systems—and thus subspaces—are *intersections of hyperplanes*. There is, however, an alternate way to describe subspaces: as *families of linear combinations*. We thus have two entirely different two modes of description for subspaces, and we call them *perps* and *spans* (or *implicit* and *explicit*), respectively.

Every subspace can be seen as either a perp or a span, and this introduces a new kind of duality into our subject. The present chapter explores both ways of generating, recognizing and describing subspaces, and how to convert back and forth between the perp (implicit) and span (explicit) points of view.

1. Basic Examples and Definitions

1.1. General subsets of $\mathbf{R}^n$. Any collection of vectors in $\mathbf{R}^n$ forms a sub*set*. Subsets may contain finitely or infinitely many vectors. They may or may not be geometrically simple or easily described. The possibilities are endless, but relatively few subsets qualify as sub*spaces*. Before we state conditions that distinguish a subspace from a mere subset, however, here are some examples of subsets that are *not* subspaces. We suspect readers can better appreciate what a subspace *is* after first seeing examples of what a subspace is *not*.

EXAMPLE 1.2. The **unit cube** is the set of all points $(x_1, x_2, \ldots, x_n) \in \mathbf{R}^n$ with $|x_i| \leq 1$ for each $i = 1, 2, \ldots, n$. Figure 1 depicts the unit cubes in $\mathbf{R}^2$ and $\mathbf{R}^3$. Cubes are subsets, but not subspaces. □

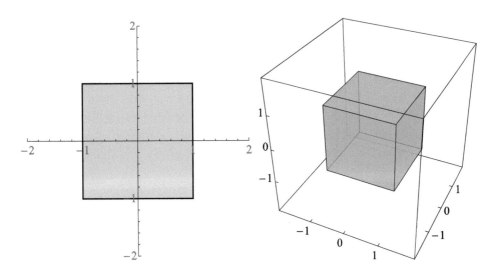

Figure 1. Unit cubes. The unit cube in $\mathbf{R}^2$ can be visualized as a square (left). The unit cube in $\mathbf{R}^3$ is depicted similarly at right.

EXAMPLE 1.3. The set of *all unit vectors* in $\mathbf{R}^n$ (all $\mathbf{x} \in \mathbf{R}^n$ with $|\mathbf{x}| = 1$) is called the **unit sphere** in $\mathbf{R}^n$. For instance, in $\mathbf{R}^2$ we get the unit circle: all (x, y) with $x^2 + y^2 = 1$. The unit sphere in $\mathbf{R}^n$, forms the boundary of the **unit ball**, which is defined by $|\mathbf{x}| \leq 1$. Balls and spheres are nice subsets of $\mathbf{R}^n$, but they are not subspaces. □

EXAMPLE 1.4. The set of all vectors in $\mathbf{R}^n$ with *integer* coordinates (for instance, vectors $(1, 3)$ or $(-5, 2)$ in $\mathbf{R}^2$) is called the **integer lattice**. Part of the (infinite) integer lattice in $\mathbf{R}^2$ is depicted on the left of Figure 2. Like cubes and spheres, lattices are subsets, but not subspaces. □

EXAMPLE 1.5. The set of all points in $\mathbf{R}^n$ with *strictly positive co-ordinates* is called the **positive orthant** in $\mathbf{R}^n$. When $n = 2$, we usually call it the positive **quadrant** (Figure 2, right). Orthants are not subspaces. □

EXAMPLE 1.6. **A non-linear solution set.** Consider the subset of $\mathbf{R}^3$ formed by all solutions (x, y, z) of the **non**-linear equation

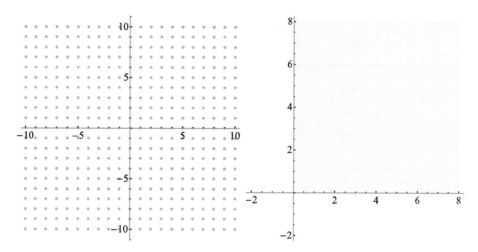

Figure 2. The dots at left are points in the integer lattice in $\mathbf{R}^2$. The shaded region at right indicates the positive quadrant.

$$xyz = 0$$

Since the product xyz vanishes if and only if one of the factors x, y or z vanishes, this subset comprises the *union* of three planes: the $z = 0$, $y = 0$, and $x = 0$ planes (Figure 3). The individual planes are subspaces, as it turns out, but taken together, their union is not. □

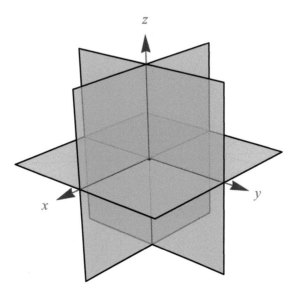

Figure 3. Any point on a coordinate plane solves $xyz = 0$. The solution set of $xyz = 0$ is the *union* of all three planes.

EXAMPLE 1.7. **Hyperplanes.** Any hyperplane is a subset, but a hyperplane that contains the *origin* (the solution set of a *homogeneous* linear equation) is a *subspace.* Hyperplanes that do not contain the origin are not subspaces, though they are *affine* subspaces, as we shall explain.   □

1.8. Subspaces of $\mathbf{R}^n$. Within the vast and diverse zoo of subsets, the *subspaces* we define next play the central role in our subject. This is because *every subspace forms a self-contained linear-algebraic microcosm in its own right.* For the definition guarantees that in a subspace, we can perform the basic operations of linear algebra—addition and scalar multiplication—and remain in that subspace.

DEFINITION 1.9 (Subspace). A non-empty subset of $S \subset \mathbf{R}^n$ is called a **subspace** if it is

- **Closed under addition**, meaning that when $\mathbf{v}, \mathbf{w} \in S$, so is $\mathbf{v} + \mathbf{w}$.

- **Closed under scalar multiplication**, meaning that when $\mathbf{v} \in S$, so is $s\,\mathbf{v}$ for any scalar s.

                                                                      □

By far, most subsets of $\mathbf{R}^n$ have neither property above, and thus do *not* qualify as subspaces. The definition above lets us test whether they do or not, however. We illustrate with more examples.

EXAMPLE 1.10. Let us verify that the plane S of solutions to $x + y + z = 0$ in $\mathbf{R}^3$ forms a subspace.

To do so using Definition 1.9, we must confirm that S is closed under addition and scalar multiplication.

Closure under addition. The defining equation for S says that a vector $\mathbf{v} = (v_1, v_2, v_3)$ belongs to S if and only if its coordinates sum to zero: $v_1 + v_2 + v_3 = 0$. We have to show that if $\mathbf{v}$ and $\mathbf{w}$ *both* pass that test, then so does their sum.

We therefore take two arbitrary vectors $\mathbf{v} = (v_1, v_2, v_3)$ and $\mathbf{w} = (w_1, w_2, w_3)$, and compute their sum:

$$\mathbf{v} + \mathbf{w} = (v_1 + w_1,\ v_2 + w_2,\ v_3 + w_3)$$

Does this sum remain in S? To find out, we again apply the test for membership in S by checking whether its coordinates add to zero. Adding them up and rearranging a bit, we get

$$v_1 + w_1 + v_2 + w_2 + v_3 + w_3 = (v_1 + v_2 + v_3) + (w_1 + w_2 + w_3)$$

This clearly *does* vanish: it equals $0 + 0$, because $\mathbf{v}, \mathbf{w} \in S$, and hence the v_i's and w_i's themselves each sum to zero. So S is closed under addition, as hoped.

Closure under scalar multiplication. We next have to show that when $\mathbf{v} \in S$, and $t \in \mathbf{R}$ is any scalar, then $t\mathbf{v}$ lies in S too. This is again straightforward. For $\mathbf{v} = (v_1, v_2, v_3)$ lies in S if and only if the sum of the v_i's vanish. But then the coordinates of $t\mathbf{v} = (tv_1, tv_2, tv_3)$ will also sum to zero, since

$$t\, v_1 + t\, v_2 + t\, v_3 = t\, (v_1 + v_2 + v_3) = t \cdot 0 = 0$$

Thus, S is closed under *both* scalar multiplication and addition. By Definition 1.9, that makes it a subspace.   □

EXAMPLE 1.11. Here are perhaps the two simplest examples of subspaces:

- $\mathbf{R}^n$ *is a subspace of itself.* If we add two vectors, in $\mathbf{R}^n$ we get another vector in $\mathbf{R}^n$, so $\mathbf{R}^n$ itself is closed under addition.

  It is also closed under scalar multiplication, since any scalar multiple of a vector in $\mathbf{R}^n$ again lies in $\mathbf{R}^n$.

- *The set* $\{\mathbf{0}\} \subset \mathbf{R}^n$ *containing only the origin is a subspace.* For if we add $\mathbf{0}$ to itself, or multiply it by any scalar, we still get $\mathbf{0}$, which stays in the set.

  □

To emphasize that a subspace $S \subset \mathbf{R}^n$ is a proper subset of $\mathbf{R}^n$ —*not* all of $\mathbf{R}^n$ —we call it a **proper** subspace.

At the other extreme, note that Definition 1.9 requires a subspace to be *non-empty*. So no subspace is *smaller* than the one-vector subspace $\{\mathbf{0}\}$ of the Example above. In fact $\{\mathbf{0}\}$ is *only* one-vector subspace, a fact we leave to the reader in Exercise 278. We often refer to $\{\mathbf{0}\}$ as the **trivial subspace**.

We now give two examples illustrating the use of Definition 1.9 to *dis*qualify certain subsets of $\mathbf{R}^n$ as subspaces. These are interesting because each satisfies one, but not *both* of the closure conditions. Definition 1.9 explicitly requires both conditions, so these sets are not subspaces.

EXAMPLE 1.12. The integer lattice in $\mathbf{R}^n$ (cf. Example 1.4 and Figure 2) is **not** a subspace because it is **not** closed under scalar multiplication.

To show a set is *not* closed, we only have to exhibit one counter-example—one vector in the lattice that goes *out* of the lattice when multiplied by one scalar. Here we can take $\mathbf{v} = (1, 1, \ldots, 1)$ as our vector, and $s = 1/2$ as our scalar. While $\mathbf{v}$ belongs to the lattice, multiplication by s now gives

$$s\mathbf{v} = \left(\tfrac{1}{2}, \tfrac{1}{2}, \ldots, \tfrac{1}{2}\right)$$

which is *not* in the lattice: lattice vectors must have integer coordinates. We thus see that the integer lattice is *not* closed under scalar multiplication, and consequently is *not* a subspace.

The integer lattice *is* closed under addition, by the way. To see that, simply note that when we take two *arbitrary* lattice points, say

$$\mathbf{a} = (a_1, a_2, \ldots, a_n) \quad \text{and} \quad \mathbf{b} = (b_1, b_2, \ldots, b_n)$$

and add them, we always get another integer lattice point. Indeed, their sum is

$$\mathbf{a} + \mathbf{b} = (a_1 + b_1, a_2 + b_2, \ldots, a_n + b_n)$$

and since we're assuming the a_i's and b_i's are integers, and sums of integers are integers, the coordinates of $\mathbf{a} + \mathbf{b}$ are integers too. So the integer lattice is closed under addition.

To summarize: *Though closed under addition, the integer lattice in $\mathbf{R}^n$ is **not** closed under scalar multiplication, hence is **not** a subspace.* $\square$

EXAMPLE 1.13. Let us show similarly that the solution set (call it Z) of the non-linear equation

$$xyz = 0$$

(compare Example 1.6) is **not** a subspace because it is *not* closed under addition.

For instance, $\mathbf{a} = (1, 1, 0)$ and $\mathbf{b} = (0, 1, 1)$ both pass the test for membership in Z. The product xyz of their coordinates vanishes for $\mathbf{a}$, and it vanishes for $\mathbf{b}$. But when we form the sum $\mathbf{a} + \mathbf{b}$, we get

$$\mathbf{a} + \mathbf{b} = (1, 1, 0) + (0, 1, 1) = (1, 2, 1)$$

whose coordinates do *not* multiply to zero. The sum of two vectors in Z therefore does *not* always belong to Z, so Z is *not* closed under addition.

On the other hand, Z *is* closed under scalar multiplication.

To see this, note that when we multiply any point $(x, y, z) \in Z$ by a scalar s, we get (sx, sy, sz), a vector whose coordinates *do* still multiply to zero:

$$(sx)(sy)(sz) = s^3(xyz) = s^3 \cdot 0 = 0$$

This puts (sx, sy, sz) in Z for *any* scalar s and *any* $(x, y, z) \in Z$. So Z is closed under scalar multiplication. Still, it failed to be closed under addition, so it is not a subspace. □

Here's a fact that often makes it easy to see that a subset is *not* a subspace:

OBSERVATION 1.14. *If a subset of* $\mathbf{R}^n$ *does* not *contain the origin, it is* not *a subspace.*

PROOF. We prove the contrapositive: *If S is a subspace, it must contain the origin.* Indeed, if S is a subspace, it is non-empty, hence contains a vector $\mathbf{v}$. Since 0 is a scalar, and S is closed under scalar multiplication, S must also contain $0 \cdot \mathbf{v} = \mathbf{0}$. □

EXAMPLE 1.15. The unit sphere in $\mathbf{R}^n$, consisting of all vectors $\mathbf{x} \in \mathbf{R}^n$ with $|\mathbf{x}| = 1$ is *not* a subspace. We can disqualify it immediately by observing that it does *not* contain the origin:

$$|\mathbf{0}| = \sqrt{0^2 + 0^2 + \cdots 0^2} = 0 \neq 1$$

□

REMARK 1.16. (Warning!) The *converse* of Observation 1.14 is *false*:

Containing the origin does not *qualify a subset as a subspace.*

The integer lattice provides a vivid illustration. It *does* contain the origin $(0, 0, \dots, 0)$, since 0 is an integer. But we showed above (Example 1.12) that it still isn't a subspace. (It isn't closed under scalar multiplication.)

REMARK 1.17. So far, we have treated linear algebra as a theory about $\mathbf{R}^n$. Increasingly, however, we shall see that the theorems of linear algebra hold equally well for *subspaces of* $\mathbf{R}^n$, since one can add and scalar multiply in them. Subspaces can be domains and ranges of linear transformations, for instance. The notion of subspace thus extends our understanding to a larger set of mathematical "universes."

One could go even further: There are linear-algebraic universes entirely *outside* $\mathbf{R}^n$. One defines these spaces, called **vector spaces**, by listing a set of axioms they must obey. Chief among these are the requirement that the space support—and be *closed* under—operations of addition and scalar multiplication.

To name just two of many possible examples, the set of all cubic polynomials forms a vector space, as do the solutions of any homogeneous linear differential equation. Two cubic polynomials, for instance, can be added to give a new cubic polynomial (we just add the coefficients). We can also scalar multiply cubic polynomials, again just by multiplying coefficients. Indeed, the analogous facts hold for polynomials of any fixed degree (compare Exercise 290). Virtually all we learn about linear algebra in this book turns out to be essential in these settings as well.

We do not explore more general vector spaces here, however, for we feel that doing so does not improve one's command of the subject much. All the linear-algebraic features of general vector spaces arise already in $\mathbf{R}^n$ and its subspaces. Indeed, $\mathbf{R}^n$ is the universal model of a vector space. Students who understand linear algebra in $\mathbf{R}^n$ and its subspaces will have little difficulty absorbing the notion of vector space if and when they encounter it.

– Practice –

278. Suppose that $S \subset \mathbf{R}^n$ is...

    a) ...the subset containing a single vector $\mathbf{v}$ and nothing else. show that S is a closed under addition if and only if $\mathbf{v} = \mathbf{0}$.

    b) ...the subset containing all vectors $\mathbf{v} \in \mathbf{R}^n$ whose first and last coordinates are equal: $v_1 = v_n$. Is S a subspace? Justify your answer.

    c) ...the subset containing all vectors $\mathbf{v} \in \mathbf{R}^n$ whose first and last coordinates both equal 1: $v_1 = v_n = 1$. Is S a subspace? Justify your answer.

279. Show that:

   a) Every subspace of $\mathbf{R}^n$ (except the trivial subspace $\{\mathbf{0}\}$) contains infinitely many vectors.

   b) When a subspace contains any pair of vectors $\mathbf{p}$ and $\mathbf{v} \neq \mathbf{0}$, it also contains every vector on the line through $\mathbf{p}$ in the $\mathbf{v}$ direction. (See Definition 4.3.)

280. Consider the set of vectors in $\mathbf{R}^n$ whose coordinates are *even* integers. Is this set closed under addition? How about the set of vectors with *odd* integer coordinates?

281. Show that:

   a) The set of all $(x, y, z) \in \mathbf{R}^3$ with $x^2 + y^2 + z^2 \leq 1$ (the unit ball) is *not* closed under addition.

   b) The unit ball in $\mathbf{R}^3$ is *not* closed under scalar multiplication.

   c) The set of all $(x, y, z) \in \mathbf{R}^3$ with $x^2 + y^2 - z^2 \leq 0$ *is* closed under scalar multiplication. Is it a subspace? [*Hint:* $(x, y, -z)$ *is in this set whenever* (x, y, z) *is.*]

282. Show that the unit cube in $\mathbf{R}^n$ (Example 1.2) is not a subspace.

283. For what values of m and b (if any) is the line with slope m and y-intercept b a subspace of $\mathbf{R}^2$?

284. Show that the graph of the absolute value function $y = |x|$ in $\mathbf{R}^2$ is closed under multiplication by non-negative scalars, but not closed under multiplication by all scalars.

285. Suppose $\mathbf{A}$ is any matrix. Show that the solutions of the homogeneous system $\mathbf{A}\mathbf{x} = \mathbf{0}$ form a subspace of $\mathbf{R}^n$. (That is, show that set of homogeneous solutions is closed under addition and scalar multiplication.)

286. Show that when $\mathbf{b} \neq \mathbf{0}$, the solution set of $\mathbf{A}\mathbf{x} = \mathbf{b}$ will be closed under *neither* addition *nor* scalar multiplication.

287. If we fix n and m, we can add and scalar multiply $n \times m$ matrices.

   a) Is the set of all $n \times m$ matrices *closed* under addition and scalar multiplication?

   b) Which of these sets of matrices are closed under both operations? Justify your answers.

   - The set of all $n \times m$ matrices with integer entries

   - The set of all *upper-triangular* $n \times n$ matrices

   - The set of all *triangular* $n \times n$ matrices

   - The set of all *symmetric* $n \times n$ matrices

   - The set of all *skew-symmetric* $n \times n$ matrices

   - The set of all $n \times m$ matrices with non-negative entries

288. Suppose $\mathbf{A}$ is an $n \times n$ matrix, and fix a scalar $\lambda \in \mathbf{R}$. Show that the set of all λ-*eigevectors* form a subspace of $\mathbf{R}^n$ (You may want to review Section 1.35 from Chapter 1.) This subspace, known as the λ-*eigenspace of* $\mathbf{A}$ will play a key role in Chapter 7.

289. Suppose $S, T \subset \mathbf{R}^n$ are two subspaces.

   a) Show that the intersection $S \cap T$ is also a subspace.

   b) We saw in Example 1.13 that the union of subspaces (for instance, several planes through the origin in $\mathbf{R}^3$) may fail to be a subspace. Give an example of two subspaces $S, T \in \mathbf{R}^3$ whose union *is* a subspace.

290. Show that the set of all quadratic polynomials $f(x) = Ax^2 + Bx + C$ is closed under addition and multiplication by scalars. Is the same true for cubic polynomials? Polynomials of degree n, where n is some fixed positive integer? *All* polynomials?

[*Note: Even though the set of quadratic polynomials is closed under both operations, they don't form a subspace of any $\mathbf{R}^n$, since polynomials aren't vectors in $\mathbf{R}^n$. They* do *form a vector space though (see Remark 1.17.]*

291. The *graph* of a mapping $F : \mathbf{R}^m \to \mathbf{R}^n$ is the subset $G_F \subset \mathbf{R}^{m+n}$ given by

$$G_F = \{(\mathbf{x}, F(\mathbf{x})) : \mathbf{x} \in \mathbf{R}^m\}$$

Show that F is linear if and only if G_F is a *subspace* of $\mathbf{R}^{n+m}$.

$- \star -$

2. Spans and Perps

In practice, every subspace of $\mathbf{R}^n$ arises in one of two dual—and beautifully complementary—ways: as a *span*, or as a *perp*. We have actually seen simple examples of each type (without saying so) already. We revisit those examples before giving the general definition.

EXAMPLE 2.1 (Simplest span: the line generated by $\mathbf{v}$). In Section 4.1 of Chapter 3, we introduced *the line generated by a vector* $\mathbf{v} \in \mathbf{R}^n$. It is simply the *union* of all scalar multiples $t\,\mathbf{v}$, as t runs over all scalars.

The line generated by $\mathbf{v}$ *is a subspace.* This is easy to check. The line is closed under addition because the sum of two multiples of $\mathbf{v}$ is still a multiple of $\mathbf{v}$. Specifically, for any two scalars $t, s \in \mathbf{R}$ we have

$$t\,\mathbf{v} + s\,\mathbf{v} = (t + s)\mathbf{v}$$

which is a scalar multiple of $\mathbf{v}$.

Similarly, the line generated by $\mathbf{v}$ is closed under scalar multiplication, since "a scalar multiple of a scalar multiple is still a scalar multiple." That is,

$$s\,(t\,\mathbf{v}) = (st)\mathbf{v}$$

again a scalar multiple of $\mathbf{v}$ for any $s, t \in \mathbf{R}$. So the line generated by $\mathbf{v} \in \mathbf{R}^n$ is indeed a subspace of $\mathbf{R}^n$. □

EXAMPLE 2.2 (Simplest perp: the hyperplane perpendicular to $\mathbf{a}$). In Section 4.15 of Chapter 3, we introduced *the hyperplane* $\mathbf{a}^\perp$. Here $\mathbf{a} \in \mathbf{R}^n$ can be any non-zero vector, and $\mathbf{a}^\perp$ denotes the solution set of the single linear equation $\mathbf{a} \cdot \mathbf{x} = 0$. Equivalently, $\mathbf{a}^\perp$ is the set of all vectors $\mathbf{x} \in \mathbf{R}^n$ that are perpendicular to $\mathbf{a}$.

The hyperplane $\mathbf{a}^\perp$ *is a subspace.* As usual, one verifies this by checking closure under addition and scalar multiplication.

To check closure under addition, we sum an arbitrary pair of vectors $\mathbf{v}, \mathbf{w} \in \mathbf{a}^\perp$. Since $\mathbf{v}$ and $\mathbf{w}$ belong to $\mathbf{a}^\perp$, they each pass the membership test:

$$\mathbf{a} \cdot \mathbf{v} = 0 \quad \text{and} \quad \mathbf{a} \cdot \mathbf{w} = 0$$

We then apply the test to their sum, and find that it too passes:

$$\mathbf{a} \cdot (\mathbf{v} + \mathbf{w}) = \mathbf{a} \cdot \mathbf{v} + \mathbf{a} \cdot \mathbf{w} = 0 + 0 = 0$$

So $\mathbf{a}^\perp$ is closed under addition, as claimed.

Similarly for scalar multiplication. If $\mathbf{v} \in \mathbf{a}^\perp$, and $t \in \mathbf{R}$ is any scalar, then $t\mathbf{v}$ remains in $\mathbf{a}^\perp$, since

$$\mathbf{a} \cdot (t\mathbf{v}) = t\,(\mathbf{a} \cdot \mathbf{v}) = t \cdot 0 = 0$$

So $\mathbf{a}^\perp$ is closed under scalar multiplication too, which makes it a subspace. □

Let us record the facts proven by these two examples:

OBSERVATION 2.3. *Given any non-zero vector* $\mathbf{a} \in \mathbf{R}^n$, *both the line generated by* $\mathbf{a}$, *and the hyperplane* $\mathbf{a}^\perp$, *are subspaces.*

The line generated by $\mathbf{v}$ is the simplest case of a *span* description of a subspace, while the hyperplane $\mathbf{a}^\perp$ is the simplest case of a *perp* description. We will eventually see that every subspace can be described both ways: as a span, or as a perp, but we first need precise definitions of these terms.

DEFINITION 2.4. Any non-empty set of vectors $\{\mathbf{a}_1, \mathbf{a}_2, \ldots, \mathbf{a}_k\} \subset \mathbf{R}^n$ determines the following two simple subsets of $\mathbf{R}^n$.

- The **span** of the set, denoted by span$\{\mathbf{a}_1, \mathbf{a}_2, \ldots, \mathbf{a}_k\}$, is the **union** of all linear combinations of the $\mathbf{a}_i$'s.

- The **perp** of the set, denoted $\{\mathbf{a}_1, \mathbf{a}_2, \ldots, \mathbf{a}_k\}^\perp$, is the **intersection** of the k hyperplanes $\mathbf{a}_1^\perp, \mathbf{a}_2^\perp, \ldots, \mathbf{a}_k^\perp$.

Occasionally, one also needs to know the span or perp (in $\mathbf{R}^n$) of the *empty* set. For that purpose, we simply declare: *The span of the empty set is* $\{\mathbf{0}\}$. *The perp of the empty set is all of* $\mathbf{R}^n$. As our theory develops, we shall see why these declarations are sensible.   □

In fact, spans and perps are always sub*spaces*—not just sub*sets*. We prove that in Proposition 2.9 below, but first, we want to point out that *we are already in the habit of using both spans and perps to describe the solution sets of homogeneous systems.*

EXAMPLE 2.5 (Homogeneous solution sets as perps). Consider the matrix

$$\mathbf{A} = \begin{bmatrix} 1 & -1 & 1 & -1 \\ 1 & 1 & -1 & -1 \end{bmatrix}$$

The homogeneous system $\mathbf{A}\mathbf{x} = \mathbf{0}$ then amounts to the following system of linear equations:

$$x_1 - x_2 + x_3 - x_4 = 0$$
$$x_1 + x_2 - x_3 - x_4 = 0$$

Taken singly, each equation here defines the simplest kind of perp: a hyperplane. Solutions of the first equation form the hyperplane $\mathbf{a}_1^\perp$, where $\mathbf{a}_1 = (1, -1, 1, -1)$, while solutions of the second equation form the hyperplane $\mathbf{a}_2^\perp$, where $\mathbf{a}_2 = (1, 1, -1, -1)$. A vector $\mathbf{x}$ solves the *system* if and only if it solves *both* equations, which puts it on *both* the hyperplanes, and hence on the *intersection*

$$\mathbf{a}_1^\perp \cap \mathbf{a}_2^\perp$$

The solution set H of the system is thus a *perp*:

$$H = \{\mathbf{a}_1, \mathbf{a}_2\}^\perp$$

$\square$

The reader can easily check that the same reasoning applies to *any* homogeneous system, thereby proving

OBSERVATION 2.6. *The solution set H of a homogeneous system $\mathbf{Ax} = \mathbf{0}$ is given by the **perp** of the rows $\mathbf{a}_i$ of $\mathbf{A}$:*

$$H = \mathbf{a}_1^\perp \cap \mathbf{a}_2^\perp \cap \cdots \cap \mathbf{a}_n^\perp$$
$$= \{\mathbf{a}_1, \mathbf{a}_2, \ldots, \mathbf{a}_n\}^\perp$$

The very same homogeneous solution set can *also* be described as a *span*, however.

EXAMPLE 2.7 (Homogeneous solution sets as spans). Let us actually *solve* the homogeneous system $\mathbf{Ax} = \mathbf{0}$ from Example 2.5 above using Gauss-Jordan elimination. To do so, we row-reduce the coefficient matrix to get

$$\text{RRE}\,(\mathbf{A}) = \begin{bmatrix} 1 & 0 & 0 & -1 \\ 0 & 1 & -1 & 0 \end{bmatrix}$$

Columns 3 and 4 are free, giving us two homogeneous generators:

$$\mathbf{h}_1 = (0, 1, 1, 0) \quad \text{and} \quad \mathbf{h}_2 = (1, 0, 0, 1)$$

The solution set H of the system is then given by all linear combinations of these two vectors, which makes it a *span*:

$$H = \text{all } s_1\mathbf{h}_2 + s_2\mathbf{h}_2$$
$$= \text{span}\,\{\mathbf{h}_1, \mathbf{h}_2\}$$

$\square$

The reader who understands these examples will see that the same reasoning applies to *any* homogeneous system. After all, the solution of any homogeneous system is the span of its homogeneous generators. Thus:

OBSERVATION 2.8. *The solution set H of a homogeneous system $\mathbf{A}x =$*
0 *is given by the **span** of the homogeneous generators we get from the*
free columns of RRE$(\mathbf{A})$ *:*

$$H = \operatorname{span}\{\mathbf{h}_1, \mathbf{h}_2, \ldots, \mathbf{h}_k\}$$

Observations 2.6 and 2.8 show that every homogeneous solution set can
be described as both a *span* and a *perp*, and we shall see that the same
holds for *every* subspace of $\mathbf{R}^n$. We can describe the same subspace
as the perp of one set, or as the span of a different set. Moreover, the
two types of description complement each other with an elegant mathematical symmetry. In this way, subspaces manifest the dual nature of
our subject.

Before going further, however, we must resolve a bit of unfinished business and prove that both spans and perps are always closed under
addition and scalar multiplication—that is, that every span and every
perp is indeed a subspace.

PROPOSITION 2.9. *Let $V = \{\mathbf{v}_1, \mathbf{v}_2, \ldots, \mathbf{v}_k\}$ be any set of k vectors in*
$\mathbf{R}^n$. *Then* span V *and* $V^\perp$ *are both subspaces of* $\mathbf{R}^n$.

PROOF. We deal first with span V, proving it closed under addition
and scalar multiplication.

To verify closure under addition, recall that a vector belongs to span V
if and only if it can be expressed as a linear combination of the $\mathbf{v}_i$'s.
So when $\mathbf{x}$ and $\mathbf{y}$ are arbitrary vectors in span V, we have

$$\mathbf{x} = c_1\mathbf{v}_1 + c_2\mathbf{v}_2 + \cdots + c_k\mathbf{v}_k$$
$$\mathbf{y} = d_1\mathbf{v}_1 + d_2\mathbf{v}_2 + \cdots + d_k\mathbf{v}_k$$

Sum these expansions and use the distributive law to collect multiples
of each $\mathbf{v}_i$ to get

$$\mathbf{x} + \mathbf{y} = (c_1 + d_1)\mathbf{v}_1 + (c_2 + d_2)\mathbf{v}_2 + \cdots + (c_k + d_k)\mathbf{v}_k$$

This is again a linear combination of the $\mathbf{v}_i$'s, so span V is closed
under addition.

An even shorter argument proves that span V is closed under scalar
multiplication, and we leave that as an exercise.

Next consider $V^\perp$. To prove closure under addition, as usual, we consider two arbitrary vectors $\mathbf{x}, \mathbf{y} \in V^\perp$. By Definition 2.4, their membership in $V^\perp$ means

$$\mathbf{x}, \mathbf{y} \in \mathbf{v}_1^\perp \cap \mathbf{v}_2^\perp \cap \cdots \cap \mathbf{v}_k^\perp$$

In particular, *both* $\mathbf{x}$ and $\mathbf{y}$ are in *each* of the hyperplanes $\mathbf{v}_i^\perp$, and we want to show that the same holds for their sum: to remain in $V^\perp$, the sum must lie in *each* hyperplane $\mathbf{v}_i^\perp$.

This holds thanks to Observation 2.3, which assures us that each hyperplane $\mathbf{v}_i^\perp$ is *itself* a subspace, and hence closed under addition and scalar multiplication. It follows that once $\mathbf{x}$ and $\mathbf{y}$ belong to $\mathbf{v}_i^\perp$, so does their sum $\mathbf{x} + \mathbf{y}$. Since this holds for each $i = 1, 2, \ldots, k$, we have

$$\mathbf{x} + \mathbf{y} \in \mathbf{v}_1^\perp \cap \mathbf{v}_2^\perp \cap \cdots \cap \mathbf{v}_k^\perp$$

This makes $V^\perp$ closed under addition. As with spans, one gets closure under scalar multiplication by a similar argument that we leave as an exercise. □

– Practice –

292. Let $\mathbf{a}_1 = (1, 0, -2)$ and $\mathbf{a}_2 = (-2, 1, 0)$ in $\mathbf{R}^3$.

   a) Find two non-zero vectors (not multiples of each other) in $\mathbf{a}_1^\perp$.

   b) Find one non-zero vector in $\{\mathbf{a}_1, \mathbf{a}_2\}^\perp$.

293. Consider the vectors $\mathbf{a}_1 = (7, -5, 4)$ and $\mathbf{a}_2 = (-2, 6, 3)$ in $\mathbf{R}^3$.

   a) Find two non-zero vectors (not multiples of each other or the $\mathbf{a}_i$'s) in $\mathrm{span}\{\mathbf{a}_1, \mathbf{a}_2\}$.

   b) Which of these three vectors belong to $\mathrm{span}\{\mathbf{a}_1, \mathbf{a}_2\}$?

$$\mathbf{b}_1 = \begin{pmatrix} 7 \\ 27 \\ 33 \end{pmatrix}, \quad \mathbf{b}_2 = \begin{pmatrix} 22 \\ -8 \\ 24 \end{pmatrix}, \quad \mathbf{b}_3 = \begin{pmatrix} 24 \\ -8 \\ 22 \end{pmatrix}$$

   [*Hint: The technique explored in Exercises 254 and 255 is useful (though not necessary) here.*]

294. Consider the vectors $\mathbf{a}_1 = (2, -6, 4)$ and $\mathbf{a}_2 = (-2, 6, 2)$ in $\mathbf{R}^3$.

   a) Find a non-zero vector in $\{\mathbf{a}_1, \mathbf{a}_2\}^\perp$.

   b) Which of these three vectors belong to $\{\mathbf{a}_1, \mathbf{a}_2\}^\perp$?

$$\mathbf{b}_1 = \begin{pmatrix} 4 \\ 0 \\ -2 \end{pmatrix}, \quad \mathbf{b}_2 = \begin{pmatrix} 0 \\ -2 \\ 6 \end{pmatrix}, \quad \mathbf{b}_3 = \begin{pmatrix} 24 \\ 8 \\ 0 \end{pmatrix}$$

295. Consider the vectors $\mathbf{a}_1 = (1, -1, 1, -1, 0)$, $\mathbf{a}_2 = (0, 1, -1, 1, -1)$, and $\mathbf{a}_3 = (-1, 0, 1, -1, 1)$ in $\mathbf{R}^5$. Find two non-zero vectors (not multiples of each other) in $\{\mathbf{a}_1, \mathbf{a}_2, \mathbf{a}_3\}^{\perp}$.

296. Consider the vectors $\mathbf{a}_1 = (1, -2, 3, 0)$, $\mathbf{a}_2 = (0, 1, -2, 3)$, and $\mathbf{a}_3 = (3, 0, 1, -2)$ in $\mathbf{R}^4$.

   a) Find three non-zero vectors (other than multiples of each other or the $\mathbf{a}_i$'s themselves) in $\mathrm{span}\{\mathbf{a}_1, \mathbf{a}_2, \mathbf{a}_3\}$.

   b) Which of these three vectors belong to $\mathrm{span}\{\mathbf{a}_1, \mathbf{a}_2, \mathbf{a}_3\}$?

$$\mathbf{b}_1 = \begin{pmatrix} 4 \\ -1 \\ 2 \\ -1 \end{pmatrix}, \quad \mathbf{b}_2 = \begin{pmatrix} 3 \\ -4 \\ 7 \\ -4 \end{pmatrix}, \quad \mathbf{b}_3 = \begin{pmatrix} -1 \\ 7 \\ 5 \\ 1 \end{pmatrix}$$

   [*Hint: The technique explored in Exercises 254 and 255 is useful (though not necessary) here.*]

297. Consider the vectors $\mathbf{a}_1 = (1, -2, 3, -4)$, and $\mathbf{a}_2 = (2, -3, 4, -5)$ in $\mathbf{R}^4$.

   a) Find two non-zero vectors (other than multiples of each other) in $\{\mathbf{a}_1, \mathbf{a}_2\}^{\perp}$.

   b) Which of these three vectors belong to $\{\mathbf{a}_1, \mathbf{a}_2\}^{\perp}$?

$$\mathbf{b}_1 = \begin{pmatrix} 2 \\ 5 \\ -4 \\ -5 \end{pmatrix}, \quad \mathbf{b}_2 = \begin{pmatrix} 2 \\ 1 \\ 4 \\ 3 \end{pmatrix}, \quad \mathbf{b}_3 = \begin{pmatrix} 4 \\ 0 \\ 0 \\ 1 \end{pmatrix}$$

298. a) Let m be any scalar. Find a vector in $\mathbf{R}^2$ whose span is the line of slope m through the origin of $\mathbf{R}^2$.

   b) Find two vectors in $\mathbf{R}^3$ whose span is the plane $x_3 = 0$ in $\mathbf{R}^3$.

   c) Find three vectors in $\mathbf{R}^4$ whose span is the hyperplane through $\mathbf{0}$ and perpendicular to $(1, 1, 1, 1)$.

299. a) Let m be any scalar. Find a vector in $\mathbf{R}^2$ whose perp is the line of slope m through the origin of $\mathbf{R}^2$.

b) Find two vectors in $\mathbf{R}^3$ whose perp is the z-axis in $\mathbf{R}^3$.

c) What is $\{\mathbf{e}_1, \mathbf{e}_2, \mathbf{e}_3\}^\perp \subset \mathbf{R}^3$? (The $\mathbf{e}_i$'s are the standard basis vectors in $\mathbf{R}^3$.)

300. Is it possible for the span of a set in $\mathbf{R}^n$ to *equal* the perp of that same set? Give an example showing it is possible, or prove that it is not possible.

301. Complete the proof of Proposition 2.9 by showing that

a) The *span* of any set $V = \{\mathbf{v}_1, \mathbf{v}_2, \ldots, \mathbf{v}_k\}$ in $\mathbf{R}^n$ is closed under scalar multiplication.

b) The *perp* of any set $V = \{\mathbf{v}_1, \mathbf{v}_2, \ldots, \mathbf{v}_k\}$ in $\mathbf{R}^n$ is closed under scalar multiplication.

$$- \star -$$

3. Nullspace

Every $n \times m$ matrix $\mathbf{A}$ gives rise in a simple, natural way to four subspaces—a span and a perp in both $\mathbf{R}^m$ and $\mathbf{R}^n$. All four carry important algebraic and geometric information about the linear transformation represented by $\mathbf{A}$.

DEFINITION 3.1. To every $n \times m$ matrix $\mathbf{A}$, we can associate the following four subspaces:

- The **column-space** of $\mathbf{A}$, denoted $\mathrm{col}(\mathbf{A})$ is the *span of the columns* of $\mathbf{A}$. It lies in $\mathbf{R}^n$.

- The **nullspace** of $\mathbf{A}$, denoted $\mathrm{nul}(\mathbf{A})$ is the *perp of the rows* of $\mathbf{A}$. It lies in $\mathbf{R}^m$.

- The **row-space** of $\mathbf{A}$, denoted $\mathrm{row}(\mathbf{A})$ is the *span of the rows* of $\mathbf{A}$. It lies in $\mathbf{R}^m$.

- The **left nullspace** of $\mathbf{A}$, denoted $\mathrm{nul}(\mathbf{A}^T)$ is the *perp of the columns* of $\mathbf{A}$. It lies in $\mathbf{R}^n$.

$\square$

3.2. Nullspace. The two *perps* associated with any matrix are its *nullspace* and its *left nullspace*. Since the left nullspace of $\mathbf{A}$ is simply the nullspace of $\mathbf{A}^T$, however, it will largely suffice to focus on $\mathrm{nul}(\mathbf{A})$.

The nullspace of $\mathbf{A}$ is defined as the perp of the rows, and Observation 2.6 then tells us that that it is also the solution set of $\mathbf{Ax} = \mathbf{0}$. This makes it easy to test vectors for membership in $\mathrm{nul}(\mathbf{A})$ —we just multiply them by $\mathbf{A}$ and see if we get $\mathbf{0}$ (hence the *null* in *nullspace*). To actually *find* vectors in $\mathrm{nul}(\mathbf{A})$, however, we have to work harder and solve the homogeneous system using Gauss–Jordan.

EXAMPLE 3.3. Suppose

$$\mathbf{A} = \begin{bmatrix} 1 & -1 & 0 \\ 0 & 1 & -1 \end{bmatrix}$$

Which (if any) of the vectors

$$\mathbf{v}_1 = \begin{pmatrix} 1 \\ 1 \\ 0 \end{pmatrix}, \quad \mathbf{v}_2 = \begin{pmatrix} 1 \\ 1 \\ 1 \end{pmatrix}, \quad \mathbf{v}_3 = \begin{pmatrix} 0 \\ 1 \\ 1 \end{pmatrix}$$

lie in $\mathrm{nul}(\mathbf{A})$? To find out, just multiply each $\mathbf{v}_i$ by $\mathbf{A}$:

$$\mathbf{Av}_1 = \begin{pmatrix} 0 \\ 1 \end{pmatrix}, \quad \mathbf{Av}_2 = \begin{pmatrix} 0 \\ 0 \end{pmatrix}, \quad \mathbf{Av}_3 = \begin{pmatrix} -1 \\ 0 \end{pmatrix}$$

These facts show that $\mathbf{v}_2$ *does* lie in $\mathrm{nul}(\mathbf{A})$, while $\mathbf{v}_1$ and $\mathbf{v}_3$ do not. But what if we sought more vectors in $\mathrm{nul}(\mathbf{A})$—how could we find them?

Since $\mathrm{nul}(\mathbf{A})$ comprises all solutions to $\mathbf{Ax} = \mathbf{0}$, we just need to solve that system. The resulting homogeneous generators will lie in $\mathrm{nul}(\mathbf{A})$, and more: their linear combinations give *all* solutions of $\mathbf{Ax} = \mathbf{0}$, and hence *all* vectors in $\mathrm{nul}(\mathbf{A})$.

In short, we can also express $\mathrm{nul}(\mathbf{A})$—the *perp* of the rows of $\mathbf{A}$— as the *span* of the homogeneous generators. To find the latter, we row-reduce $\mathbf{A}$ in the familiar way, getting

$$\mathrm{RRE}(\mathbf{A}) = \begin{bmatrix} 1 & 0 & -1 \\ 0 & 1 & -1 \end{bmatrix}$$

The last column (only) is free, yielding one generator:

$$\mathbf{h} = (1, 1, 1)$$

So nul(**A**) is the span of $(1, 1, 1)$. Geometrically, it is *the line generated by* $(1, 1, 1)$.

Note that we now have two descriptions of this line: as the *span* of $(1, 1, 1)$, and as the *perp* of the rows of **A**. Figure 4 illustrates.

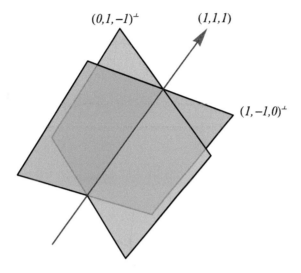

Figure 4. The black arrow indicates the line of solutions to the homogeneous system in Example 3.3. It can be seen as a *span* or a *perp*: the span of $(1, 1, 1)$, or the perp of $\{(1, -1, 0), (0, 1, -1)\}$ (i.e., the intersection of the planes).

$\square$

EXAMPLE 3.4. Which of these vectors

$$\mathbf{v}_1 = \begin{pmatrix} 1 \\ 1 \\ 1 \end{pmatrix}, \quad \mathbf{v}_2 = \begin{pmatrix} 3 \\ 3 \\ 1 \end{pmatrix}, \quad \mathbf{v}_3 = \begin{pmatrix} 3 \\ 2 \\ 1 \end{pmatrix}, \quad \mathbf{v}_4 = \begin{pmatrix} 1 \\ 2 \\ 1 \end{pmatrix}$$

belongs to the nullspace of the one-rowed matrix

$$\mathbf{A} = \begin{bmatrix} 2 & -4 & 6 \end{bmatrix}$$

Again, we can answer this easily by multiplying **A** with each vector to see whether we get **0** or not. Here, multiplication by **A** merely amounts to dotting with $(2, -4, 6)$, and gives

$$\mathbf{A}\mathbf{v}_1 = 4, \quad \mathbf{A}\mathbf{v}_2 = 0, \quad \mathbf{A}\mathbf{v}_3 = 8, \quad \mathbf{A}\mathbf{v}_4 = 0$$

which shows that $\mathbf{v}_2$ and $\mathbf{v}_4$ lie in nul(**A**). The other two vectors do not.

To find *other* vectors in nul($\mathbf{A}$), we again solve the homogeneous system, here just the single equation $2x - 4y + 6z = 0$. The solutions of a linear equation in $\mathbf{R}^3$ form a plane; we may visualize nul($\mathbf{A}$) here as the plane orthogonal to $(2, -4, 6)$. Algebraically, we find the points in this plane in the usual way: we start by row-reducing $\mathbf{A}$ to get

$$\text{RRE}(\mathbf{A}) = \begin{bmatrix} 1 & -2 & 3 \end{bmatrix}$$

The last two columns are free, and our usual procedure supplies two homogeneous generators:

$$\mathbf{h}_1 = \begin{pmatrix} 2 \\ 1 \\ 0 \end{pmatrix} \quad \text{and} \quad \mathbf{h}_2 = \begin{pmatrix} -3 \\ 0 \\ 1 \end{pmatrix}$$

The nullspace of $\mathbf{A}$ is then given by the span (all linear combinations) of these two generators   □

The examples above show that when a subspace $S \subset \mathbf{R}^n$ arises as the nullspace of a matrix $\mathbf{A}$, we can easily test any vector $\mathbf{v}$ for membership in S by computing $\mathbf{Av}$.

They also show that when S arises as a nullspace, we do *not* directly get any explicit vector in S. For that we must *solve* the homogeneous system, which then allows us to describe S as a *span*—the span of the homogeneous generators. This involves more work than the easier "membership test" above.

3.5. Summary. Here is a list of some key properties of the nullspace of an $n \times m$ matrix $\mathbf{A}$. When we reach the corresponding summary for column-space, in the next section, the reader should come back to this and compare the two lists. They exhibit a satisfying complementarity.

- *The nullspace of $\mathbf{A}_{n \times m}$ is a subspace of $\mathbf{R}^m$ (not $\mathbf{R}^n$).*
- *It is easy to test any vector for membership in* nul($\mathbf{A}$) *(just multiply by $\mathbf{A}$ and see if you get $\mathbf{0}$).*
- *To actually* find *vectors in* nul($\mathbf{A}$), *we must solve* $\mathbf{Ax} = \mathbf{0}$.
- *The nullspace of $\mathbf{A}$ is orthogonal to every* row *of $\mathbf{A}$ (it is the perp of the rows).*
- *The nullspace of $\mathbf{A}$ is trivial if and only if* RRE($\mathbf{A}$) *has a pivot in every column.*
- *The nullspace of $\mathbf{A}$ cannot be trivial if $m > n$.*

– Practice –

302. Answer these questions about the nullspace:

a) What is the nullspace of the identity matrix $\mathbf{I}_n$?

b) What is the nullspace of the $n \times m$ zero matrix $\mathbf{0}_{n \times m}$?

c) Which $n \times m$ matrices have all of $\mathbf{R}^m$ for their nullspace?

d) Which $n \times n$ matrices have only $\{\mathbf{0}\}$ for their nullspace?

303. Find two vectors that span the nullspace of this matrix. What is its *left* nullspace?

$$\begin{bmatrix} 1 & -1 & 0 & 1 \\ 0 & 1 & -1 & 0 \end{bmatrix}$$

304. Consider the matrix

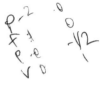

$$\mathbf{A} = \begin{bmatrix} 1 & -2 & 0 & 0 \\ 0 & 0 & 2 & 1 \end{bmatrix}$$

a) Which of these vectors lie in the nullspace of $\mathbf{A}$?

$$\mathbf{x} = \begin{pmatrix} -2 \\ 1 \\ 0 \\ 0 \end{pmatrix}, \quad \mathbf{y} = \begin{pmatrix} -4 \\ 2 \\ -2 \\ 1 \end{pmatrix}, \quad \mathbf{z} = \begin{pmatrix} 0 \\ -2 \\ 1 \\ 0 \end{pmatrix}, \quad \mathbf{w} = \begin{pmatrix} 0 \\ -2 \\ 2 \\ 0 \end{pmatrix}$$

b) Find all vectors in $\text{nul}(\mathbf{A})$.

305. Consider this matrix:

$$\mathbf{B} = \begin{bmatrix} 2 & 1 & 3 \\ 4 & 2 & 6 \\ 2 & 1 & 3 \\ 6 & 3 & 9 \end{bmatrix}$$

a) Which vectors below lie in the nullspace of $\mathbf{B}$?

$$\mathbf{x} = \begin{pmatrix} 2 \\ 2 \\ -2 \end{pmatrix}, \quad \mathbf{y} = \begin{pmatrix} 3 \\ -6 \\ 3 \end{pmatrix}, \quad \mathbf{z} = \begin{pmatrix} 4 \\ -2 \\ -2 \end{pmatrix}, \quad \mathbf{w} = \begin{pmatrix} -2 \\ 4 \\ -2 \end{pmatrix}$$

b) Find *all* vectors in $\text{nul}(\mathbf{B})$.

306. Let $a, b, c \in \mathbf{R}$ be any scalars that are not all zero. What then is the nullspace of this matrix?

$$\begin{bmatrix} 0 & c & -b \\ -c & 0 & a \\ b & -a & 0 \end{bmatrix}$$

307. Consider matrices $\mathbf{A}$ and $\mathbf{B}$ below.

a) Which columns of $\mathbf{B}$ lie in $\mathrm{nul}(\mathbf{A})$?

b) Which columns of $\mathbf{A}$ lie in $\mathrm{nul}(\mathbf{B})$?

$$\mathbf{A} = \begin{bmatrix} 1 & 1 & 1 & 1 \\ 1 & 0 & 0 & -1 \\ 1 & -1 & 1 & -1 \\ 0 & 0 & 0 & 0 \end{bmatrix} \qquad \mathbf{B} = \begin{bmatrix} 1 & 1 & 1 & 1 \\ -1 & 1 & 1 & -1 \\ -1 & 1 & -1 & 1 \\ 1 & 1 & -1 & -1 \end{bmatrix}$$

308. Show that:

a) The nullspace and left nullspace of a *symmetric* matrix are always the same.

b) Does this hold for *skew*-symmetric matrices?

c) The nullspace and left nullspace of this matrix are the same, though it is neither symmetric nor skew-symmetric:

$$\begin{bmatrix} 1 & 2 & 3 \\ 4 & 5 & 6 \\ 7 & 8 & 9 \end{bmatrix}$$

309. Prove true or give a counterexample:

a) If an $n \times m$ matrix has more rows than columns, then its nullspace contains *only* the origin.

b) If an $n \times m$ matrix has two identical columns, then its nullspace contains more than just the origin.

c) If an $n \times m$ matrix has two identical rows, then its nullspace contains more than just the origin.

d) If every row of an $n \times m$ matrix $\mathbf{A}$ lies in the nullspace of that same matrix $\mathbf{A}$, then all entries of $\mathbf{A}$ vanish.

310. When $\mathbf{A}$ and $\mathbf{B}$ are two matrices of the same size, the nullspace of $\mathbf{A} + \mathbf{B}$ bears little or no relation to that of $\mathrm{nul}(\mathbf{A})$ and $\mathrm{nul}(\mathbf{B})$.

    a) Can you find non-zero square matrices $\mathbf{A}$ and $\mathbf{B}$, each having a non-trivial nullspace, such that $\mathrm{nul}(\mathbf{A})$ and $\mathrm{nul}(\mathbf{B})$ are both *proper* subsets of $\mathrm{nul}(\mathbf{A} + \mathbf{B})$?

    b) Can you find non-zero square matrices $\mathbf{A}$ and $\mathbf{B}$, each having a non-trivial nullspace, and yet $\mathrm{nul}(\mathbf{A} + \mathbf{B}) = \{\mathbf{0}\}$?

311. Show that:

    a) The nullspace of a product $\mathbf{AB}$ always contains that of $\mathbf{B}$. Why is this true?

    b) The nullspace of $\mathbf{AB}$ may *not* contain the nullspace of $\mathbf{A}$. Use the matrices

$$\mathbf{A} = \begin{bmatrix} 1 & 1 \\ 2 & 2 \end{bmatrix} \quad \text{and} \quad \mathbf{B} = \begin{bmatrix} 0 & -1 \\ 1 & 0 \end{bmatrix}$$

    [*Hint: Find a vector that lies in* $\mathrm{nul}(\mathbf{A})$ *but not in* $\mathrm{nul}(\mathbf{B})$.]

4. Column-Space

The nullspace and left nullspace are the two perps associated with a matrix. The two *spans* are its *column-space* and *row-space*—the spans of the columns and rows, respectively. We focus on the column-space here, since the row-space of $\mathbf{A}$ is also the column-space of $\mathbf{A}^T$. Still, the row-space has its own distinct identity and utility, especially with regard to orthogonality, the subject of Chapter 6, and we'll have more to say about it there.

Where $\mathrm{nul}(\mathbf{A})$ was already familiar to us as the solution set of $\mathbf{Ax} = \mathbf{0}$, $\mathrm{col}(\mathbf{A})$ is likewise familiar. For by definition, a vector $\mathbf{b}$ lies in $\mathrm{col}(\mathbf{A})$ if and only if it is a linear combination of the columns:

$$\mathbf{b} = x_1 \mathbf{c}_1(\mathbf{A}) + x_2 \mathbf{c}_2(\mathbf{A}) + \cdots + x_m \mathbf{c}_m(\mathbf{A})$$

Recalling that *linear combination equals matrix/vector multiplication,* however, we can recognize the expansion of $\mathbf{b}$ above as the familiar matrix/vector product $\mathbf{Ax}$, where $\mathbf{x} = (x_1, x_2, \ldots, x_m)$.

This simple insight is important enough to record for future reference.

OBSERVATION 4.1. *The column-space of a matrix* $\mathbf{A}_{n \times m}$ *is the same as the set of products* $\mathbf{Ax}$ *for all* $\mathbf{x} \in \mathbf{R}^m$. *In symbols,*

$$\mathrm{col}(\mathbf{A}) = \{\mathbf{Ax} \colon \mathbf{x} \in \mathbf{R}^m\}$$

Equivalently, $\mathrm{col}(\mathbf{A})$ *is the* image *of the linear transformation represented by* $\mathbf{A}$.

The last sentence here is true because the *outputs* of a transformation T, which make up its image, are all possible $T(\mathbf{x})$'s, as $\mathbf{x}$ runs over its domain. When $\mathbf{A}$ represents T this is the same as all possible $\mathbf{Ax}$'s which, as we have just observed, is the same as $\mathrm{col}(\mathbf{A})$.

To make this more concrete, consider a specific example.

EXAMPLE 4.2. Let

$$\mathbf{A} = \begin{bmatrix} 1 & 0 & -1 \\ -1 & -1 & 0 \\ -1 & 2 & 3 \end{bmatrix}$$

What are some vectors in $\mathrm{col}(\mathbf{A})$? Well, the columns of $\mathbf{A}$ themselves obviously lie in $\mathrm{col}(\mathbf{A})$. Moreover, given any $\mathbf{x} \in \mathbf{R}^3$, $\mathbf{Ax} \in \mathrm{col}(\mathbf{A})$, by Observation 4.1. For instance, with $\mathbf{x} = (1, 2, -3)$, we find that

$$\mathbf{Ax} = \begin{pmatrix} 4 \\ -3 \\ -6 \end{pmatrix} \quad \text{lies in} \quad \mathrm{col}(\mathbf{A})$$

Obviously, it would be equally easy to generate as many other vectors in $\mathrm{col}(\mathbf{A})$ simply by choosing different $\mathbf{x}$'s.

On the other hand, suppose we are *given* a vector $\mathbf{b}$, and we want to *test* it for membership in $\mathrm{col}(\mathbf{A})$. For instance: Do either of the vectors $(1, 1, 1)$ or $(2, 2, -10)$ lie in $\mathrm{col}(\mathbf{A})$? Observation 4.1 above shows that to answer this question, we must determine whether the system

$$\mathbf{Ax} = \mathbf{b}$$

has a solution when $\mathbf{b}$ is either of the vectors $(1, 1, 1)$ or $(2, 2, -10)$ of interest.

This is something we have done many times. It requires us to row-reduce the augmented matrix $[\mathbf{A}|\mathbf{b}]$. Here, where $\mathbf{b} = (1, 1, 1)$, that produces

$$\text{RRE}\left(\begin{bmatrix} \mathbf{A} \mid \mathbf{b} \end{bmatrix}\right) = \begin{bmatrix} 1 & 0 & -1 & \mid & 0 \\ 0 & 1 & 1 & \mid & 0 \\ 0 & 0 & 0 & \mid & 1 \end{bmatrix}$$

The non-zero entry in the lower-right corner signals that we *cannot* solve $\mathbf{A}\mathbf{x} = (1, 1, 1)$, and hence $(1, 1, 1)$ does *not* lie in $\text{col}(\mathbf{A})$.

Doing the same experiment with $\mathbf{b} = (2, 2, -10)$ however, we get

$$\text{RRE}\left(\begin{bmatrix} \mathbf{A} \mid \mathbf{b} \end{bmatrix}\right) = \begin{bmatrix} 1 & 0 & -1 & \mid & 2 \\ 0 & 1 & 1 & \mid & -4 \\ 0 & 0 & 0 & \mid & 0 \end{bmatrix}$$

Here we see that we *can* solve $\mathbf{A}\mathbf{x} = (2, 2, -10)$, and hence that $(2, 2, -10)$ *does* lie in $\text{col}(\mathbf{A})$. Indeed, we can read off the particular solution $\mathbf{x}_p = (2, -4, 0)$, which tells us

$$\begin{pmatrix} 2 \\ 2 \\ -10 \end{pmatrix} = 2 \begin{pmatrix} 1 \\ -1 \\ -1 \end{pmatrix} - 4 \begin{pmatrix} 0 \\ -1 \\ 2 \end{pmatrix} + 0 \begin{pmatrix} -1 \\ 0 \\ 3 \end{pmatrix}$$

The reader will easily verify that this is true. □

Note the perfect complementarity between this example and the null-space Example 3.3, for instance. When we have a nullspace, it is *easy to test* vectors for membership (just multiply by $\mathbf{A}$ and see if you get $\mathbf{0}$), but *hard to find* vectors *in* the nullspace (we have to solve the homogeneous system).

The column-space story is exactly the opposite: it is *hard to test* vectors for membership (we have to solve an inhomogeneous system), but *easy to find* vectors *in* the column-space (just multiply $\mathbf{A}$ by any $\mathbf{x}$).

We will return to this beautiful duality in the next section.

To conclude this section, however, we present two important theoretical results.

The first is an extremely useful solvability criterion, and *we strongly recommend that students contemplate it and its simple proof thoroughly.* In fact, we call it the *V.I.P.* for *Very Important Proposition* to signal its import and to enable future reference. It follows almost trivially from Observation 4.1 above, which was itself quite transparent, but there is something subtle and unexpected about it.

PROPOSITION 4.3 (V.I.P.). *The system* $\mathbf{A}\mathbf{x} = \mathbf{b}$ *has a solution if and only if* $\mathbf{b} \in \text{col}(\mathbf{A})$.

PROOF. $\mathbf{Ax} = \mathbf{b}$ has a solution if and only if $\mathbf{b}$ can be written as $\mathbf{Ax}$ for some $\mathbf{x}$. By Observation 4.1, this is the same as saying $\mathbf{b} \in \mathrm{col}(\mathbf{A})$. □

When $\mathbf{A}$ is *square*, this criterion has an immediate interesting consequence. For when $\mathbf{A}$ is $n \times n$, solvability of $\mathbf{Ax} = \mathbf{b}$ for all $\mathbf{b} \in \mathbf{R}^n$ is equivalent to *invertibility* (Conclusion (d) of Theorem 5.7 in Chapter 4). By the proposition just proven, however, it is also equivalent to having $\mathbf{b} \in \mathrm{col}(\mathbf{A})$ for every $\mathbf{b} \in \mathbf{R}^n$. This gives a new condition equivalent to invertibility. We have accumulated several since Theorem 5.7, so we now collect them in a new version of that result:

THEOREM 4.4. *For any $n \times n$ matrix $\mathbf{A}$, the following are equivalent:*

a) $\mathbf{A}$ *is invertible.*

b) *The homogeneous system $\mathbf{Ax} = \mathbf{0}$ has only the trivial solution.*

c) $\mathrm{RRE}(\mathbf{A}) = \mathbf{I}_n$.

d) *Given any $\mathbf{b} \in \mathbf{R}^n$, we can solve $\mathbf{Ax} = \mathbf{b}$.*

e) $\det(\mathbf{A}) \neq 0$.

f) $\mathrm{nul}(\mathbf{A}) = \{\mathbf{0}\}$.

g) $\mathrm{col}(\mathbf{A}) = \mathbf{R}^n$.

PROOF. It suffices to show that conditions (b)–(g) are all equivalent to invertibility. Conditions (b), (c), and (d) here come directly from Theorem 5.7 of Chapter 4, where we first saw that equivalence. The determinant condition (e) here is equivalent to invertibility by the third conclusion of Corollary 6.8 in Chapter 4. Observaton 2.6 shows that condition (f) above is just another way of stating (b), since the perp of the rows of $\mathbf{A}$ is, by definition, the same as $\mathrm{nul}(\mathbf{A})$. Finally, condition (g) above is equivalent to invertibility because by our V.I.P. (Proposition 4.3) $\mathrm{col}(\mathbf{A}) = \mathbf{R}^n$ means we can solve $\mathbf{Ax} = \mathbf{b}$ for every $\mathbf{b} \in \mathbf{R}^n$, and hence (g) is equivalent to (b). □

REMARK 4.5. Note the complementarity of the relationships of $\mathrm{col}(\mathbf{A})$ and $\mathrm{nul}(\mathbf{A})$ to invertibility in the Theorem just proven. Conclusion (g) links invertibility of $\mathbf{A}$ to the *maximality* of $\mathrm{col}(\mathbf{A})$, while Conclusion (f) links it to the *minimality* of $\mathrm{nul}(\mathbf{A})$. □

4.6. "Span" as a verb. In Definition 2.4, we defined *span* as a *noun*. It is also makes a useful **verb**: Instead of saying that a subspace S *is* the span of a set V, we often say that V *spans* S, or equivalently that the vectors in V span S.

In short, $S = \text{span } V$ and "*V* spans S" mean the same thing: the linear combinations of the vectors in V fill the subspace S.

EXAMPLE 4.7. In $\mathbf{R}^3$, the plane $z = 0$, also known as $\mathbf{e}_3^\perp$, is the *span of* the first two standard basis vectors $\{\mathbf{e}_1, \mathbf{e}_2\}$. There, "span" functions as a noun.

We can make the statement using it as a verb, however, by saying that $\mathbf{e}_1$ and $\mathbf{e}_2$ in $\mathbf{R}^3$ *span* the $z = 0$ plane.

The verbal form is often useful with the spans associated to a matrix too. For instance, we defined $\text{col}(\mathbf{A})$ and $\text{row}(\mathbf{A})$ as the *span of* the columns and rows of $\mathbf{A}$, respectively. To make the same definitions using *span* as a verb, we could say

The columns of $\mathbf{A}$ ***span*** *its column-space. The rows of* $\mathbf{A}$ ***span*** *its row-space.*

Similarly, *the homogeneous generators* ***span*** *the nullspace of* $\mathbf{A}$. □

4.8. Summary. Now we list of some key properties of the column-space of an $n \times m$ matrix $\mathbf{A}$, just as we did for nullspace in the previous section. The reader should compare and contrast the column-space list here with the nullspace list there.

- *The columnspace of* $\mathbf{A}_{n \times m}$ *is a subspace of* $\mathbf{R}^n$ *(not* $\mathbf{R}^m$ *).*

- *It* is *easy to* find *vectors in* $\text{col}(\mathbf{A})$—*simply multiply* $\mathbf{A}$ *by any vector* $\mathbf{x} \in \mathbf{R}^m$.

- *To test a vector* $\mathbf{b} \in \mathbf{R}^n$ *for membership in* $\text{col}(\mathbf{A})$, *however, we must solve* $\mathbf{Ax} = \mathbf{b}$. $\mathbf{b} \in \text{col}(\mathbf{A})$ *if and only if a solution exists.*

- *Every* column *of* $\mathbf{A}$ *is orthogonal to the left nullspace of* $\mathbf{A}$, *which is the perp of the columns.*

- *The column-space of* $\mathbf{A}$ *is all of* $\mathbf{R}^n$ *if and only if* RRE$(\mathbf{A})$ *has a pivot in every column.*

- *The column-space of* $\mathbf{A}$ *cannot be all of* $\mathbf{R}^n$ *if* $n > m$.

– Practice –

312. Answer these questions about the column-space:

    a) What is the column-space of the identity matrix I_n?

    b) What is the column-space of the zero matrix $0_{n \times m}$?

    c) Which $n \times m$ matrices have only $\{0\}$ for their column-space?

    d) Which $n \times n$ matrices have all of $\mathbf{R}^n$ for their column-space?

313. Consider the matrix

$$A = \begin{bmatrix} 1 & -2 & 3 \\ -2 & 1 & -2 \\ 3 & -2 & 1 \end{bmatrix}$$

    a) Find four vectors (not multiples of single columns) in $\mathrm{col}(A)$.

    b) Which vectors below lie in $\mathrm{col}(A)$? [*Hint: Though not necessary, the technique of Exercises 254 and 255 can help.*]

$$\mathbf{x} = \begin{pmatrix} 2 \\ -3 \\ 2 \end{pmatrix}, \quad \mathbf{y} = \begin{pmatrix} 6 \\ -5 \\ 6 \end{pmatrix}, \quad \mathbf{z} = \begin{pmatrix} 2 \\ -5 \\ 2 \end{pmatrix}, \quad \mathbf{w} = \begin{pmatrix} 2 \\ -6 \\ 6 \end{pmatrix}$$

    c) Which of the vectors in part (b) lie in the *row-space* of A?

314. Consider the matrix

$$L = \begin{bmatrix} 1 & -4 \\ -2 & 3 \\ -3 & 2 \\ 4 & -1 \end{bmatrix}$$

    a) Find three vectors (not multiples of single columns) that lie in $\mathrm{col}(L)$.

b) Which vectors below lie in col($\mathbf{L}$)? [*Hint: Though not neces-sary, the technique of Exercises 254 and 255 can help.*]

$$\mathbf{x} = \begin{pmatrix} -2 \\ -7 \\ -8 \\ 7 \end{pmatrix}, \quad \mathbf{y} = \begin{pmatrix} 3 \\ -1 \\ -1 \\ 3 \end{pmatrix}, \quad \mathbf{z} = \begin{pmatrix} 3 \\ -1 \\ 1 \\ -3 \end{pmatrix}, \quad \mathbf{w} = \begin{pmatrix} 2 \\ -6 \\ 6 \\ 2 \end{pmatrix}$$

c) Do either (or both) of the standard basis vectors $\mathbf{e}_1, \mathbf{e}_2 \subset \mathbf{R}^2$ lie in the *row-space* of $\mathbf{L}$?

315. For each set V given below, either show that span $V = \mathbf{R}^3$, or find a vector *not* in span V. In the latter case, describe span V as a point, line, or plane. All four possibilities occur here.

a) $V = \{(1, 1, 0), \ (0, 1, 1), \ (1, 0, 1)\}$

b) $V = \{(1, -1, 0), \ (0, 1, -1), \ (1, 0, -1)\}$

c) $V = \{(1, -1, 0), \ (-1, 1, 0), \ (0, 0, 0)\}$

d) $V = \{(1, 1, 0), (2, 2, 0), (3, 3, 0)\}$

e) $V = \{(0, 0, 0), \ (0, 0, 0), \ (0, 0, 0)\}$

316. Use the V.I.P. (Proposition 4.3) to decide if the column-space of each $n \times n$ matrix is all of $\mathbf{R}^n$ or not.

a) $\begin{bmatrix} 2 & -1 & 0 \\ 0 & 1 & -2 \\ 1 & 0 & -1 \end{bmatrix}$ b) $\begin{bmatrix} 1 & 2 & -3 & 0 \\ 0 & 1 & 2 & -3 \\ -3 & 0 & 1 & 2 \\ 2 & -3 & 0 & 1 \end{bmatrix}$

c) $\begin{bmatrix} 1 & 1 & 1 & 1 \\ 1 & -1 & 1 & -1 \\ 1 & 1 & -1 & -1 \\ 1 & -1 & -1 & 1 \end{bmatrix}$

317. Determine whether the *row*-space of each matrix in Exercise 316 is all of $\mathbf{R}^n$ or not.

318. Determine in each case whether **b** lies in the column-space of **A**:

a) **b** $= (1, 1, 1)$, $\mathbf{A} = \begin{bmatrix} 1 & -1 & 0 \\ 0 & 1 & -1 \\ -1 & 0 & 1 \end{bmatrix}$

b) **b** $= (1, 2, 3)$, $\mathbf{A} = \begin{bmatrix} 1 & 2 & 3 \\ 0 & 1 & 2 \\ 0 & 0 & 1 \end{bmatrix}$

c) **b** $= (1, 4, 1)$, $\mathbf{A} = \begin{bmatrix} 0 & 2 & 1 \\ -2 & 0 & 2 \\ -1 & -2 & 0 \end{bmatrix}$

d) **b** $= (1, 1, 1)$, $\mathbf{A} = \begin{bmatrix} 0 & 2 & 1 \\ -2 & 0 & 2 \\ -1 & -2 & 0 \end{bmatrix}$

319. Show that:

a) When **A** has m columns and **B** has m rows, we always have $\mathrm{col}(\mathbf{AB}) \subset \mathrm{col}(\mathbf{A})$.

b) Contrastingly, $\mathrm{col}(\mathbf{AB})$ generally neither contains, nor lies in, $\mathrm{col}(\mathbf{B})$. Demonstrate this by considering

$$\mathbf{A} = \begin{bmatrix} 0 & 1 \\ 1 & 0 \end{bmatrix} \quad \text{and} \quad \mathbf{B} = \begin{bmatrix} 1 & 1 \\ 2 & 2 \end{bmatrix}$$

c) Use the results of (a) and (b) above to prove: *When* **A** *has* m *columns and* **B** *has* m *rows, we always have* $\mathrm{row}(\mathbf{AB}) \subset \mathrm{row}(\mathbf{B})$, *but in general,* $\mathrm{row}(\mathbf{AB})$ *neither contains, nor lies in,* $\mathrm{row}(\mathbf{A})$.

320. Let

$$\mathbf{A} = \begin{bmatrix} 2 & 4 & 0 & 6 \\ 4 & 8 & 2 & 6 \end{bmatrix}$$

a) Find a matrix **B** whose column-space equals $\mathrm{nul}(\mathbf{A})$.

b) Find a matrix **C** whose row-space is the left nullspace of **A**.

321. Find a 2×2 matrix whose nullspace and column-space both lie in the span of $(1, 1)$.

322. When $\mathbf{A}$ and $\mathbf{B}$ are both $n \times n$, the column-space of $\mathbf{AB}$ bears little relation to that of $\mathbf{A}$ and $\mathbf{B}$ separately.

    a) Can you find 2×2 matrices $\mathbf{A}$ and $\mathbf{B}$ that both have nontrivial column-spaces, and yet $\text{col}(\mathbf{A} + \mathbf{B}) = \{\mathbf{0}\}$?

    b) Can you find 2×2 matrices $\mathbf{A}$ and $\mathbf{B}$, neither of which have all of $\mathbf{R}^2$ as their column-space, and yet $\text{col}(\mathbf{A} + \mathbf{B}) = \mathbf{R}^2$?

323. Find (different) matrices whose (a) nullspace, (b) column-space, (c) row-space, and (d) left nullspace, is spanned by $(0, 0, 1, 4)$ and $(1, 2, 0, -3)$.

5. Perp/Span Conversion

At the beginning of Section 2, we asserted that every subspace $S \in \mathbf{R}^n$ can be described *both* ways: as the span of a set $V \subset \mathbf{R}^n$, and as the perp of some other set $W \subset \mathbf{R}^n$. The obvious question then arises as to whether, and how, we might convert from a span description to a perp description and vice-versa:

    i) Given a spanning set V for a subspace S, can we *find* a set W whose perp is S?

    ii) Conversely, given that S is the *perp* of a set $W \subset \mathbf{R}^n$, can we *find* a *spanning* set V for S?

Algorithms for both conversions do exist. One is already familiar, and we will soon present the other. First, however, we compare the strengths and weaknesses of the two types of subspace description. They serve different purposes, and complement each other with an elegant mathematical symmetry.

To facilitate our discussion and connect to other areas of mathematics, we also use the terms *explicit* and *implicit*, respectively for the span and perp descriptions of a subspace.

DEFINITION 5.1 (Explicit and Implicit). An **explicit** description of a subspace $S \subset \mathbf{R}^n$ is a set V of vectors that *span* S.

An **implicit** description of a subspace $S \subset \mathbf{R}^n$ is a set W of vectors whose *perp* is S:

$$\text{Explicit description:} \quad S = \operatorname{span} V$$

$$\text{Implicit description:} \quad S = W^{\perp}$$

□

The terms *span* and *perp* need not actually appear for us to recognize a description as explicit or implicit. For instance, $S = \operatorname{col}(\mathbf{A})$ describes S *explicitly*, because $\operatorname{col}(\mathbf{A})$ is a *span*—the span of the columns of $\mathbf{A}$. Contrastingly, $S = \operatorname{nul}(\mathbf{A})$, describes S *implicitly*, since $\operatorname{nul}(\mathbf{A})$ is a perp—the *perp* of the rows of $\mathbf{A}$.

These terms make sense because a span description of a subspace S *gives an explicit formula* for every vector in S—as a linear combination. For instance, if $S = \operatorname{span}\{\mathbf{v}_1, \mathbf{v}_2, \ldots, \mathbf{v}_k\}$, the formula

$$c_1 \mathbf{v}_1 + c_2 \mathbf{v}_2 + \cdots + c_k \mathbf{v}_k$$

gives every vector in S. By choosing different coefficients $c_1, c_2 \ldots, c_k$, we can *explicitly* generate as many vectors in S as we like.

When we describe a subspace S as the **perp** of a set, on the other hand, we do **not** explicitly present *any* vector in S. Instead, we describe S by giving a *test for membership*. A vector $\mathbf{x}$ belongs to $W^{\perp}$ if and only $\mathbf{a} \cdot \mathbf{x} = \mathbf{0}$ for every vector $\mathbf{a}$ in W. Since W does not list any vectors in S, but only provides a test for membership, it makes sense to call the description implicit.

The two descriptions therefore complement each other perfectly:

- An **explicit** (span) description of a subspace S makes it
  - Easy to produce vectors in S, but
  - Hard to test whether a given vector belongs to S.
- The **implicit** (perp) description of a subspace S is just the opposite, making it
  - Easy to test whether a given vector belongs to S, but
  - Hard to produce actual vectors in S

The terms *easy* and *hard* have specific meanings here:

Easy means we need only form a linear combination—or equivalently, perform a matrix/vector multiplication.

For example, it is "easy" to produce a member of a subspace that is explicitly given as $\mathrm{col}(\mathbf{A})$ for some matrix $\mathbf{A}$ with m columns. By the V.I.P., we just choose any vector $\mathbf{x} \in \mathbf{R}^m$ and compute $\mathbf{A}\mathbf{x}$ (Proposition 4.3).

Similarly, to *test* whether a vector $\mathbf{x}$ lies in a subspace given implicitly as $\mathrm{nul}(\mathbf{A})$, we again compute $\mathbf{A}\mathbf{x}$. If it vanishes, then $\mathbf{x} \in \mathrm{nul}(\mathbf{A})$, otherwise it does not.

Hard, on the other hand, means we have to solve a linear system, and hence row-reduce a matrix.

For instance, to *find* vectors in $\mathrm{nul}(\mathbf{A})$ we must *solve* $\mathbf{A}\mathbf{x} = \mathbf{0}$. The homogeneous generators and their linear combinations then give us a span formula for vectors in $\mathrm{nul}(\mathbf{A})$. Similarly, to *test* whether a vector $\mathbf{b}$ lies in $\mathrm{col}(\mathbf{A})$, we must *solve* $\mathbf{A}\mathbf{x} = \mathbf{b}$. The system has a solution if and only if $\mathbf{b} \in \mathrm{col}(\mathbf{A})$ (again, Proposition 4.3, our V.I.P.).

The terms *easy* and *hard* are relative, of course, but in the present context, one can actually quantify them. The number of scalar additions and multiplications required by a typical $n \times n$ matrix/vector multiplication is proportional to n^2. The number of operations required in *solving* an $n \times n$ system, on the other hand, is proportional to n^3. For even moderately large n, of course, n^2 is *much* smaller than n^3, making matrix/vector multiplication quite a bit easier than system-solving.

REMARK 5.2 (Image/pre-image). Another way to see the span/perp, explicit/implicit duality is this:

A span *description presents a subspace as the **image** of a linear map, while the* perp *description presents it as the **pre-image** of* $\mathbf{0}$ *under some other linear map.*

Indeed, the span of a set $V = \{\mathbf{v}_1, \mathbf{v}_2, \ldots, \mathbf{v}_m\}$ is, by definition, all linear combinations of the $\mathbf{v}_i$'s. But our mantra, *linear combination equals matrix/vector multiplication* (Remark 1.22 from Chapter 1), shows that the linear combinations of the $\mathbf{v}_i$'s form the *image* of the linear map represented by the matrix $\mathbf{V}$ having the $\mathbf{v}_i$'s as columns:

$$\mathbf{V} = \begin{bmatrix} \mathbf{v}_1 & \mathbf{v}_2 & \cdots & \mathbf{v}_k \\ | & | & \cdots & | \end{bmatrix}$$

Thus, $\mathrm{span}\, V$ is always an *image*: the image of the linear transformation represented by $\mathbf{V}$.

Similarly, *perps are pre-images.* For as noted in Observation 2.6, the perp of a set

$$A = \{\mathbf{a}_1, \mathbf{a}_2, \ldots, \mathbf{a}_n\} \subset \mathbf{R}^m$$

is the same as $\mathrm{nul}(\mathbf{A})$, where

$$\mathbf{A} = \begin{bmatrix} \mathbf{a}_1 & \rule{2cm}{0.4pt} \\ \mathbf{a}_2 & \rule{2cm}{0.4pt} \\ \vdots & \vdots \\ \mathbf{a}_n & \rule{2cm}{0.4pt} \end{bmatrix}$$

But $\mathrm{nul}(\mathbf{A})$ can also be seen as the pre-image of $\mathbf{0}$ under the linear map represented by $\mathbf{A}$.

Thus, $A^\perp$ is always a *pre-image*: the pre-image of $\mathbf{0}$ under the linear transformation represented by $\mathbf{A}$.   □

Since the span and perp descriptions have complementary strengths and weaknesses, we often want to know a subspace *explicitly* when it's given *implicitly*, or vice-versa. In other words, we want to *convert* one mode of subspace description into the other:

- Convert an *implicit* (perp) subspace description to an *explicit* (span) description.
- Convert an *explicit* (span) subspace description to an *implicit* (perp) description.

5.3. Converting perp to span. To recast a perp as a span, we simply solve a homogeneous system.

Indeed, we saw in Observation 2.6 (and just mentioned again in Remark 5.2) that a subspace defined implicitly (as the perp of a set $\{\mathbf{a}_1, \mathbf{a}_2, \ldots, \mathbf{a}_n\}$) is nothing but the solution set of $\mathbf{Ax} = \mathbf{0}$, where row i of $\mathbf{A}$ is given by $\mathbf{a}_i$ for each $i = 1, 2, \ldots, n$. This solution set is of course the *span of the homogeneous generators* we get from the free columns of $\mathrm{RRE}(\mathbf{A})$. In short,

Converting a perp description of a subspace to a span description of that same subspace, we always solve a homogeneous system.

EXAMPLE 5.4. Let

$$\mathbf{A} = \begin{bmatrix} 1 & 2 & 3 & 0 \\ 0 & 1 & 2 & 3 \end{bmatrix}$$

and find a matrix $\mathbf{B}$ with $\mathrm{col}(\mathbf{B}) = \mathrm{nul}(\mathbf{A})$.

Here we have a perp description (the nullspace of $\mathbf{A}$ is the perp of its rows), and we want to convert it to a span description as $\mathrm{col}(\mathbf{B})$ (the span of the columns of a matrix $\mathbf{B}$). We want to do an implicit-to-explicit conversion.

Specifically, we need the columns of $\mathbf{B}$ to span the nullspace of $\mathbf{A}$. What spans the nullspace of $\mathbf{A}$? The homogeneous generators $\mathbf{h}_i$ that we get from $\text{RRE}(\mathbf{A})$, of course. So all we need to do is find the $\mathbf{h}_i$'s and set them up as columns of a matrix $\mathbf{B}$.

Accordingly, we compute

$$\text{RRE}(\mathbf{A}) = \begin{bmatrix} 1 & 0 & -1 & -6 \\ 0 & 1 & 2 & 3 \end{bmatrix}$$

write down the homogeneous generators

$$\mathbf{h}_1 = \begin{pmatrix} 1 \\ -2 \\ 1 \\ 0 \end{pmatrix}, \qquad \mathbf{h}_2 = \begin{pmatrix} 6 \\ -3 \\ 0 \\ 1 \end{pmatrix}$$

and arrange them as columns to get our answer:

$$\mathbf{B} = \begin{bmatrix} 1 & 6 \\ -2 & -3 \\ 1 & 0 \\ 0 & 1 \end{bmatrix}$$

Now $\text{col}(\mathbf{B}) = \text{nul}(\mathbf{A})$, as desired. □

In sum: *To convert the perp of $\{\mathbf{a}_1, \mathbf{a}_2, \ldots, \mathbf{a}_k\}$ to a span, we:*

    i) *Construct a matrix $\mathbf{A}$ having the $\mathbf{a}_i$'s as rows, and*

    ii) *Solve $\mathbf{A}\mathbf{x} = \mathbf{0}$ to find its homogeneous generators $\mathbf{h}_1, \mathbf{h}_2, \ldots, \mathbf{h}_l$.*

We can then rewrite

$$\{\mathbf{a}_1, \mathbf{a}_2, \ldots, \mathbf{a}_k\}^{\perp} = \text{span}\,\{\mathbf{h}_1, \mathbf{h}_2, \ldots, \mathbf{h}_l\}$$

The perp of the $\mathbf{a}_i$'s has become the span of the $\mathbf{h}_j$'s.

5.5. Converting span to perp. To recast a span as a perp requires a new idea. Here, we show how to do it using Proposition 4.3—the V.I.P. again. We give an alternative and more elegant procedure for doing this in Section 1.13 of Chapter 6, but we're not yet in a position to explain that method.

We can best explain the one we have in mind here using examples. We give two, after which we'll summarize.

EXAMPLE 5.6. Consider the subspace $S \subset \mathbf{R}^3$ spanned by $(1, -1, 0)$, $(0, 1, -1)$, and $(-1, 0, 1)$. This is an explicit (span) description. How can we describe this same subspace *implicitly* (as a perp)?

The answer has us realize S as the column-space of a matrix $\mathbf{B}$. The V.I.P. (Proposition 4.3), will then help us find a matrix $\mathbf{A}$ with

$$\mathrm{col}(\mathbf{B}) = \mathrm{nul}(\mathbf{A})$$

On the left, we have our original span description of S, but on the right, get an implicit, perp description of the same subspace.

The first step, realizing S as a column-space is easy: Set the given vectors up as columns of a matrix $\mathbf{B}$. Proposition 4.3 (the V.I.P.) then tells us that a vector $\mathbf{b}$ lies in $S = \mathrm{col}(\mathbf{B})$ if and only if the *in*homogeneous system $\mathbf{Bx} = \mathbf{b}$ has a solution. To see if it does, we form the usual augmented matrix $[\mathbf{B} \,|\, \mathbf{b}]$, with an *indeterminate* $\mathbf{b} = (x, y, z)$:

$$\left[\begin{array}{ccc|c} 1 & 0 & -1 & x \\ -1 & 1 & 0 & y \\ 0 & -1 & 1 & z \end{array} \right]$$

Row-reduction takes this to

$$\left[\begin{array}{ccc|c} 1 & 0 & -1 & x \\ 0 & 1 & -1 & y + x \\ 0 & -1 & 1 & z \end{array} \right]$$

and then

$$\left[\begin{array}{ccc|c} 1 & 0 & -1 & x \\ 0 & 1 & -1 & y + x \\ 0 & 0 & 0 & z + y + x \end{array} \right]$$

The key observation is this: the latter matrix represents a system with *no* solution, *unless* the final entry in the bottom row is *zero*, that is, $z + y + x = 0$. This single homogeneous equation, however, is equivalent to the perp condition $(1, 1, 1) \cdot (x, y, z) = 0$, or in other words,

$$\mathbf{Ab} = \mathbf{0}$$

where $\mathbf{A} = [1\ 1\ 1]$.

Thus, $\mathbf{b} \in \mathrm{col}(\mathbf{B})$ if and only if $\mathbf{b} \in \mathrm{nul}(\mathbf{A})$. We have converted the span description to a perp description. □

EXAMPLE 5.7. Find an implicit (perp) description for the span of $(1,1,1,1)$ and $(1,-1,1,-1)$ in $\mathbf{R}^4$.

As above, we set up a matrix with the given spanning vectors as the columns of a matrix $\mathbf{B}$, and augment it with a general vector $\mathbf{b}$:

$$[\mathbf{B}\,|\,\mathbf{b}] = \begin{bmatrix} 1 & 1 & | & b_1 \\ 1 & -1 & | & b_2 \\ 1 & 1 & | & b_3 \\ 1 & -1 & | & b_4 \end{bmatrix}$$

Subtract the first row from each row below it to get

$$\begin{bmatrix} 1 & 1 & | & b_1 \\ 0 & -2 & | & b_2 - b_1 \\ 0 & 0 & | & b_3 - b_1 \\ 0 & -2 & | & b_4 - b_1 \end{bmatrix}$$

and subtract the second row from the last to get

$$\begin{bmatrix} 1 & 1 & | & b_1 \\ 0 & -2 & | & b_2 - b_1 \\ 0 & 0 & | & b_3 - b_1 \\ 0 & 0 & | & b_4 - b_2 \end{bmatrix}$$

We don't need to go any further: the system clearly has a solution if and only if the last two right-hand entries vanish—that is, if and only if $\mathbf{b}$ solves the homogeneous system

$$\begin{aligned} b_3 - b_1 &= 0 \\ b_4 - b_2 &= 0 \end{aligned}$$

The coefficient matrix of this new system is clearly

$$\mathbf{A} = \begin{bmatrix} -1 & 0 & 1 & 0 \\ 0 & -1 & 0 & 1 \end{bmatrix}$$

We have thus converted an explicit, span description to an implicit, perp one. The subspace spanned by $(1,1,1,1)$ and $(1,-1,1,-1)$, that is, the *column-space* of the original matrix $\mathbf{B}$ is now the *nullspace* of a new matrix, namely $\mathbf{A}$ above. □

Here's a summary of the method used by these two examples.

To convert a span description of a subspace $S \subset \mathbf{R}^n$ to a perp description,

i) *Set up the spanning vectors for S as the columns of a matrix $\mathbf{B}$.*

ii) *Row-reduce the augmented matrix $[\mathbf{B} \,|\, \mathbf{b}]$, with an undetermined "variable" vector $\mathbf{b} = (b_1, b_2, \ldots, b_n)$.*

iii) *Each row of zeros at the bottom of $\mathrm{RRE}(\mathbf{B})$ now ends with a linear combination of the b_i's in the last column.*

iv) *Set each of these linear combinations equal to zero, and write the resulting homogeneous system in vector form as $\mathbf{A}\mathbf{x} = \mathbf{0}$.*

The procedure starts with the span description $S = \mathrm{col}(\mathbf{B})$, and converts it to the perp description $S = \mathrm{nul}(\mathbf{A})$.

– Practice –

324. Find a span description for

   a) The plane $y = 0$ in $\mathbf{R}^3$.

   b) The y-axis in $\mathbf{R}^3$.

   c) The set of all vectors in $\mathbf{R}^5$ whose last two coordinates vanish.

   d) All of $\mathbf{R}^n$.

325. Find a perp description for

   a) The subspace spanned by $(0, 1, 0)$ (the y-axis) in $\mathbf{R}^3$.

   b) The set of all vectors in $\mathbf{R}^5$ whose last two coordinates vanish.

   c) The trivial subspace $\{\mathbf{0}\} \subset \mathbf{R}^n$.

326. Find an explicit (span) description for the nullspace of

$$\mathbf{A} = \begin{bmatrix} 2 & 4 & -2 & 4 \\ 3 & 6 & 3 & -6 \end{bmatrix}$$

327. Find an explicit (span) description for

   a) The plane through $\mathbf{0}$ and orthogonal to $(1, -3, 2) \in \mathbf{R}^3$.

b) The hyperplane through $\mathbf{0}$ and orthogonal to $(2, -3, 4, -5, 1)$ in $\mathbf{R}^5$.

328. Find an implicit (perp) description for

    a) The plane spanned by $(2, 0, 1)$ and $(0, 2, -1)$ in $\mathbf{R}^3$.

    b) The subspace spanned by $(2, -4, 2, -6)$ and $(2, 4, 2, 6)$ in $\mathbf{R}^4$.

329. Find an implicit (perp) description for

    a) The line spanned by $(1, 2, 3)$ in $\mathbf{R}^3$. (Figure 4 depicts a similar situation.)

    b) The line spanned by $(2, -1, -2, 1)$ in R^4.

330. Find

    a) A span description for the intersection of the hyperplanes $x_1 + x_3 + x_5 = 0$ and $x_2 + x_4 = 0$ in $\mathbf{R}^5$.

    b) A perp description for the subspace spanned by $(1, 0, 1, 0, 1)$ and $(0, 1, 0, 1, 0)$ in $\mathbf{R}^5$.

331. Find a matrix whose nullspace equals $\operatorname{col}(\mathbf{A})$, where

$$\mathbf{A} = \begin{bmatrix} 3 & -1 & 5 \\ 0 & 0 & 0 \\ -3 & 1 & -5 \end{bmatrix}$$

332. Consider the matrix

$$\mathbf{A} = \begin{bmatrix} 1 & 1 \\ 2 & 1 \\ -2 & 2 \\ 1 & -2 \end{bmatrix}$$

    a) Find a matrix $\mathbf{X}$ with $\operatorname{col}(\mathbf{X}) = \operatorname{nul}(\mathbf{A}^T)$.

    b) Find a matrix $\mathbf{N}$ with $\operatorname{nul}(\mathbf{N}) = \operatorname{col}(\mathbf{A})$.

333. Find an *implicit* description for $\operatorname{row}(\mathbf{A})$, where

$$\mathbf{A} = \begin{bmatrix} 1 & 2 & -2 & 1 \\ 0 & 1 & 1 & 0 \end{bmatrix}$$

334. Suppose $V = \{\mathbf{v}_1, \mathbf{v}_2, \ldots, \mathbf{v}_k\}$ is a finite set of non-zero vectors in $\mathbf{R}^n$.

    a) Can $\operatorname{span} V \cap V^\perp$ ever be completely empty? Why or why not?

    b) Can the span of V ever be the *same* as the perp of $\mathbf{V}$? Why or why not?

    c) Can any non-zero vector ever lie in *both* the span of V and the perp of V? Why or why not?

335. Recall that the *row-space* of a matrix is the span of its rows.

    a) Show that swapping rows doesn't change the row-space of a matrix. (The same holds for the other elementary row operations, but you needn't show it here.)

    b) Swapping rows usually *does* change the column-space of a matrix. Exhibit a 2-by-2 matrix whose column-space gets changed by that operation. Find a vector in the original column-space that no longer lies in the changed column-space.

$$- \star -$$

6. Independence

6.1. Reducing a spanning set. Suppose we have a spanning set V for a subspace $S \subset \mathbf{R}^n$, and we create a larger set V' by adjoining one or more additional vectors *from* S into V. Then $V' \supset V$, so $\operatorname{span} V' \supset \operatorname{span} V = S$. On the other hand, S is closed under addition and scalar multiplication, so $\operatorname{span} V' \subset \operatorname{span} V = S$. The new set V' is *larger* than V, *but has the same span* as V.

This shows that spanning sets for a subspace come in different sizes, and in particular, that we can easily make a spanning set *bigger* without changing its span. But how *small* can we make it? Specifically,

Question: *If V spans S, which vectors (if any) can we delete from V, without changing its span S?*

We will answer this after proving Proposition 6.3 below.

First, however, we clarify a key point of terminology.

DEFINITION 6.2. Suppose $\mathbf{A}$ is a matrix—*not* necessarily in reduced row-echelon form. Henceforth, when we refer to the **pivot** or **free columns** of $\mathbf{A}$, we mean the columns that *end up* with or without pivots *after* row-reduction.

Thus, to *find* the pivot or free columns of any matrix, we row-reduce, and see which columns end up with pivots. The corresponding columns of the *original* matrix are then its pivot (or free) columns. □

PROPOSITION 6.3. *Each **free** column of a matrix can be written as a linear combination of its **pivot** columns.*

PROOF. For simplicity, let us assume that columns 1 through r are pivot columns, while the remaining columns $r+1$ through m are free. (In general the pivot and free columns intermingle, but that doesn't really change the argument—it just complicates the notation.)

Now let $j \geq r+1$ be the index of any free column. We get the corresponding homogeneous generator $\mathbf{h}$ by putting a 1 in position j, and setting all other free variables to zero. Let p_i denote the coordinate of $\mathbf{h}$ in each "pivot" position $i \leq r$. Since $\mathbf{h}$ solves the homogeneous system, we have

$$\mathbf{Ah} = \mathbf{0}$$

Expand the matrix/vector product using our mantra: *matrix/vector multiplication is a linear combination*. Since all the free coordinates of $\mathbf{h}$ except the jth vanish, $\mathbf{Ah}$, as a linear combination, will involve only column j and the pivot columns (1 through r) of $\mathbf{A}$, and hence will look like this:

$$p_1\,\mathbf{c}_1(\mathbf{A}) + p_2\,\mathbf{c}_2(\mathbf{A}) + \cdots + p_r\,\mathbf{c}_r(\mathbf{A}) + 1\,\mathbf{c}_j(\mathbf{A}) = \mathbf{0}$$

Solve this for the free column $\mathbf{c}_j(\mathbf{A})$ to express it as a linear combination of the pivot columns:

$$\mathbf{c}_j(\mathbf{A}) = -p_1\,\mathbf{c}_{j_1}(\mathbf{A}) - p_2\,\mathbf{c}_{j_2}(\mathbf{A}) - \cdots - p_r\,\mathbf{c}_{j_r}(\mathbf{A})$$

This is what we wanted! We have shown how to express any free column as a linear combination of pivot columns. □

COROLLARY 6.4. *The pivot columns of a matrix span its entire column-space. Deleting the free columns doesn't change the column-space.*

PROOF. Let $\mathbf{A}$ be any matrix, and suppose $\mathbf{v} \in \operatorname{col}(\mathbf{A})$. Then by definition, $\mathbf{v}$ is a linear combination of the columns of $\mathbf{A}$. If any *free* column appears in that linear combination, however, we can substitute a linear combination of *pivot* columns by the Proposition above. Replacing all free columns by linear combinations of pivot columns in this way, we express $\mathbf{v}$ as a linear combination of pivot columns only. Since we can do this for any $\mathbf{v} \in \operatorname{col}(\mathbf{A})$, the pivot columns span all of $\operatorname{col}(\mathbf{A})$.   □

EXAMPLE 6.5. The matrix

$$\mathbf{A} = \begin{bmatrix} 1 & -1 & 0 \\ 0 & 1 & -1 \\ -1 & 0 & 1 \end{bmatrix}$$

quickly row-reduces to

$$\operatorname{RRE}(\mathbf{A}) = \begin{bmatrix} 1 & 0 & -1 \\ 0 & 1 & -1 \\ 0 & 0 & 0 \end{bmatrix}$$

So the first two columns of $\mathbf{A}$ are its pivot columns. By the Corollary just proven, the first two columns of $\mathbf{A}$ span all of $\operatorname{col}(\mathbf{A})$. In particular, column 3 should be a linear combination of columns one and two—and it is. In fact, the homogeneous generator we get from it is $\mathbf{h} = (1, 1, 1)$, which tells us

$$1 \cdot \mathbf{c}_1(\mathbf{A}) + 1 \cdot \mathbf{c}_2(\mathbf{A}) + 1 \cdot \mathbf{c}_3(\mathbf{A}) = \mathbf{0}$$

Solve this for $\mathbf{c}_3(\mathbf{A})$ to write the latter as a linear combination of the first two columns:

$$\mathbf{c}_3(\mathbf{A}) = -\mathbf{c}_1(\mathbf{A}) - \mathbf{c}_2(\mathbf{A})$$

Caution: Following Definition 6.2, we selected the first two columns of $\mathbf{A}$ *itself* here. The first two columns of the *reduced* matrix $\operatorname{RRE}(\mathbf{A})$ do **not** span $\operatorname{col}(\mathbf{A})$. In fact, they're not even *in* $\operatorname{col}(\mathbf{A})$ (check it)! □

Before we proceed from our discussion of spanning sets, we prove a fact—especially useful later on—relating span and orthogonality.

PROPOSITION 6.6. *If $\mathbf{0} \neq \mathbf{w} \in \mathbf{R}^n$, and $\mathbf{w}$ is orthogonal to each vector in a set V, then $\mathbf{w}$ is not in the span of V.*

PROOF. We will prove the contrapositive: *If a vector $\mathbf{w}$ is in the span of V, and is orthogonal to every vector in V, then $\mathbf{w} = \mathbf{0}$.*

To do this, label the vectors in V as $\mathbf{v}_1, \mathbf{v}_2, \ldots, \mathbf{v}_k$. Then $\mathbf{w} \in \operatorname{span} V$ would mean we could write

$$(37) \qquad \mathbf{w} = c_1 \mathbf{v}_1 + c_2 \mathbf{v}_2 + \cdots + c_k \mathbf{v}_k$$

In this case, dotting $\mathbf{w}$ with itself and applying the distributive rule for dot products (Corollary 1.30 of Chapter 1) would give

$$\begin{aligned}
|\mathbf{w}|^2 &= \mathbf{w} \cdot \mathbf{w} \\
&= \mathbf{w} \cdot (c_1 \mathbf{v}_1 + c_2 \mathbf{v}_2 + \cdots + c_k \mathbf{v}_k) \\
&= c_1 (\mathbf{w} \cdot \mathbf{v}_1) + c_2 (\mathbf{w} \cdot \mathbf{v}_2) + \cdots + c_k (\mathbf{w} \cdot \mathbf{v}_k) \\
&= 0
\end{aligned}$$

since we assume $\mathbf{w}$ orthogonal to each of the $\mathbf{v}_i$'s. In short, a vector $\mathbf{w}$ **in** the span of the $\mathbf{v}_i$'s, **and** orthogonal to them, must be $\mathbf{0}$. This proves the contrapositive, and hence the Proposition. $\qquad \square$

Exercises 347 and 344 develop some useful first consequences of this fact. We'll appeal to it again in the discussion leading up to Theorem 8.2—perhaps the deepest result of this chapter.

6.7. Independence. The pivot columns of $\mathbf{A}$ don't just span $\operatorname{col}(\mathbf{A})$—they span *efficiently*, in the sense that we need *every* pivot column to span $\operatorname{col}(\mathbf{A})$; if we discard any pivot column, the column-space shrinks. A central concept in linear algebra, *independence*, expresses this fact.

DEFINITION 6.8. A non-empty set of V of vectors in $\mathbf{R}^n$ is **linearly independent** (or simply **independent**) if no proper subset of V spans all of $\operatorname{span} V$. If a set of vectors is not *independent*, we call it **dependent**.

This definition of independence has the advantage of showing clearly why independence is desirable: Every vector in an independent set is *essential*—none can be discarded without losing span.

On the other hand, Definition 6.8 provides no practical *test* for independence. We remedy that lack with Proposition 6.9 below, which equates independence to more easily testable properties. Conclusion (c) in particular yields a simple and concrete independence test.

PROPOSITION 6.9 (Independence). *The following six statements are equivalent:*

   a) *The columns of a matrix* $\mathbf{A}$ *are independent.*

   b) *No column of* $\mathbf{A}$ *is a linear combination of the others.*

   c) *All columns of* $\mathbf{A}$ *are pivot columns (no free columns).*

   d) $\mathbf{Ax} = \mathbf{0}$ *has only the trivial solution.*

   e) $\mathrm{nul}(\mathbf{A}) = \{\mathbf{0}\}.$

   f) *No linear combination of the columns* $\mathbf{c}_i(\mathbf{A})$ *vanishes unless each of its coefficients vanish. That is,*

$$a_1 \, \mathbf{c}_1(\mathbf{A}) + a_2 \, \mathbf{c}_2(\mathbf{A}) + \cdots + a_m \, \mathbf{c}_m(\mathbf{A}) = \mathbf{0}$$

   only when $a_1 = a_2 = \cdots = a_m = 0.$

PROOF. As usual, it will suffice to prove each of the following six implications:

$$(\mathrm{a}) \Rightarrow (\mathrm{b}) \Rightarrow (\mathrm{c}) \Rightarrow (\mathrm{d}) \Rightarrow (\mathrm{e}) \Rightarrow (\mathrm{f}) \Rightarrow (\mathrm{a})$$

Individually, most of these have very short proofs.

(a)⇒ **(b):** We need to show that when the columns of $\mathbf{A}$ are independent, no column is a linear combination of the others. We do so via the contrapositive: *If one column is a linear combination of the others, the remaining columns—a proper subset—will still span* $\mathrm{col}(\mathbf{A})$, *making the full set of columns* **dependent**. For instance, suppose the last column $c_m(\mathbf{A})$ is a linear combination of the first $m - 1$:

$$c_m(\mathbf{A}) = a_1 \, c_1(\mathbf{A}) + a_2 \, c_2 + \cdots a_{m-1} \, c_{m-1}(\mathbf{A})$$

Then in any linear combination of *all* columns, we can replace $c_m(\mathbf{A})$ by the right-hand side above, thereby rewriting the original linear combination using the first $m-1$ columns only. The full set of m columns would then be **dependent**, by our definition, because the first $m - 1$ columns alone—a proper subset—suffice to span the full column space.

(b)⇒ **(c):** Again, we argue the contrapositive: *If* $\mathbf{A}$ *has a free column, then some column is a linear combination of the others.* Indeed, Proposition 6.3 guarantees that the free column itself is a linear combination of the other columns—in fact, of the remaining pivot columns alone.

(c)⇒ **(d):** Here we argue directly. Since each column has a pivot, there are no homogeneous generators, and hence the particular solution is the only solution. Since $\mathbf{0}$ solves the homogeneous system, it must be that unique particular solution.

(d)$\Rightarrow$ (e): This one is trivial: by definition, nul($\mathbf{A}$) *equals* the solution set of $\mathbf{Ax} = \mathbf{0}$.

(e)$\Rightarrow$ (f): Also trivial: If we define $\mathbf{x} = (a_1, a_2, \ldots, a_m)$, the linear combination in (f) just gives the column description of $\mathbf{Ax} = \mathbf{0}$. So (f) merely restates that nul($\mathbf{A}$) contains only the origin.

(f)$\Rightarrow$(a): We argue the contrapositive again: *If the columns of $\mathbf{A}$ are **dependent**, then some **non**-trivial linear combination of them vanishes.* Indeed, suppose the columns of $\mathbf{A}$ are dependent. Then some subset of the columns—the first $m - 1$, say—span the entire column-space. In that case, the last column is a linear combination of the $m - 1$ columns that precede it:

$$\mathbf{c}_m(\mathbf{A}) = a_1\,\mathbf{c}_1(\mathbf{A}) + a_2\,\mathbf{c}_2(\mathbf{A}) + \cdots + a_{m-1}\,\mathbf{c}_{m-1}(\mathbf{A})$$

Subtracting $\mathbf{c}_m(\mathbf{A})$ from both sides then yields

$$\mathbf{0} = a_1\,\mathbf{c}_1(\mathbf{A}) + a_2\,\mathbf{c}_2(\mathbf{A}) + \cdots + a_{m-1}\,\mathbf{c}_{m-1}(\mathbf{A}) - \mathbf{c}_m(\mathbf{A})$$

Some of the a_i's might vanish here, but they can't *all* be zero, since the coefficient of $\mathbf{c}_m(\mathbf{A})$ is $-1 \neq 0$. So the right-hand side is a non-trivial, linear combination of the columns of $\mathbf{A}$ that *vanishes*. This proves the contrapositive, as desired. $\qquad\square$

Traditionally, one takes Conclusion (f) above as the *definition* of independence. We derive it as a consequence of Definition 6.8, because we find the latter easier for students to understand and remember.

Conclusion (c), as mentioned above, amounts to an easy *test* for the independence (or dependence) of a set of vectors. We restate it here for emphasis:

COROLLARY 6.10 (Independence test). *To see if a set of vectors is independent, set its vectors up as columns of a matrix, and row-reduce. If every column gets a pivot, the vectors are independent. If not, they're dependent.*

EXAMPLE 6.11. Are $(1, 2, 3)$, $(2, 1, 3)$, and $(2, -3, -1)$ independent?

To answer this using the test above, we create a matrix $\mathbf{A}$ having these vectors as columns, and row-reduce:

$$\mathbf{A} = \begin{bmatrix} 1 & 2 & 2 \\ 2 & 1 & -3 \\ 3 & 3 & -1 \end{bmatrix} \longrightarrow \text{RRE}(\mathbf{A}) = \begin{bmatrix} 1 & 0 & 1 \\ 0 & 1 & 1 \\ 0 & 0 & 0 \end{bmatrix}$$

Column three of the reduced matrix has no pivot, so by Conclusion (c) of the Proposition, the given set is **dependent**. The third vector is a linear combination of the first two (by Proposition 6.3), and can thus be discarded without losing span. □

EXAMPLE 6.12. Are the vectors $(1, -2, 3)$, $(2, -3, 1)$, and $(3, -1, 2)$ independent?

Here the answer is "yes," since a routine row-reduction takes the matrix $\mathbf{A}$ having these vectors as columns, to $\text{RRE}(\mathbf{A}) = \mathbf{I}_n$, which has a pivot in every column. □

Before going any further, we state another important corollary to our Independence Proposition 6.9. It isn't quite as obvious as one might first think.

COROLLARY 6.13. *The pivot columns of any matrix are independent.*

PROOF. Proposition 6.9 tells us that the columns of a matrix are independent if they all reduce to pivot columns. What we need to show here is that if we *discard* the free columns of a matrix $\mathbf{A}$, the new matrix $\mathbf{A}'$ that remains will have *no* free columns. In other words, no pivot column of $\mathbf{A}$ will become a free column of $\mathbf{A}'$.

This holds because the Gauss–Jordan algorithm *skips over* free columns—it never executes a row operation based on a free-column entry, for if we don't find a pivot in some column, we move on to the next. *Exactly the same* sequence of row operations thus row-reduces *both* $\mathbf{A}$ and $\mathbf{A}'$. To put it another way, deleting free columns and row-reducing commute!

We thus get $\text{RRE}(\mathbf{A}')$ either way, which means the columns of $\text{RRE}(\mathbf{A}')$ are the *pivot* columns of $\text{RRE}(\mathbf{A})$. It follows that $\text{RRE}(\mathbf{A}')$ has no free columns, making the columns of $\mathbf{A}'$, and thus the pivot columns of $\mathbf{A}$, independent by Conclusion (c) of Proposition 6.9. □

We conclude our introduction to independence with a more specialized test that has two immediate, but important corollaries.

PROPOSITION 6.14. *If* $\mathbf{A}$ *is a matrix, and each* column *of* $\mathbf{A}$ *has a non-zero entry in a* row *that has zeros everywhere else, the* columns *of* $\mathbf{A}$ *are independent.*

PROOF. Suppose $\mathbf{A}$ has m columns. For each $j = 1, 2, \ldots, m$, divide the row with the non-zero entry in column j by that entry to make it a one. Then swap that row into the jth position. The first m rows of $\mathbf{A}$ now form an $m \times m$ identity matrix. So far we have only used elementary row operations, and we can now use the leading one in each column to zero out all entries in any rows below the first m rows to put the matrix in RRE form, with a pivot in every column.

Proposition 6.9(c) now certifies that the columns of $\mathbf{A}$ are independent, as claimed.   □

COROLLARY 6.15. *The non-zero rows of a matrix in* RRE *form are always independent.*

PROOF. The non-zero rows of a matrix in RRE form each have a pivot in some column that is zeros above and below.

Now delete any *zero* rows from the bottom of the matrix, and then transpose. The resulting matrix satisfies hypotheses of Proposition 6.14. Each *column* now has a 1 in a *row* that is otherwise zero. Its columns—the non-zero *rows* of the original matrix—are therefore independent.   □

COROLLARY 6.16. *The homogeneous generators we get when solving a homogeneous system are independent.*

PROOF. We construct the jth homogeneous generator by putting a "1" in the jth free variable position, while putting zeros in that same position in every other homogeneous generator. If we now set the homogeneous generators up as columns of a matrix $\mathbf{H}$, that matrix satisfies the conditions of Proposition 6.14.   □

– **Practice** –

336. Find two vectors whose span equals the span of this set in $\mathbf{R}^3$:

$$\left\{ \begin{pmatrix} 1 \\ -2 \\ 1 \end{pmatrix}, \begin{pmatrix} 2 \\ -1 \\ -1 \end{pmatrix}, \begin{pmatrix} 1 \\ 1 \\ -2 \end{pmatrix} \right\}$$

337. Find three vectors whose span equals the span of this set in $\mathbf{R}^4$:

$$\left\{ \begin{pmatrix} 2 \\ 0 \\ -2 \\ 0 \end{pmatrix}, \begin{pmatrix} 0 \\ 2 \\ 0 \\ -2 \end{pmatrix}, \begin{pmatrix} 1 \\ 2 \\ -1 \\ -2 \end{pmatrix}, \begin{pmatrix} 2 \\ 1 \\ -2 \\ 1 \end{pmatrix} \right\}$$

338. Determine whether each set is independent. If not, write one vector in the set as a linear combination of the other two.

a) $\left\{ \begin{pmatrix} 1 \\ 2 \\ 3 \end{pmatrix}, \begin{pmatrix} 2 \\ 3 \\ 4 \end{pmatrix}, \begin{pmatrix} 3 \\ 4 \\ 5 \end{pmatrix} \right\}$

b) $\left\{ \begin{pmatrix} 1 \\ 2 \\ 3 \end{pmatrix}, \begin{pmatrix} 2 \\ 3 \\ 5 \end{pmatrix}, \begin{pmatrix} 3 \\ 4 \\ 5 \end{pmatrix} \right\}$

339. Determine whether this set is independent. If not, write one vector in the set as a linear combination of the other three:

$$\left\{ \begin{pmatrix} -1 \\ 2 \\ 2 \\ 1 \end{pmatrix}, \begin{pmatrix} -5 \\ 5 \\ -3 \\ 5 \end{pmatrix}, \begin{pmatrix} 1 \\ -1 \\ 0 \\ -1 \end{pmatrix}, \begin{pmatrix} 4 \\ -1 \\ 9 \\ -4 \end{pmatrix} \right\}$$

340. Which of the following sets (if either) in $\mathbf{R}^3$ is dependent? Which is independent?

a) $\left\{ \begin{pmatrix} 1 \\ 2 \\ 3 \end{pmatrix}, \begin{pmatrix} 3 \\ 4 \\ 5 \end{pmatrix}, \begin{pmatrix} 5 \\ 6 \\ 7 \end{pmatrix} \right\}$

b) $\left\{ \begin{pmatrix} 1 \\ 2 \\ 3 \end{pmatrix}, \begin{pmatrix} 4 \\ 4 \\ 4 \end{pmatrix}, \begin{pmatrix} 3 \\ 2 \\ 1 \end{pmatrix} \right\}$

341. Let

$$\mathbf{v}_1 = (1, 0, 1, 0, 1), \ \mathbf{v}_2 = (1, 1, 0, 1, 1), \ \mathbf{v}_3 = (1, 0, 0, 0, 1)$$

$$\mathbf{v}_4 = (0, 1, 0, 1, 0), \ \mathbf{v}_5 = (0, 0, 1, 0, 0), \ \mathbf{v}_6 = (1, 1, 1, 0, 0)$$

Which of the following sets in $\mathbf{R}^5$ is dependent? Which is independent?

    a) $\{\mathbf{v}_1, \mathbf{v}_2, \mathbf{v}_3\}$ b) $\{\mathbf{v}_1, \mathbf{v}_2, \mathbf{v}_4\}$ c) $\{\mathbf{v}_2, \mathbf{v}_5, \mathbf{v}_6\}$

    d) $\{\mathbf{v}_1, \mathbf{v}_2, \mathbf{v}_3, \mathbf{v}_5, \mathbf{v}_6\}$ e) $\{\mathbf{v}_1, \mathbf{v}_2, \mathbf{v}_3, \mathbf{v}_6\}$

342. Do any *five* of the vectors $\mathbf{v}_i$ in Exercise 341 form an independent subset? If so, which five? If not, how do you know?

343. Show that a set of k vectors in $\mathbf{v}_1, \mathbf{v}_2, \ldots, \mathbf{v}_k \in \mathbf{R}^n$ cannot be *independent* if $k > n$. [*Hint: Regard the $\mathbf{v}_i$'s as matrix columns.*]

344. Show that a set of k vectors $\mathbf{v}_1, \mathbf{v}_2, \ldots, \mathbf{v}_k \in \mathbf{R}^n$ cannot *span* $\mathbf{R}^n$ if $k < n$. [*Hint: Regard the $\mathbf{v}_i$'s as matrix **rows**, argue that there exists a vector $\mathbf{h} \neq \mathbf{0}$ orthogonal to all of them, and use Proposition 6.6.*]

345. Suppose $S \subset \mathbf{R}^n$ is a subspace.

    a) If a set with only three vectors *spans* S, can S contain an *independent* set with four vectors? Why or why not?

    b) (Hard) If S contains an *independent* set of four vectors, can a set with only three vectors *span* it? Why or why not?

346. Suppose $T : \mathbf{R}^m \to \mathbf{R}^n$ is linear, and we have an *independent* set $\{\mathbf{u}_1, \ldots, \mathbf{u}_k\}$ in $\mathbf{R}^m$.

    a) Give a simple example showing that the image set

$$\{T(\mathbf{u}_1), \ T(\mathbf{u}_2), \ldots, T(\mathbf{u}_k)\}$$

    in the range need not be independent.

    b) Show that, even so, $\{T(\mathbf{u}_1), \ldots, T(\mathbf{u}_k)\}$ *will* be independent if T is one-to-one.

    c) Show that $\{T(\mathbf{u}_1), \ldots, T(\mathbf{u}_k)\}$ will also be independent, whether or not T is one-to-one, if the $\mathbf{u}_i$'s are independent vectors in the *row-space* of $\mathbf{A}$.

347. We call a set $V \subset \mathbf{R}^n$ **orthogonal** if every vector in V is orthogonal to every other vector in V (that is, $\mathbf{v}_i \cdot \mathbf{v}_j = 0$ whenever $\mathbf{v}_i$ and $\mathbf{v}_j$ are distinct vectors in V.)

Show that an orthogonal set *not* containing $\mathbf{0}$ must be independent. [*Hint: Proposition 6.6.*]

$$- \star -$$

7. Basis

Suppose a set B both *spans* a subspace $S \subset \mathbf{R}^n$, and is *independent*.

In this case, including *more* vectors from S won't increase the span, since B already spans S, which is closed under addition and scalar multiplication. The new set *will* be *dependent* though, since the smaller, original set B already spanned S.

On the other hand, suppose we *delete* one or more vectors from B. In this case, the new set will no longer *span* all of S, since that would mean the original set B—which we assumed independent—was *dependent*.

In short, when an independent set B spans a subspace S, we can't add a vector from S without making it dependent, and we can't delete a vector without reducing its span. In this sense, an independent spanning subset of a subspace is optimal: it's a *maximal* independent set, and a *minimal* spanning set at the same time. We call such sets *bases*:

DEFINITION 7.1. A **basis** for a subspace $S \subset \mathbf{R}^n$ is a subset of S that both

- *Spans S, and is*
- *Independent.*

$\square$

EXAMPLE 7.2 (Standard basis). As a first example, let us verify that the standard basis for $\mathbf{R}^n$ actually is a basis for $\mathbf{R}^n$ according to Definition 7.1. The criteria are that it must *span* $\mathbf{R}^n$, and it must be independent.

The standard basis does span $\mathbf{R}^n$. Indeed, we saw this way back in Observation 1.12 of Chapter 1, which says we can express any $\mathbf{x} = (x_1, x_2, \ldots, x_n) \in \mathbf{R}^n$ as a linear combination of the standard basis vectors in an obvious way. The standard basis vectors thus span $\mathbf{R}^n$, as needed.

They are also independent. For when we set them up as columns of a matrix, we get the $n \times n$ *identity* matrix, whose columns obviously all have pivots, and are hence independent, by Proposition 6.9 (c).   □

OBSERVATION 7.3. *The columns of any invertible matrix form a basis for* $\mathbf{R}^n$.

PROOF. We must show that the columns of any invertible matrix both *span* $\mathbf{R}^n$, and are *independent*.

When $\mathbf{A}$ is invertible, we can solve $\mathbf{Ax} = \mathbf{b}$ with any $\mathbf{b} \in \mathbf{R}^n$ (just take $\mathbf{x} - \mathbf{A}^{-1}\mathbf{b}$). So col($\mathbf{A}$) (the span of the columns) is all of $\mathbf{R}^n$ by the V.I.P. (Proposition 4.3).

Since an invertible matrix row-reduces to $\mathbf{I}_n$, every column is a pivot column too. So by Proposition 6.9 (b), they're independent. This completes the proof.   □

Contrastingly, the columns of a general $n \times m$ matrix may or may not span $\mathbf{R}^n$. Likewise, they may or may not be independent. The *pivot* columns of a matrix span its column-space (Corollary 6.4), however, and they're also independent, by Corollary 6.13. So we immediately get an easy (and useful) conclusion:

PROPOSITION 7.4. *The **pivot** columns of a matrix* $\mathbf{A}$ *always form a basis for* col($\mathbf{A}$).

This Proposition is what gives us a practical way to *find* a basis—an independent spanning set—for any subspace given *explicitly*—as a span. We take a spanning set, set up its vectors as columns of a matrix, row-reduce to distinguish pivot columns from free columns, and discard the latter.

EXAMPLE 7.5. Find a basis for the span of $(1, -3, 5)$, $(3, -9, 15)$, and $(2, 2, -2)$.

To do this, set the given vectors up as columns in a matrix $\mathbf{A}$. By Proposition 7.4, the pivot columns of $\mathbf{A}$ will then provide a basis.

We therefore row-reduce the matrix

$$\mathbf{A} = \begin{bmatrix} 1 & 3 & 2 \\ -3 & -9 & 2 \\ 5 & 15 & -2 \end{bmatrix}$$

This gives

$$\mathrm{RRE}(\mathbf{A}) = \begin{bmatrix} 1 & 3 & 0 \\ 0 & 0 & 1 \\ 0 & 0 & 0 \end{bmatrix}$$

Pivots appear in the first and third columns, so columns 1 and 3 of $\mathbf{A}$ form a basis B for $\mathrm{col}(\mathbf{A})$:

$$B = \left\{ \begin{pmatrix} 1 \\ -3 \\ 5 \end{pmatrix}, \begin{pmatrix} 2 \\ 2 \\ -2 \end{pmatrix} \right\}$$

$\square$

This example shows how to extract a basis from any *span*. But how do we find a basis for a *perp*? Since the nullspace of a matrix is the perp of its rows, the answer follows easily from Corollary 6.16, as we now show:

PROPOSITION 7.6. *The homogeneous generators for* $\mathrm{nul}(\mathbf{A})$ *always form a* basis *for* $\mathrm{nul}(\mathbf{A})$.

PROOF. We must show that the homogeneous generators both *span* $\mathrm{nul}(\mathbf{A})$ and are also *independent*.

Spanning is immediate: the nullspace of $\mathbf{A}$ *is* the set of solutions to $\mathbf{Ax} = \mathbf{0}$ which, we know, *is* the span of the homogeneous generators.

Independence is even quicker: it is exactly the statement of Corollary 6.16. $\square$

EXAMPLE 7.7. Find bases for **both** $\mathrm{nul}(\mathbf{A})$ and $\mathrm{col}(\mathbf{A})$, given

$$\mathbf{A} = \begin{bmatrix} 2 & 2 & 3 & 3 \\ -3 & -3 & 2 & 2 \\ 2 & 2 & 1 & 1 \end{bmatrix}$$

To get either basis, we start by row-reducing $\mathbf{A}$, which yields

$$\mathrm{RRE}(\mathbf{A}) = \begin{bmatrix} 1 & 1 & 0 & 0 \\ 0 & 0 & 1 & 1 \\ 0 & 0 & 0 & 0 \end{bmatrix}$$

The pivot columns are columns 1 and 3, so the first and third columns of $\mathbf{A}$ form a basis for the column-space:

$$\text{Basis for col}(\mathbf{A}) = \left\{ \begin{pmatrix} 2 \\ -3 \\ 2 \end{pmatrix}, \begin{pmatrix} 3 \\ 2 \\ 1 \end{pmatrix} \right\}$$

By inspecting the two *free* columns, we can also write generators for the nullspace, namely $\mathbf{h}_1 = (-1, 1, 0, 0)$ and $\mathbf{h}_2 = (0, 0, -1, 1)$. Thus,

$$\text{Basis for nul}(\mathbf{A}) = \left\{ \begin{pmatrix} -1 \\ 1 \\ 0 \\ 0 \end{pmatrix}, \begin{pmatrix} 0 \\ 0 \\ -1 \\ 1 \end{pmatrix} \right\}$$

$\square$

These examples show how to find a basis for any subspace $S \subset \mathbf{R}^n$. If S comes to us as a span, we realize it as the column space of a matrix. The pivot columns of that matrix then give us a basis for S. If, on the other hand, we encounter S as a perp, we realize it as the nullspace of a matrix, whose homogeneous generators then give us a basis for S.

7.8. Linear interpolation. One reason bases are important is this: *A linear transformation is completely determined by what it does to any basis for its domain.* In a way, this fact, called *linear interpolation* is just linear algebra's generalization of the familiar statement that "two points determine a line."

As a special case of linear interpolation, we saw (in Observation 5.5 of Chapter 1) that any linear map $T: \mathbf{R}^m \to \mathbf{R}^n$ is completely determined by its effect on the *standard* basis. In fact, we saw that the m outputs $T(\mathbf{e}_j)$ form the columns of the matrix $\mathbf{A}$ that represents T, and using that matrix, we can then compute $T(\mathbf{x})$ for *any* input $\mathbf{x}$ by the simple formula $T(\mathbf{x}) = \mathbf{A}\mathbf{x}$.

We now extend that principle from the standard basis to *any* basis for the domain $\mathbf{R}^m$ of T. Specifically, suppose we have a basis $B = \{\mathbf{b}_1, \mathbf{b}_2, \ldots, \mathbf{b}_m\}$ for $\mathbf{R}^m$, and we know the outputs

$$\mathbf{w}_j = T(\mathbf{b}_j), \quad j = 1, 2, \ldots, m$$

Then we can *still* construct the matrix $\mathbf{A}$ representing T, and from that, compute $T(\mathbf{x}) = \mathbf{A}\mathbf{x}$ for any input $\mathbf{x} \in \mathbf{R}^m$.

To see this, let $\mathbf{A}$ denote the (as yet unknown) matrix that represents T. Then for each basis vector $\mathbf{b}_j \in B$, we have

$$\mathbf{A}\mathbf{b}_j = T(\mathbf{b}_j) = \mathbf{w}_j$$

But if $\mathbf{Ab}_j = \mathbf{w}_j$ for each $j = 1, 2, \ldots, m$, then (since matrix multiplication works column-by-column), we actually have

(38) $$\mathbf{AB} = \mathbf{W}$$

where $\mathbf{B}$ is the matrix whose jth column is $\mathbf{b}_j$, while $\mathbf{W}$ is the matrix whose jth column is $\mathbf{w}_j$. Since we're given the basis vectors $\mathbf{b}_j$ in $\mathbf{R}^m$, and the corresponding outputs $\mathbf{w}_j$ in $\mathbf{R}^n$, we can easily construct both $\mathbf{B}$ and $\mathbf{W}$.

Now $\mathbf{A}$ is still unknown, but: *Since the m columns of $\mathbf{B}$ form a* **basis** *for* $\mathbf{R}^m$, *the matrix* $\mathbf{B}$ *is* **invertible**, *by Observation 7.3.* We can therefore solve (38) for $\mathbf{A}$:

(39) $$\mathbf{A} = \mathbf{W}\mathbf{B}^{-1}$$

and this gives us the desired formula for the matrix $\mathbf{A}$ that represents T. As mentioned above, once we have $\mathbf{A}$, we can compute $T(\mathbf{x}) = \mathbf{Ax}$ for any input $\mathbf{x} \in \mathbf{R}^m$, and T is thus completely determined, as claimed.

Let us demonstrate:

EXAMPLE 7.9. Suppose $T \colon \mathbf{R}^2 \to \mathbf{R}^3$ is linear, with

$$T\begin{pmatrix} 1 \\ 3 \end{pmatrix} = \begin{pmatrix} 0 \\ 1 \\ -2 \end{pmatrix} \quad \text{and} \quad T\begin{pmatrix} 2 \\ 5 \end{pmatrix} = \begin{pmatrix} 3 \\ 0 \\ 3 \end{pmatrix}$$

Since $(1, 3)$ and $(2, 5)$ form a basis for the domain $\mathbf{R}^2$ and we're given their images under T, we have exactly the situation discussed above, with

$$\mathbf{b}_1 = \begin{pmatrix} 1 \\ 3 \end{pmatrix}, \ \mathbf{b}_2 = \begin{pmatrix} 2 \\ 5 \end{pmatrix}, \quad \mathbf{w}_1 = \begin{pmatrix} 0 \\ 1 \\ -1 \end{pmatrix}, \ \mathbf{w}_2 = \begin{pmatrix} 3 \\ 0 \\ 3 \end{pmatrix}$$

These vectors form the columns of the $\mathbf{B}$ and $\mathbf{W}$ in (38), but we want to use (39) to compute $\mathbf{A}$, so we must first find the *inverse* of $\mathbf{B}$. Using either the formula from Exercise 235 of Chapter 4, or by row-reducing $[\mathbf{B} \,|\, \mathbf{I}]$, we find that

$$\mathbf{B}^{-1} = \begin{bmatrix} 1 & 2 \\ 3 & 5 \end{bmatrix}^{-1} = \begin{bmatrix} -5 & 2 \\ 3 & -1 \end{bmatrix}$$

so that T is represented by

$$\mathbf{A} = \mathbf{W}\mathbf{B}^{-1} = \begin{bmatrix} 0 & 3 \\ 1 & 0 \\ -2 & 3 \end{bmatrix} \begin{bmatrix} -5 & 2 \\ 3 & -1 \end{bmatrix}$$

$$= \begin{bmatrix} 9 & -3 \\ -5 & 2 \\ 19 & -7 \end{bmatrix}$$

We emphasize again the key point: Once we have $\mathbf{A}$, we can find $T(\mathbf{x})$ for *any* input vector $\mathbf{x}$. For instance,

$$T(2,2) = \begin{bmatrix} 9 & -3 \\ -5 & 2 \\ 19 & -7 \end{bmatrix} \begin{pmatrix} 2 \\ 2 \end{pmatrix} = \begin{pmatrix} 12 \\ -6 \\ 24 \end{pmatrix}$$

$\square$

– Practice –

348. Find a basis for the column-space. Then find one for the row-space.

$$\mathbf{A} = \begin{bmatrix} 1 & 2 & 3 & 5 \\ 2 & 4 & 8 & 12 \\ 3 & 6 & 7 & 13 \end{bmatrix}$$

349. Find a basis for the column-space. Then find one for the row-space:

$$\mathbf{A} = \begin{bmatrix} 1 & 1 & 1 & 0 & 0 & 0 \\ 1 & 0 & 0 & 1 & 1 & 0 \\ 0 & 1 & 0 & 1 & 0 & 1 \\ 0 & 0 & 1 & 0 & 1 & 1 \end{bmatrix}$$

350. Find a basis for $\{(0,1,0,1),(1,0,1,0)\}^{\perp}$ in $\mathbf{R}^4$.

351. Find a basis for the perp of the rows of the matrix in Exercise 349. Then find a basis for the perp of the *columns*.

352. Let $V = \{\mathbf{v}_1, \mathbf{v}_2, \ldots, \mathbf{v}_6\}$, where the $\mathbf{v}_i$'s are the vectors in Exercise 341.

    a) Find a basis for the span of V.

    b) Find a basis for the perp of V.

353. The columns of any invertible $n \times n$ matrix form a basis for $\mathbf{R}^n$. Is the converse true? That is, if we take any basis for $\mathbf{R}^n$ and make its vectors the columns of an $n \times n$ matrix, do we always get an invertible matrix? Justify your answer with either a proof or a counterexample.

354. Suppose $\mathbf{v}_1$ and $\mathbf{v}_2$ form a basis for a subspace $S \subset \mathbf{R}^n$. Let $\mathbf{w}_1 = \mathbf{v}_1 + \mathbf{v}_2$ and $\mathbf{w}_2 = \mathbf{v}_1 - \mathbf{v}_2$. Show that $\mathbf{w}_1$ and $\mathbf{w}_2$ form a basis for S too.

355. Use Exercise 335 and Corollary 6.15 to prove

    The pivot rows of RRE$(\mathbf{A})$ *form a basis for the row-space of* $\mathbf{A}$.

356. Let

$$\mathbf{A} = \begin{bmatrix} 0 & 1 & 2 & 3 \\ 4 & 5 & 6 & 7 \\ 8 & 9 & 10 & 11 \\ 12 & 13 & 14 & 15 \end{bmatrix}$$

Find bases for each of the subspaces row$(\mathbf{A})$, col$(\mathbf{A})$, nul$(\mathbf{A})$, and nul$(\mathbf{A}^T)$.

357. Suppose

$$U = \{\mathbf{u}_1, \mathbf{u}_2, \ldots, \mathbf{u}_k\}$$

is an **independent** set in $\mathbf{R}^n$, and $\mathbf{z} \in \mathbf{R}^n$ is **not** in the **span** of U. Show that if we adjoin $\mathbf{z}$ to U, the new set

$$U' = \{\mathbf{u}_1, \mathbf{u}_2, \ldots, \mathbf{u}_k, \mathbf{z}\}$$

will still be independent. [*Hint: What happens if we row reduce* $[\mathbf{U}|\mathbf{z}]$, *where the columns of* $\mathbf{U}$ *are the vectors in U?*]

358. If $T\colon \mathbf{R}^2 \to \mathbf{R}^2$ maps $(3, -1)$ to $(1, -4)$, and maps $(5, -2)$ to $(-3, 2)$, what matrix represents T? Where does T send the standard basis vectors $(1, 0)$ and $(0, 1)$?

359. Find a 2-by-2 matrix for which $(4, 5)$ is an eigenvector with eigenvalue $\lambda = -1$, and $(3, 4)$ is an eigenvector with eigenvalue $\lambda = 2$. (See Definition 1.36 of Chapter 1 to review the meanings of *eigenvalue* and *eigenvector*.) [*Hint: Use linear interpolation.*]

360. Find a 3-by-3 matrix for which $(1, 0, 0)$ is an eigenvector with eigenvalue $\lambda = 2$, $(1, 1, 0)$ is an eigenvector with eigenvalue $\lambda = -1$, and $(1, 1, 1)$ is an eigenvector with eigenvalue $\lambda = 3$. [*Hint: Use linear interpolation.*]

361. What matrix represents the linear transformation $Q \colon \mathbf{R}^4 \to \mathbf{R}^2$, with

$$
\begin{aligned}
Q(1, 1, 1, 1) &= (1, 1) \\
Q(1, -1, 1, -1) &= (1, -1) \\
Q(1, -1, -1, 1) &= (-1, 1) \\
Q(1, 1, -1, -1) &= (-1, -1)
\end{aligned}
$$

362. Suppose $B = \{\mathbf{b}_1, \mathbf{b}_2, \ldots, \mathbf{b}_n\}$ is a basis for $\mathbf{R}^n$. Suppose further that $S \colon \mathbf{R}^n \to \mathbf{R}^n$ is a linear transformation that sends $\mathbf{b}_i$ to $-\mathbf{b}_i$ for each $i = 1, 2, \ldots, n$. What $n \times n$ matrix represents S?

$$- \star -$$

8. Dimension and Rank

8.1. Dimension. We now know how to extract, from any spanning set for a subspace S, an *independent* spanning set—a basis—for S. Each vector in such a basis is then *essential*: we can't delete it without losing span. In that way, bases for S are *minimal* spanning sets.

We may still wonder, however: *Is it possible that some bases for a given subspace S have fewer vectors than others, making them even "more minimal"?*

Experience suggests otherwise. For instance, Observation 7.3, tells us that the columns of any invertible $n \times n$ matrix form a basis for $\mathbf{R}^n$. That gives many bases with exactly n vectors.

It's easy to see that no basis for $\mathbf{R}^n$ can have *more* than n vectors. For a matrix with more than n columns in $\mathbf{R}^n$ would have only n rows, hence a free column, violating the independence we require any basis to have.

Similarly, no basis for $\mathbf{R}^n$ can have *fewer* than n vectors, for if we set them up as *rows* of a matrix, that matrix too would have more columns than rows, and hence a free column, giving us a homogeneous generator—a non-zero vector in the perp of the rows. By Proposition 6.6, that puts it outside the span of the rows, violating the requirement that a basis for $\mathbf{R}^n$ must *span* all of $\mathbf{R}^n$.

We now want to show that this same fact holds, not just for $\mathbf{R}^n$, but for all its *subspaces* too.

The following theorem, one of the most important in the subject, completely settles the question.

THEOREM 8.2 (Dimension). *All bases for a given subspace* $S \subset \mathbf{R}^n$ *contain the same number of vectors.*

Before we prove this, we briefly explore the major consequence for which it is named. Specifically, it assigns each subspace $S \subset \mathbf{R}^n$ two characteristic integers $d, c \geq 0$ called the *dimension* and *codimension* of S, respectively:

DEFINITION 8.3 (Dimension). The number of vectors common to every basis for a subspace $S \subset \mathbf{R}^n$ is called the **dimension** of S, denoted by $\dim(S)$. The difference in dimension between $\mathbf{R}^n$ and S, namely $n - \dim(S)$, is called the **codimension** of S, which we abbreviate as $\operatorname{codim}(S)$.   □

Dimension is perhaps the *most basic feature of a subspace*. Here are some examples.

EXAMPLE 8.4. As noted above, every basis for $\mathbf{R}^n$, like the standard basis, contains n vectors. The dimension of $\mathbf{R}^n$ is thus n, while its codimension is $n - n = 0$.   □

EXAMPLE 8.5 (Hyperplanes). Let $\mathbf{a} = (a_1, a_2, \ldots, a_n) \in \mathbf{R}^n$ be any non-zero vector. Then the hyperplane $\mathbf{a}^\perp$ has dimension $n - 1$. For we know that $\mathbf{a}^\perp$ is also the nullspace of the matrix

$$\mathbf{A} = \begin{bmatrix} a_1 & a_2 & \cdots & a_n \end{bmatrix}$$

Such a matrix must have exactly $n - 1$ free columns, hence $n - 1$ homogeneous generators. By Proposition 7.6, these form a basis for $\operatorname{nul}(\mathbf{A}) = \mathbf{a}^\perp$. It follows from the Dimension Theorem that *every* basis for $\mathbf{a}^\perp$ contains exactly $n - 1$ vectors, and hence $\mathbf{a}^\perp$ has dimension $n - 1$ and codimension $n - (n - 1) = 1$.

This gives meaning to our intuitive sense that hyperplanes in $\mathbf{R}^n$ are "copies" of $\mathbf{R}^{n-1}$. Both have dimension $n - 1$. For instance, every plane in $\mathbf{R}^3$ has dimension two, as we might expect. $\qquad\square$

EXAMPLE 8.6 (Lines). The line generated by any non-zero vector $\mathbf{v} \in \mathbf{R}^n$ has dimension 1, and hence codimension $n - 1$. For $\{\mathbf{v}\}$ spans the line, and is independent, and thus forms a 1-element basis for the line. By the Dimension Theorem 8.2, *every* basis for the line also contains exactly one vector. Lines are thus 1-dimensional according to our definition of dimension. $\qquad\square$

Let us now turn toward the proof of the Dimension Theorem (8.2) itself. Most of the work goes into Lemma 8.7 below, a fact very useful in its own right. To facilitate the proof, we formalize a naming convention we have already used informally:

When we use uppercase *without* bold to name a set of vectors, like this:

$$V = \{\mathbf{v}_1, \mathbf{v}_2, \cdots, \mathbf{v}_k\}$$

we use the *same* uppercase letter—but *in* bold—to name the *matrix* having those vectors as columns:

$$\mathbf{V} = \begin{bmatrix} \mathbf{v}_1 & \mathbf{v}_2 & \cdots & \mathbf{v}_k \\ | & | & \cdots & | \end{bmatrix}$$

With this notation in hand, here's the key lemma:

LEMMA 8.7. *Suppose $S \subset \mathbf{R}^n$ has a basis with exactly k vectors. Then*

   i) *Any subset of S with **more** than k vectors is **dependent**, and*

   ii) *No subset of S with **fewer** than k vectors can **span** S.*

PROOF. By assumption, S has a basis B with exactly k vectors. Call this basis

$$U = \{\mathbf{u}_1, \mathbf{u}_2, \ldots, \mathbf{u}_k\}$$

To deduce conclusion (i), we must show that any subset of S with $k' > k$ vectors must be dependent. Let

$$V = \{\mathbf{v}_1, \mathbf{v}_2, \ldots, \mathbf{v}_{k'}\}$$

Since U is a basis, the $\mathbf{u}_j$'s span S, so if we set them up as columns of a matrix $\mathbf{U}$, we have

$$S = \mathrm{col}(\mathbf{U})$$

Now the $\mathbf{v}_j$'s *also lie in* S, so each $\mathbf{v}_j$ is a linear combination of the $\mathbf{u}_i$'s. Linear combination is the same as matrix/vector multiplication, however, so for each $j = 1, \ldots, k'$, we can write

$$\mathbf{v}_j = \mathbf{U}\mathbf{x}_j$$

for some $\mathbf{x}_j$. Moreover, since matrix multiplication works column-by-column, we now have

(40) $$\mathbf{V} = \mathbf{U}\mathbf{X}$$

where $\mathbf{X}$ and $\mathbf{V}$ are the matrices with columns $\mathbf{x}_j$ and $\mathbf{v}_j$, respectively.

Now $\mathbf{V}$ is $n \times k'$, and $\mathbf{U}$ is $n \times k$. So $\mathbf{X}$ must be $k \times k'$, which means $\mathbf{X}$ *has more columns than rows, and hence a* free *column*.

That gives us a homogeneous generator $\mathbf{h} \neq \mathbf{0}$ with $\mathbf{X}\mathbf{h} = \mathbf{0}$. But if $\mathbf{X}\mathbf{h} = \mathbf{0}$, then by (40) above, we have

$$\mathbf{V}\mathbf{h} = \mathbf{U}\mathbf{X}\mathbf{h} = \mathbf{U}(\mathbf{X}\mathbf{h}) = \mathbf{U}\mathbf{0} = \mathbf{0}$$

So $\mathbf{V}\mathbf{h} = \mathbf{0}$. But *this makes the columns of* $\mathbf{V}$ *—the* $\mathbf{v}_j$*'s—dependent*, by conclusion (d) of Proposition 6.9, thereby proving conclusion (i) of our Lemma.

Conclusion (ii) then follows easily from (i). For if S contained a *spanning* set V with $k' < k$ vectors, then V (or some independent subset of it), would form a *basis* for S with fewer than k vectors. Our *original* basis U would then be a spanning set with *more* vectors than this new basis, violating conclusion (i).   □

This Lemma makes the Dimension Theorem 8.2 (every basis for a subspace S has the same number of vectors) easy to prove:

PROOF OF DIMENSION THEOREM. Take any two bases B and B' for a subspace $S \subset \mathbf{R}^n$, with k and k' vectors in B and B', respectively. As bases, B and B' are both independent. By conclusion (i) of the Lemma, neither has more vectors than the other. This forces $k' = k$, and we're done: any two bases for S have the same number of vectors.   □

Another implication of the Lemma above often proves useful:

COROLLARY 8.8. *If* $S \subset \mathbf{R}^n$ *is a subspace of dimension* k, *then* any *set of* k independent *vectors in* S *forms a basis for* S. *Similarly, any* spanning *set of* k *vectors forms a basis for* S.

In other words, if we know that $\dim(S) = k$, and we have a set of k vectors in S, we only have to verify *one* of the defining conditions for a basis. The remaining condition (independence or spanning) follows automatically when the number of vectors equals the dimension.

PROOF OF COROLLARY 8.8. Since S has dimension k, it has a basis U with k vectors. Let $V \subset S$ be some other subset with k vectors:

$$V = \{\mathbf{v}_1, \mathbf{v}_2, \ldots, \mathbf{v}_k\}$$

If V is independent, it must also *span* S. For otherwise, we could find a vector $\mathbf{v}_{k+1}$ in S, but outside the span of V. Adjoining $\mathbf{v}_{k+1}$ to V would give us a new set, V' with $k+1$ vectors, and V' would be then *independent*, by Exercise 357. But since V' has more than $k = \dim(S)$ vectors, this contradicts Lemma 8.7. So if V was independent, it also spans, making it a basis for S.

Similarly, if V *spans* S, it must also be independent. For if it were *dependent*, we would get at least one free column on row-reducing the matrix $\mathbf{V}$ having the $\mathbf{v}_j$'s as columns. Keeping only the pivot columns, we would then have a spanning set with fewer than $k = \dim(S)$ vectors. This again contradicts the Lemma. ∎

8.9. Rank and nullity.

The dimensions of the column-space and nullspace of a matrix are among the most useful fact we can know about it. Both have names.

DEFINITION 8.10. The **rank** of a matrix $\mathbf{A}$, denoted $\operatorname{rank}(\mathbf{A})$, is the dimension of its column-space. The dimension of $\operatorname{nul}(\mathbf{A})$ is called its **nullity**, written $\operatorname{nullity}(\mathbf{A})$. ∎

PROPOSITION 8.11. *The rank and nullity of a matrix respectively equal its number of pivot and free columns. If $\mathbf{A}$ has m columns, this means*

$$\operatorname{rank}(\mathbf{A}) + \operatorname{nullity}(\mathbf{A}) = m$$

PROOF. By definition, the rank of $\mathbf{A}$ is the dimension of $\operatorname{col}(\mathbf{A})$. Since the pivot columns of $\mathbf{A}$ always form a basis for the column-space, $\operatorname{rank}(\mathbf{A})$ is the number of pivot columns.

Similarly, $\operatorname{nullity}(\mathbf{A})$ is the dimension of its nullspace, and by Proposition 7.6, the generating vectors $\mathbf{h}_i$ form a basis for it. Since each free column gives rise to one homogeneous generator and vice-versa, the nullity of $\mathbf{A}$ equals the number of free columns.

The Proposition's last statement is now obvious, since the number of pivot columns plus the number of free columns equals the total number of columns.   □

We will continue to encounter rank and nullity going forward but for now, we content ourselves with one more useful fact that, at first, seems almost magical:

PROPOSITION 8.12. *Let* $\mathbf{A}$ *be any matrix. Then*

$$\operatorname{rank}(\mathbf{A}) = \operatorname{rank}(\mathbf{A}^T)$$

Note that the pivot columns form a basis for $\operatorname{col}(\mathbf{A})$, so the number of pivots is $\dim(\operatorname{col}(\mathbf{A}))$, the rank of $\mathbf{A}$. When we row-reduce $\mathbf{A}^T$, on the other hand, the number of pivots we get is the dimension of $\operatorname{col}(\mathbf{A}^T) = \operatorname{row}(\mathbf{A})$. Thus, the Proposition tells us that

$$\dim(\operatorname{col}(\mathbf{A})) = \dim(\operatorname{row}(\mathbf{A}))$$

even though $\operatorname{col}(\mathbf{A}) \subset \mathbf{R}^n$ and $\operatorname{row}(\mathbf{A}) \subset \mathbf{R}^m$ typically lie in entirely different numeric vector spaces. More operationally, it says we get the same number of pivots whether we row-reduce $\mathbf{A}$ or $\mathbf{A}^T$, even though the row-reduction processes for the two will be completely different!

PROOF OF PROPOSITION 8.12. The rank of $\mathbf{A}$ is the dimension of $\operatorname{col}(\mathbf{A})$, and the rank of $\mathbf{A}^T$ is the dimension of $\operatorname{col}(\mathbf{A}^T)$. But $\operatorname{col}(\mathbf{A}^T) = \operatorname{row}(\mathbf{A})$, so it suffices to show that

$$\dim(\operatorname{col}(\mathbf{A})) = \dim(\operatorname{row}(\mathbf{A}))$$

By Exercise 355, the pivot *rows* of RRE$(\mathbf{A})$ form a basis for $\operatorname{row}(\mathbf{A})$, and the dimension of $\operatorname{row}(\mathbf{A})$ is therefore the number of pivots in RRE$(\mathbf{A})$. By Proposition 7.4, however, the pivot *columns* of $\mathbf{A}$ form a basis for $\operatorname{col}(\mathbf{A})$, so the dimension of $\operatorname{col}(\mathbf{A})$ is *also* the number of pivots in RRE$(\mathbf{A})$. Thus, $\dim(\operatorname{col}(\mathbf{A})) = \dim(\operatorname{row}(\mathbf{A}))$, just as we needed.   □

8.13. Affine subspaces. The solution set of the *homogeneous* system $\mathbf{Ax} = \mathbf{0}$ is always a subspace. We can describe it as $\operatorname{nul}(\mathbf{A})$, a perp, or, as the span of the homogeneous generators we get from the free columns.

The solution set of an *inhomogeneous* system $\mathbf{Ax} = \mathbf{b}$, on the other hand, can *never* be a subspace since it cannot contain the origin (Proof: $\mathbf{A0} = \mathbf{0} \neq \mathbf{b}$).

Geometrically, however, the solution set of $\mathbf{Ax} = \mathbf{b}$ "looks" very much like a subspace. Indeed, we get it by adding just one vector—a particular solution $\mathbf{x}_p$—to each vector in $\mathrm{nul}(\mathbf{A})$, which *is* a subspace. As we have know since Theorem 4.1 of Chapter 2, the solution set for $\mathbf{Ax} = \mathbf{b}$ takes the form

$$\mathbf{x}_p + \mathrm{nul}(\mathbf{A})$$

Here $\mathbf{x}_p$ is one ("particular") solution of the inhomogeneous system, and of course $\mathrm{nul}(\mathbf{A})$ comprises all solutions of the homogeneous system.

Geometrically, $\mathbf{x}_p + \mathrm{nul}(\mathbf{A})$ is the set we get by translating $\mathrm{nul}(\mathbf{A})$ away from the origin by the vector $\mathbf{x}_p$ (Figure 5). It is no longer a subspace, but we shall signal its resemblance to one by calling it an *affine* subspace:

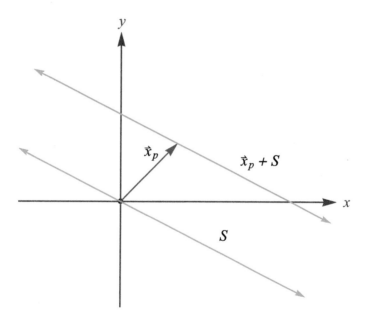

Figure 5. Affine subspace. When we displace a subspace S by an amount $\mathbf{x}_p$, we get the *affine subspace* $\mathbf{x}_p + S$. Note that $\mathbf{x}_p + S$ will not be a subspace, unless $\mathbf{x}_p \in S$. (Compare Figure 5, Chapter 3.)

DEFINITION 8.14. A **k-dimensional affine subspace** of $\mathbf{R}^n$ is a subset having the form

$$\mathbf{a} + S = \{\mathbf{a} + \mathbf{s} \colon \mathbf{s} \in S\}$$

where $S \subset \mathbf{R}^n$ is a k-dimensional subspace. An affine subspace $\mathbf{a} + S$ is *not* an actual subspace, unless $\mathbf{a} \in S$ (Exercise 373).   □

We call 1-dimensional affine subspaces of $\mathbf{R}^n$ **lines**. A **plane** in $\mathbf{R}^n$ is any 2-dimensional affine subspace.

Note that the solution set of a single inhomogeneous equation $\mathbf{a} \cdot \mathbf{x} = b$ (a hyperplane parallel to $\mathbf{a}^\perp$; see Definition 4.16 in Chapter 3) is always an $(n-1)$-dimensional affine subspace—a **hyperplane**. For we can always write it as $\mathbf{x}_p + \mathbf{a}^\perp$, where $\mathbf{x}_p$ is a particular solution to $\mathbf{a} \cdot \mathbf{x} = b$.

Conversely, $(n-1)$-dimensional affine subspaces are always hyperplanes (Exercise 374).

– Practice –

363. For each set of vectors below, determine (i) Whether it is independent, and (ii) The dimension and codimension of its span.

a) $\{(1, 1, 1, -3), (-3, 1, 1, 1), (1, -3, 1, 1), (1, 1, -3, 1)\}$

b) $\{(1, 3, 5), (3, 5, 1), (5, 1, 3)\}$

c) $\{(6, 3, 0), (0, 3, 2), (6, -3, -4), (9, 0, -3)\}$

364. Compute the dimensions of all four subspaces (column-space, nullspace, row-space, and left nullspace) for each matrix below.

a) $\begin{bmatrix} 2 & -4 & 8 \\ 0 & 3 & -3 \end{bmatrix}$ b) $\begin{bmatrix} 1 & 5 & -1 & 5 \\ 5 & -2 & 1 & 4 \\ -1 & 1 & 3 & 3 \end{bmatrix}$

365. Compute the **co**dimensions of all four subspaces (column-space, nullspace, row-space, and left nullspace) for each matrix below.

a) $\begin{bmatrix} 1 & -3 \\ -3 & 1 \\ 1 & -3 \end{bmatrix}$ b) $\begin{bmatrix} 2 & -3 & 5 \\ 7 & 2 & -3 \\ 5 & 7 & 2 \\ -3 & 5 & 7 \end{bmatrix}$

366. Find the dimension of the affine subspace of solutions to each of these four-variable systems:

a)
$$\begin{aligned} x + y + z &= 2 \\ y + z + w &= 1 \\ x + z + w &= 3 \end{aligned}$$

b)
$$\begin{aligned} x - 2y + z - 2w &= 3 \\ x + y - 2z &= 2 \\ y + z - 2w &= 1 \end{aligned}$$

367. Suppose we have an inhomgeneous linear system of n equations in m unknowns.

   a) If $n = 7$ and $m = 12$, and the augmented matrix for the system reduces with 7 pivots, what are the dimension and codimension of the solution set (as an affine subspace)?

   b) If $n = 12$ and $m = 7$, and the augmented matrix for the system reduces with 7 pivots, what are the dimension and codimension of the solution set (as an affine subspace)?

   c) if $n = 12$ and $m = 7$, the augmented matrix has eight columns. If they are all pivot columns, what does that say about the solution set?

368. Suppose a 3-by-4 matrix $\mathbf{A}$ represents a linear mapping that is *onto*. What will the dimension of the solution set to $\mathbf{Ax} = \mathbf{0}$ be? Explain your answer.

369. Explain why each statement below about the rank r of an $n \times m$ matrix $\mathbf{A}$ is true, or give an example showing it to be false.

   a) $\mathbf{A}$ is invertible if and only if $m = n = r$.

   b) If $r < m$, the columns of $\mathbf{A}$ are dependent.

   c) If $r < n$, the columns of $\mathbf{A}$ are dependent.

   d) If $r < m$, the rows of $\mathbf{A}$ are dependent.

   e) If $r < n$, the rows of $\mathbf{A}$ are dependent.

370. Suppose we have k vectors $\mathbf{v}_1, \mathbf{v}_2, \ldots, \mathbf{v}_k$ in a subspace $S \subset \mathbf{R}^n$.

   a) If the $\mathbf{v}_i$'s *span* S, what can you deduce about the dimension and codimension of S?

   b) If the $\mathbf{v}_i$'s are *independent*, what can you deduce about the dimension and codimension of S?

   c) If the $\mathbf{v}_i$'s *both* span S *and* are independent, what can you deduce about the dimension and codimension of S?

371. Suppose V and W are subspaces of $\mathbf{R}^n$, and define their **sum** to be
$$V + W := \{\mathbf{v} + \mathbf{w} : \mathbf{v} \in V \text{ and } \mathbf{w} \in W\}.$$
Show that

    a) $V + W$ is a subspace.

    b) $\dim(V + W) \le \dim(V) + \dim(W)$.

    c) Give an example in which the inequality in (b) is actually strict.

    d) Give an example in which the inequality in (b) is actually an equality.

372. Suppose $\mathbf{A}$ is an $n \times m$ matrix.

    a) Express $\operatorname{rank}(\mathbf{A})$ in terms of $\operatorname{nullity}(\mathbf{A}^T)$.

    b) Express $\operatorname{nullity}(\mathbf{A}^T)$ in terms of $\operatorname{nullity}(\mathbf{A})$.

373. Prove: *If $S \subset \mathbf{R}^n$ is a subspace and $\mathbf{a} \notin S$, the affine subspace $\mathbf{a} + S$ doesn't contain the origin, hence isn't a subspace.*

374. Prove that every $(n - 1)$-dimensional affine subspace $S \subset \mathbf{R}^n$ is a hyperplane (the solution set of a single inhomogeneous equation). [*Hint: Find a vector orthogonal to each vector in a basis for S.*]

375. Show that when $\mathbf{b}$ is *in* the affine subspace $\mathbf{a} + S$, then $\mathbf{b} + S = \mathbf{a} + S$ (they are the same affine subspace).

376. Show that the intersection of two affine subspaces (when non-empty) is again an affine subspace. [*Hint: Express each subspace as the solution set of a linear system. Then express their intersection the same way. Alternatively, use Exercises 289 and 375.*]

$-\star-$

CHAPTER 6

Orthogonality

In the first two sections of this chapter, we use orthogonality to complete our understanding of subspaces—especially the relationship between the span and perp of the same set. In the last two sections, we explore other topics that use orthogonality: First, we focus on *orthonormal bases*: bases comprised of mutually orthogonal unit vectors. Such bases are particularly useful, and they also have special geometric significance, for when they form the columns of a matrix, that matrix represents a *distortion-free* linear transformation, as we shall see. Lastly, we explain the Gram–Schmidt algorithm, which systematically replaces the vectors of any basis for a subspace $S \subset \mathbf{R}^n$ by those of an orthonormal basis.

1. Orthocomplements

The nullspace of a matrix is the perp of its rows—we defined it that way in Chapter 5. So a vector lies in $\mathrm{nul}(\mathbf{A})$ if and only if it is orthogonal to (dots to zero with) each row of $\mathbf{A}$.

A vector orthogonal to every row, however, will also be orthogonal to any *linear combination* of rows, since the dot product distributes over vector addition. It follows that each vector in $\mathrm{nul}(\mathbf{A})$ is actually orthogonal to the entire *row-space* of $\mathbf{A}$. Similarly, each row is orthogonal to the entire *nullspace* of $\mathbf{A}$.

In short, the nullspace and row-space of an $n \times m$ matrix $\mathbf{A}$ enjoy a mutual perpendicularity relationship. The nullspace consists precisely of all vectors orthogonal to the row-space.

DEFINITION 1.1. We say a vector $\mathbf{w} \in \mathbf{R}^n$ is **orthogonal to a subspace** $S \subset \mathbf{R}^n$ if it is orthogonal to *every* vector in S.

The set of *all* vectors orthogonal to S is called the **orthogonal complement** (or **orthocomplement**) of S, denoted by $S^\perp$. In symbols,

$$S^\perp = \{\mathbf{w} \in \mathbf{R}^n : \mathbf{w} \cdot \mathbf{v} = 0 \text{ for all } \mathbf{v} \in S\}$$

See Figure 1. □

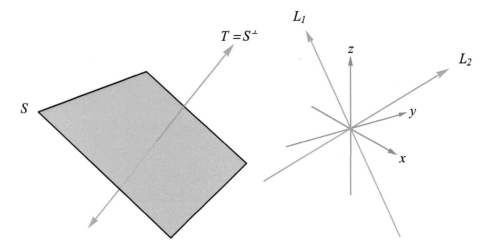

Figure 1. Orthocomplements (and not). The orthocomplement of a plane $S \subset \mathbf{R}^3$ is the perpendicular line T
through $\mathbf{0}$, because that line contains *every* vector orthogonal to the plane. The blue lines at right are orthogonal to
each other, but they are **not** orthocomplements, since L_2
does **not** contain **every** vector orthogonal to L_1.

Definition 1.1 doesn't certify $\mathbf{w}$ as orthogonal to a subspace S until it
is orthogonal to *every* vector in S. Fortunately, we don't have to test
it on every vector—it suffices to test on a spanning set:

LEMMA 1.2. *If* $S = \mathrm{span}\{\mathbf{v}_1, \mathbf{v}_2, \ldots, \mathbf{v}_k\}$, *then* $\mathbf{w}$ *is orthogonal to* S
if and only if it is orthogonal to each $\mathbf{v}_i$.

PROOF. By definition, $\mathbf{w}$ is orthogonal to S if it dots to zero with
every vector in S. In that case, it certainly dots to zero with each of
the $\mathbf{v}_i$'s. Conversely, if $\mathbf{w}$ dots to zero with each of the $\mathbf{v}_i$'s, it dots
to zero with any linear combination of the $\mathbf{v}_i$'s (since the dot product
distributes) and hence is orthogonal to S. $\square$

It follows immediately from this lemma that the set of *all* $\mathbf{w}$ orthogonal
to $S = \mathrm{span}\{\mathbf{v}_1, \mathbf{v}_2, \ldots, \mathbf{v}_k\}$—the orthocomplement of S—is simply
the set of all $\mathbf{w}$ orthogonal to *each* $\mathbf{v}_i$. But we already have a name
for that—it's the *perp* of $\{\mathbf{v}_1, \mathbf{v}_2, \ldots, \mathbf{v}_k\}$. This proves:

COROLLARY 1.3. *The orthocomplement of the* ***span*** *of a set is the*
perp *of that* same *set: If* $S = \mathrm{span}\, V$, *then* $S^\perp = V^\perp$.

REMARK 1.4. Note that the "perp" superscript $\perp$ has a slightly different meaning depending on whether we apply it to a *set* (as in $V^\perp$
above) or a *subspace* (as in $S^\perp$). Technically, the perp of a set consists

of everything orthogonal to each vector in the set. The Corollary shows that the perp of a set is also the perp of the entire *span* of that set. □

EXAMPLE 1.5. Consider the line L spanned by $(1,0,0)$ in $\mathbf{R}^3$—the x-axis. What is the orthocomplement $L^\perp$?

By Corollary 1.3, the orthocomplement $L^\perp$ is simply $(1,0,0)^\perp$, the plane perpendicular to $(1,0,0)$, also known as the y, z-coordinate plane, or the $x = 0$ plane. □

EXAMPLE 1.6. We just saw that when L is the x-axis, $L^\perp$ is the $x = 0$ plane. Now we show that, at least for this example, it works the other way too: when P is the $x = 0$ plane, $P^\perp$ is the x-axis.

To see this, we need to show that the x-axis contains *all* vectors orthogonal to the $x = 0$ plane. Corollary 1.3 makes this easy.

For the $x = 0$ coordinate plane is spanned by the standard basis vectors $\mathbf{e}_2 = (0,1,0)$ and $\mathbf{e}_3 = (0,0,1)$. So by the corollary, $P^\perp = \{\mathbf{e}_1, \mathbf{e}_2\}^\perp$, and hence $P^\perp = \text{nul}(\mathbf{A})$, where

$$\mathbf{A} = \begin{bmatrix} 0 & 1 & 0 \\ 0 & 0 & 1 \end{bmatrix}$$

This matrix is already in RRE form, and yields one homogeneous generator $\mathbf{h} = (1,0,0)$. So $P^\perp = \text{nul}(\mathbf{A}) = \text{span}\{\mathbf{h}\}$, which is the x-axis. In short, the x-axis forms the orthocomplement of the $x = 0$ plane, as claimed. □

The examples above illustrate a fundamental fact:

If T is the orthocomplement of S, then S is the orthocomplement of T. "Orthocomplement" is a reciprocal relationship.

This may not be surprising, but it isn't easy to prove—it does not follow directly from the definition. Indeed, its proof is somewhat subtle. We will derive it as one of several corollaries to Proposition 1.7 below. Most of the work goes into proving that proposition, which, stated informally, says that the nullspace and row-space of a matrix are always *orthocomplements of each other.*

PROPOSITION 1.7. *For any matrix $\mathbf{A}$, we have*

$$\text{row}(\mathbf{A})^\perp = \text{nul}(\mathbf{A})$$
$$\text{nul}(\mathbf{A})^\perp = \text{row}(\mathbf{A})$$

The row-space and nullspace of a matrix are always orthocomplements of each other.

PROOF. The definition of row-space (Definition 3.1 in Chapter 5) says that $\mathrm{nul}(\mathbf{A})$ is the perp of the rows, so by Corollary 1.3, it is also the perp of the row-*space*, thereby proving the first identity. The second requires a subtler argument, however, which starts with the observation that $\mathrm{nul}(\mathbf{A})^{\perp}$ certainly *contains* $\mathrm{row}(\mathbf{A})$:

$$(41) \qquad \mathrm{nul}(\mathbf{A})^{\perp} \supset \mathrm{row}(\mathbf{A})$$

Indeed, each row of $\mathbf{A}$ dots to zero with any solution $\mathbf{x}$ of $\mathbf{Ax} = \mathbf{0}$ since we can compute $\mathbf{Ax}$ by dotting $\mathbf{x}$ with each row of $\mathbf{A}$. The same then goes for any linear combination of rows—anything in $\mathrm{row}(\mathbf{A})$— since dot products distribute over vector addition. So any vector in $\mathrm{row}(\mathbf{A})$ is orthogonal to everything in $\mathrm{nul}(\mathbf{A})$, which confirms (41).

To conclude that $\mathrm{row}(\mathbf{A})$ *equals* $\mathrm{nul}(\mathbf{A})^{\perp}$, however, we must also be sure that $\mathrm{nul}(\mathbf{A})^{\perp}$ contains *only* $\mathrm{row}(\mathbf{A})$. We argue that by contradiction, assuming that, to the contrary, there were some

$$(42) \qquad \mathbf{z} \in \mathrm{nul}(\mathbf{A})^{\perp} \quad \text{with} \quad \mathbf{z} \notin \mathrm{row}(\mathbf{A})$$

To contradict this, observe that if we had such a vector $\mathbf{z}$, we could append it to a basis $\{\mathbf{u}_1, \mathbf{u}_2, \ldots, \mathbf{u}_k\}$ for $\mathrm{row}(\mathbf{A})$ to get a new set

$$B = \{\mathbf{u}_1, \mathbf{u}_2, \ldots, \mathbf{u}_k, \mathbf{z}\}$$

Since $\mathbf{z} \notin \mathrm{row}(\mathbf{A})$ however, this set is *independent* (Exercise 357). In fact, we may assume that $\mathbf{z}$ *is a **unit** vector orthogonal to each of the* $\mathbf{u}_i$*'s* (Exercise 389). As a basis for $\mathrm{row}(\mathbf{A})$, the $\mathbf{u}_i$'s *span* it, so by Lemma 1.2, we have $\mathbf{z} \in \mathrm{row}(\mathbf{A})^{\perp}$. By the first identity of our Propostion, this means $\mathbf{z} \in \mathrm{nul}(\mathbf{A})$.

But wait: we now have *both* $\mathbf{z} \in \mathrm{nul}(\mathbf{A})$ *and by* (41), $\mathbf{z} \in \mathrm{nul}(\mathbf{A})^{\perp}$. So $\mathbf{z}$ is both **in** and **orthogonal to** $\mathrm{nul}(\mathbf{A})$, which makes it orthogonal to *itself*, forcing

$$0 = \mathbf{z} \cdot \mathbf{z} = |\mathbf{z}|^2$$

Since we assumed above that $\mathbf{z}$ was a *unit* vector, this is a contradiction. So (42) must not be impossible, which makes (41) an *equality*, not just a containment. In short, Proposition's second identity holds. $\square$

Now consider an arbitrary subspace $S \subset \mathbf{R}^n$. Take any basis for S, and make its vectors the *rows* of a matrix $\mathbf{A}$. Then $S = \mathrm{row}(\mathbf{A})$, and hence $S^{\perp} = \mathrm{nul}(\mathbf{A})$ by the Proposition just proven. Take orthocomplements on both sides of this equation and apply the Proposition again, to get

$$\left(S^{\perp}\right)^{\perp} = \mathrm{nul}(\mathbf{A})^{\perp} = \mathrm{row}(\mathbf{A}) = S$$

This proves the corollary we promised:

COROLLARY 1.8. *If $S \subset \mathbf{R}^n$ is any subspace, then*

$$\left(S^\perp\right)^\perp = S$$

It follows immediately that, as mentioned earlier, *If $T = S^\perp$, then $T^\perp = S$.* For when $T = S^\perp$, we can take the orthocomplement of both sides to get $T^\perp = \left(S^\perp\right)^\perp$, which equals S, by the Corollary.

More generally, Proposition 1.7 lets us find the orthocomplement of *any* subspace S, whether we encounter it as span or perp. We state this as an additional corollary:

COROLLARY 1.9. *Let $V = \{\mathbf{v}_1, \mathbf{v}_2, \dots, \mathbf{v}_k\} \subset \mathbf{R}^n$ be any set of vectors. Then:*

    i) *If S is the span of V, then $S^\perp$ is the perp of V.*

    ii) *If S is the perp of V, then $S^\perp$ is the span of V.*

PROOF. When S is the span of V, we have $S = \text{row}\left(\mathbf{V}^T\right)$, where, as usual, $\mathbf{V}$ is the matrix having the $\mathbf{v}_i$'s as columns. The Proposition's first identity then yields $S^\perp = \text{nul}\left(\mathbf{V}^T\right)$. By definition, the latter is the perp of the rows of $\mathbf{V}^T$, which makes it the perp of V. This proves (i).

To prove (ii), we essentially just switch *span* with *perp*, and *row* with *nul* above: when S is the perp of V, we have $S = \text{nul}\left(\mathbf{V}^T\right)$. The Proposition's second identity then yields $S^\perp = \text{row}\left(\mathbf{V}^T\right)$, which makes it the span of V. $\square$

REMARK 1.10. Note that conclusion (i) above is not really new—it restates our earlier Corollary 1.3. We include it above to highlight the beautiful duality between (i) and (ii), one aspect of the broader duality between spans and perps, row-spaces and nullspaces. $\square$

We illustrate the basic use of Proposition 1.7 and its corollaries with two simple examples:

EXAMPLE 1.11. Find the orthocomplement $P^\perp$ of the plane P that $(1, 2, -2, -1)$ and $(0, 1, 1, 0)$ span in $\mathbf{R}^4$.

Since P arises here as the span of two vectors, Corollary 1.9 (i) tells us that $P^\perp$ is just the perp of those same two vectors. If we create a matrix $\mathbf{A}$ having those two vectors as rows, then $P = \text{row}(\mathbf{A})$, and hence, by Proposition 1.7 (i), $P^\perp = \text{nul}(\mathbf{A})$:

$$P^{\perp} = \text{nul} \begin{bmatrix} 1 & 2 & -2 & -1 \\ 0 & 1 & 1 & 0 \end{bmatrix}$$

This gives $P^{\perp}$ *implicitly*, which could be exactly what we want. If we preferred an *explicit* description instead, we simply do a perp-to-span conversion by row-reducing $\mathbf{A}$ to get its homogeneous generators. The reader can easily calculate them to conclude that

$$P^{\perp} = \text{span} \left\{ \begin{pmatrix} 4 \\ -1 \\ 1 \\ 0 \end{pmatrix}, \begin{pmatrix} 1 \\ 0 \\ 0 \\ 1 \end{pmatrix} \right\}$$

$\square$

We just saw how to find the orthocomplement of *span*, both implicitly and explicitly. What about finding the orthocomplement of a *perp*? For that, we use Proposition 1.7 (ii).

EXAMPLE 1.12. Let

$$\mathbf{A} = \begin{bmatrix} 1 & 2 & -2 & 1 \\ 0 & 1 & 1 & 0 \end{bmatrix}$$

Find the orthocomplement of $S = \text{nul}(\mathbf{A})$.

In a sense, Proposition 1.7(ii) makes this trivial, immediately giving $S^{\perp} = \text{row}(\mathbf{A})$. In short, the rows of $\mathbf{A}$ span the orthocomplement of $\text{nul}(\mathbf{A})$, giving us an explicit description of $\text{nul}(\mathbf{A})^{\perp}$ with no real work.

Suppose, however, that we wanted an *implicit* description of $\text{nul}(\mathbf{A})^{\perp} = \text{row}(\mathbf{A})$. This is trickier, but one way to do it is to *first* convert $\text{nul}(\mathbf{A})$ (an implicit description) to an explicit one by finding the homogeneous generators $\mathbf{h}_i$ for $\mathbf{A}$. Conveniently, we already found these in the last example: they are $(4, -1, 1, 0)$ and $(1, 0, 0, 1)$. So we can write

$$\text{nul}(\mathbf{A}) = \text{span} \left\{ \begin{pmatrix} 4 \\ -1 \\ 1 \\ 0 \end{pmatrix}, \begin{pmatrix} 1 \\ 0 \\ 0 \\ 1 \end{pmatrix} \right\}$$

and hence, by Corollary 1.9(i),

$$\mathrm{nul}(\mathbf{A})^{\perp} = \mathrm{perp}\left\{ \begin{pmatrix} 4 \\ -1 \\ 1 \\ 0 \end{pmatrix}, \begin{pmatrix} 1 \\ 0 \\ 0 \\ 1 \end{pmatrix} \right\} = \mathrm{nul}\begin{bmatrix} 4 & -1 & 1 & 0 \\ 1 & 0 & 0 & 1 \end{bmatrix}$$

This indeed gives an implicit description of $S^{\perp} = \mathrm{row}(\mathbf{A})$. □

1.13. Span-to-perp conversion. The last part of Example 1.12 is a bit intricate, but if we review its logic, we get something we promised at the beginning of Section 5.5 of Chapter 5: a more elegant method for converting from span to perp.

In the first part of the example, we had an *explicit* (span) description $S = \mathrm{row}(\mathbf{A})$, but we wanted an *implicit* (perp) description of that same subspace S. To get it, we took three steps, two of which involved no effort:

i) First, we went to the orthocomplement $S^{\perp} = \mathrm{nul}(\mathbf{A})$. (This is just a shift of focus—no calculation involved.)

ii) Second, we computed an *explicit* description of $S^{\perp} = \mathrm{nul}(\mathbf{A})$ as the span of the homogeneous generators $\mathbf{h}_i$. (This requires calculation, but of a familiar kind.)

iii) Finally, we took the matrix having the $\mathbf{h}_i$'s as columns (call it $\mathbf{H}$) *transposed* it, so that now $S^{\perp} = \mathrm{row}(\mathbf{H}^T)$, and paired Proposition 1.7 with Corollary 1.9 to get the result:

$$S = (S^{\perp})^{\perp} = \mathrm{row}(\mathbf{H}^T)^{\perp} = \mathrm{nul}(\mathbf{H}^T)$$

We started out with the span description $S = \mathrm{row}(\mathbf{A})$ and ended up with the perp description $S = \mathrm{nul}(\mathbf{H}^T)$. We summarize this as another corollary to Proposition 1.7:

COROLLARY 1.14. *Let $\mathbf{A}$ be any matrix, and make its homogeneous generators the columns of a new matrix $\mathbf{H}$. Then*

$$\mathrm{nul}(\mathbf{A}) = \mathrm{row}(\mathbf{H}^T)$$
$$\mathrm{row}(\mathbf{A}) = \mathrm{nul}(\mathbf{H}^T)$$

PROOF. The first identity is obvious, since the rows of $\mathbf{H}^T$ are the homogeneous generators of $\mathbf{A}$. We argued the second identity in Example 1.12. It boils down to this:

$$\text{row}(\mathbf{A}) = \text{nul}(\mathbf{A})^{\perp} \qquad\qquad \text{Proposition 1.7(ii)}$$
$$= \text{row}(\mathbf{H}^{T})^{\perp}$$
$$= \text{nul}(\mathbf{H}^{T}) \qquad\qquad \text{Proposition 1.7(i)}$$

This proves the result.   □

The Corollary's second identity gives the elegant span-to-perp method we promised, since $\text{row}(\mathbf{A})$ is a span, while $\text{nul}(\mathbf{H}^{T})$ is a perp. The only real work involved is the familiar task of computing the homogeneous generators of $\mathbf{A}$, which then form the *rows* of $\mathbf{H}^{T}$.

We diagram this three-step span-to-perp algorithm below (Figure 2), where it amounts to going from the upper-left to the upper-right of the diagram by going down, across, and up, instead of straight across.

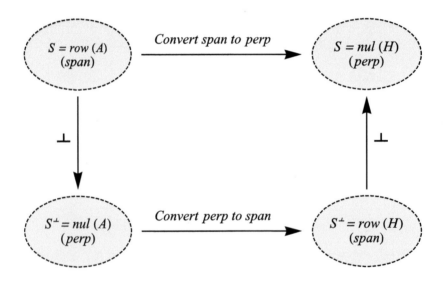

Figure 2. Instead of converting explicit to implicit directly as in Section 1.13 (upper arrow), we can take an orthocomplement (left arrow), convert implicit to explicit (lower arrow), then take another orthocomplement (right arrow).

Two examples will serve to demonstrate.

EXAMPLE 1.15. Find a perp description for the span of $(1, -1, 0)$, $(0, 1, -1)$, and $(-1, 0, 1)$ in $\mathbf{R}^{3}$.

Note that this is indeed a span-to-perp conversion problem. We show how to solve it using the second identity of Corollary 1.14, that is, by going left-to-right across the top of the diagram in Figure 2.

We must first write our span S as a row-space, by writing the given vectors as matrix rows:

$$S = \text{row}(\mathbf{A}) \quad \text{where} \quad \mathbf{A} = \begin{bmatrix} 1 & -1 & 0 \\ 0 & 1 & -1 \\ -1 & 0 & 1 \end{bmatrix}$$

Corollary 1.14 then says we can also realize S as a perp by writing

$$S = \text{nul}(\mathbf{H}^T)$$

Here the rows of $\mathbf{H}^T$ are the homogeneous generators of $\mathbf{A}$. Finding them is the only work we really have to do. A familiar row-reduction shows that $\text{RRE}(\mathbf{A})$ has only one free column, and hence one homogeneous generator $\mathbf{h} = (1, 1, 1)$. So

$$S = \text{nul}(\begin{bmatrix} 1 & 1 & 1 \end{bmatrix})$$

This is the implicit description we sought. Equivalently, we could say that S is the plane $x + y + z = 0$, or the plane $(1, 1, 1)^\perp$.   □

EXAMPLE 1.16. Find an implicit description for the span of $(1, 1, 1, 1)$ and $(1, -1, 1, -1)$ in $\mathbf{R}^4$.

As in the previous example, we write the spanning vectors as rows to get

$$S = \text{row} \begin{bmatrix} 1 & 1 & 1 & 1 \\ 1 & -1 & 1 & -1 \end{bmatrix}$$

The matrix on the right reduces to

$$\begin{bmatrix} 1 & 0 & 1 & 0 \\ 0 & 1 & 0 & 1 \end{bmatrix}$$

Columns 3 and 4 are free, and give rise to homogeneous generators $(-1, 0, 1, 0)$ and $(0, -1, 0, 1)$, respectively. The second identity of Corollary 1.14 now says we can realize S as a perp by writing these vectors as matrix rows, and then taking the nullspace:

$$S = \text{nul} \begin{bmatrix} -1 & 0 & 1 & 0 \\ 0 & -1 & 0 & 1 \end{bmatrix}$$

□

REMARK 1.17. Many different matrices have the same column-space or nullspace. So if we want, for instance, to describe a subspace S as the nullspace of a matrix $\mathbf{A}$, there will always be (infinitely) many correct answers. Indeed, elementary row operations don't change the row-space or the nullspace of a matrix, so once we have $S = \mathrm{nul}(\mathbf{A})$, we also have $S = \mathrm{nul}(\mathbf{A}')$ for every matrix $\mathbf{A}'$ obtainable by applying elementary row operations to $\mathbf{A}$.

Similarly, if we start with a perp description $S = \mathrm{nul}(\mathbf{A})$, convert to the explicit description $\mathrm{row}(\mathbf{H}^T) = \mathrm{nul}(\mathbf{A})$, and then convert *back* by finding the homogeneous generators of $\mathbf{H}^T$ and making *them* the rows of a new matrix $\mathbf{A}'$, we *don't* expect to get back our original matrix $\mathbf{A}$ (we will *not* generally have $\mathbf{A} = \mathbf{A}'$). While $\mathbf{A}$ and $\mathbf{A}$ *will* have the same nullspace, they will *not* generally be the same matrix.

Likewise, if we start with $S = \mathrm{row}(\mathbf{A})$, convert to the implicit description $\mathrm{row}(\mathbf{A}) = \mathrm{nul}(\mathbf{H}^T)$, and then convert *back* to explicit by making the nullspace generators of $\mathbf{H}^T$ the rows of a new matrix $\mathbf{A}'$, we *will* have $\mathrm{row}(\mathbf{A}') = \mathrm{row}(\mathbf{A})$, even though $\mathbf{A} \neq \mathbf{A}'$ in general.   □

We can demonstrate this using Example 1.16 above. We started with the span description

$$S = \mathrm{row} \begin{bmatrix} 1 & 1 & 1 & 1 \\ 1 & -1 & 1 & -1 \end{bmatrix}$$

and converted to the perp description

$$S = \mathrm{nul} \begin{bmatrix} -1 & 0 & 1 & 0 \\ 0 & -1 & 0 & 1 \end{bmatrix}$$

If we now convert *back* to explicit in the usual way, by computing the homogenous generators of this last matrix—they are $(1,0,1,0)$ and $(0,1,0,1)$ —we get

$$S = \mathrm{row} \begin{bmatrix} 1 & 0 & 1 & 0 \\ 0 & 1 & 0 & 1 \end{bmatrix}$$

If we call the first and third matrices above $\mathbf{A}$ and $\mathbf{A}'$, respectively, we can see by inspection that they have the same row-spaces: the sum and difference of the rows of the third give the rows of the first, while conversely, half the sum and difference of the rows in the first give those of the third. But they are obviously not the same matrix.

We conclude this section with one more important consequence of Proposition 1.7:

COROLLARY 1.18. *If $S \subset \mathbf{R}^n$ is any subspace, then*

$$\dim(S^\perp) = \operatorname{codim}(S)$$

PROOF. Take any spanning set for S and make its vectors the columns of a matrix $\mathbf{A}$. Then

$$S = \operatorname{col}(\mathbf{A}) = \operatorname{row}(\mathbf{A}^T)$$

The first identity of Proposition 1.7 then gives

$$S^\perp = \operatorname{nul}(\mathbf{A}^T)$$

So $\dim(S)^\perp$ is the nullity of $\mathbf{A}^T$, which has n columns. Since rank plus nullity equals total number of columns (Proposition 8.11 of Chapter 5), we now know that

$$\operatorname{rank}(\mathbf{A}^T) + \dim(S)^\perp = n$$

But $\operatorname{rank}(\mathbf{A}^T) = \operatorname{rank}(\mathbf{A})$, by Proposition 8.12 of the same chapter, and $\operatorname{rank}(\mathbf{A}) = \dim(\operatorname{col}(\mathbf{A})) = \dim(S)$ by definition of rank. The desired result now follows easily:

$$\dim(S)^\perp = n - \dim(S) = \operatorname{codim}(S)$$

$\square$

REMARK 1.19. In Exercise 371 of Chapter 5, we introduced the *sum* $V + W$ of subspaces $V, W \subset \mathbf{R}^n$, which is just all possible sums $\mathbf{v} + \mathbf{w}$, where $\mathbf{v} \in V$ and $\mathbf{w} \in W$. The *intersection* $V \cap W$, of course, is all vectors common to both V *and* W.

Exercise 390 below, we explore some basic inequalities satisfied by the *dimensions* of the sum and intersection:

$$\dim(V + W) \leq \dim(V) + \dim(W)$$
$$\operatorname{codim}(V \cap W) \leq \operatorname{codim}(V) + \operatorname{codim}(W)$$

Both follow easily from one beautiful identity:

(43) $$\dim(V + W) = \dim(V) + \dim(W) - \dim(V \cap W)$$

To prove (43), just take a basis $\{\mathbf{u}_1, \ldots, \mathbf{u}_r\}$ for $V \cap W$, and extend it to bases for V and W, respectively, of the form

$$\{\mathbf{u}_1, \ldots, \mathbf{u}_r, \mathbf{v}_1, \ldots, \mathbf{v}_k\}$$

and

$$\{\mathbf{u}_1, \ldots, \mathbf{u}_r, \mathbf{w}_1, \ldots, \mathbf{w}_l\}$$

It is easy to show that the $\mathbf{u}_i$'s , $\mathbf{v}_i$'s, and $\mathbf{w}_i$'s form an independent set that spans (hence forms a basis for) $V + W$. Identity (43) can then be verified by simply counting basis vectors, and the dimension inequality above follows trivially.

To get the codimension inequality, convert (43) into a statement about codimension, by subtracting each side from the dimension n of the space containing V and W (using $n = n + n - n$ on the right) whence it becomes

$$\operatorname{codim}(V + W) = \operatorname{codim}(V) + \operatorname{codim}(W) - \operatorname{codim}(V \cap W)$$

$\square$

– Practice –

377. Let P denote the plane spanned by $(1, 4, -1, -1)$ and $(1, 4, 1, 4)$ in $\mathbf{R}^4$

    a) Find a basis for $P^\perp$.

    b) Find an implicit description of $P^\perp$.

378. Consider the plane $2x - 3y + z = 0$ in $\mathbf{R}^3$.

    a) Find a basis for its orthocomplement.

    b) Find an implicit description of its orthocomplement.

379. Find a basis for the orthocomplement of the line spanned by $(1, 1, 1, 1, 1) \subset \mathbf{R}^5$.

380. Suppose $(a, b) \in \mathbf{R}^2$, and $(a, b) \neq \mathbf{0}$. Derive a formula, involving a and b, for a vector spanning the orthocomplement of the 1-dimensional subspace spanned by (a, b).

381. The *cross product* $\mathbf{u} \times \mathbf{v}$ of two vectors $\mathbf{u}, \mathbf{v} \in \mathbf{R}^3$, is defined in Exercise 63 of Chapter 1. Show that when $\mathbf{u} \neq \mathbf{0} \neq \mathbf{v}$, the cross-product $\mathbf{u} \times \mathbf{v}$ forms a basis for the orthocomplement of $\operatorname{span}\{\mathbf{u}, \mathbf{v}\}$.

382. Letting S denote the column-space of the matrix below, find bases for S and $S^{\perp}$.

$$A = \begin{bmatrix} 1 & 3 & 3 \\ 2 & -2 & 2 \\ 3 & -7 & 1 \\ -2 & 2 & -2 \\ 1 & 3 & 3 \end{bmatrix}$$

383. Letting S denote the nullspace of the matrix below, find bases for S and $S^{\perp}$.

$$B = \begin{bmatrix} 4 & 2 & 4 & 1 \\ -2 & 4 & 2 & 1 \\ 1 & 2 & 2 & 1 \\ 2 & 2 & 2 & 2 \end{bmatrix}$$

384. Find an explicit description for the perp of the set

$$\left\{ \begin{pmatrix} 1 \\ -2 \\ 3 \\ -2 \\ 1 \end{pmatrix}, \begin{pmatrix} 2 \\ 2 \\ 2 \\ 2 \\ 2 \end{pmatrix} \right\} \subset \mathbf{R}^5 \}$$

385. Find an implicit description for the span of the set in Exercise 384.

386. Given any two independent vectors $\mathbf{R}^3$, the cross product (Exercises 63, 381) produces a third vector orthogonal to both. If you were given $n-1$ independent vectors in $\mathbf{R}^n$, how could you find an nth vector orthogonal to all of them?

387. Suppose $S \subset \mathbf{R}^m$ is a subspace, and we have independent sets in $U \subset S$ and $V \subset S^{\perp}$. Prove that $U \cup V$ will be independent in $\mathbf{R}^m$.

388. Supposing that $S \subset \mathbf{R}^n$ is a subspace, prove:

a) $S \cap S^\perp = \{\mathbf{0}\}$.

b) $\dim(S) + \dim(S^\perp) = n$. [*Hint: Proposition 8.11 of Chapter 5.*]

389. Suppose $B = \{\mathbf{u}_1, \mathbf{u}_2, \ldots, \mathbf{u}_k\}$ is a basis for a subspace $S \subset \mathbf{R}^n$, and that $\mathbf{z} \in \mathbf{R}^n$ is *not* in S.

a) Explain why $\mathbf{Bx} = \mathbf{z}$ has no solution.

b) Show that $\mathbf{B}^T\mathbf{Bx} = \mathbf{B}^T\mathbf{z}$ *does* have a solution. [*Hint: Show that* $\text{nul}(\mathbf{B}^T\mathbf{B})$ *is trivial, so that* $\mathbf{B}^T\mathbf{B}$ *is invertible. Use the conclusion of Corollary 2.8:* $\mathbf{Ax} \cdot \mathbf{Ax} = \mathbf{x} \cdot \mathbf{A}^T\mathbf{Ax}$.]

c) Show that when $\mathbf{x}$ solves $\mathbf{B}^T\mathbf{Bx} = \mathbf{B}^T\mathbf{z}$ (as in (b) above), $\mathbf{z} - \mathbf{Bx}$ is non-zero, and lies in $S^\perp$.

If we normalize $\mathbf{z} - \mathbf{Bx}$, we produce a unit vector perpendicular to each of the $\mathbf{u}_i$'s. This justifies the "we may assume" statement in the proof of Proposition 1.7 above.

390. Suppose V and W are subspaces of $\mathbf{R}^n$. You may want to review Remark 1.19 above, and Exercises 289 and 371 from Chapter 5.

a) Show that $(V + W)^\perp = V^\perp \cap W^\perp$.

b) Show that $(V \cap W)^\perp = V^\perp + W^\perp$.

c) Show that $\text{codim}(V \cap W) \leq \text{codim}(V) + \text{codim}(W)$.

[*Hint: Write* V *and* W *as nullspaces.*]

$$- \star -$$

2. Four Subspaces, 16 Questions

We conclude our focus on subspaces here by summarizing what we have so-far learned about the four subspaces associated with any matrix. In particular, we consider four basic questions one needs to answer in trying to understand these, or any other, subspaces.

Any $n \times m$ matrix $\mathbf{A}$ presents us with four fundamental subspaces:

- its **column-space** $\mathrm{col}(\mathbf{A})$

- its **nullspace** $\mathrm{nul}(\mathbf{A})$

- its **row-space** $\mathrm{row}(\mathbf{A})$

- its **left nullspace** $\mathrm{nul}(\mathbf{A}^T)$.

To claim we "know" any subspace, we should be able to answer four questions about it:

- i) Where does S lie (e.g., $\mathbf{R}^n$ or $\mathbf{R}^m$)?

- ii) How do we find a *basis* for S (an efficient *span* description)?

- iii) What are the dimension and codimension of S?

- iv) How do we find a basis for $S^\perp$ (an efficient *perp* description for S itself)?

We now review the answers to each of these questions for each of the four subspaces associated with a matrix $\mathbf{A}_{n \times m}$.

Column-space:

- i) *Where does $S = \mathrm{col}(\mathbf{A})$ lie?*

  Since $\mathbf{A}$ has n rows, each column of $\mathbf{A}$ has n entries. The columns of $\mathbf{A}$, and thus $\mathrm{col}(\mathbf{A})$, therefore lie in $\mathbf{R}^n$.

- ii) *How do we find a basis for S?*

  The pivot columns of $\mathbf{A}$ form a basis for $\mathrm{col}(\mathbf{A})$.

- iii) *What are the dimension and codimension of S?*

  Since the pivot columns form a basis, the dimension of $\mathrm{col}(\mathbf{A})$ is the number of pivot columns—the *rank* of $\mathbf{A}$.

- iv) *How do we find a basis for $S^\perp$?*

  Since $\mathrm{col}(\mathbf{A}) = \mathrm{row}(\mathbf{A}^T)$, we have

$$\mathrm{col}(\mathbf{A})^\perp = \mathrm{row}(\mathbf{A}^T)^\perp = \mathrm{nul}(\mathbf{A}^T)$$

  the orthocomplement of $\mathrm{col}(\mathbf{A})$ is thus $\mathrm{nul}(\mathbf{A}^T)$, so the homogeneous generators for $\mathbf{A}^T$ yield a basis for $\mathrm{col}(\mathbf{A})^\perp$.

Nullspace:

   i) *Where does* $S = \mathrm{nul}(\mathbf{A})$ *lie?*

      Since $\mathbf{A}$ has m columns, each rows has m entries. To form the product $\mathbf{Ax}$ (whose vanishing defines $\mathrm{nul}(\mathbf{A})$) $\mathbf{x}$ must also have m entries, so $\mathrm{nul}(\mathbf{A}) \subset \mathbf{R}^m$.

   ii) *How do we find a basis for* S?

      The homogeneous generators $\mathbf{h}_i$ we get from $\mathrm{RRE}(\mathbf{A})$ form a basis for $\mathrm{nul}(\mathbf{A})$ by Proposition 7.6.

   iii) *What are the dimension and codimension of* S?

      The dimension of $\mathrm{nul}(\mathbf{A})$ is by definition $\mathrm{nullity}(\mathbf{A})$. According to Proposition 8.11 of Chapter 5, we have

$$\dim(\mathrm{nul}(\mathbf{A})) = \mathrm{nullity}(\mathbf{A}) = m - r$$

      where $r = \mathrm{rank}(\mathbf{A})$ is the number of pivots in $\mathrm{RRE}(\mathbf{A})$.

   iv) *How do we find a basis for* $S^\perp$?

      By Proposition 1.7, $\mathrm{nul}(\mathbf{A})^\perp = \mathrm{row}(\mathbf{A})$. Any basis for $\mathrm{row}(\mathbf{A})$ will therefore serve our purpose. By Exercise 355, the pivot *rows* of $\mathrm{RRE}(\mathbf{A})$ provide such a basis. The pivot *columns* of $\mathbf{A}^T$ also provide a (different) basis.

REMARK 2.1. We can answer any question about the row-space or left nullspace of a matrix $\mathbf{A}$ by answering the corresponding questions for the column-space and nullspace of $\mathbf{A}^T$. We prefer to face $\mathrm{row}(\mathbf{A})$ on its own terms, though, since this allows us to highlight some facts we might otherwise miss.   □

Row-space:

   i) *Where does* $S = \mathrm{row}(\mathbf{A})$ *lie?*

      Since $\mathbf{A}$ has m columns, each row of $\mathbf{A}$ has m entries. So the rows of $\mathbf{A}$, and thus $\mathrm{row}(\mathbf{A})$, lie in $\mathbf{R}^m$.

   ii) *How do we find a basis for* S?

      By Exercise 355, the *pivot rows* of $\mathrm{RRE}(\mathbf{A})$ form a basis for $\mathrm{row}(\mathbf{A})$. Note that these are *not* typically rows of $\mathbf{A}$ itself. The pivot *columns* of $\mathbf{A}^T$, which *are* rows of $\mathbf{A}$, yield another (generally different) basis.

   iii) *What are the dimension and codimension of* S?

The pivot rows of $\mathrm{RRE}(\mathbf{A})$ form a basis for $\mathrm{row}(\mathbf{A})$, so its dimension is the number of pivots, which equals the rank r of $\mathbf{A}$. So the dimension of $\mathrm{row}(\mathbf{A})$, like that of $\mathrm{col}(\mathbf{A})$, is r.

iv) *How do we find a basis for $S^{\perp}$?*

By Proposition 1.7 above, the orthocomplement of $\mathrm{row}(\mathbf{A})$ is $\mathrm{nul}(\mathbf{A})$. The homogeneous generators $\mathbf{h}_i$ for $\mathbf{A}$ therefore give the desired basis, by Proposition 7.6.

Left nullspace:

We handle the left nullspace by transposing and answering the four questions for the nullspace of $\mathbf{A}^T$. We address the four questions for the left nullspace here mostly for completeness; not much is really new.

i) *How do we find a basis for $S = \mathrm{nul}(\mathbf{A}^T)$?*

By the same reasoning we used above to deduce $\mathrm{nul}(\mathbf{A}) \subset \mathbf{R}^m$, we see that $\mathrm{nul}(\mathbf{A}^T) \subset \mathbf{R}^n$.

ii) *Where does $S = \mathrm{nul}(\mathbf{A})$ lie?*

The homogeneous generators for $\mathbf{A}^T$ form a basis for $\mathrm{nul}(\mathbf{A}^T)$ by Proposition 7.6.

iii) *What are the dimension and codimension of S?*

The homogeneous generators for $\mathbf{A}^T$ form a basis for the left nullspace, and we get one generator for each free column of $\mathbf{A}^T$. The dimension of $\mathrm{nul}(\mathbf{A}^T)$ therefore equals $n - r$, the total number of columns of $\mathbf{A}^T$ (namely n) minus the number of pivot columns. The latter number is r because rank of $\mathbf{A}^T$ equals the rank of $\mathbf{A}$ by Proposition 8.12.

iv) *How do we find a basis for $S^{\perp}$?*

The orthogonal complement of $\mathrm{nul}(\mathbf{A}^T)$ is $\mathrm{row}(\mathbf{A}^T) = \mathrm{col}(\mathbf{A})$ by Proposition 1.7. The pivot columns of $\mathbf{A}$ therefore form a basis for the orthocomplement of $\mathrm{nul}(\mathbf{A})$.

– **Practice** –

391. Consider the matrix

$$\mathbf{A} = \begin{bmatrix} 1 & 5 & -3 \\ 1 & -3 & 5 \end{bmatrix}$$

a) Find sets of vectors R and C that span $\mathrm{row}(\mathbf{A})$ and $\mathrm{col}(\mathbf{A})$, respectively.

b) Find a set V such that $\mathrm{row}(\mathbf{A}) = V^{\perp}$.

c) What is the dimension of $\mathrm{col}(\mathbf{A})^{\perp}$?

392. Consider the matrix

$$\mathbf{A} = \begin{bmatrix} 2 & 4 & 4 & -2 \\ 1 & 2 & 4 & -3 \\ 1 & 2 & 3 & -2 \end{bmatrix}$$

Find bases for each of the four subspaces associated with $\mathbf{A}$

393. Find the dimension and codimension of each of the four subspaces associated with

$$\begin{bmatrix} -5 & -4 & -3 \\ 1 & 2 & 3 \\ -2 & -1 & 0 \\ 4 & 5 & 6 \end{bmatrix}$$

394. Consider the matrix

$$\mathbf{B} = \begin{bmatrix} 1 & -1 \\ 2 & -2 \\ -1 & 1 \end{bmatrix}$$

For each of the four subspaces associated with $\mathbf{B}$,

a) Find an implicit (perp) description for that subspace.

b) Determine the dimension of the subspace.

395. Consider the matrix

$$\mathbf{C} = \begin{bmatrix} 3 & -3 & 0 \\ 0 & 1 & -1 \\ -3 & 0 & 3 \end{bmatrix}$$

For each of the four subspaces associated with $\mathbf{C}$,

a) Find a basis for its orthocomplement.

b) Find an implicit description of its orthocomplement.

396. A certain matrix $\mathbf{A}$ has a 5-dimensional left nullspace, and a 6-dimensional row-space.

a) How rows does $\mathbf{A}$ have?

b) What's the minimum number of columns $\mathbf{A}$ can have? Is there a maximum?

397. Give an example of a matrix $\mathbf{A}$ with the given properties, or explain why none exists.

a) $\dim(\mathrm{nul}(\mathbf{A})) = 3$, $\dim(\mathrm{col}(\mathbf{A})) = 2$, and $\dim\left(\mathrm{nul}\left(\mathbf{A}^T\right)\right) = 1$

b) $\dim(\mathrm{row}(\mathbf{A})) = 3$, $\dim\left(\mathrm{nul}\left(\mathbf{A}^T\right)\right) = 0$, and $\dim(\mathrm{nul}(\mathbf{A})) = 2$

c) $\dim(\mathrm{row}(\mathbf{A})) = 3$, $\dim(\mathrm{nul}(\mathbf{A})) = 1$, and $\dim(\mathrm{col}(\mathbf{A})) = 2$

398. Consider the general 3×3 *skew-symmetric* matrix

$$\mathbf{X} = \begin{bmatrix} 0 & a & -b \\ -a & 0 & c \\ b & -c & 0 \end{bmatrix}$$

Here $a, b, c \in \mathbf{R}$ can be any scalars.

a) Explain why $\mathrm{row}(\mathbf{X}) = \mathrm{col}(\mathbf{X})$ and $\mathrm{nul}(\mathbf{X}) = \mathrm{nul}\left(\mathbf{X}^T\right)$.

b) Show that the nullity of such a matrix is always at *least* one.

c) Give examples (i.e., choose a, b, c) for which $\mathrm{nullity}(\mathbf{X}) = 1$, $\mathrm{nullity}(\mathbf{X}) = 2$, and $\mathrm{nullity}(\mathbf{X}) = 3$.

399. What are the answers to the sixteen questions of this section in the case where $\mathbf{A}_{n \times n}$ is *invertible*?

400. Discuss the answers to the sixteen questions of this section in the case where $\mathbf{A}_{n \times n}$ is *diagonal*. (The answers depend on how many (if any) diagonal entries vanish.)

— ★ —

3. Orthonormal Bases

In Chapter 3, we defined the *length* of a vector $\mathbf{v} = (v_1, v_2, \ldots, v_n)$ as

$$|\mathbf{v}| = \sqrt{\mathbf{v} \cdot \mathbf{v}} = \sqrt{v_1^2 + v_2^2 + \cdots v_n^2}$$

Definition 3.25 of that same Chapter makes vectors $\mathbf{v}, \mathbf{w} \in \mathbf{R}^n$ orthogonal to each other if and only if $\mathbf{v} \cdot \mathbf{w} = 0$.

To state our next definition, we use the **Kronecker delta** δ_{ij}, which denotes the *row i, column j entry of the identity matrix:*

$$\delta_{ij} = \begin{cases} 1, & \text{if } i = j \\ 0, & \text{if } i \neq j \end{cases}$$

DEFINITION 3.1. A set $\{\mathbf{u}_1, \mathbf{u}_2, \ldots, \mathbf{u}_k\} \subset \mathbf{R}^n$ is **orthonormal** if

- Each $\mathbf{u}_i$ has **unit length**, so that $|\mathbf{u}_i|^2 = 1$ for every $i = 1, 2, \ldots, k$, and

- Each $\mathbf{u}_i$ is **orthogonal to every other** $\mathbf{u}_j$. That is, $\mathbf{u}_i \cdot \mathbf{u}_j = 0$, whenever $i \neq j$.

Since $|\mathbf{u}_i|^2 = \mathbf{u}_i \cdot \mathbf{u}_i$, we can express both these conditions as a single identity using the Kronecker delta:

A set $\{\mathbf{u}_1, \mathbf{u}_2, \ldots, \mathbf{u}_k\}$ is orthonormal if and only if

$$(44) \qquad\qquad \mathbf{u}_i \cdot \mathbf{u}_j = \delta_{ij} \quad \text{for all } i, j$$

$\square$

We can say this even more compactly using matrix notation. Set the $\mathbf{u}_i$'s up as the columns of a matrix $\mathbf{U}$, and observe that by the dot-product method of matrix multiplication, $\mathbf{u}_i \cdot \mathbf{u}_j$ gives the i, j entry of $\mathbf{U}^T \mathbf{U}$. Since δ_{ij} is the i, j entry of the identity matrix, identity (44) above reveals a remarkable property of any orthonormal set U: the transpose of $\mathbf{U}$ is a *left inverse* for $\mathbf{U}$ itself:

PROPOSITION 3.2. *A set $U = \{\mathbf{u}_1, \mathbf{u}_2, \ldots, \mathbf{u}_k\}$ is orthonormal in $\mathbf{R}^n$ if and only if*

$$\mathbf{U}^T \mathbf{U} = \mathbf{I}_k$$

PROOF. When (44) holds, the i, j entry of $\mathbf{U}^T \mathbf{U}$ is

$$\mathbf{r}_i\left(\mathbf{U}^T\right) \cdot \mathbf{c}_j\left(\mathbf{U}\right) = \mathbf{u}_i \cdot \mathbf{u}_j = \delta_{ij}$$

which is the i, j entry of the identity matrix.

Conversely, when $\mathbf{U}^T\mathbf{U} = \mathbf{I}_k$, we have $\mathbf{u}_i \cdot \mathbf{u}_j = \delta_{ij}$, which makes U orthonormal. $\qquad\square$

REMARK 3.3. In Proposition 3.2 of Chapter 4, we showed that for *square* matrices, $\mathbf{AB} = \mathbf{I}$ forces $\mathbf{BA} = \mathbf{I}$. While orthonormality ensures $\mathbf{U}^T\mathbf{U} = \mathbf{I}_k$, however, it does *not* generally imply $\mathbf{UU}^T = \mathbf{I}_n$; it is *not* generally a *right* inverse for $\mathbf{U}$, because *unless $k = n$, $\mathbf{U}$ is not square.*

The square case $k = n$ is especially important, however, and we explore it in Section 3.15 below. $\qquad\square$

We will shortly give interesting examples of orthonormal sets, but first, we record a useful corollary to Proposition 3.2:

COROLLARY 3.4. *Any orthonormal set is independent.*

PROOF. By Conclusion (d) of Proposition 6.9 in Chapter 5, $U = \{\mathbf{u}_1, \mathbf{u}_2, \dots, \mathbf{u}_k\}$ in $\mathbf{R}^n$ is independent if and only if $\mathbf{Ux} = \mathbf{0}$ has only the trivial solution. If U is orthonormal, however, then $\mathbf{Ux} = \mathbf{0}$ implies $\mathbf{U}^T\mathbf{Ux} = \mathbf{0}$ too, which Proposition 3.2 above then rewrites as

$$\mathbf{I}_k\mathbf{x} = \mathbf{0}$$

which clearly forces $\mathbf{x} = \mathbf{0}$. So any solution of $\mathbf{Ux} = \mathbf{0}$ has to be trivial. As noted at the outset, this makes U independent. $\qquad\square$

EXAMPLE 3.5. It is easy to check that the columns of this symmetric matrix

$$\mathbf{U} = \begin{bmatrix} \frac{1}{\sqrt{2}} & \frac{1}{\sqrt{2}} \\ \frac{1}{\sqrt{2}} & -\frac{1}{\sqrt{2}} \end{bmatrix}$$

form an orthonormal set. Since they are independent by Corollary 3.4, and there are two of them, they actually form an orthonormal *basis* for $\mathbf{R}^2$ (Corollary 8.8 of Chapter 5). It is easy to check that

$$\mathbf{U}^T\mathbf{U} = \begin{bmatrix} \frac{1}{\sqrt{2}} & \frac{1}{\sqrt{2}} \\ \frac{1}{\sqrt{2}} & -\frac{1}{\sqrt{2}} \end{bmatrix} \begin{bmatrix} \frac{1}{\sqrt{2}} & \frac{1}{\sqrt{2}} \\ \frac{1}{\sqrt{2}} & -\frac{1}{\sqrt{2}} \end{bmatrix} = \mathbf{I}_2$$

just as the Proposition predicts. $\qquad\square$

EXAMPLE 3.6. The vectors

$$\tfrac{1}{2}(1,1,1,1) \quad \text{and} \quad \tfrac{1}{2}(1,1,-1,-1)$$

are orthonormal in $\mathbf{R}^4$, as is easily checked by computing their dot products with themselves and each other. These same dot products arise when we compute

$$\begin{bmatrix} 1/2 & 1/2 & 1/2 & 1/2 \\ 1/2 & 1/2 & -1/2 & -1/2 \end{bmatrix} \begin{bmatrix} 1/2 & 1/2 \\ 1/2 & 1/2 \\ 1/2 & -1/2 \\ 1/2 & -1/2 \end{bmatrix} = \mathbf{I}_2$$

which also proves orthonormality, by Proposition 3.2.   □

In Chapter 1, we noted that the linear system $\mathbf{Ax} = \mathbf{b}$ can be seen as a *column problem*: How do we express the right-hand side $\mathbf{b}$ as a linear combination of the *columns* of $\mathbf{A}$?

Proposition 3.2 makes it especially easy to solve this problem when the columns involved are orthonormal. For when the columns of $\mathbf{U}$ are orthonormal, it allows us to solve

$$(45) \qquad\qquad \mathbf{Ux} = \mathbf{b}$$

by simply left-multiplying both sides of the equation by $\mathbf{U}^T$:

$$\mathbf{U}^T\mathbf{Ux} = \mathbf{U}^T\mathbf{b}$$

By Proposition 3.2, the left-hand side reduces to $\mathbf{I}_k\mathbf{x} = \mathbf{x}$, and we're done:

$$\mathbf{x} = \mathbf{U}^T\mathbf{b}$$

Note that we actually *compute* each coordinate x_i of $\mathbf{U}^T\mathbf{b}$ by dotting $\mathbf{b}$ with each $\mathbf{u}_i$ in turn:

$$x_i = \mathbf{u}_i \cdot \mathbf{b}, \quad i = 1, 2, \ldots, n$$

We record this fact as follows for easy reference.

OBSERVATION 3.7. *If U is an orthonormal basis for a subspace $S \subset \mathbf{R}^n$, and $\mathbf{b} \in S$, then*

$$\mathbf{b} = x_1\mathbf{u}_1 + x_2\mathbf{u}_2 + \cdots x_k\mathbf{u}_k$$

where $x_i = \mathbf{u}_i \cdot \mathbf{b}$ for each $i = 1, 2, \ldots, n$.

EXAMPLE 3.8. The vectors $\mathbf{v} = (2, 2, -3, -3)$ and $\mathbf{w} = (0, 0, 4, 4)$, belong to the plane spanned by these two vectors:

$$\mathbf{u}_1 = \tfrac{1}{2}(1, 1, 1, 1) \quad \text{and} \quad \mathbf{u}_2 = \tfrac{1}{2}(1, 1, -1, -1)$$

Because the $\mathbf{u}_i$'s are orthonormal, we can easily expand any vector in their span (like $\mathbf{v}$ and $\mathbf{w}$ above) as linear combinations of them. Indeed, Observation 3.7 shows how to do so by computing a couple of dot products in each case.

$$\begin{aligned} \mathbf{v} &= (\mathbf{u}_1 \cdot \mathbf{v})\mathbf{u}_1 + (\mathbf{u}_2 \cdot \mathbf{v})\mathbf{u}_2 \\ &= -\mathbf{u}_1 + 5\,\mathbf{u}_2 \end{aligned}$$

$$\begin{aligned} \mathbf{w} &= (\mathbf{u}_1 \cdot \mathbf{w})\mathbf{u}_1 + (\mathbf{u}_2 \cdot \mathbf{w})\mathbf{u}_2 \\ &= 2\mathbf{u}_1 - 2\,\mathbf{u}_2 \end{aligned}$$

$\qquad\qquad\qquad\qquad\qquad\qquad\qquad\qquad\qquad\qquad\qquad\qquad\square$

EXAMPLE 3.9. By an easy exercise, these three vectors form an orthonormal basis for $\mathbf{R}^3$:

$$\mathbf{u}_1 = \frac{1}{\sqrt{2}}\begin{pmatrix} 1 \\ 0 \\ -1 \end{pmatrix}, \quad \mathbf{u}_2 = \frac{1}{\sqrt{6}}\begin{pmatrix} 1 \\ -2 \\ 1 \end{pmatrix}, \quad \mathbf{u}_3 = \frac{1}{\sqrt{3}}\begin{pmatrix} 1 \\ 1 \\ 1 \end{pmatrix}$$

By Observation 3.7, then, we can easily expand any vector in $\mathbf{R}^3$ as a linear combination of them. For instance, let $\mathbf{b} = (1, 2, 3)$. Then

$$\begin{aligned} \mathbf{b} &= (\mathbf{b} \cdot \mathbf{u}_1)\,\mathbf{u}_1 + (\mathbf{b} \cdot \mathbf{u}_2)\,\mathbf{u}_2 + (\mathbf{b} \cdot \mathbf{u}_3)\,\mathbf{u}_3 \\ &= \frac{1}{\sqrt{2}}(1 - 3)\,\mathbf{u}_1 + \frac{1}{\sqrt{6}}(1 - 4 + 3)\,\mathbf{u}_2 + \frac{1}{\sqrt{3}}(1 + 2 + 3)\,\mathbf{u}_3 \\ &= -\sqrt{2}\,\mathbf{u}_1 + 2\sqrt{3}\,\mathbf{u}_3 \end{aligned}$$

The reader will easily check that this is indeed correct: $(1, 2, 3) = -\sqrt{2}\,\mathbf{u}_1 + 2\sqrt{3}\,\mathbf{u}_3$. $\qquad\qquad\qquad\qquad\qquad\qquad\qquad\qquad\square$

3.10. Distortion-free maps. We saw above that having an orthonormal basis U for a subspace makes it easy to solve the *column* problem (i.e., express a given vector as a linear combination of the basis vectors) in that subspace. This followed from Proposition 3.2, which asserts that when the columns of $\mathbf{U}$ are orthonormal, we have

$$\mathbf{U}^T\mathbf{U} = \mathbf{I}_k$$

This same identity has a strong *geometric* interpretation, however, when we regard the matrix $\mathbf{U}$ as representing a linear transformation. Then it says that when $\mathbf{U}$ has orthonormal columns, it represents a *distortion-free* linear map, meaning that multiplication by $\mathbf{U}$ leaves lengths and angles unchanged. To say this precisely, we have

DEFINITION 3.11 (Distortion-free mapping). A linear map $T : \mathbf{R}^m \to \mathbf{R}^n$ is **distortion-free** or **distortionless** if *it preserves dot products*, so that for all inputs $\mathbf{v}, \mathbf{w} \in \mathbf{R}^m$, we have

$$T(\mathbf{v}) \cdot T(\mathbf{w}) = \mathbf{v} \cdot \mathbf{w}$$

□

Because we used dot products to define lengths and angles, this simple-looking condition forces T to preserve the lengths and angles in the following sense:

OBSERVATION 3.12. *If* $T : \mathbf{R}^m \to \mathbf{R}^n$ *is a distortion-free linear map, then for all inputs* $\mathbf{v}, \mathbf{w} \in \mathbf{R}^m$, *we have*

$$|T(\mathbf{v})| = |\mathbf{v}| \quad and \quad \angle\, T(\mathbf{v}), T(\mathbf{w}) = \angle\, \mathbf{v}, \mathbf{w}$$

PROOF. We prove the second identity and leave the first as an exercise.

Let $\mathbf{v}, \mathbf{w} \in \mathbf{R}^m$ be any two non-zero vectors. Since T is distortion-free, we have $T(\mathbf{v}) \cdot T(\mathbf{w}) = \mathbf{v} \cdot \mathbf{w}$, and, assuming the first identity of the Observation, $|T(\mathbf{v})|\,|T(\mathbf{w})| = |\mathbf{v}|\,|\mathbf{w}|$. So

$$\frac{T(\mathbf{v}) \cdot T(\mathbf{w})}{|T(\mathbf{v})|\,|T(\mathbf{w})|} = \frac{\mathbf{v} \cdot \mathbf{w}}{|\mathbf{v}|\,|\mathbf{w}|}$$

By Proposition 3.18 of Chapter 3, the two sides here give the cosines of $\angle\, T(\mathbf{v}), T(\mathbf{w})$ and $\angle\, \mathbf{v}, \mathbf{w}$, respectively. Angles between vectors are completely determined by their cosines, so this proves the Observation's second identity. □

EXAMPLE 3.13 (Two distortion-free maps). The mapping $T : \mathbf{R}^2 \to \mathbf{R}^2$ given by

$$T(x, y) = (-y, x)$$

(a 90° rotation of the plane) is distortion-free. Indeed, suppose $\mathbf{v} = (v_1, v_2)$ and $\mathbf{w} = (w_1, w_2)$ are any two vectors in $\mathbf{R}^2$. Then

$$T(\mathbf{v}) \cdot T(\mathbf{w}) = (-v_2, v_1) \cdot (-w_2, w_1) = v_2 w_2 + v_1 w_1 = \mathbf{v} \cdot \mathbf{w}$$

So T preserves dot products—it is distortion-free.

As a second example, consider the transformation $T : \mathbf{R}^2 \to \mathbf{R}^3$ represented by

$$\mathbf{A} = \frac{1}{\sqrt{6}} \begin{bmatrix} \sqrt{2} & \sqrt{3} \\ \sqrt{2} & 0 \\ \sqrt{2} & -\sqrt{3} \end{bmatrix}$$

A routine calculation shows that $\mathbf{A}^T\mathbf{A} = \mathbf{I}_2$, so the columns of $\mathbf{A}$ are orthonormal. To see that it represents a distortion-free transformation, we must show that multiplication by $\mathbf{A}$ preserves dot products.

For that, take two arbitrary vectors $\mathbf{x}$ and $\mathbf{x}'$, say, in the domain, and note that by definition of T

$$T(\mathbf{x}) \cdot T(\mathbf{x}') = \mathbf{A}\mathbf{x} \cdot \mathbf{A}\mathbf{x}'$$

Now use Corollary 2.8 from Chapter 4, which lets us write

$$\mathbf{A}\mathbf{x} \cdot \mathbf{A}\mathbf{x}' = \mathbf{x} \cdot \mathbf{A}^T\mathbf{A}\mathbf{x}' = \mathbf{x} \cdot \mathbf{I}_2\,\mathbf{x}' = \mathbf{x} \cdot \mathbf{x}'$$

as hoped: multiplication by $\mathbf{A}$ preserves dot products, so T is distortionless. $\qquad\qquad\square$

In the example just discussed, distortion freedom ultimately rested on the identity $\mathbf{A}^T\mathbf{A} = \mathbf{I}_2$—a property familiar from our discussion of matrices with orthonormal columns. Indeed, Proposition 3.2 says that $\mathbf{A}^T\mathbf{A} = \mathbf{I}$ *if and only if* $\mathbf{A}$ has orthonormal columns. By the exact same argument used in the example above, all such matrices represent distortion-free transformations. Specifically,

PROPOSITION 3.14. *A linear transformation* $\mathbf{R}^m \to \mathbf{R}^n$ *is distortion-free if and only if it is represented by a matrix with orthonormal columns.*

PROOF. If $T : \mathbf{R}^m \to \mathbf{R}^n$ is distortion-free, then by Theorem 5.6 of Chapter 1, T is represented by a matrix $\mathbf{A}$ in the sense that

$$T(\mathbf{x}) = \mathbf{A}\mathbf{x} \quad \text{for every input } \mathbf{x} \in \mathbf{R}^m$$

Moreover, by the column description of matrix/vector multiplication, we have

$$\mathbf{c}_j(\mathbf{A}) = \mathbf{A}\mathbf{e}_j$$

($\mathbf{c}_j(\mathbf{A})$ denotes column j of $\mathbf{A}$, while $\mathbf{e}_j$ is the jth standard basis vector in $\mathbf{R}^m$.) But then for each $i, j = 1, 2, \ldots, m$, we have

$$\mathbf{c}_i(\mathbf{A}) \cdot \mathbf{c}_j(\mathbf{A}) = \mathbf{A}\mathbf{e}_i \cdot \mathbf{A}\mathbf{e}_j = T(\mathbf{e}_i) \cdot T(\mathbf{e}_j) = \mathbf{e}_i \cdot \mathbf{e}_j = \delta_{ij}$$

The second equals sign holds because $\mathbf{A}$ represents T, and the next-to-last because T is distortion-free. In short, we get

$$\mathbf{c}_i(\mathbf{A}) \cdot \mathbf{c}_j(\mathbf{A}) = \delta_{ij}$$

which says the columns of $\mathbf{A}$ are orthonormal. This proves the "only if" half of the Proposition.

The "if" half follows almost immediately because when $\mathbf{A}$ has orthonormal columns, we have $\mathbf{A}^T\mathbf{A} = \mathbf{I}_m$ by Proposition 3.2. That being the case, we can apply Corollary 2.8 from Chapter 4 just as in Example 3.13 above, deducing that for any two inputs $\mathbf{v}, \mathbf{w} \in \mathbf{R}^m$, we have

$$T(\mathbf{v}) \cdot T(\mathbf{w}) = \mathbf{A}\mathbf{v} \cdot \mathbf{A}\mathbf{w} = \mathbf{v} \cdot \mathbf{A}^T\mathbf{A}\mathbf{w} = \mathbf{v} \cdot \mathbf{I}_m\mathbf{w} = \mathbf{v} \cdot \mathbf{w}$$

So T preserves dot products, which makes it distortion-free.   □

3.15. Orthogonal matrices. When U is an orthonormal basis for *all* of $\mathbf{R}^n$, we get a *square* $(n \times n)$ matrix when we set its vectors up as columns Proposition 3.2 then says that $\mathbf{U}^T$ actually *inverts* $\mathbf{U}$. Since transposition is vastly easier than row-reduction, this yields a truly valuable short-cut—when it applies.

DEFINITION 3.16. A square matrix $\mathbf{A}$ is **orthogonal** if it can be inverted by mere transposition. In other words, $\mathbf{A}$ is orthogonal iff

$$\mathbf{A}^T = \mathbf{A}^{-1}$$

□

With this terminology at our disposal, we can state some basic facts about orthogonal matrices.

OBSERVATION 3.17. *The following are equivalent.*

  a) $\mathbf{A}$ *is an orthogonal* $n \times n$ *matrix.*

  b) $\mathbf{A}^T$ *is an orthogonal* $n \times n$ *matrix.*

  c) *The columns of* $\mathbf{A}$ *form an orthonormal basis for* $\mathbf{R}^n$.

  d) *The rows of* $\mathbf{A}$ *form an orthonormal basis for* $\mathbf{R}^n$.

PROOF. **(a)** $\iff$ **(b)**. By Definition 3.16, we have orthogonality of $\mathbf{A}$ if and only if $\mathbf{A}^T = \mathbf{A}^{-1}$. Take the inverse on both sides of this equation to get

$$\left(\mathbf{A}^T\right)^{-1} = \mathbf{A}$$

Since $\mathbf{A} = \left(\mathbf{A}^T\right)^T$, we can write this as

$$\left(\mathbf{A}^T\right)^{-1} = \left(\mathbf{A}^T\right)^T$$

read backwards, this says that $\mathbf{A}^T$ is orthogonal too. If we read the whole chain of identities backwards, we get the converse: If the transpose is orthogonal, so is the matrix.

(a) $\iff$ **(c)**. Once more, we start with the orthogonality of $\mathbf{A}$, which means

$$\mathbf{A}^T \mathbf{A} = \mathbf{I}_n$$

Since the entry in row i, column j of $\mathbf{I}_n$ is δ_{ij}, this implies

$$\mathbf{r}_i \left(\mathbf{A}^T\right) \cdot \mathbf{c}_j \left(\mathbf{A}\right) = \delta_{ij}$$

and hence, since the rows of $\mathbf{A}^T$ are the columns of $\mathbf{A}$,

$$\mathbf{c}_i \left(\mathbf{A}\right) \cdot \mathbf{c}_j \left(\mathbf{A}\right) = \delta_{ij}$$

The columns of $\mathbf{A}$ are thus orthonormal, as (c) claims. The converse follows immediately from Proposition 3.2.

(b) $\iff$ **(d)**. Orthonormality of the rows of $\mathbf{A}$ is equivalent to orthonormality of the columns of $\mathbf{A}^T$. Since (c) is equivalent to (a), this holds if and only if $\mathbf{A}^T$ orthogonal. In short, (b) $\iff$ (d) as claimed.

Schematically, we have now proven

$$(d) \iff (b) \iff (a) \iff (c)$$

so that all four statements are equivalent as claimed. $\qquad\square$

Orthogonal matrices have other remarkable properties too.

PROPOSITION 3.18. *The product of two orthogonal matrices is again orthogonal. The determinant of an orthogonal matrix is always ± 1.*

PROOF. The inverse of a product is the product of the inverses in reverse order (Proposition 3.10 of Chapter 4) and the same rule governs products of transposes (Proposition 2.7 of Chapter 4). Since we also have orthogonality of $\mathbf{A}$ and $\mathbf{B}$, meaning that $\mathbf{A}^{-1} = \mathbf{A}^T$ and $\mathbf{B}^{-1} = \mathbf{B}^T$, we get

$$(\mathbf{AB})^{-1} = \mathbf{B}^{-1}\mathbf{A}^{-1} = \mathbf{B}^T\mathbf{A}^T = (\mathbf{AB})^T$$

This shows that $\mathbf{AB}$ is orthogonal, as promised.

To see that the determinant of an orthogonal matrix is ± 1, note first the obvious fact that $\det \mathbf{I}_n = 1$, and recall that the determinant of a product is the product of the determinants (Theorem 6.10 of Chapter 4). It follows that we always have

$$1 = \det \mathbf{I}_n = \det\left(\mathbf{A}^{-1}\mathbf{A}\right) = \det\left(\mathbf{A}^{-1}\right)\det\left(\mathbf{A}\right)$$

Here, however, $\mathbf{A}$ is orthogonal, so $\mathbf{A}^{-1} = \mathbf{A}^T$, and hence

$$1 = \det\left(\mathbf{A}^T\right)\det\left(\mathbf{A}\right) = (\det \mathbf{A})^2$$

because $\det \mathbf{A}^T = \det \mathbf{A}$ by Proposition 6.15 of Chapter 4. In short, we have $(\det \mathbf{A})^2 = 1$, and hence $\det \mathbf{A} = \pm 1$, as claimed. $\square$

REMARK 3.19. Orthogonal matrices play a major role in many fields of mathematics, and we may interpret the first conclusion of Proposition 3.18 as saying that that orthogonal matrices form a closed algebraic system of their own—a system important enough to have a well-established name: the $n \times n$ **orthogonal group**, denoted by $\mathrm{O}(n)$. It plays truly fundamental roles in both geometry and physics. $\square$

REMARK 3.20. As discussed in Exercise 410, all 2-by-2 orthogonal matrices take the form

$$\begin{bmatrix} \cos\theta & -\sin\theta \\ \sin\theta & \cos\theta \end{bmatrix} \quad \text{or} \quad \begin{bmatrix} \cos\theta & \sin\theta \\ \sin\theta & -\cos\theta \end{bmatrix}$$

Matrices of the first type correspond to *rotation* of the plane counterclockwise by θ. More precisely, when we multiply a vector $\mathbf{v} \in \mathbf{R}^2$ by such a matrix, the corresponding geometric vector in the plane rotates counterclockwise by the angle θ (radians).

Similarly, matrices of the second type correspond to *reflection* across the line that makes an angle of $\theta/2$ with the positive x-axis.

It is geometrically clear that both rotation and reflection are distortion-free, changing neither lengths of vectors nor angles between them. $\square$

– Practice –

401. Show that the columns of the following matrices yield orthonormal bases for $\mathbf{R}^2$ and $\mathbf{R}^3$, respectively:

a) $\begin{bmatrix} 0.6 & -0.8 \\ 0.8 & 0.6 \end{bmatrix}$
b) $\begin{bmatrix} 0.36 & 0.48 & 0.8 \\ 0.48 & 0.64 & -0.6 \\ -0.80 & 0.60 & 0.0 \end{bmatrix}$

402. Let

$$U = \frac{1}{3} \begin{bmatrix} 1 & 2 & -2 \\ 2 & 1 & 2 \\ 2 & -2 & -1 \end{bmatrix}$$

a) Compute $\mathbf{U}^T\mathbf{U}$ to show that the columns of this matrix form an orthonormal basis U for $\mathbf{R}^3$.

b) Use Observation 3.7 to express each of the vectors $(1, -2, 0)$, $(0, 1, -2)$ and $(1, 0, -2)$ as linear combinations of the columns of $\mathbf{U}$.

403. Let

$$U = \frac{1}{7} \begin{bmatrix} 2 & 3 & -6 \\ 6 & 2 & 3 \\ 3 & -6 & -2 \end{bmatrix}$$

a) Compute $\mathbf{U}^T\mathbf{U}$ to show that the columns of this matrix form an orthonormal basis U for $\mathbf{R}^3$.

b) Use Observation 3.7 to express each of the vectors $(1, -2, 0)$, $(0, 1, -2)$, and $(1, 0, -2)$ as linear combinations of the columns of $\mathbf{U}$.

404. Consider all matrices of the form

$$\mathbf{U}(\theta) := \begin{bmatrix} \cos\theta & 0 & \sin\theta \\ 0 & 1 & 0 \\ \sin\theta & 0 & -\cos\theta \end{bmatrix}, \quad \theta \in \mathbf{R}$$

a) Show that for any fixed θ, the columns of $\mathbf{U}(\theta)$ form an orthonormal basis for $\mathbf{R}^3$.

b) Expand $(1, -1, 0)$ and $(0, 1, -1) \in \mathbf{R}^3$ as a linear combination of the columns of $\mathbf{U}(\theta)$. Use Observation 3.7 (no need to row-reduce anything).

405. Define

$$
J = \begin{bmatrix} 0 & 1 & 0 & 0 \\ -1 & 0 & 0 & 0 \\ 0 & 0 & 0 & 1 \\ 0 & 0 & -1 & 0 \end{bmatrix}, \quad K = \begin{bmatrix} 0 & 0 & 0 & -1 \\ 0 & 0 & -1 & 0 \\ 0 & 1 & 0 & 0 \\ 1 & 0 & 0 & 0 \end{bmatrix},
$$

$$
L = \begin{bmatrix} 0 & 0 & 1 & 0 \\ 0 & 0 & 0 & -1 \\ -1 & 0 & 0 & 0 \\ 0 & 1 & 0 & 0 \end{bmatrix}
$$

Show that for any $\mathbf{v} \neq \mathbf{0}$ in $\mathbf{R}^4$, the set

$$
\left\{ \frac{\mathbf{v}}{|\mathbf{v}|}, \frac{J\mathbf{v}}{|\mathbf{v}|}, \frac{K\mathbf{v}}{|\mathbf{v}|}, \frac{L\mathbf{v}}{|\mathbf{v}|} \right\}
$$

forms an orthonormal basis.

406. Consider the vector $\mathbf{u}_1 = (1, \sqrt{3})/2$ in $\mathbf{R}^2$, and the vectors $\mathbf{v}_1 = (1, 1, 1)/\sqrt{3}$, $\mathbf{v}_2 = (1, 0, -1)/\sqrt{2}$ in $\mathbf{R}^3$.

   a) Find a vector $\mathbf{u}_2$ that makes $\{\mathbf{u}_1, \mathbf{u}_2\}$ an orthonormal basis for $\mathbf{R}^2$. How many possibilities are there for such a vector?

   b) Find a vector $\mathbf{v}_3$ that makes $\{\mathbf{v}_1, \mathbf{v}_2, \mathbf{v}_3\}$ an orthonormal basis for $\mathbf{R}^3$. How many possibilities are there for such a vector?

407. Find an orthonormal basis for the plane spanned by these two vectors in $\mathbf{R}^4$:

$$
\mathbf{u}_1 = (1, -2, -2, 1), \quad \mathbf{u}_2 = (1, -2, 6, -3)
$$

[*Hint: Start by normalizing* $\mathbf{u}_1$. *Then seek a second vector in the span of* $\mathbf{u}_1$ *and* $\mathbf{u}_2$, *but orthogonal to* $\mathbf{u}_1$.]

408. Prove that distortion-free linear mappings preserve *lengths* of vectors (the first identity of Observation 3.12).

409. Which of the following represent distortion-free linear transformations? Justify your answers.

a) $\dfrac{1}{10} \begin{bmatrix} 6 & -8 \\ 8 & 6 \end{bmatrix}$ b) $\dfrac{1}{9} \begin{bmatrix} 1 & -4 \\ 4 & -7 \\ 8 & 4 \end{bmatrix}$ c) $\dfrac{1}{9} \begin{bmatrix} 1 & 4 & 8 \\ -4 & -7 & 4 \end{bmatrix}$

410. Show that every 2×2 orthogonal matrix takes one of the two forms displayed in Remark 3.20 above. [*Hint: If $a^2 + b^2 = 1$, then (a, b) is on the unit circle in $\mathbf{R}^2$, and hence takes the form $(\cos \theta, \sin \theta)$ for some angle θ.*]

411. Prove the claims made by Remark 3.20. Specifically show that when we multiply a unit vector of the form $(\cos \phi, \sin \phi)$ (or any vector that makes an angle ϕ with the x-axis) by a matrix of the first or second type, we respectively get

$$(\cos(\phi + \theta), \sin(\phi + \theta)), \quad \text{or} \quad (\cos(\theta - \phi), \sin(\theta - \phi))$$

(The latter results from reflecting $(\cos \phi, \sin \phi)$ across the line generated by $(\cos \frac{1}{2}\theta, \sin \frac{1}{2}\theta)$, a less obvious fact.)

412. Proposition 3.18 says that the product of two orthogonal matrices is again orthogonal. Verify this in the 2-by-2 case by computing products of pairs of orthogonal 2-by-2 matrices of the forms displayed in Remark 3.20. For instance, show that the product of two matrices of the first type, e.g.,

$$\begin{bmatrix} \cos \alpha & -\sin \alpha \\ \sin \alpha & \cos \alpha \end{bmatrix} \begin{bmatrix} \cos \beta & -\sin \beta \\ \sin \beta & \cos \beta \end{bmatrix}$$

is orthogonal. There are three other combinations of the two types; verify their products are orthogonal too. [*Hint: The addition and subtraction identities for sine and cosine can be helpful here.*]

413. Which linear transformations below are distortion-free? Justify your answer.

    a) $T : \mathbf{R}^3 \to \mathbf{R}^3$, given by $T(x, y, z) = (z, x, y)$

    b) $P : \mathbf{R}^4 \to \mathbf{R}^3$, given by $P(x_1, x_2, x_3, x_4) = (x_1, x_2, x_3)$

    c) $Q : \mathbf{R}^2 \to \mathbf{R}^4$, given by $Q(x, y) = (x + y, \ x - y, \ x - y, \ x + y)$

    d) $S : \mathbf{R}^2 \to \mathbf{R}^4$, given by $S(x, y) = \frac{1}{2}(x + y, \ x - y, \ x - y, \ x + y)$

414. Show: *A linear map $T : \mathbf{R}^m \to \mathbf{R}^n$ is distortion-free if it preserves lengths*—that is, if $|T(\mathbf{v})| = |\mathbf{v}|$ for every $\mathbf{v} \in \mathbf{R}^m$.
[*Hint: To show that $T(\mathbf{v}) \cdot T(\mathbf{w}) = \mathbf{v} \cdot \mathbf{w}$, argue that*

$$|T(\mathbf{v}) + T(\mathbf{w})|^2 = |T(\mathbf{v} + \mathbf{w})|^2 = |\mathbf{v} + \mathbf{w}|^2,$$

then expand out both ends of the equality and do some cancelling.]

415. Length preservation alone makes a linear map distortion-free, by Exercise 414. Does *angle* preservation make a linear map distortion-free? Find out by considering the map given by $T(\mathbf{x}) = 2\mathbf{x}$. Show that for any $\mathbf{v}, \mathbf{w}$ in the domain of T, we do have angle-preservation:

$$\angle T(\mathbf{v}), T(\mathbf{w}) = \angle \mathbf{v}, \mathbf{w}$$

Does T also preserve dot products?

$$- \star -$$

4. The Gram–Schmidt Algorithm

4.1. Introduction. In Section 6 of Chapter 5, we learned three ways to find a basis for a subspace S.

- If S is described as a *span*, we have a set U of vectors that span S, and the pivot columns of the matrix $\mathbf{U}$ having those vectors as columns give us a basis.

- Alternatively, we can transpose $\mathbf{U}$ and row-reduce. The pivot *rows* of $\text{RRE}(\mathbf{U})$ then form a basis for S.

- Finally, if we have an *implicit* description for S as $\text{nul}(\mathbf{A})$ for some matrix $\mathbf{A}$, then the homogeneous generators we get from the free columns of $\text{RRE}(\mathbf{A})$ give us a basis.

The methods listed above seldom produce *orthonormal* bases, however, and yet (as the previous section shows) orthonormal bases have especially desirable qualities.

Fortunately, there is an algorithm, known as the **Gram–Schmidt process**, that transforms *any* basis for a subspace S into an *orthonormal* basis for S, and does it in a particularly nice way by preserving the spans as it proceeds. Before we explain what we mean by this and present Gram–Schmidt in detail, we consider a simple example that reveals the basic idea behind it.

EXAMPLE 4.2. The vectors $\mathbf{u}_1 = (1, -1, 0)$ and $\mathbf{u}_2 = (0, 1, -1)$ form a basis for the perp of $(1, 1, 1)$—the plane P defined implicitly by $x + y + z = 0$. This basis is not orthonormal, though. Neither $\mathbf{u}_1$ nor $\mathbf{u}_2$ have unit length. They aren't orthogonal to each other either.

We can use them to *construct* an orthonormal basis for P, however, in two main steps. First, we modify the original basis $\{\mathbf{u}_1, \mathbf{u}_2\}$ to get an *orthogonal* basis $\{\mathbf{v}_1, \mathbf{v}_2\}$ for P. Then we normalize the $\mathbf{v}_i$'s.

The second step will be trivial: we just divide each $\mathbf{v}_i$ by its length. The first step, however—modifying the original basis $\{\mathbf{u}_1, \mathbf{u}_2\}$ to get an *orthogonal* basis $\{\mathbf{v}_1, \mathbf{v}_2\}$—is subtler. It starts easily, for we simply set

$$\mathbf{v}_1 \;:=\; \mathbf{u}_1 \;=\; (1, -1, 0)$$

The next step is harder, but this is where the main idea behind Gram–Schmidt comes in. To get $\mathbf{v}_2$, we "correct" $\mathbf{u}_2$, accomplishing two goals in one operation: *We make it perpendicular to* $\mathbf{v}_1$, *while still keeping it within the span of* $\{\mathbf{u}_1, \mathbf{u}_2\}$. This is done by subtracting off a multiple of the previous vector $\mathbf{v}_1$, that is, by setting

$$(46) \qquad\qquad \mathbf{v}_2 \;:=\; \mathbf{u}_2 - c\,\mathbf{v}_1$$

where c is a scalar we have yet to determine. Notice that this automatically formula keeps $\mathbf{v}_2$ in the span of $\mathbf{u}_1 = \mathbf{v}_1$ and $\mathbf{u}_2$. The key point now is to choose c so that $\mathbf{v}_2$ will also be *orthogonal* to $\mathbf{v}_1$. This means we want

$$\begin{aligned}
0 &= \mathbf{v}_2 \cdot \mathbf{v}_1 \\
&= (\mathbf{u}_2 - c\mathbf{v}_1) \cdot \mathbf{v}_1 \\
&= \mathbf{u}_2 \cdot \mathbf{v}_1 - c\,(\mathbf{v}_1 \cdot \mathbf{v}_1)
\end{aligned}$$

which we simply solve for c to get

$$c = \frac{\mathbf{u}_2 \cdot \mathbf{v}_1}{\mathbf{v}_1 \cdot \mathbf{v}_1}$$

Putting this value into (46) yields

$$(47) \qquad\qquad \mathbf{v}_2 = \mathbf{u}_2 - \left(\frac{\mathbf{u}_2 \cdot \mathbf{v}_1}{\mathbf{v}_1 \cdot \mathbf{v}_1} \right) \mathbf{v}_1$$

which now has the properties we need: Together with $\mathbf{v}_1$, it spans P, *and* it is orthogonal to $\mathbf{v}_1$.

Plugging in the actual vectors $\mathbf{u}_1$ and $\mathbf{u}_2$ we started with, we have

$$c = \frac{\mathbf{u}_2 \cdot \mathbf{v}_1}{\mathbf{v}_1 \cdot \mathbf{v}_1} = \frac{(0, 1, -1) \cdot (1, -1, 0)}{(1, -1, 0) \cdot (1, -1, 0)} = -\frac{1}{2}$$

Inserting this into (47), we get

$$\mathbf{v}_2 = \begin{pmatrix} 0 \\ 1 \\ -1 \end{pmatrix} - \left(-\frac{1}{2}\right)\begin{pmatrix} 1 \\ -1 \\ 0 \end{pmatrix}$$

$$= \begin{pmatrix} 1/2 \\ 1/2 \\ -1 \end{pmatrix}$$

We are not yet done: we still need to normalize the $\mathbf{v}_i$'s to get *unit* vectors that we shall call $\mathbf{w}_i$'s. Accordingly, we set

$$\mathbf{w}_1 := \frac{\mathbf{v}_1}{|\mathbf{v}_1|} = \frac{(1,-1,0)}{\sqrt{1+1+0}} = \frac{(1,-1,0)}{\sqrt{2}}$$

$$\mathbf{w}_2 = \frac{\mathbf{v}_2}{|\mathbf{v}_2|} = \frac{(1,1,-2)}{\sqrt{1+1+4}} = \frac{(1,1,-2)}{\sqrt{6}}$$

Notice that we got $\mathbf{w}_2$ by normalizing $2\mathbf{v}_2 = (1,1,-2)$, instead of normalizing $\mathbf{v}_2$ itself. That's because *normalizing* $k\mathbf{v}$ *gives the same result as normalizing* $\mathbf{v}$ *for any positive scalar* $k > 0$ (Exercise 155). It's often easier to normalize a vector *after* clearing all denominators with a large scalar. In any case, we now have the desired orthonormal basis for P:

$$U = \left\{ \frac{1}{\sqrt{2}}\begin{pmatrix} 1 \\ -1 \\ 0 \end{pmatrix}, \frac{1}{\sqrt{6}}\begin{pmatrix} 1 \\ 1 \\ -2 \end{pmatrix} \right\}$$

$\square$

4.3. General case. The ideas used above allow us to *orthonormalize* any independent set $U = \{\mathbf{u}_1, \mathbf{u}_2, \ldots, \mathbf{u}_k\}$ to get an *orthonormal* set $W = \{\mathbf{w}_1, \mathbf{w}_2, \ldots, \mathbf{w}_k\}$ with precisely the same span as U itself. In fact, the span of the first j $\mathbf{w}_i$'s always equals that of the first j $\mathbf{u}_i$'s, so the procedure *preserves span* along the way. It goes as follows:

Gram–Schmidt algorithm

Step 1. *Set* $\mathbf{v}_1 = \mathbf{u}_1$.

Step 2. *Set* $\mathbf{v}_2 = \mathbf{u}_2 - \left(\dfrac{\mathbf{u}_2 \cdot \mathbf{v}_1}{\mathbf{v}_1 \cdot \mathbf{v}_1} \right) \mathbf{v}_1$.

Step 3. *Set* $\mathbf{v}_3 = \mathbf{u}_3 - \left(\dfrac{\mathbf{u}_3 \cdot \mathbf{v}_1}{\mathbf{v}_1 \cdot \mathbf{v}_1} \right) \mathbf{v}_1 - \left(\dfrac{\mathbf{u}_3 \cdot \mathbf{v}_2}{\mathbf{v}_2 \cdot \mathbf{v}_2} \right) \mathbf{v}_2$.

$$\vdots$$

Step i. *For each remaining* $i = 4, \ldots, k$, *produce a vector* $\mathbf{v}_i$ *perpendicular to all previous* $\mathbf{v}_j$'s, *and lying in the span of* $\{\mathbf{u}_1, \ldots, \mathbf{u}_i\}$ *via the formula*

$$\mathbf{v}_i := \mathbf{u}_i - \sum_{j=1}^{i} \left(\frac{\mathbf{u}_i \cdot \mathbf{v}_j}{\mathbf{v}_j \cdot \mathbf{v}_j} \right) \mathbf{v}_j$$

$$\vdots$$

Step k + 1. *The set* $\{\mathbf{v}_1, \mathbf{v}_2, \ldots, \mathbf{v}_k\}$ *is now* **orthogonal,** *with the same span as the original* $\mathbf{u}_i$'s. *To complete the Gram–Schmidt process, simply normalize the* $\mathbf{v}_i$s, *setting*

$$\mathbf{w}_i = \frac{\mathbf{v}_i}{|\mathbf{v}_i|}, \quad i = 1, \ldots, k$$

Now $W = \{\mathbf{w}_1, \mathbf{w}_2, \ldots, \mathbf{w}_k\}$ *yields an orthonormal basis for the span of the original set* U.

The prescription for Gram–Schmidt above is a learnable routine, but it can be tedious. For sets of two or three vectors, one can apply it by hand without too much effort, but for larger sets, computer assistance is nice.

EXAMPLE 4.4. Use Gram–Schmidt to orthonormalize this basis for $\mathbf{R}^3$:

$$\mathbf{B} = \left\{ \begin{pmatrix} 1 \\ -1 \\ 0 \end{pmatrix}, \begin{pmatrix} 0 \\ 1 \\ -1 \end{pmatrix}, \begin{pmatrix} 1 \\ 0 \\ 1 \end{pmatrix} \right\}$$

Step 1. Define $\mathbf{v}_1 = \mathbf{u}_1 = (1, -1, 0)$.

Step 2. Compute $\mathbf{v}_2$ as described above:

$$\mathbf{v}_2 = \mathbf{u}_2 - \left(\frac{\mathbf{u}_2 \cdot \mathbf{v}_1}{\mathbf{v}_1 \cdot \mathbf{v}_1} \right) \mathbf{v}_1$$

This is exactly the same calculation we did in the previous example, and with the very same vectors. So again,

$$\mathbf{v}_2 = \begin{pmatrix} 1/2 \\ 1/2 \\ -1 \end{pmatrix}$$

As noted there, however, we can scalar multiply this vector by 2 to clear the denominators without changing its direction (i.e., leaving it perpendicular to $\mathbf{v}_1$). Doing so gives a cleaner result:

$$\mathbf{v}_2 = \begin{pmatrix} 1 \\ 1 \\ -2 \end{pmatrix}$$

Step 3. Now compute $\mathbf{v}_3$ using the "Step i" formula above with $i = 3$:

$$\mathbf{v}_3 = \mathbf{u}_3 - \left(\frac{\mathbf{u}_3 \cdot \mathbf{v}_1}{\mathbf{v}_1 \cdot \mathbf{v}_1} \right) \mathbf{v}_1 - \left(\frac{\mathbf{u}_3 \cdot \mathbf{v}_2}{\mathbf{v}_2 \cdot \mathbf{v}_2} \right) \mathbf{v}_2$$

To organize the calculation, first compute

$$\frac{\mathbf{u}_3 \cdot \mathbf{v}_1}{\mathbf{v}_1 \cdot \mathbf{v}_1} = \frac{(1,0,1) \cdot (1,-1,0)}{(1,-1,0) \cdot (1,-1,0)} = \frac{1}{2},$$

$$\frac{\mathbf{u}_3 \cdot \mathbf{v}_2}{\mathbf{v}_2 \cdot \mathbf{v}_2} = \frac{(1,0,1) \cdot (1,1,-2)}{(1,1,-2) \cdot (1,1,-2)} = \frac{-1}{6}$$

Now put these coefficients into the formula for $\mathbf{v}_3$ above to get

$$
\begin{aligned}
\mathbf{v}_3 \;&=\; \mathbf{u}_3 - \frac{1}{2}\,\mathbf{v}_1 + \frac{1}{6}\,\mathbf{v}_2 \\[2mm]
&=\; \begin{pmatrix} 1 \\ 0 \\ 1 \end{pmatrix} - \frac{1}{2}\begin{pmatrix} 1 \\ -1 \\ 0 \end{pmatrix} + \frac{1}{6}\begin{pmatrix} 1 \\ 1 \\ -2 \end{pmatrix} \\[2mm]
&=\; \begin{pmatrix} 1 - \frac{1}{2} + \frac{1}{6} \\[1mm] 0 + \frac{1}{2} + \frac{1}{6} \\[1mm] 1 - \frac{2}{6} \end{pmatrix} \\[2mm]
&=\; \frac{2}{3}\begin{pmatrix} 1 \\ 1 \\ 1 \end{pmatrix}
\end{aligned}
$$

Once more, we can ignore the positive scalar $2/3$ here (i.e., multiply by $3/2$) because we're about to...

Step 4. ...finish up by normalizing:

$$
\begin{aligned}
\mathbf{w}_1 \;&=\; \frac{\mathbf{v}_1}{|\mathbf{v}_1|} = \frac{(1,-1,0)}{\sqrt{1+1+0}} = \frac{1}{\sqrt{2}}\,(1,-1,0)\,, \\[3mm]
\mathbf{w}_2 \;&=\; \frac{\mathbf{v}_2}{|\mathbf{v}_2|} = \frac{(1,1,-2)}{\sqrt{1+1+4}} = \frac{1}{\sqrt{6}}\,(1,1,-2)\,, \\[3mm]
\mathbf{w}_3 \;&=\; \frac{\mathbf{v}_3}{|\mathbf{v}_3|} = \frac{(1,1,1)}{\sqrt{1+1+1}} = \frac{1}{\sqrt{3}}\,(1,1,1)
\end{aligned}
$$

Simple inspection reveals that these $\mathbf{w}_i$'s are indeed mutually perpendicular unit vectors. Our new orthonormal basis for $\mathbf{R}^3$ is thus

$$
W = \left\{ \frac{1}{\sqrt{2}}\begin{pmatrix} 1 \\ -1 \\ 0 \end{pmatrix},\ \frac{1}{\sqrt{6}}\begin{pmatrix} 1 \\ 1 \\ -2 \end{pmatrix},\ \frac{1}{\sqrt{3}}\begin{pmatrix} 1 \\ 1 \\ 1 \end{pmatrix} \right\}
$$

$\square$

– **Practice** –

416. Use Gram–Schmidt to convert $\{(1,1),(1,2)\}$ into an orthonormal basis for $\mathbf{R}^2$.

417. Find an orthonormal basis for the plane spanned by $(1,2,3)$ and $(3,2,1)$ in $\mathbf{R}^3$.

418. Find a basis for the plane $x + 2y - z = 0$ in $\mathbf{R}^3$. Then use Gram–Schmidt to make it orthonormal.

419. Use Gram–Schmidt to turn the given (ordered) set into an orthonormal basis for $\mathbf{R}^3$.

$$\text{a)} \quad U = \left\{ \begin{pmatrix} 1 \\ 0 \\ 0 \end{pmatrix}, \begin{pmatrix} 1 \\ 1 \\ 0 \end{pmatrix}, \begin{pmatrix} 1 \\ 1 \\ 1 \end{pmatrix} \right\}$$

$$\text{b)} \quad U = \left\{ \begin{pmatrix} 1 \\ 1 \\ 1 \end{pmatrix}, \begin{pmatrix} 1 \\ 1 \\ 0 \end{pmatrix}, \begin{pmatrix} 1 \\ 0 \\ 0 \end{pmatrix} \right\}$$

Notice that while the sets in (a) and (b) are the same (as sets), Gram–Schmidt yields different results for the two cases because the vectors come in a different order.

420. Find a basis for $\mathbf{R}^2$ that is orthonormal, *and* whose first vector is a multiple of $(3,4)$.

421. Find a basis for $\mathbf{R}^3$ that is orthonormal, *and* whose first two vectors span the plane $2x - 4y = 0$. [*Hint: Inspection is all you need to find a vector orthogonal to the given plane.*]

422. Find a basis for $\mathbf{R}^3$ that is orthonormal, *and* whose first two vectors lie in the span of $(1, -2, 2)$ and $(2, -1, 2)$. [*Hint: The cross product of (Exercises 63 and 381) may be useful here.*]

423. Consider the vectors $\mathbf{u}_1 = (1, 2, 3, 4)$, $\mathbf{u}_2 = (-2, 1, -4, 3)$, and $\mathbf{u}_3 = (-4, -3, 2, 1)$.

　　a) Verify that the $\mathbf{u}_i$'s are mutually orthogonal, and all have the same length.

b) Find a fourth vector orthogonal to all three $\mathbf{u}_i$'s.

c) Normalize the resulting set of four vectors to produce an orthonormal basis for $\mathbf{R}^4$.

424. Find an orthonormal basis for the nullspace of

$$\mathbf{B} = \begin{bmatrix} 1 & 6 & 5 & -2 \\ 3 & 2 & 7 & -2 \end{bmatrix}$$

425. Find a basis for the hyperplane $x_1 + x_2 + x_3 + x_4 = 0$ in $\mathbf{R}^4$. Then use Gram–Schmidt to make it orthonormal.

 $-\star-$

CHAPTER 7

Linear Transformation

1. Kernel and Image

In this final chapter, we explore the *geometric structure* of linear transformations. Our findings will address the remaining unanswered mapping questions from Chapter 1. Specifically, we shall understand the geometry of images and pre-images of linear maps, and quantify the ways that linear maps distort subsets in the domain.

Virtually all we have learned so far will come into play, and we need to go deeper into a topic we introduced just briefly in Chapter 1: eigenvalues and eigenvectors.

As always, the *computational* tool for exploring a linear transformation T is the matrix that represents it; that is, the matrix $\mathbf{A}$ that calculates $T(\mathbf{x})$ as the matrix/vector product $\mathbf{A}\mathbf{x}$.

By Observation 5.5 of Chapter 1, we can always construct $\mathbf{A}$ from T if we know how T affects the standard basis vectors in the domain:

$$\mathbf{c}_j(\mathbf{A}) = T(\mathbf{e}_j)$$

Here $\mathbf{c}_j(\mathbf{A})$ and $\mathbf{e}_j$ are the jth column of $\mathbf{A}$, and the jth standard basis vector in $\mathbf{R}^m$, respectively.

We warm up with a simple example. Though very basic, it illustrates the kind of structural understanding we develop in this chapter.

EXAMPLE 1.1. Consider the transformation T represented by

$$\mathbf{A} = \begin{bmatrix} 1 & -2 \\ -1 & 2 \end{bmatrix}$$

Since $\mathbf{A}$ is 2×2, the domain and range of T are both $\mathbf{R}^2$:

$$T \colon \mathbf{R}^2 \to \mathbf{R}^2$$

To understand the effect of T, we view $\mathbf{R}^2$ geometrically, using numeric/geometric duality.

First consider the image. A vector $\mathbf{w} \in \mathbf{R}^2$ lies in the image of T if and only if

$$\mathbf{w} = T(\mathbf{x}) = \mathbf{A}\mathbf{x}$$

for some $\mathbf{x}$ in the domain. Reading this backwards, we see that $\mathbf{w}$ lies in the image of T if and only if the system

$$\mathbf{A}\mathbf{x} = \mathbf{w}$$

has a solution. By Proposition 4.3 of Chapter 5, however, a solution exists if and only if $\mathbf{w} \in \text{col}(\mathbf{A})$. So we see that *the image of T is precisely the same as* $\text{col}(\mathbf{A})$. To visualize it geometrically, we compute a basis for $\text{col}(\mathbf{A})$ by row-reducing:

$$\text{RRE}(\mathbf{A}) = \begin{bmatrix} 1 & -2 \\ 0 & 0 \end{bmatrix}$$

The single pivot in the first column means $\mathbf{c}_1(\mathbf{A}) = (1, -1)$ spans $\text{col}(\mathbf{A})$, and hence spans the image of T. Geometrically, this means the image of T is the line generated by $(1, -1)$. That line is *not* all of $\mathbf{R}^2$, so T is not onto.

At the same time, the *free* column in $\text{RRE}(\mathbf{A})$ means that when $\mathbf{w} = (w_1, w_2)$ *does* lie in the image, it will have *infinitely* many pre-images, namely solutions of $\mathbf{A}\mathbf{x} = \mathbf{w}$. We can read them off after reducing the *augmented* matrix $[\mathbf{A} \,|\, \mathbf{w}]$ to get

$$\text{RRE}\left([\mathbf{A} \,|\, \mathbf{w}]\right) = \begin{bmatrix} 1 & -2 & w_1 \\ 0 & 0 & 0 \end{bmatrix}$$

Column two is free, yielding the single homogeneous generator $(2, 1)$, so $T^{-1}(\mathbf{w})$ is given by the image of the mapping

$$\mathbf{x}(s) = \begin{pmatrix} w_1 \\ 0 \end{pmatrix} + s \begin{pmatrix} 2 \\ 1 \end{pmatrix}$$

This too is a line: the line through $(w_1, 0)$ in the $(2, 1)$ direction. Just as the image of T is a line in the range, the pre-image of $\mathbf{w}$ is a line in the domain, and T sends every point on it to $\mathbf{w}$. Since $\mathbf{w}$ was an *arbitrary* vector in the image, we see that the pre-image of *any* $\mathbf{w}$ in the image is a translate of $\text{nul}(\mathbf{A})$—a line in the $(2, 1)$ direction (Figure 1).

Our analysis here reveals that T is neither onto, nor one-to-one. It fails to be onto because its image does *not* include the entire range $\mathbf{R}^2$. It fails to be one-to-one because each $\mathbf{w}$ in the image of T has

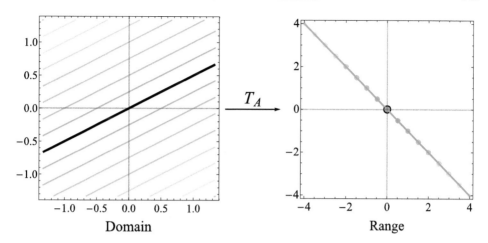

Figure 1. Pre-image and image. The blue line at right shows the *image* of T in the range $\mathbf{R}^2$. Points on the line are outputs of T; other points in $\mathbf{R}^2$ are not. Each output (e.g., each *dot* on the right) has an entire line of pre-images (same-hued *line* on the left). The black line at left is $\mathrm{nul}(\mathbf{A})$, the pre-image of $\mathbf{0}$. All other pre-images are parallel copies of that one.

multiple pre-images, a result of the fact that $\mathrm{nul}(\mathbf{A})$ contains multiple vectors (here, an entire line). $\square$

Guided by this example, we now take up the general case.

1.2. Kernel and image. Two key aspects of the example above hold for *every* linear map:

PROPOSITION 1.3. *Let* $T \colon \mathbf{R}^m \to \mathbf{R}^n$ *be any linear transformation. Let* $\mathbf{A}$ *be the matrix that represents it.*

  a) *The image of* T *in* $\mathbf{R}^n$ *is the column-space of* $\mathbf{A}$. *In particular, the image of* T *is always a subspace of* $\mathbf{R}^n$.

  b) *The pre-image of each* $\mathbf{w} \in \mathrm{col}(\mathbf{A})$ *is a translate of the null-space of* $\mathbf{A}$, *an affine subspace of the form* $\mathbf{x}_p + \mathrm{nul}(\mathbf{A})$.

PROOF. By definition (and as in Example 1.1 above), $\mathbf{w}$ lies in the image of T if and only if $T(\mathbf{x}) = \mathbf{w}$ for some $\mathbf{x}$ in the domain $\mathbf{R}^m$. Since $\mathbf{A}$ represents T, this in turn holds if and only if we can solve

$$(48) \qquad\qquad \mathbf{A}\mathbf{x} = \mathbf{w}$$

which is true if and only if $\mathbf{w} \in \mathrm{col}(\mathbf{A})$, by Proposition 4.3 of Chapter 5. So the image of T equals $\mathrm{col}(\mathbf{A})$, as our first assertion claims, and since $\mathrm{col}(\mathbf{A})$ is subspace of $\mathbf{R}^n$, so is the image of T.

To get the second assertion, observe that by the same reasoning, the *pre*-image of any $\mathbf{w} \in \mathrm{col}(\mathbf{A})$ is exactly the solution set of the inhomogeneous system (48). But Theorem 4.1 of Chapter 2 (our Structure Theorem for pre-images of linear maps) tells us we can always express that solution set as the affine subspace

$$\mathbf{x}_p + \mathrm{nul}(\mathbf{A})$$

where $\mathbf{x}_p$ is any particular solution of (48). This completes the proof. $\square$

We sometimes denote the **image** of T by $\mathrm{im}(T)$.

The nullspace of the representing matrix $\mathbf{A}$ is itself a pre-image: it is $T^{-1}(\mathbf{0})$, and as we just saw, every other pre-image is a parallel copy of it, obtained by simply adding a particular solution. When discussing the linear map T represented by $\mathbf{A}$, we usually this crucial subspace the *kernel* of T:

DEFINITION 1.4. The **kernel** of a linear transformation $T \colon \mathbf{R}^m \to \mathbf{R}^n$, denoted $\ker(T)$, is the pre-image of $\mathbf{0} \in \mathbf{R}^n$. In short, $\ker(T) = T^{-1}(\mathbf{0})$.

When $\mathbf{A}$ represents T, we have $\mathbf{x} \in \ker(T)$ if and only if $T(\mathbf{x}) = \mathbf{A}\mathbf{x} = \mathbf{0}$. So just as Proposition 1.3 equates the image of T to $\mathrm{col}(\mathbf{A})$, we can equate the *kernel* of T to $\mathrm{nul}(\mathbf{A})$. For future reference, we state these facts as follows:

OBSERVATION 1.5. *We always have* $\ker(T) = \mathrm{nul}(\mathbf{A})$ *and* $\mathrm{im}(T) = \mathrm{col}(\mathbf{A})$. *In particular,* $\ker(T) \subset \mathbf{R}^m$ *and* $\mathrm{im}(T) \subset \mathbf{R}^n$ *are subspaces of the domain and range, respectively.*

We give the dimensions of these subspaces obvious names:

DEFINITION 1.6. The **nullity** of a linear transformation is the dimension of its kernel. The **rank** of the transformation is the dimension of its image. Observation 1.5 makes it obvious that the rank and nullity of a linear *map* are the same as the rank and nullity of the *matrix* representing it. $\square$

For any $n \times m$ matrix, we have $\mathrm{rank} + \mathrm{nullity} = m$ (Proposition 8.11; "pivot columns plus free columns equal all columns"). Since every linear map from $\mathbf{R}^m$ to $\mathbf{R}^n$ is represented by an $n \times m$ matrix, we immediately get the corresponding fact for linear maps:

OBSERVATION 1.7. *For any linear map* $T : \mathbf{R}^n \to \mathbf{R}^m$, *we have*

$$\mathrm{rank}(T) + \mathrm{nullity}(T) = \text{dimension of domain} = m$$

To practice with these terms, we study a few more examples.

EXAMPLE 1.8. Suppose $T : \mathbf{R}^3 \to \mathbf{R}^2$ is the linear transformation represented by

$$\mathbf{A} = \begin{bmatrix} 1 & 2 & 1 \\ 2 & 5 & 1 \end{bmatrix}$$

Row-reducing, we get

$$\mathrm{RRE}([\mathbf{A}]) = \begin{bmatrix} 1 & 0 & 3 \\ 0 & 1 & -1 \end{bmatrix}$$

Clearly $\mathbf{A}$, and hence T, has rank 2, which means the image of T forms a 2-dimensional subspace of the range $\mathbf{R}^2$. That makes it *all* of $\mathbf{R}^2$, so T is onto.

Meanwhile, the single free column tells us T has nullity 1. The kernel is therefore a *line* in the domain $\mathbf{R}^3$, spanned by the single homogeneous generator $\mathbf{h} = (-3, 1, 1)$. We again have a case where $\ker(T)$ contains multiple points, so T is *not* one-to-one. It is onto, but not one-to-one. $\square$

EXAMPLE 1.9. Consider the transformation $G : \mathbf{R}^2 \to \mathbf{R}^3$ represented by the *transpose* of the matrix $\mathbf{A}$ above, so that G is represented by

$$\mathbf{B} = \begin{bmatrix} 1 & 2 \\ 2 & 5 \\ 1 & 1 \end{bmatrix} \quad \text{and} \quad \mathrm{RRE}(\mathbf{B}) = \begin{bmatrix} 1 & 0 \\ 0 & 1 \\ 0 & 0 \end{bmatrix}$$

The representing matrix still has rank 2, so the image of G is again 2-dimensional. Now, however, the range is $\mathbf{R}^3$, so *the image is not the entire range*: G is *not* onto.

On the other hand, it *is* one-to-one. For the nullity—the number of free columns—is now zero. There are no homogeneous generators, so $G^{-1}(\mathbf{0})$ —and hence every pre-image—contains exactly one point. $\square$

The reasoning in these examples applies quite generally. We leave it as an exercise for the reader to prove:

PROPOSITION 1.10. *Let* $T: \mathbf{R}^m \to \mathbf{R}^n$ *be a linear map. Then*

　　a) T *is onto if and only if it has rank* n.

　　b) T *is one-to-one if and only if it has nullity zero.*

– Practice –

426. For each matrix, sketch a diagram like that in Figure 1 depicting the kernel, image, and pre-images of the linear map $T: \mathbf{R}^2 \to \mathbf{R}^2$ it represents.

　a)　$\mathbf{A} = \begin{bmatrix} 1 & 2 \\ 2 & 4 \end{bmatrix}$　　b)　$\mathbf{B} = \begin{bmatrix} 3 & 0 \\ 0 & 0 \end{bmatrix}$　　b)　$\mathbf{B} = \begin{bmatrix} 0 & 2 \\ 0 & 0 \end{bmatrix}$

427. For each matrix below, sketch a diagram like that in Figure 1 depicting the kernel, image, and pre-images of the linear map $T: \mathbf{R}^2 \to \mathbf{R}^2$ it represents. These sketches will be qualitatively different that those in Figure 1 and Exercise 426. Why?

　　a)　$\mathbf{A} = \begin{bmatrix} 1 & 2 \\ 3 & 4 \end{bmatrix}$　　b)　$\mathbf{B} = \begin{bmatrix} 0 & 0 \\ 0 & 0 \end{bmatrix}$

428. Suppose $T: \mathbf{R}^2 \to \mathbf{R}^3$ is a linear transformation, with

$$T(1,0) = (1,-1,0), \quad \text{and} \quad T(0,1) = (0,1,-1)$$

What matrix represents T? What are the kernel and image of T? (Give bases.) Is T one-to-one and/or onto?

429. Suppose $T: \mathbf{R}^3 \to \mathbf{R}^3$ is a linear transformation, with

$$T(1,0,0) = (-1,0), \quad T(0,1,0) = (1,1), \quad \text{and} \quad T(0,0,1) = (1,-1)$$

What matrix represents T? What are the kernel and image of T? (Give bases.) Is T one-to-one and/or onto?

430. Let
$$\mathbf{A} = \begin{bmatrix} 0 & 2 & 0 \\ 0 & 0 & -3 \end{bmatrix}$$

a) Name the domain and range of the linear map T represented by $\mathbf{A}$. What is the image of T? Is T onto? Is it one-to-one?

b) Name the domain and range of the linear map S represented by $\mathbf{B} = \mathbf{A}^T$. What is the image of S? Is S onto? Is it one-to-one?

431. Suppose $(d_1, d_2, d_3) \in \mathbf{R}^3$, and consider the diagonal matrix
$$\mathbf{D} = \begin{bmatrix} d_1 & 0 & 0 \\ 0 & d_2 & 0 \\ 0 & 0 & d_3 \end{bmatrix}$$

Explain why the linear map represented by D will be *both* onto and one-to-one unless one of the d_i's vanishes, in which case it will be *neither* onto *nor* one-to-one.

432. Suppose $T \colon \mathbf{R}^4 \to \mathbf{R}^4$ is the linear map represented by
$$\mathbf{A} = \begin{bmatrix} 1 & 2 & 3 & 4 \\ 5 & 6 & 7 & 8 \\ 9 & 10 & 11 & 12 \\ 13 & 14 & 15 & 16 \end{bmatrix}$$

a) What are the dimensions of the image and kernel of T?

b) Is T one-to-one, onto, both, or neither?

c) How would you describe $T^{-1}(\mathbf{w})$ when $\mathbf{w}$ lies in the image of T? What is $T^{-1}(\mathbf{w})$ when $\mathbf{w}$ does *not* lie in the image of T?

433. Suppose $T \colon \mathbf{R}^3 \to \mathbf{R}^2$ is a linear transformation, with
$$T(-1, 1, 0) = (1, 0)$$
$$T(1, -1, 1) = (1, 1)$$
$$T(0, 1, -1) = (0, 1)$$

a) What matrix represents T?

b) Find the kernel and image of T. Is T one-to-one and/or onto?

434. Prove Proposition 1.10.

435. Prove true, or give a counterexample:

    a) *The transformation represented by a matrix* **A** *is **onto** if and only if the transformation represented by* $\mathbf{A}^T$ *is **one-to-one**.*

    b) *The transformation represented by a matrix* **A** *is **one-to-one** if and only if the transformation represented by* $\mathbf{A}^T$ *is **onto**.*

436. Prove: *Every linear transformation is one-to-one on the row-space.*

More precisely, for any **w** in the image of a transformation T represented by a matrix **A**, the pre-image $T^{-1}(\mathbf{w})$ has at most one point in row(**A**). [*Hint: If there were two such points, their difference would lie in both* row(**A**) *and* nul(**A**).]

437. (*Linear transformations map subspaces to subspaces.*) Suppose $T \colon \mathbf{R}^m \to \mathbf{R}^n$ is a linear map, and $S \subset \mathbf{R}^m$ is a subspace.

    a) Show that T maps S to a *subspace* of $\mathbf{R}^n$. In other words, show that

$$T(S) := \{T(\mathbf{x}) \colon \mathbf{x} \in S\} \subset \mathbf{R}^n$$

    is a subspace.

    b) Similarly, show that T maps *affine* subspaces of $\mathbf{R}^m$ to affine subspaces of $\mathbf{R}^n$ (see Section 8.13, Chapter 5).

2. The Linear Rank Theorem

2.1. The prototypical map. Suppose $r \leq m, n$, and consider the rank-r linear transformation $U_r \colon \mathbf{R}^m \to \mathbf{R}^n$ given by

$$U_r\,(x_1,\, x_2,\, \ldots,\, x_m) = (x_1,\, x_2,\, \ldots,\, x_r,\, 0,\, 0,\, \ldots,\, 0)$$

The first r coordinates of the input become the first r coordinates of the output, which are followed by $n - r \geq 0$ zeros. Applying U_r to the standard basis vectors in the domain $\mathbf{R}^m$, we get

$$U_r(\mathbf{e}_j) = \begin{cases} \mathbf{e}_j & \text{if } j \leq r \\ \mathbf{0} & \text{if } j > r \end{cases}$$

so U_r is represented by the matrix

$$\mathbf{U}_r := \begin{bmatrix} \mathbf{I}_r & 0 \\ \hline 0 & 0 \end{bmatrix}$$

which has $m-r$ *columns* of zeros at the right, and $n-r$ *rows* of zeros at the bottom. This matrix is already in RRE form, with r pivot rows, and $m-r$ free columns, so we have

$$\text{rank}(U_r) = r \quad \text{and} \quad \text{nullity}(U_r) = m - r$$

It now follows from Proposition 1.10 above that U_r is one-to-one or onto if and only if $r = m$ or $r = n$, respectively.

When $r < n$, one or more rows lack pivots, so U_r is *not* onto. Its r-dimensional image—the column-space of the matrix above—is spanned by the *first* r standard basis vectors. Implicitly, it is the perp of the *last* $n - r$ standard basis vectors; it consists of all $\mathbf{y} \in \mathbf{R}^n$ whose last $n - r$ coordinates vanish.

The homogeneous generators for $\mathbf{U}_r$ are easily found to be the *last* $m - r$ standard basis vectors in the *domain* $\mathbf{R}^m$. They span $\ker(U_r)$, which consequently has dimension $m - r$. The kernel can also be described as the perp of the first r standard basis vectors in the domain, consisting of all $\mathbf{x} \in \mathbf{R}^m$ whose first r coordinates vanish.

We depict the situation schematically in Figure 2.

What do the image and pre-images "look like" for U_r?

To answer this, consider any $\mathbf{b} = (b_1, \ldots, b_r, 0, \ldots, 0)$ in the image of U_r. The *pre*-image of $\mathbf{b}$ clearly contains *any* vector in the domain whose first r coordinates are $b_1, b_2, \ldots, b_r$. One such vector is clearly

$$\mathbf{x}_p = (b_1, b_2, \ldots, b_r, 0, 0, \ldots, 0) \in \mathbf{R}^m$$

So by our Structure Theorem for linear pre-images (Theorem 4.1 of Chapter 2) $U_r^{-1}(\mathbf{b})$ is the $(m - r)$-dimensional affine subspace

$$\mathbf{x}_p + \ker(U_r)$$

Note that $\mathbf{U}_r$ is already reduced, and its homogeneous generators are simply the last $m - r$ standard basis vectors in $\mathbf{R}^m$, so we get the following geometric picture:

The domain $\mathbf{R}^m$ is filled by $(m - r)$ dimensional affine subspaces parallel to the "vertical" kernel, the span of the last *$m - r$ standard basis vectors in the domain $\mathbf{R}^m$. Each of these affine subspaces gets sent to*

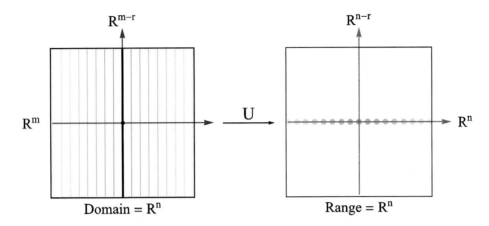

Figure 2. The "prototypical" linear map is neither 1:1 nor onto. The dotted line at right represents the r-dimensional *image* of U in the range $\mathbf{R}^n$. The pre-image of each point (blue dot) on this line is an affine subspace on the left, represented by a vertical line of the same color. The black line through $\mathbf{0}$ in the domain is the kernel.

a different point in the r-dimensional image, which is the span of the first r *standard basis vectors in* $\mathbf{R}^n$.

We do allow the possibility that $r = n$ (so that U is onto), and/or $r = m$, which makes U one-to-one. In the onto case, the dotted line in our range picture fattens out to cover the whole square—in effect, the whole square is filled with colored dots; every point in the entire range has a pre-image.

When $r = m$, the map is one-to-one and the kernel becomes 0-dimensional—a single point. Still, every pre-image is a copy of the kernel, so in this case, *all* pre-images are single points.

If we have *both* $m = n = r$, then U is the simplest rank-r linear isomorphism: the identity map.

Once we understand Figure 2 as allowing all these possibilities, we understand the geometry of *all* linear transformations in a certain sense. The similarity between Figures 1 and 2 is no accident: the Structure and Rank theorems we explain and prove next predict it.

2.2. Rank and structure. The *rank* of a linear transformation is the dimension of its image. Given the dimension of the domain, the rank tells us the dimension of the kernel (by Observation 1.7), and hence that of all pre-images, which are just parallel copies of the kernel: $\ker(T) + \mathbf{x}_p$. A single number—the rank—therefore gives a detailed structural picture of a linear map from $\mathbf{R}^m$ to $\mathbf{R}^n$. The following easy result describes this picture. It holds for *every* linear mapping.

THEOREM 2.3 (Structure theorem for linear maps). *If a linear map* $T \colon \mathbf{R}^m \to \mathbf{R}^n$ *has rank* r, *then*

    a) *The image of* T *is an* r-*dimensional subspace of the range,*

    b) *The kernel of* T *is an* $(m - r)$-*dimensional subspace of the domain,*

    c) *The pre-image of each point in* $\mathrm{im}(T)$ *is a translated copy of* $\ker(T)$, *and*

    d) *These pre-images completely fill the domain.*

PROOF. Conclusion (a) holds by definition of *rank*. Conclusion (b) then follows, because $\mathrm{rank}(T) + \mathrm{nullity}(T) = m$ (Observation 1.7).

If we write $\mathbf{A}$ for the matrix representing T, then Conclusion (c) simply restates something we have known since Chapter 2: The pre-image of any point $\mathbf{b} \in \mathrm{im}(T)$ amounts to nothing more nor less than the solution set of the system

$$\mathbf{A}\mathbf{x} = \mathbf{b}$$

which, by Theorem 4.1 of Chapter 2, comprises all vectors of the form

$$\mathbf{x}_p + \mathrm{nul}(\mathbf{A})$$

Since $\mathrm{nul}(\mathbf{A}) = \ker(T)$, each pre-image is a parallel copy of $\ker(T)$, as Conclusion (c) claims.

Finally, for Conclusion (d), we note that *every* vector $\mathbf{x}$ in the domain maps to *some* $\mathbf{b} \in \mathrm{im}(T)$, and $\mathbf{x}$ then lies in the pre-image of that $\mathbf{b}$. So every $\mathbf{x} \in \mathbf{R}^m$ lies on a pre-image. This shows that (d) holds. $\square$

Roughly speaking, the Structure Theorem says that the geometry of a rank-r linear mapping from $\mathbf{R}^m$, to $\mathbf{R}^n$ always looks much like that of the rank-r archetype $U_r : \mathbf{R}^m \to \mathbf{R}^n$ of Section 2.1 above.

2.4. The Linear Rank Theorem. Theorem 2.5 below makes the universality of our prototypical map U_r above even more precise. It says that if we *precede* T with an isomorphism of the domain, and then *follow it* with an isomorphism of the range, we get *exactly* U_r.

THEOREM 2.5 (Linear Rank Theorem). *Suppose* $\mathbf{A}$ *represents a rank-r linear map* $T: \mathbf{R}^m \to \mathbf{R}^n$. *Then there are invertible matrices* $\mathbf{B}$ *and* $\mathbf{C}$ *for which*

$$\mathbf{C}^{-1}\mathbf{A}\,\mathbf{B} = \mathbf{U_r}$$

PROOF OF RANK THEOREM. Since T has nullity $m-r$, there are $m-r$ nullspace generators $\{\mathbf{h}_1, \ldots, \mathbf{h}_{m-r}\}$ for $\mathrm{nul}(\mathbf{A})$. Take a basis for the *row-space* of $\mathbf{A}$, say $\{\mathbf{u}_1, \ldots, \mathbf{u}_r\}$, and include it with the $\mathbf{h}_i$'s to form the set

$$B = \{\mathbf{u}_1, \ldots, \mathbf{u}_r, \mathbf{h}_1, \ldots, \mathbf{h}_{m-r}\}$$

Since the row-space and nullspace of $\mathbf{A}$ are orthocomplements, and the $\mathbf{u}_i$'s and $\mathbf{h}_j$'s were independent, it is easy to show (Exercise 387) that the union of the two sets is also independent, and hence forms a *basis* for the domain $\mathbf{R}^m$.

Set the vectors in B up as the columns of a matrix $\mathbf{B}$. Since $\mathbf{A}\mathbf{h}_i = \mathbf{0}$ for each $i = 1, 2, \ldots, r$, and matrix multiplication operates column by column, we will then have

$$\mathbf{A}\mathbf{B} = \begin{bmatrix} \mathbf{v}_1 & \cdots & \mathbf{v}_r & 0 & \cdots & 0 \\ \Big| & \cdots & \Big| & \vdots & \cdots & \vdots \\ & \cdots & & 0 & \cdots & 0 \end{bmatrix}$$

where $\mathbf{v}_i = \mathbf{A}\mathbf{u}_i = T(\mathbf{u}_i)$ for $i = 1, 2, \ldots, r$. Since the $\mathbf{u}_i$'s are independent in $\mathrm{row}(\mathbf{A})$, their images, the $\mathbf{v}_i$'s, will be independent too (Exercise 346).

The $\mathbf{v}_i$'s lie in the range $\mathbf{R}^n$ of our map. Augment them with additional independent vectors $\mathbf{v}_{r+1}, \ldots, \mathbf{v}_n$ to get a basis C for $\mathbf{R}^n$. Since $\mathbf{C}^{-1}\mathbf{C} = \mathbf{I}$ and matrix multiplication works column-by-column, we will have

$$\mathbf{C}^{-1} \begin{bmatrix} \mathbf{v}_1 & \cdots & \mathbf{v}_r \\ \Big| & \cdots & \Big| \\ & \cdots & \end{bmatrix} = \begin{bmatrix} \mathbf{I}_r \\ - - - \\ \mathbf{0}_{(n-r)\times r} \end{bmatrix}$$

That is, the first r columns of the $n \times n$ identity matrix. It follows immediately that

$$\mathbf{C}^{-1}\mathbf{A}\,\mathbf{B} = \mathbf{U_r}$$

exactly as desired. $\square$

REMARK 2.6. Notice that if we multiply on the right by $\mathbf{B}^{-1}$ and on the left by $\mathbf{C}$, the identity $\mathbf{C}^{-1}\mathbf{A}\mathbf{B} = \mathbf{U}_r$ turns into

$$\mathbf{C}\,\mathbf{U}_r\mathbf{B}^{-1} = \mathbf{A}$$

This says that any rank-r linear transformation (represented by $\mathbf{A}$) is accomplished by $\mathbf{U}_r$, after we "prepare" the domain (using B^{-1}) and then "adjust" the range (using $\mathbf{C}$).

EXAMPLE 2.7. The matrix

$$\mathbf{A} = \begin{bmatrix} 1 & 0 & -1 & 0 \\ 0 & 1 & 0 & -1 \\ 1 & 2 & -1 & -2 \end{bmatrix}$$

row reduces to

$$\text{RRE}(\mathbf{A}) = \begin{bmatrix} 1 & 0 & -1 & 0 \\ 0 & 1 & 0 & -1 \\ 0 & 0 & 0 & 0 \end{bmatrix}$$

so $\mathbf{A}$ has rank 2. By the Linear Rank Theorem, there should consequently exist invertible matrices $\mathbf{B}$ and $\mathbf{C}$ for which

(49)
$$\mathbf{C}^{-1}\mathbf{A}\,\mathbf{B} = \begin{bmatrix} 1 & 0 & 0 & 0 \\ 0 & 1 & 0 & 0 \\ 0 & 0 & 0 & 0 \end{bmatrix} = \mathbf{U}$$

Indeed, the proof of that Theorem says we can construct $\mathbf{B}$ by making its first two columns a basis for $\text{row}(\mathbf{A})$, and its last two columns a basis for $\text{nul}(\mathbf{A})$. We can thus take the first two rows of $\text{RRE}(\mathbf{A})$ above as the first two columns of $\mathbf{B}$, and the two homogeneous generators as the last two to get

$$\mathbf{B} = \begin{bmatrix} 1 & 0 & 1 & 0 \\ 0 & 1 & 0 & 1 \\ -1 & 0 & 1 & 0 \\ 0 & -1 & 0 & 1 \end{bmatrix}$$

We then compute

$$\mathbf{A}\mathbf{B} = \begin{bmatrix} 2 & 0 & 0 & 0 \\ 0 & 2 & 0 & 0 \\ 2 & 4 & 0 & 0 \end{bmatrix}$$

The vanishing of the last two columns here (and independence of the first two) is exactly as the proof of the Rank Theorem predicts.

To get $\mathbf{C}$, take the first two columns of $\mathbf{AB}$, and augment them with one independent vector since we're trying to construct a basis for the range, $\mathbf{R}^3$. In general, any vector in $\text{nul}\big((\mathbf{AB})^T\big)$ will be independent of $\text{col}(\mathbf{AB})$ since these two subspaces are orthocomplements. So we could row-reduce $(\mathbf{AB})^T$ and take a homogeneous generator as our independent third vector. In the present case however, $(0,0,1)$ is obviously independent of $\text{col}(\mathbf{AB})$, and has the virtue of simplicity. So we take

$$\mathbf{C} = \begin{bmatrix} 2 & 0 & 0 \\ 0 & 2 & 0 \\ 2 & 4 & 1 \end{bmatrix}$$

Computing its inverse, we get

$$\mathbf{C}^{-1} = \begin{bmatrix} \frac{1}{2} & 0 & 0 \\ 0 & \frac{1}{2} & 0 \\ -1 & -2 & 1 \end{bmatrix}$$

Checking the product verifies that as predicted, we have (49):

$$\mathbf{C}^{-1}\mathbf{A}\mathbf{B} = \begin{bmatrix} 1 & 0 & 0 & 0 \\ 0 & 1 & 0 & 0 \\ 0 & 0 & 0 & 0 \end{bmatrix}$$

The isomorphisms $\mathbf{B}$ of the domain, and $\mathbf{C}^{-1}$ of range that we have constructed here distort domain and range in precisely the right way to turn multiplication by $\mathbf{A}$ into the prototypical rank-2 from transformation from $\mathbf{R}^4$ to $\mathbf{R}^3$, namely $U_r(x_1, x_2, x_3, x_4) = (x_1, x_2, 0)$. $\quad\square$

– Practice –

438. Find bases for the kernel and image of this map $T\colon \mathbf{R}^4 \to \mathbf{R}^4$.

$$T(x,y,z,w) = (x - z,\ y - w,\ z - x,\ w - y)$$

439. Find bases for the *orthocomplements* of the kernel and image of the transformation $S\colon \mathbf{R}^4 \to \mathbf{R}^3$ given by

$$S(x,y,z,w) = (x - 2y + z - 2w,\ y - 2z + w - 2x,\ x + y + z + w)$$

440. Suppose $T: \mathbf{R}^6 \to \mathbf{R}^8$ is a linear transformation.

   a) Given no other information about T, how small could the dimension of $\ker(T)$ be?

   b) If the pre-image of one point in $\mathbf{R}^8$ has dimension 3, what is the dimension of $\operatorname{im}(T)$?

   c) Suppose a point $\mathbf{b}$ lies in the image of T, and that $\operatorname{im}(T)$ has codimension 6. What can you say about the pre-image of $\mathbf{b}$? Be as precise as you can.

441. Suppose $T: \mathbf{R}^7 \to \mathbf{R}^5$ is a linear transformation.

   a) Given no other information about T, how large could the codimension of $\ker(T)$ be?

   b) If the pre-image of one point in $\mathbf{R}^5$ has codimension 4, what is the codimension of $\operatorname{im}(T)$?

   c) Suppose a point $\mathbf{b}$ lies in the image of T, and that $\operatorname{im}(T)$ has codimension 3. What can you say about the pre-image of $\mathbf{b}$? Be as precise as you can.

442. Find a linear transformation with the stated properties:

   a) The domain is $\mathbf{R}^4$, the range is $\mathbf{R}^2$, and the kernel is spanned by $(1, 2, -3, 0)$ and $(0, 1, 2, -3)$.

   b) The domain is $\mathbf{R}^2$, the range is $\mathbf{R}^4$, and the image is spanned by $(1, 2, -3, 0)$ and $(0, 1, 2, -3)$.

443. Find a linear transformation with the stated properties:

   a) The domain and range are both $\mathbf{R}^3$, and the kernel is orthogonal to both $(3, -1, 3)$ and $(1, -3, 1)$.

   b) The domain and range are both $\mathbf{R}^3$, and the image is orthogonal to both $(3, -1, 3)$ and $(1, -3, 1)$.

444. Let $T: \mathbf{R}^2 \to \mathbf{R}^2$ be the linear map represented by

$$\mathbf{A} = \begin{bmatrix} 1 & 2 \\ 2 & 4 \end{bmatrix}$$

The image and kernel of T are both lines in $\mathbf{R}^2$. Carefully sketch the image on one set of axes, and the kernel (along with a few non-empty pre-images) on another. Your sketches should agree with the predictions of Theorem 2.3.

445. Let $X \colon \mathbf{R}^2 \to \mathbf{R}^2$ be the linear map represented by

$$\mathbf{X} = \begin{bmatrix} 1 & -1 \\ 2 & -2 \\ 3 & -3 \end{bmatrix}$$

The image and kernel of X are lines in $\mathbf{R}^3$ and $\mathbf{R}^2$, respectively.

a) Find a vector that generates the image line in $\mathbf{R}^3$.

b) Carefully sketch the kernel and several non-empty pre-images in the (x, y)-plane.

446. The matrix

$$\mathbf{A} = \begin{bmatrix} 1 & 2 \\ 2 & 4 \end{bmatrix}$$

has rank 1. According to the Linear Rank Theorem, there consequently exist invertible 3×3 matrices $\mathbf{B}$ and $\mathbf{C}$ for which

$$\mathbf{C}^{-1}\mathbf{A}\,\mathbf{B} = \begin{bmatrix} 1 & 0 \\ 0 & 0 \end{bmatrix}$$

Find matrices $\mathbf{B}$ and $\mathbf{C}$ that accomplish this, and verify that they do. (Compare Example 2.7.)

447. The matrix

$$\mathbf{A} = \begin{bmatrix} 1 & 2 & 3 \\ 4 & 5 & 6 \\ 7 & 8 & 9 \end{bmatrix}$$

has rank 2. According to the Linear Rank Theorem, there consequently exist invertible 3×3 matrices $\mathbf{B}$ and $\mathbf{C}$ for which

$$\mathbf{C}^{-1}\mathbf{A}\,\mathbf{B} = \begin{bmatrix} 1 & 0 & 0 \\ 0 & 1 & 0 \\ 0 & 0 & 0 \end{bmatrix}$$

Find matrices $\mathbf{B}$ and $\mathbf{C}$ that accomplish this, and verify that they do.

448. Suppose $\mathbf{A}$ is an $n \times m$ matrix of rank r.

a) Show that when $\mathbf{B}_{m \times m}$ is invertible, $\mathbf{AB}$ will always have the same rank and nullity as $\mathbf{A}$.

b) Show that when $\mathbf{C}_{n \times n}$ is invertible, $\mathbf{CA}$ will always have the same rank and nullity as $\mathbf{A}$.

c) Deduce from (a) and (b) that when $\mathbf{C}$ and $\mathbf{B}$ are both invertible matrices of the appropriate sizes, $\mathbf{CAB}$ will always have the same rank and nullity as $\mathbf{A}$.

[*Hint: Show that* $\mathrm{col}(\mathbf{AB}) = \mathrm{col}(\mathbf{A})$ *and that* $\mathrm{nul}(\mathbf{CA}) = \mathrm{nul}(\mathbf{A})$. *Be sure to use the invertibility of* $\mathbf{B}$ *and* $\mathbf{C}$. *These statements fail without that.*]

$$- \star -$$

3. Eigenspaces

The Linear Rank Theorem gives us a good structural picture of the general rank-r linear mapping $T \colon \mathbf{R}^m \to \mathbf{R}^n$. We can get a far more detailed picture, however, when domain and range coincide: when $m = n$. For when images and pre-images—outputs and inputs—occupy the *same* numeric vector space, we may assess their geometric relation to each other. For example, we can measure the angle between an input $\mathbf{x}$ and its image $T(\mathbf{x})$. Mappings from $\mathbf{R}^n$ to *itself* thus enjoy extra structure and finer detail. We call such mappings *operators*.

DEFINITION 3.1. Linear mappings that have the *same* domain and range are called **linear operators**. A linear operators T has an **inverse operator** S if and only if ST is the identity map: $S(T(\mathbf{x})) = T(S(\mathbf{x})) = \mathbf{x}$ for every $\mathbf{x} \in \mathbf{R}^n$. □

A linear operator T on $\mathbf{R}^n$ is represented by an $n \times n$ matrix, and of course, the operator is invertible if and only if its representing matrix is invertible. Recall that invertible linear maps are also called *isomorphisms* (Definition 6.14 of Chapter 2).

OBSERVATION 3.2. *If* $\mathbf{T}$ *and* $\mathbf{S}$ *represent operators* T *and* S *on* $\mathbf{R}^n$, *respectively, then* $\mathbf{ST}$ *represents* ST. *If* T *is an isomorphism, then the inverse matrix* $\mathbf{T}^{-1}$ *represents the inverse isomorphism.*

PROOF. Exercise 459. □

The ability to make direct comparisons between images and pre-images allows us, in particular, to ask whether they are ever *collinear*—whether $T(\mathbf{x})$ is ever a scalar multiple of $\mathbf{x}$ so that $T(\mathbf{x}) = \lambda \mathbf{x}$ for some scalar λ. If this happens, and $\mathbf{A}$ is the matrix that represents T, we clearly have $\mathbf{Ax} = \lambda \mathbf{x}$, so that $\mathbf{x}$ is an *eigenvector* of $\mathbf{A}$. In this case we also call it an eigenvector of T. We introduced eigenvalues and eigenvectors of a *matrix* in Definition 1.36 from Chapter 1). The definition below adds the notion of *eigenspace* to that vocabulary.

To prepare for a careful definition, let $I : \mathbf{R}^n \to \mathbf{R}^n$ denote the **identity operator**, given by $I(\mathbf{x}) = \mathbf{x}$ for all $\mathbf{x} \in \mathbf{R}^n$. The identity operator is of course represented by the identity matrix $\mathbf{I}$.

Now notice that whenever $T : \mathbf{R}^n \to \mathbf{R}^n$ is a linear operator, we can construct a new operator for any scalar $\lambda \in \mathbf{R}$, by subtracting λI from T. The resulting operator, $T - \lambda I$ simply applies T, then subtracts off λ times the input:

$$(T - \lambda I)(\mathbf{x}) := T(\mathbf{x}) - \lambda(\mathbf{x})$$

DEFINITION 3.3 (Eigenspace). Suppose $T : \mathbf{R}^n \to \mathbf{R}^n$ is a linear operator. Then for each scalar $\lambda \in \mathbf{R}$, the *kernel* of $T - \lambda I$ is called the λ-**eigenspace** of T. We typically write E_λ for the λ-eigenspace. □

The first thing to note here is that, for each scalar λ, the λ-eigenspace $E_\lambda \subset \mathbf{R}^n$ is a *subspace*. We noted this in Exercise 288 of Chapter 5, but we can now see it as an immediate consequence of our defining E_λ as the *kernel* of $T - \lambda I$. For the kernel of *any* linear map is a subspace (Observation 1.5).

Recall too that every subspace contains the origin. So we always have $\{\mathbf{0}\} \subset E_\lambda$. In fact, for almost any value of λ we might choose, E_λ is *exactly* $\{\mathbf{0}\}$—it is the trivial subspace, and holds no real interest for us.

Only for rare values of λ does E_λ contain *more* than just the origin. In that case, as a subspace of positive dimension, it contains *infinitely many* vectors. Only then, when $\dim(E_\lambda) > 0$, do we find λ interesting.

DEFINITION 3.4 (Eigenvalues and Eigenvectors). When $T : \mathbf{R}^n \to \mathbf{R}^n$ is a linear operator, λ is a scalar, and E_λ has *positive* dimension, we call λ an **eigenvalue** of T, and we call each $\mathbf{v} \in E_\lambda$ an **eigenvector** (or more specifically, a λ-**eigenvector**) of T.

The set of all eigenvalues of T is called the **spectrum** of T, denoted $\mathrm{spec}(T)$. □

While stated somewhat abstractly here, *eigenvalue* and *eigenvector* are actually very geometric—and familiar—concepts. For when $\dim(E_\lambda) > 0$, as discussed above, there are infinitely many vectors $\mathbf{v} \neq \mathbf{0}$ in $E_\lambda = \ker(T - \lambda I)$. Such a vector $\mathbf{v}$ then satisfies $T(\mathbf{v}) - \lambda\mathbf{v} = \mathbf{0}$, or equivalently,

$$T(\mathbf{v}) = \lambda\mathbf{v}$$

In other words, T sends $\mathbf{v} \neq \mathbf{0}$ to *a multiple of itself.* This is *exactly* the condition we used to define eigenvalues and eigenvectors of a *matrix* $\mathbf{A}$ in Definition 1.36 of Chapter 1. There, we wrote

$$\mathbf{A}\mathbf{v} = \lambda\,\mathbf{v}$$

Of course, the two conditions become one when $\mathbf{A}$ represents T. Either way, the geometric meanings of *eigenvalue* and *eigenvector* are plain and simple:

A vector $\mathbf{v} \neq 0$ *in* $\mathbf{R}^n$ *is an* **eigenvector** *of an operator* T *on* $\mathbf{R}^n$ *if and only if* T *maps it to a scalar multiple of itself:* $T(\mathbf{v}) = \lambda\,\mathbf{v}$. *We then call the scalar* λ *an* **eigenvalue** *of* T.

3.5. Kernel as eigenspace. The 0-*eigenspace* E_0 of a linear operator has long been familiar to us: it is none other than the *kernel* of the operator. For,

$$E_0 = \ker(T - 0\,I) = \ker(T)$$

Geometrically, vectors in the kernel are eigenvectors because they get mapped to scalar multiples of themselves: *zero* times themselves.

Note in particular that the presence or absence of the *scalar* 0 in the spectrum tells us whether or not T has a non-trivial kernel—a key piece of information about any linear operator.

Since the kernel of T is the same as the nullspace of the matrix representing it, and a square matrix is invertible if and only if its nullspace is trivial (Theorem 4.4 of Chapter 5), we immediately have:

OBSERVATION 3.6. *The spectrum of an operator contains* $\lambda = 0$ *if and only if the operator is singular.*

Arguably, this makes E_0 the most important eigenspace. It tells us whether the operator is invertible or not. As we shall see, however, other eigenspaces carry rich information too.

REMARK 3.7. While the *scalar* 0 holds great interest as an eigen*value*, the *vector* $\mathbf{0}$ is of virtually **no** interest as an eigen*vector*, because it lies in *every* eigenspace. Indeed,

$$T(\mathbf{0}) = \lambda\,\mathbf{0}$$

for *every* scalar λ. This holds for *every* linear map, so it tells us nothing about T in particular. For this reason, we don't call λ an eigenvalue of T unless $T(\mathbf{v}) = \lambda\,\mathbf{v}$ for some *non-zero* vector $\mathbf{v}$. $\square$

To begin seeing the importance of eigenspaces other than $E_0 = \ker(T)$, and to give all this *eigen*-vocabulary more meaning, we study some examples. They are all rank 2 operators on $\mathbf{R}^2$, so from the perspective of the Linear Rank Theorem, they are all structurally equivalent to the identity map $U_2(\mathbf{x}) = I(\mathbf{x}) = \mathbf{x}$. This isn't very enlightening.

By focusing on their eigenvalues and eigenvectors though, we can detect fundamental geometric differences between them. The number of *independent* eigenvectors for each example will play a special role, and the corresponding eigenvalues reveal an even finer level of detail.

EXAMPLE 3.8 (Homothety). When we multiply every $\mathbf{x} \in \mathbf{R}^n$ by a fixed scalar λ, we get an operator called **homothety** H_λ. Homothety is the simplest kind of linear operator. As a scalar multiple of the identity map, H_λ sends every vector to λ times itself,

$$H_\lambda(\mathbf{x}) = \lambda\,\mathbf{x}$$

so *every* vector is an eigenvector with eigenvalue λ. To put it another way, $H_\lambda - \lambda I = \mathbf{0}$, and hence $E_\lambda = \ker(H_\lambda - \lambda I)$ is *all* of $\mathbf{R}^2$.

In short, H_λ has *one* non-trivial eigenspace, namely E_λ, which is all of $\mathbf{R}^2$, and exactly one eigenvalue, namely λ, so the spectrum of H_λ contains only λ.

$$\operatorname{spec}(H_\lambda) = \{\lambda\}$$

Note too that H_λ is represented by the matrix

$$\lambda\,\mathbf{I} = \begin{bmatrix} \lambda & 0 \\ 0 & \lambda \end{bmatrix}$$

clearly displaying the spectrum along its diagonal.

When, $\lambda = -2$, for instance, Figure 3 depicts the geometric effect of H_λ. There, and in all future diagrams of this type, each point on the figure should be seen as the tip of an arrow emanating from the origin. □

EXAMPLE 3.9 (Diagonal dilatation). The operator on $\mathbf{R}^2$ represented by

$$\mathbf{A} = \begin{bmatrix} 3 & 0 \\ 0 & -1 \end{bmatrix}$$

has two eigenvalues, and the standard basis vectors are both eigenvectors, as we can verify by direct calculation:

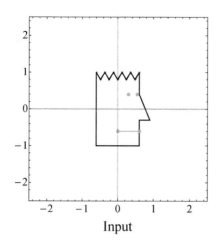

Input

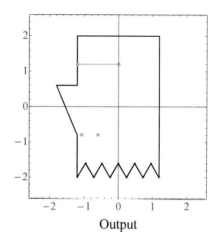

Output

Figure 3. Homothety with $\lambda = -2$. Homothety with factor $\lambda = -2$ magnifies all arrows (and hence any figure) in the plane by a factor of 2, then reflects it through the origin (reverses every vector). *All* vectors are $\lambda = -2$ eigenvectors.

$$\mathbf{A}\,\mathbf{e}_1 = \begin{bmatrix} 3 & 0 \\ 0 & -1 \end{bmatrix} \begin{pmatrix} 1 \\ 0 \end{pmatrix} = \begin{pmatrix} 3 \\ 0 \end{pmatrix} = 3\,\mathbf{e}_1$$

$$\mathbf{A}\,\mathbf{e}_2 = \begin{bmatrix} 3 & 0 \\ 0 & -1 \end{bmatrix} \begin{pmatrix} 0 \\ 1 \end{pmatrix} = \begin{pmatrix} 0 \\ -1 \end{pmatrix} = -\,\mathbf{e}_2$$

Since eigenspaces are *subspaces*, it follows that the entire *line* generated by $\mathbf{e}_1$ —the x-axis—lies in E_3, while similarly, the y-axis lies in E_{-1}. In fact, the eigenspaces E_3 and E_{-1} *are* the x- and y-axes respectively, by an easy calculation (Exercise 457). Equally easily, one can show that these are the *only* eigenspaces, which means

$$\operatorname{spec}(\mathbf{A}) = \{3, -1\}$$

Geometrically, this information paints the picture suggested by Figure 4. The operator represented by $\mathbf{A}$ stretches vectors by a factor of 3 along E_3 (the x-axis) and flips it along E_{-1} (the y-axis).

□

More generally, an operator represented by a diagonal matrix of *any* size behaves very much like the example just given. If λ_i is the ith diagonal entry, the operator scalar multiplies the ith coordinate of every input by λ_i. It thus stretches (when $|\lambda_i| > 0$), shrinks, (when $|\lambda_i| < 0$) or collapses ($\lambda_i = 0$). When $\lambda_i < 0$, it also reverses direction along that axis.

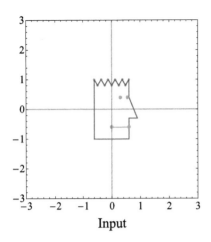
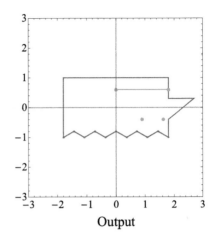

Input   Output

Figure 4. A diagonal map. Diagonal matrices represent linear operators that stretch or shrink the input along the axes.

EXAMPLE 3.10 (Non-diagonal dilatation). Now consider the transformation represented by

$$\mathbf{B} = \begin{bmatrix} 1 & -2 \\ -2 & 1 \end{bmatrix}$$

We claim $\mathbf{B}$ has the same spectrum as $\mathbf{A}$ in the previous example, and like $\mathbf{A}$, has two eigenspaces E_3 and E_{-1} that form mutually perpendicular lines. The difference is that here, *the eigenspaces don't line up with the axes*. Instead, they are generated by $(1,1)$ and $(1,-1)$, as the following calculation reveals:

$$\mathbf{B} \begin{pmatrix} 1 \\ 1 \end{pmatrix} = \begin{bmatrix} 1 & -2 \\ -2 & 1 \end{bmatrix} \begin{pmatrix} 1 \\ 1 \end{pmatrix} = \begin{pmatrix} -1 \\ -1 \end{pmatrix} = - \begin{pmatrix} 1 \\ 1 \end{pmatrix}$$

$$\mathbf{B} \begin{pmatrix} 1 \\ -1 \end{pmatrix} = \begin{bmatrix} 1 & -2 \\ -2 & 1 \end{bmatrix} \begin{pmatrix} 1 \\ -1 \end{pmatrix} = \begin{pmatrix} 3 \\ -3 \end{pmatrix} = 3 \begin{pmatrix} 1 \\ -1 \end{pmatrix}$$

This shows that $\lambda = -1$ and $\lambda = 3$ are eigenvalues, with corresponding eigenvectors $(1,1)$ and $(1,-1)$, respectively. As in the previous example, it follows that the lines they generate form the eigenspaces E_3 and E_{-1} respectively, and that $\mathrm{spec}(\mathbf{B})$ contains no other eigenvalues, so here again,

$$\mathrm{spec}(\mathbf{B}) = \{3, -1\}$$

The geometry of this operator thus mimics that of the diagonal operator in Example 3.9 above, except that the stretching and flipping don't line up with the axes, but rather with the new, "tilted" eigenspaces. For this reason, if we start with the same input figure we used in Figure 4, the geometry is harder to understand (Figure 5).

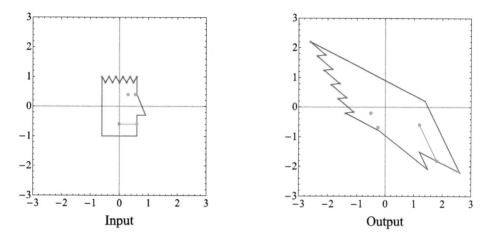

Figure 5. A poor way to picture the effect of **B**. How would you describe the way the input picture gets distorted?

The trick is to *align the input picture with the eigenspaces*. The geometric similarity between **A** and **B** then gets much clearer (Figure 6) as we see the picture stretching by a factor of 3 along the $(1, -1)$ direction, while flipping along the $(1, 1)$ direction.

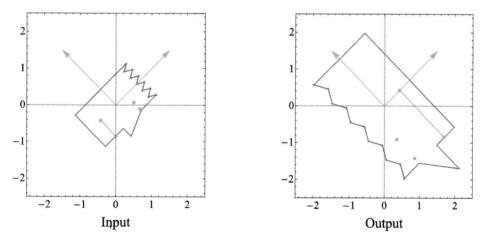

Figure 6. To see the effect of **B** better, first align the input picture with the eigenvector directions (blue arrows). The tripling along the northwest-pointing eigenvector then shows up clearly, as does the flip along the other eigenvector.

□

EXAMPLE 3.11 (Rotation). The transformation represented by

$$\mathbf{R} = \begin{bmatrix} 0 & -1 \\ 1 & 0 \end{bmatrix}$$

has **no** eigenvalues. Indeed, $\mathbf{Rx}$ can *never* be a multiple of a non-zero input $\mathbf{x}$ because $\mathbf{Rx}$ is always *perpendicular* to its input $\mathbf{x}$. To see this, just compute

$$\mathbf{R} \begin{pmatrix} x \\ y \end{pmatrix} = \begin{bmatrix} 0 & -1 \\ 1 & 0 \end{bmatrix} \begin{pmatrix} x \\ y \end{pmatrix} = \begin{pmatrix} -y \\ x \end{pmatrix}$$

This output clearly dots to zero with the input (x, y), showing that for this transformation, output is always orthogonal to—and consequently cannot be a scalar multiple of—the input. Geometrically, $\mathbf{R}$ *rotates* vectors by $\pi/2$ (90°) counterclockwise, as shown in Figure 7, preventing any non-zero input from mapping to a multiple of itself: $\mathbf{R}$ cannot have any eigenvectors. All eigenspaces are trivial and $\mathrm{spec}(\mathbf{R}) = \emptyset$ (the empty set).

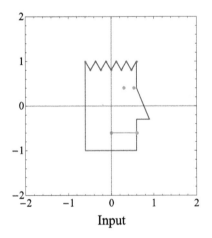
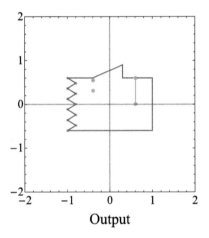

Input   Output

Figure 7. No eigenvalues. An operator that rotates the plane by 90° maps every vector perpendicular to itself, hence has no eigenvalues, and no non-trivial eigenvectors.

□

EXAMPLE 3.12 (Shear). Finally, we consider the operator represented by this triangular matrix:

$$\mathbf{S} = \begin{bmatrix} 1 & 1 \\ 0 & 1 \end{bmatrix}$$

This operator has just one eigenvalue, $\lambda = 1$, and the corresponding eigenspace E_1 turns out to be the x-axis. We can easily verify that E_1 *contains* the x-axis:

$$\mathbf{S} \begin{pmatrix} x \\ 0 \end{pmatrix} = \begin{bmatrix} 1 & 1 \\ 0 & 1 \end{bmatrix} \begin{pmatrix} x \\ 0 \end{pmatrix} = \begin{pmatrix} x \\ 0 \end{pmatrix}$$

For now, we ask the reader to accept our claim that these are actually the *only* eigenvectors, so that

$$\mathrm{spec}(\mathbf{S}) = \{1\}$$

Geometrically, this tells us that $\mathbf{S}$ leaves the x-axis (the line of eigenvectors) fixed (each vector maps to 1 times itself). The rest of the plane, on the other hand contains no eigenvectors: the operator tilts vectors above the x-axis to the right; vectors below it tilt ("shear") to the left. (Figure 8).

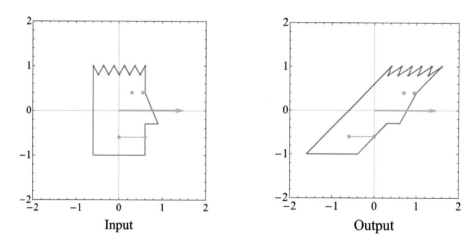

Input   Output

Figure 8. A shear in the plane. A shear preserves the eigenvector direction (blue arrow), while tilting the perpendicular direction.

$\square$

The examples above show that a linear operator on $\mathbf{R}^2$ may have two eigenvalues, one, or none at all. They suggest that different eigenvalues always correspond to linearly independent eigenvectors. Example 3.8 (homothety) shows that a single eigenspace can have dimension bigger than one. The existence of eigenvalues, their signs and magnitudes, and the dimensions of the eigenspaces are details that reveal much about an operator. In the next two sections, we will learn (a) how to find all eigenvalues and eigenspaces, and (b) how to interpret the information they provide.

– Practice –

449. Let T be the linear operator on $\mathbf{R}^2$ represented by this matrix:

$$\begin{bmatrix} -43 & -60 \\ 30 & 42 \end{bmatrix}$$

Show that $(-3, 2)$ and $(-4, 3)$ are eigenvectors of T. Find the corresponding eigenvalues and sketch the corresponding eigenspaces.

450. Let G be the linear operator on $\mathbf{R}^4$ represented by this matrix:

$$\begin{bmatrix} 2 & -1 & 0 & 1 \\ -1 & 2 & 1 & 0 \\ 0 & 1 & 2 & -1 \\ 1 & 0 & -1 & 2 \end{bmatrix}$$

Show that $(1, -1, -1, 1)$, $(0, 1, 0, 1)$, $(1, 0, 1, 0)$, and $(-1, -1, 1, 1)$ are all eigenvectors of T. If every eigenspace of G contains at least one of these vectors, what is $\mathrm{spec}(T)$?

451. Show that the standard basis vectors in $\mathbf{R}^2$ are eigenvectors for the (diagonal) operator represented by

$$\begin{bmatrix} 5 & 0 \\ 0 & -5 \end{bmatrix}$$

What is its spectrum? Sketch a version of Figure 4 to depict the effect of this operator. Your sketch should have the same input figure, but a different output figure.

452. Show that $(2, 1)$ and $(-1, 2)$ are eigenvectors for the operator represented by

$$\begin{bmatrix} 3 & 4 \\ 4 & -3 \end{bmatrix}$$

What is its spectrum? Sketch a version of Figure 6 to depict the effect of this operator. The input and output "faces" should line up with the given eigenvectors, but have the same shape and size as the ones in Exercise 451.

453. Reasoning directly from the definitions (Definition 3.4), find all eigenvalues and eigenvectors of the following operators on $\mathbf{R}^n$?

   a) The *zero* mapping $Z(\mathbf{x}) \equiv \mathbf{0}$

   b) The *identity* mapping $I(\mathbf{x}) = \mathbf{x}$

   c) The rank-r *projection* mapping $P_r : \mathbf{R}^n \to \mathbf{R}^n$ given by

   $$P(x_1, x_2, \ldots, x_n) = (x_1, x_2, \ldots, x_r, 0, \ldots, 0)$$

454. (Compare Exercise 59.) Given non-zero vectors $\mathbf{u} = (u_1, u_2, \ldots u_m)$ and $\mathbf{v} = (v_1, v_2, \ldots, v_n)$ in $\mathbf{R}^m$ and $\mathbf{R}^n$, respectively, consider the rank-1 mapping $T \colon \mathbf{R}^m \to \mathbf{R}^n$ given by

$$T(\mathbf{x}) = \mathbf{u}^*(\mathbf{x})\mathbf{v}$$

   a) Show that $\operatorname{im}(T)$ is the line spanned by $\mathbf{v}$.

   b) Show that $\ker(T)$ is the hyperplane $\mathbf{u}^\perp$.

   c) Show that the only eigenspaces of T are its image and kernel.

   d) What is $\operatorname{spec}(T)$?

While it is instructive to write down the matrix representing T (as in Exercise 59), the Exercise above can be solved by reasoning directly from the formula, without recourse to the matrix.

455. Suppose that $\lambda, \mu \in \operatorname{spec}(T)$ are two *different* eigenvalues of an operator $T : \mathbf{R}^n \to \mathbf{R}^n$. Show that while the λ- and μ-eigenspaces both contain $\mathbf{0}$, they have no other vector in common.

456. Show:

   a) No skew-symmetric 2-by-2 matrix (except $\mathbf{0}_{2\times2}$) has an eigenvalue. [*Hint: Compare Example 3.11.*]

   b) Every skew-symmetric 3-by-3 matrix has an eigenvalue. [*Hint: Does the matrix below have a nullspace?*]

$$\begin{bmatrix} 0 & a & -b \\ -a & 0 & c \\ b & -c & 0 \end{bmatrix}$$

457. Show that the x- and y-axes are the *only* eigenspaces of the matrix $\mathbf{A}$ explored in Example 3.9. (Show that (x, y) never maps to a multiple of itself unless $x = 0$ or $y = 0$.)

458. Use our mantra (Remark 1.22 of Chapter 1) to show that whenever we have an $n \times n$ *diagonal* matrix

$$\mathbf{D} = \begin{bmatrix} \lambda_1 & 0 & \cdot & 0 \\ 0 & \lambda_2 & \cdots & 0 \\ \vdots & \vdots & \ddots & \vdots \\ 0 & 0 & \cdots & \lambda_n \end{bmatrix}$$

each standard basis vector in $\mathbf{R}^n$ will be an eigenvector, with eigenvalue λ_i.

459. Prove Observation 3.2.

460. Suppose λ is an eigenvalue of $\mathbf{A}$, and p is a positive integer. Show that

- a) Any eigenvector of $\mathbf{A}$ is also an eigenvector of $\mathbf{A}^p$ ($\mathbf{A}$ multiplied by itself p times).

- b) λ^p is an eigenvalue of $\mathbf{A}^p$.

- c) Can $\mathbf{A}^2$ have an eigenvector that is *not* an eigenvector of $\mathbf{A}$? Give an example where this happens, or prove that it cannot happen. [*Hint: Consider Example 3.11.*]

$$- \star -$$

4. Eigenvalues and Eigenspaces: Calculation

We now tackle the job of actually *finding* the eigenvalues and non-trivial eigenspaces of a linear operator T on $\mathbf{R}^n$. The strategy we describe is standard and straightforward, though it can be tedious in high dimensions. It has two basic stages:

- 1) First, find the eigenvalues of T. This is a fundamentally *non*-linear problem, and can be difficult. In essence, it asks us to factor a degree-n polynomial.

- 2) Second, for each eigenvalue, find a basis for the corresponding eigenspace. This stage is straightforward, reducing to a familiar task: finding homogeneous generators for a nullspace.

4.1. Characteristic polynomial and eigenvalues. To find the eigenvalues of an operator $T \colon \mathbf{R}^n \to \mathbf{R}^n$, we work with the $n \times n$ matrix $\mathbf{A}$ that represents it. Recall that a scalar λ is an *eigenvalue* of T if and only if

$$(50) \qquad \ker(T - \lambda I) = \mathrm{nul}(\mathbf{A} - \lambda \mathbf{I})$$

has dimension *greater than zero*, which means

$$(51) \qquad (\mathbf{A} - \lambda \mathbf{I})\,\mathbf{x} = \mathbf{0}$$

must have a non-trivial solution.

We know this happens if and only if the coefficient matrix $\mathbf{A} - \lambda \mathbf{I}$ has a free column. But $\mathbf{A} - \lambda \mathbf{I}$ is a *square* matrix, so a free column means $\mathbf{A} - \lambda \mathbf{I}$ is *not* invertible, and hence has determinant zero.

So: *The eigenvalues of T are the scalars λ for which*

$$(52) \qquad \det(\mathbf{A} - \lambda \mathbf{I}) = 0$$

Now it follows quite easily from the general determinant formula (see (33) in Chapter 4, or (62) of Appendix A) that for any $n \times n$ matrix $\mathbf{A}$, the determinant above is a degree-n polynomial in λ. In fact, the left side of (52) can always be expanded as

$$(53) \quad \det(\mathbf{A} - \lambda \mathbf{I})$$
$$= \pm \lambda^n + a_{n-1}\lambda^{n-1} + a_{n-2}\lambda^{n-1} + \cdots + a_1 \lambda + a_0$$

DEFINITION 4.2 (Characteristic polynomial). We call the degree-n polynomial given by (53) the **characteristic polynomial** of T (or of the matrix $\mathbf{A}$ that represents it). Setting it equal to zero, we get the **characteristic equation**

$$\det(\mathbf{A} - \lambda \mathbf{I}) = 0$$

$\square$

A degree-n polynomial has at most n roots.[1] So we may summarize the facts and definitions above as follows:

PROPOSITION 4.3. *The eigenvalues of the operator represented by an $n \times n$ matrix $\mathbf{A}$ are the roots of its characteristic polynomial*

$$\det(\mathbf{A} - \lambda \mathbf{I})$$

which is a degree-n polynomial in λ. A linear operator on $\mathbf{R}^n$ therefore has at most n eigenvalues.

---

[1]This basic fact about polynomials is contained in the so-called Fundamental Theorem of Algebra.

Finding the roots of a degree-n polynomial by hand can be difficult or even impossible. The case $n = 2$, however, arising when we deal with 2×2 matrices, is always manageable, since the characteristic equation in that case is quadratic, and we can either factor it, or if necessary, apply the quadratic formula.

EXAMPLE 4.4. Find the eigenvalues of the linear operator $T \colon \mathbf{R}^2 \to \mathbf{R}^2$ represented by

$$\mathbf{A} = \begin{bmatrix} 2 & -3 \\ -3 & 2 \end{bmatrix}$$

By Proposition 4.3 above, the eigenvalues of T are the roots of the characteristic polynomial

$$\det (\mathbf{A} - \lambda \mathbf{I}) = \begin{vmatrix} 2 - \lambda & -3 \\ -3 & 2 - \lambda \end{vmatrix}$$

$$= (2 - \lambda)^2 - (-3)^2$$

$$= \lambda^2 - 4\lambda - 5$$

$$= (\lambda - 5)(\lambda + 1)$$

The roots (and hence eigenvalues) are now clear: we have $\mathrm{spec}(T) = \{5, -1\}$.

Given the eigenvalues, the eigen*spaces* of T are easy to find, but we temporarily leave that aside. We will return to it in Example 4.9. □

EXAMPLE 4.5. Find the eigenvalues of the linear operator represented by

$$\mathbf{B} = \begin{bmatrix} 2 & 2 \\ 3 & 4 \end{bmatrix}$$

As in the previous example, we start by computing the characteristic polynomial

$$\det\left(\mathbf{B} - \lambda\mathbf{I}\right) = \begin{vmatrix} 2 - \lambda & 2 \\ 3 & 4 - \lambda \end{vmatrix}$$

$$= (2 - \lambda)(4 - \lambda) - 2 \cdot 3$$

$$= \lambda^2 - 6\lambda + 2$$

This characteristic polynomial doesn't have integer roots, so we can't easily factor it like the previous example. It does have two roots, though, since its discriminant is positive:

$$b^2 - 4ac = 6^2 - 4 \cdot 1 \cdot 2 = 28 > 0$$

The eigenvalues of $\mathbf{B}$ are consequently given by the quadratic formula:

$$\lambda = \frac{-b \pm \sqrt{b^2 - 4ac}}{2a} = 3 \pm \sqrt{7}$$

We may therefore conclude that

$$\text{spec}(\mathbf{B}) = \{3 + \sqrt{7},\ 3 - \sqrt{7}\}$$

$\square$

EXAMPLE 4.6. As a final 2×2 example, consider the operator represented by

$$\mathbf{C} = \begin{bmatrix} 2 & -2 \\ 3 & 4 \end{bmatrix}$$

Here, the characteristic polynomial is

$$\det\left(\mathbf{C} - \lambda\mathbf{I}\right) = \begin{vmatrix} 2 - \lambda & -2 \\ 3 & 4 - \lambda \end{vmatrix}$$

$$= (2 - \lambda)(4 - \lambda) + 2 \cdot 3$$

$$= \lambda^2 - 6\lambda + 14$$

This has a *negative* discriminant:

$$b^2 - 4ac = 36 - 56 = -20 < 0$$

The characteristic equation therefore has *no* roots, and hence **C** has no eigenvalues: $\text{spec}(\mathbf{C}) = \emptyset$. □

For a 3×3 matrix, the characteristic polynomial will be cubic. There *is* a formula for the roots of a cubic, but it is difficult to remember and harder to use. The situation for 4×4 matrices is even worse, and for larger matrices, many characteristic polynomials *cannot* be factored exactly by *any* algorithm. Fortunately, approximation methods (Newton's method, for instance) often work well, especially with computer assistance.

Still, we can sometimes factor the characteristic polynomials of larger matrices directly, making it possible to identify their eigenvalues without too much work.

EXAMPLE 4.7. Let us find the spectrum of the operator T on $\mathbf{R}^3$ represented by

$$\mathbf{T} = \begin{bmatrix} 1 & -1 & 0 \\ -1 & 2 & -1 \\ 0 & -1 & 1 \end{bmatrix}$$

It takes some care to find the characteristic polynomial by our usual methods, but it can be done:

$$\det(\mathbf{T} - \lambda\mathbf{I}) = \begin{vmatrix} 1-\lambda & -1 & 0 \\ -1 & 2-\lambda & -1 \\ 0 & -1 & 1-\lambda \end{vmatrix}$$

$$= (1 - \lambda) \begin{vmatrix} 1 & -\frac{1}{1-\lambda} & 0 \\ -1 & 2 - \lambda & -1 \\ 0 & -1 & 1 - \lambda \end{vmatrix}$$

$$= (1 - \lambda) \begin{vmatrix} 1 & -\frac{1}{1-\lambda} & 0 \\ 0 & \frac{\lambda^2 - 3\lambda + 1}{1-\lambda} & -1 \\ 0 & -1 & 1 - \lambda \end{vmatrix}$$

$$= (\lambda^2 - 3\lambda + 1) \begin{vmatrix} 1 & \frac{1}{1-\lambda} & 0 \\ 0 & 1 & -\frac{1-\lambda}{\lambda^2 - 3\lambda + 1} \\ 0 & -1 & 1 - \lambda \end{vmatrix}$$

$$= (\lambda^2 - 3\lambda + 1) \begin{vmatrix} 1 & -\frac{1}{1-\lambda} & 0 \\ 0 & 1 & -\frac{1-\lambda}{\lambda^2 - 3\lambda + 1} \\ 0 & 0 & 1 - \lambda - \frac{1-\lambda}{\lambda^2 - 3\lambda + 1} \end{vmatrix}$$

Since the determinant of a triangular matrix is the product of its diagonal entries, we can now extract the characteristic polynomial of T with just a little algebra:

$$\det (\mathbf{T} - \lambda \mathbf{I}) = (1 - \lambda)(\lambda^2 - 3\lambda)$$

Though cubic, we can factor this by inspection, identify its roots, and deduce that $\text{spec}(T) = \{0, 1, 3\}$.   □

4.8. Finding eigenspaces. Once we actually know an eigenvalue λ, we can find a basis for the eigenspace E_λ (the subspace of all λ-eigenvectors) with relative ease.

Indeed, the λ-eigenspace E_λ of T is by definition the *kernel* of $T - \lambda I$. To get a basis for it, we simply take the matrix $\mathbf{A}$, that represents T, and compute the homogeneous generators for $\mathbf{A} - \lambda \mathbf{I}$.

EXAMPLE 4.9. In Example 4.4, we found that when T is represented by

$$\mathbf{A} = \begin{bmatrix} 2 & -3 \\ -3 & 2 \end{bmatrix}, \quad \text{we have} \quad \text{spec}(T) = \{5, -1\}$$

Let us now find bases for the eigenspaces E_5 and E_{-1} in turn.

$\boxed{E_5:}$ The $\lambda = 5$ eigenspace is the nullspace of

$$\mathbf{A} - 5\mathbf{I} = \begin{bmatrix} 2-5 & -3 \\ -3 & 2-5 \end{bmatrix} = \begin{bmatrix} -3 & -3 \\ -3 & -3 \end{bmatrix}$$

Clearly,

$$\text{RRE}\,(\mathbf{A} - 5\,\mathbf{I}) = \begin{bmatrix} 1 & 1 \\ 0 & 0 \end{bmatrix}$$

has one free column, and the $\lambda = 5$ eigenspace is therefore spanned by the homogeneous generator

$$\mathbf{v}_5 := \begin{pmatrix} -1 \\ 1 \end{pmatrix}$$

$\boxed{E_{-1}:}$ The $\lambda = -1$ eigenspace is the nullspace of

$$\mathbf{A} - \lambda\mathbf{I} = \mathbf{A} + \mathbf{I} = \begin{bmatrix} 2+1 & -3 \\ -3 & 2+1 \end{bmatrix} = \begin{bmatrix} 3 & -3 \\ -3 & 3 \end{bmatrix}$$

so now

$$\text{RRE}\,(\mathbf{A} + \mathbf{I}) = \begin{bmatrix} 1 & -1 \\ 0 & 0 \end{bmatrix}$$

Again we have one free column, so E_{-1} is spanned by the single homogeneous generator

$$\mathbf{v}_{-1} := \begin{pmatrix} 1 \\ 1 \end{pmatrix}$$

These facts reveal the geometry of T: It is just like the diagonal operator

$$\begin{bmatrix} 5 & 0 \\ 0 & -1 \end{bmatrix}$$

except that, instead of magnifying by a factor of 5 along the x-axis and flipping the y-axis, it magnifies by a factor of 5 along the direction spanned by $\mathbf{v}_5 = (-1, 1)$, while flipping the direction spanned by $\mathbf{v}_{-1} = (1, 1)$, as suggested by Figure 9.

$\square$

EXAMPLE 4.10. Find the spectrum and eigenspaces of the operator on $\mathbf{R}^4$ represented by

$$\mathbf{P} = \begin{bmatrix} 0 & 0 & 2 & 0 \\ 0 & 0 & 0 & 2 \\ 2 & 0 & 0 & 0 \\ 0 & 2 & 0 & 0 \end{bmatrix}$$

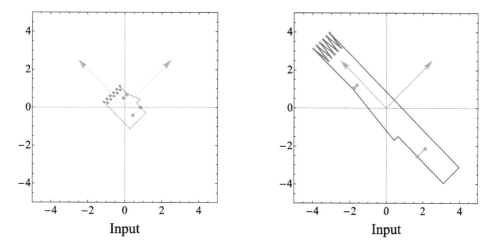

Figure 9. The operator in Examples 4.4 and 4.9 is a dilatation with eigenvalues 5 and −1. Blue arrows show the eigenvector directions.

As always, we first find the eigenvalues, given by the roots of the characteristic polynomial

$$\det\left(\mathbf{P}-\lambda\mathbf{I}\right) = \begin{vmatrix} -\lambda & 0 & 2 & 0 \\ 0 & -\lambda & 0 & 2 \\ 2 & 0 & -\lambda & 0 \\ 0 & 2 & 0 & -\lambda \end{vmatrix}$$

$$= \lambda^2 \begin{vmatrix} 1 & 0 & -2/\lambda & 0 \\ 0 & 1 & 0 & -2/\lambda \\ 2 & 0 & -\lambda & 0 \\ 0 & 2 & 0 & -\lambda \end{vmatrix}$$

$$= \begin{vmatrix} 1 & 0 & -2/\lambda & 0 \\ 0 & 1 & 0 & -2/\lambda \\ 0 & 0 & 4-\lambda^2 & 0 \\ 0 & 0 & 0 & 4-\lambda^2 \end{vmatrix}$$

$$= \left(4-\lambda^2\right)^2$$

Though it has degree 4, this characteristic polynomial is already partially factored, and its roots are clearly ±2. So $\operatorname{spec}(\mathbf{P}) = \{-2, 2\}$. In particular, $\mathbf{P}$ has two non-trivial eigenspaces. Let us find them.

$\boxed{E_2:}$ The $\lambda = 2$ eigenspace is the nullspace of

$$\mathbf{P} - 2\,\mathbf{I} = \begin{bmatrix} -2 & 0 & 2 & 0 \\ 0 & -2 & 0 & 2 \\ 2 & 0 & -2 & 0 \\ 0 & 2 & 0 & -2 \end{bmatrix}$$

This reaches RRE form after just a few operations:

$$\text{RRE}\,(\mathbf{P} - 2\,\mathbf{I}) = \begin{bmatrix} 1 & 0 & -1 & 0 \\ 0 & 1 & 0 & -1 \\ 0 & 0 & 0 & 0 \\ 0 & 0 & 0 & 0 \end{bmatrix}$$

The two free columns reveal that E_2, the nullspace of this matrix, is 2-dimensional—a plane. We easily get a basis for it by writing the homogeneous generators

$$\mathbf{v}_2 = \begin{pmatrix} 1 \\ 0 \\ 1 \\ 0 \end{pmatrix}, \quad \mathbf{w}_2 = \begin{pmatrix} 0 \\ 1 \\ 0 \\ 1 \end{pmatrix}$$

Multiplying these vectors by $\mathbf{P}$ doubles them both (try it), showing that they are indeed $\lambda = 2$ eigenvectors.

$\boxed{E_{-2}:}$ Repeating the same procedure with $\lambda = -2$, we get

$$\text{RRE}\,(\mathbf{P} + 2\,\mathbf{I}) = \begin{bmatrix} 1 & 0 & 1 & 0 \\ 0 & 1 & 0 & 1 \\ 0 & 0 & 0 & 0 \\ 0 & 0 & 0 & 0 \end{bmatrix}$$

Again we find two free columns, so E_{-2} is plane just like E_2, and again we get a basis by writing the homogeneous generators, namely

$$\mathbf{v}_{-2} = \begin{pmatrix} -1 \\ 0 \\ 1 \\ 0 \end{pmatrix}, \quad \mathbf{w}_{-2} = \begin{pmatrix} 0 \\ -1 \\ 0 \\ 1 \end{pmatrix}$$

Multiplication by $\mathbf{P}$ sends each to *minus* its double, confirming that both are $\lambda = -2$ eigenvectors.

Note that the eigenspaces E_2 and E_{-2} here turn out to be orthocomplements of each other. This fact is related to the symmetry of $\mathbf{P}$, as we shall see from the Spectral Theorem 6.14 (next section).   □

We conclude this section by summarizing the procedure for finding the eigenvalues and the eigenspace bases for any linear operator.

4.11. Summary: Finding eigenvalues and eigenvectors. To find the spectrum and eigenspaces of a linear operator T on $\mathbf{R}^n$, we use a 4-step procedure.

1) Find the $n \times n$ matrix $\mathbf{A}$ that represents T.

2) Find the *characteristic polynomial* of T by simplifying

$$\det \mathbf{A} - \lambda \mathbf{I}$$

It will always have degree n.

3) The roots (if any) of the characteristic polynomial form the **eigenvalues** of T. There will be at most n of them.

Note: $\lambda = 0$ *is a root of the characteristic polynomial if and only if T has a non-trivial kernel.*

4) The eigenspace E_λ for each eigenvalue λ is the *nullspace* of $\mathbf{A} - \lambda \mathbf{I}$. We therefore get a basis for E_λ by finding the homogeneous generators of the latter matrix.

– Practice –

461. Find the spectrum of each matrix.

a) $\begin{bmatrix} 3 & 2 \\ 2 & 1 \end{bmatrix}$ b) $\begin{bmatrix} 3 & -2 \\ 2 & 1 \end{bmatrix}$ c) $\begin{bmatrix} 3 & -2 \\ 0 & 3 \end{bmatrix}$

462. Find the spectrum of the operator in Example 4.7 using the 3-by-3 determinant trick from Section 6.5 of Chapter 4.

463. Find the spectrum of each matrix.

a) $\begin{bmatrix} 1 & -2 & 0 \\ -2 & 3 & -2 \\ 0 & -2 & 1 \end{bmatrix}$ b) $\begin{bmatrix} 1 & -2 & 0 \\ 0 & 1 & -2 \\ 0 & -2 & 4 \end{bmatrix}$ c) $\begin{bmatrix} 1 & -2 & 0 \\ 0 & 1 & -2 \\ 0 & 0 & 4 \end{bmatrix}$

464. Compute the dimension of the λ-eigenspace for $\mathbf{A}$ when

a) $\lambda = 3$, $\mathbf{A} = \begin{bmatrix} 2 & 0 & 1 \\ 0 & 3 & 0 \\ 1 & 0 & 2 \end{bmatrix}$ b) $\lambda = -2$, $\mathbf{A} = \begin{bmatrix} -2 & 1 & -3 \\ 0 & -2 & 1 \\ 0 & 0 & -2 \end{bmatrix}$

465. Compute the dimension of the $\lambda = -1$ eigenspace for $\mathbf{A}$ when

a) $\mathbf{A} = \begin{bmatrix} 0 & 0 & 0 & 1 \\ 0 & 0 & 1 & 0 \\ 0 & 1 & 0 & 0 \\ 1 & 0 & 0 & 0 \end{bmatrix}$ b) $\mathbf{A} = \begin{bmatrix} -1 & 1 & 1 & 1 \\ 1 & -1 & 1 & 1 \\ 1 & 1 & -1 & 1 \\ 1 & 1 & 1 & -1 \end{bmatrix}$

466. Find bases for the eigenspaces of each matrix in Exercise 461. In each case, determine whether $\mathbf{R}^2$ has a basis made of eigenvectors for that operator.

467. Find bases for the eigenspaces of each matrix in Exercise 463. In each case, determine whether $\mathbf{R}^{3t}$ has a basis made of eigenvectors for that operator.

468. Find the spectrum of the operator represented by each matrix below, and bases for all eigenspaces.

a) $\begin{bmatrix} 1 & 2 & 0 & 0 \\ 2 & 4 & 0 & 0 \\ 0 & 0 & 1 & 2 \\ 0 & 0 & 2 & 4 \end{bmatrix}$ b) $\begin{bmatrix} 1 & 2 & 0 & 1 \\ 2 & 4 & 0 & 0 \\ 0 & 0 & 1 & 2 \\ 0 & 0 & 2 & 4 \end{bmatrix}$

469. Find bases for all eigenspaces of these triangular matrices:

a) $\begin{bmatrix} 1 & 0 \\ 2 & 3 \end{bmatrix}$ b) $\begin{bmatrix} 1 & 2 & 4 \\ 0 & 4 & 3 \\ 0 & 0 & 5 \end{bmatrix}$ c) $\begin{bmatrix} 1 & -1 & 1 & -1 \\ 0 & 1 & -1 & 1 \\ 0 & 0 & 1 & -1 \\ 0 & 0 & 0 & 1 \end{bmatrix}$

470. Give a proof, if true, or a counterexample, if false: *The eigenvalues of a triangular matrix are the same as its diagonal entries.*

471. Give a proof, if true, or a counterexample, if false: *A square matrix and its transpose always have the same spectrum.*

472. Give examples of:

   a) A 2-by-2 matrix whose spectrum contains -1 and 4.

   b) A non-diagonal 2-by-2 whose spectrum contains -1 and 4.

   c) A 2-by-2 matrix that has $(1, 2)$ as an eigenvector with eigenvalue -1, and $(2, 3)$ as an eigenvector with eigenvalue 4.

473. Give examples of:

   a) A 3-by-3 matrix whose spectrum contains $1, 2$, and 3.

   b) A non-diagonal 3-by-3 whose spectrum contains $1, 2$, and 3.

   c) A 3-by-3 matrix that has $(1, 0, 0)$ as an eigenvector with eigenvalue 1, $(1, -1, 0)$ as an eigenvector with eigenvalue 2, and $(0, 1, -1)$ as an eigenvector with eigenvalue 3.

$$- \star -$$

5. Eigenvalues and Eigenspaces: Similarity

More than any other tool we have, eigenvalues and eigenvectors expose the geometry of linear operators. Every operator on $\mathbf{R}^n$ belongs to one of a relatively small set of geometrically meaningful classes. The most important of these classes can be understood completely through their eigenvalues and eigenspaces.

5.1. Similarity. The classes we have in mind are defined by the mathematical notion of *similarity*, also known as *conjugacy*.

DEFINITION 5.2 (Similarity). Linear operators T and S on $\mathbf{R}^n$ are **similar** or **conjugate**, written $T \sim S$, if there is an isomorphism $B : \mathbf{R}^n \to \mathbf{R}^n$ such that

$$T = B S B^{-1}$$

Equivalently, operators T and S are similar if the matrices $\mathbf{T}$ and $\mathbf{S}$ that represent them satisfy

$$\mathbf{T} = \mathbf{B} \mathbf{S} \mathbf{B}^{-1}$$

for some invertible $n \times n$ matrix $\mathbf{B}$.   □

REMARK 5.3. The operator formulation $T = BSB^{-1}$ is equivalent to the matrix formulation $\mathbf{T} = \mathbf{B}\mathbf{S}\mathbf{B}^{-1}$ by Observation 1.19 and Exercise 262 of Chapter 4. (The Observation asserts that the operator we get by composing two or more other operators (doing them in succession) is represented by the product of their representing matrices. The Exercise says that when $\mathbf{B}$ represents an isomorphism, $\mathbf{B}^{-1}$ represents the inverse isomorphism.) Together, these facts ensure that when $\mathbf{S}$ and $\mathbf{B}$ represent S and B, respectively, $\mathbf{B}\mathbf{S}\mathbf{B}^{-1}$ represents BSB^{-1}. □

Based on the definition above, similarity may seem like a purely algebraic and abstract relationship. But it actually expresses a basic idea that arises in many situations, in and out of mathematics. To convey its meaning, we present two informal metaphors.

Let T represent the operation of *Tom doing his laundry*, while S is that of of *Tom's mother doing her laundry*. These operations are actually "similar" in roughly the same way that *similarity* is defined above: we can relate them to each other by writing

$$T = BSB^{-1}$$

What is B^{-1} here? It is the invertible operation of *delivering Tom's laundry to his mother*. Indeed, if he does B^{-1} in this sense—bring his laundry home to mom—then she does hers (S), and finally, Tom does B—bringing the clean laundry back to his house (the inverse of bringing it to her), the effect is essentially the same as T, that is, Tom doing his own laundry. Symbolically, $BSB^{-1} = T$.

Here's a more geometric example. Tina can draw *vertical* arrows, but no other kind. Call that T. Sue can draw only *horizontal* arrows. Call that S. Here again we can write

$$T = BSB^{-1}$$

this time by taking B^{-1} to be the invertible operation of rotating a sheet of paper by 90°. After all, if we rotate a sheet of paper by 90°, have Sue draw a horizontal arrow on it, then invert—undo—the 90° rotation, we get a vertical arrow, just as if Tina had drawn it.

Now consider an example that models the Tina/Sue story in a linear-algebraic form.

EXAMPLE 5.4. Suppose T and S denote the operators on $\mathbf{R}^2$ represented by

$$\mathbf{T} = \begin{bmatrix} 1 & 0 \\ 0 & 2 \end{bmatrix} \quad \text{and} \quad \mathbf{S} = \begin{bmatrix} 2 & 0 \\ 0 & 1 \end{bmatrix}$$

Geometrically, T stretches the plane by a factor of 2 vertically, leaving horizontal distances unchanged, while S stretches the *horizontal* direction by a factor of 2 while leaving vertical distances fixed.

These operators are clearly different, but when we rotate the plane by $90°$, do S, and then rotate back by the same amount, the result is indistinguishable from T.

We can verify this algebraically, recalling Example 3.11, which shows that rotation by $90°$ (counterclockwise) is the linear operator represented by

$$\mathbf{R} = \begin{bmatrix} 0 & 1 \\ -1 & 0 \end{bmatrix} \quad \text{with inverse} \quad \mathbf{R}^{-1} = \begin{bmatrix} 0 & -1 \\ 1 & 0 \end{bmatrix}$$

A simple calculation (check it!) verifies that $\mathbf{RSR}^{-1}$ indeed equals $\mathbf{T}$:

$$\begin{bmatrix} 0 & 1 \\ -1 & 0 \end{bmatrix} \begin{bmatrix} 2 & 0 \\ 0 & 1 \end{bmatrix} \begin{bmatrix} 0 & -1 \\ 1 & 0 \end{bmatrix} = \begin{bmatrix} 1 & 0 \\ 0 & 2 \end{bmatrix}$$

$\square$

Among the many important features of similarity, we should highlight three right away:

PROPOSITION 5.5. *The similarity relation between operators on* $\mathbf{R}^n$ *has these properties:*

a) *(Reflexivity) Every linear operator is similar to itself.*

b) *(Symmetry) If T is similar to S, then S is similar to T.*

c) *(Transitivity) If T is similar to S and S is similar to R, then T is similar to R.*

PROOF. We prove (b), leaving (a) and (c) to Exercise 481. Let $\mathbf{T}$ and $\mathbf{S}$ be the matrices that represent similar operators T and S, respectively. By the definition of similarity, we can write

$$\mathbf{T} = \mathbf{BSB}^{-1}$$

for some invertible $\mathbf{B}$. Multiply this on the left by $\mathbf{B}^{-1}$ and on the right by $\mathbf{B}$ to get

$$\mathbf{S} = \mathbf{B}^{-1}\mathbf{T}\mathbf{B} = \mathbf{B}^{-1}\mathbf{T}(\mathbf{B}^{-1})^{-1} = \mathbf{CTC}^{-1}$$

where $\mathbf{C} = \mathbf{B}^{-1}$. Since $\mathbf{C}$ is invertible, this makes S is similar to T as claimed. $\square$

DEFINITION 5.6. The set of all linear operators *similar* to a given linear operator T is called the **similarity class** of T. □

The similarity class of T will typically (though not always) contain many operators beside T itself. When it does, all of them will be similar to *each other*, by Proposition 5.5.

Note that different similarity classes don't overlap at all. For if the classes of T and S shared some operator R, then by conclusion (c) of Proposition 5.5, R would be similar to both of them, making them similar to each other (by transitivity), thus forcing S, T, and R all into a *single* class.

What interests us here is that all operators in a given class share certain basic geometric features. To some extent, *if you've seen one (operator in the class), you've seen them all*. This makes it fruitful to seek the defining algebraic and/or geometric traits common to *all* operators in a given class.

To describe the nature of *every* class would take us beyond (but not far beyond) our present abilities. We will therefore content ourselves with understanding the most basic and important classes: those that contain *diagonal* matrices. Our approach focuses on the spectrum, which correlates closely, though not exactly, with similarity classes. For when T and S are similar, they always have the same spectrum; the spectrum is a *similarity invariant*, as we will show in Proposition 5.14 below. The *converse* is not true, however, since operators with the same spectrum are not always similar. Here's an example.

EXAMPLE 5.7. The operators represented by

$$2\mathbf{I} = \begin{bmatrix} 2 & 0 \\ 0 & 2 \end{bmatrix} \quad \text{and} \quad \mathbf{A} = \begin{bmatrix} 2 & 1 \\ 0 & 2 \end{bmatrix}$$

have the same spectrum: $\operatorname{spec}(2\mathbf{I}) = \operatorname{spec}(\mathbf{A}) = \{2\}$, as the reader will easily check. But $2\mathbf{I}$ is *not* similar to $\mathbf{A}$.

Indeed, the similarity class of $2\mathbf{I}$ contains *only* $2\mathbf{I}$. Indeed, any matrix *similar* to $2\mathbf{I}$ must actually *equal* $\mathbf{I}$, because for *any* invertible matrix $\mathbf{B}$, we have

$$\mathbf{B}\,(2\mathbf{I})\,\mathbf{B}^{-1} = 2\mathbf{B}\mathbf{I}\mathbf{B}^{-1} = 2\mathbf{B}\mathbf{B}^{-1} = 2\mathbf{I}$$

So even though both matrices have the same spectrum, $\mathbf{A} \not\sim \mathbf{B}$. For, $\mathbf{A} \neq \mathbf{B}$. □

In the example just given, we showed that $2\mathbf{I}_2$ is the *only* matrix in its similarity class. The argument generalizes easy to higher dimensions and other scalars (Exercise 484).

OBSERVATION 5.8. *The similarity class of* $\lambda \mathbf{I}_n$, *for any* n, *and any scalar* λ, *contains* only $\lambda \mathbf{I}$.

The multiples of $\mathbf{I}$ addressed by Observation 5.8 are the simplest kinds of operators: they just scalar multiply every vector in $\mathbf{R}^n$ by λ. For instance, $2\mathbf{I}$ doubles every vector, while $-(1/4)\mathbf{I}$ shrinks every vector to a quarter of its length and reverses its direction. Such operators are called *homotheties*. We encountered the 2×2 case in Example 3.8.

DEFINITION 5.9 (Homothety). Scalar multiplication by λ in $\mathbf{R}^n$ is represented by the matrix $\lambda \mathbf{I}$, and is called the n-dimensional **homothety with factor** λ.

Observation 5.8 above says that each homothety forms a one-operator similarity class. One can strengthen this statement with a strong converse:

PROPOSITION 5.10. *If the similarity class of* T *contains* T *alone, then* T *is a homothety.*

PROOF. If T is alone in its similarity class, then for *every* invertible $n \times n$ matrix $\mathbf{B}$, we have

$$\mathbf{B}\mathbf{T}\mathbf{B}^{-1} = \mathbf{T} \quad \text{or equivalently,} \quad \mathbf{B}\mathbf{T} - \mathbf{T}\mathbf{B} = 0$$

where $\mathbf{T}$ is the matrix that represents T. This identity holds in particular when $\mathbf{B} = \mathbf{E}_{ij}$, the matrix we get by changing the i, j entry of $\mathbf{0}_{n \times n}$ from a zero to a 1. If one carefully analyzes

$$\mathbf{E}_{ij}\mathbf{T} - \mathbf{T}\mathbf{E}_{ij}$$

however, one finds that it vanishes if and only if $\mathbf{T}$ is a multiple of the identity matrix, as claimed (Exercise 485).   □

Proposition 5.10 tells all there is to know about the *smallest* similarity classes: *The classes that contain only one operator are the homothety classes.*

On the other hand, Example 5.7 and Exercise 484 show that many operators have the same *spectrum* as a homothety without actually being *similar* to it.

Fortunately, this vexing situation, where an operator has the same spectrum as T but is *not* similar to T, can *only* arise when T has *repeated eigenvalues*—a statistically rare occurrence.

To see what we mean by this, recall that T has a particular scalar r as an eigenvalue if and only if r is a root of the characteristic polynomial $\det(T - \lambda I)$. Of course, when r is a root, the characteristic polynomial has a factor of the form $(\lambda - r)^k$ for some integer $k > 0$.

DEFINITION 5.11. An eigenvalue r of T has **algebraic multiplicity** k when the characteristic polynomial has a factor of the form $(\lambda - r)^k$, but not $(\lambda - r)^{k+1}$. When r has multiplicity $k > 1$, we call it a **repeated eigenvalue**. The **geometric** multiplicity of r is the dimension of the r-eigenspace.

Algebraic and geometric multiplicity are both important—and they don't always agree:

EXAMPLE 5.12. The matrices

$$3I = \begin{bmatrix} 3 & 0 \\ 0 & 3 \end{bmatrix} \quad \text{and} \quad A = \begin{bmatrix} 3 & -2 \\ 0 & 3 \end{bmatrix}$$

both have the same characteristic polynomial: $(\lambda - 3)^2$. So 3 is a twice-repeated eigenvalue in both cases. This means $\lambda = 3$ has the *same* algebraic multiplicity, namely 2, for both matrices.

The *geometric* multiplicity of $\lambda = 3$ is *different* in each case, however. When we subtract $\lambda = 3$ off the diagonal of $3I$, we get the zero matrix, whose nullspace is all of $\mathbf{R}^2$. It follows that the 3-eigenspace of $3I$ is $\mathbf{R}^2$, which is 2-dimensional. So for $3I$, the geometric multiplicity of $\lambda = 3$ is *two*.

Constrastingly, the geometric multiplicity of $\lambda = 3$ for A is only *one*, since we only get one free column when we subtract $\lambda = 3$ off the diagonal of A.

Here's another example: Consider the triangular matrix

$$B = \begin{bmatrix} -1 & -1 & 0 & 0 \\ 0 & 5 & 1 & 0 \\ 0 & 0 & 5 & 0 \\ 0 & 0 & 0 & -1 \end{bmatrix}$$

which clearly has characteristic polynomial

$$(\lambda + 1)^2 (\lambda - 5)^2$$

Its spectrum is $\{-1, 5\}$, and *both* eigenvalues are twice-repeated, so both $\lambda = -1$ and $\lambda = 5$ have algebraic multiplicity 2.

They have different geometric multiplicities, on the other hand.

When we subtract -1 and 5, respectively off the diagonal here, we get

$$
\mathbf{B} + \mathbf{I} = \begin{bmatrix} 0 & -1 & 0 & 0 \\ 0 & 6 & 1 & 0 \\ 0 & 0 & 6 & 0 \\ 0 & 0 & 0 & 0 \end{bmatrix} \quad \text{and} \quad \mathbf{B} - 5\mathbf{I} = \begin{bmatrix} -6 & -1 & 0 & 0 \\ 0 & 0 & 1 & 0 \\ 0 & 0 & 0 & 0 \\ 0 & 0 & 0 & -6 \end{bmatrix}
$$

It's easy to see that (after row-reduction) the first will have two free columns, while the second will only have one. It follows that the geometric multiplicity of $\lambda = -1$ is two (equalling its algebraic multiplicity), but the geometric multiplicity of $\lambda = 5$ is only *one*, which is *less* than its algebraic multiplicity.   □

REMARK 5.13. *The geometric multiplicity of an eigenvalue never exceeds its algebraic multiplicity.* We aren't in a good position to prove that here, however.   □

We have now seen that operators with the same spectrum may not be similar, and that the geometric multiplicity of an eigenvalue may be less than its algebraic multiplicity. These facts do make it harder to understand the geometry and similarity classes of linear operators, and we will consider them further in the next section. Let us conclude this one, however, with a useful positive result:

PROPOSITION 5.14. *Similar operators have the same spectrum, and each eigenvalue has the same algebraic and geometric multiplicity in both operators.*

PROOF. We first prove that similarity gives T and S the *same characteristic polynomial*. It follows immediately that they have same spectrum, and that each eigenvalue has the same algebraic multiplicity in both operators.

As usual, we employ the matrices $\mathbf{T}$ and $\mathbf{S}$ that represent T and S, respectively. By definition of similarity, there exists some invertible matrix $\mathbf{B}$ such that

(54) $$\mathbf{T} = \mathbf{B}\mathbf{S}\mathbf{B}^{-1}$$

Now use the trivial identity $\mathbf{I} = \mathbf{B}\mathbf{B}^{-1} = \mathbf{B}\mathbf{I}\mathbf{B}^{-1}$ to deduce

$$\det\left(\mathbf{T} - \lambda\,\mathbf{I}\right) - \det\left(\mathbf{B}\mathbf{S}\mathbf{D}^{-1} - \lambda\,\mathbf{I}\right)$$
$$= \det\left(\mathbf{B}\mathbf{S}\mathbf{B}^{-1} - \lambda\,\mathbf{B}\mathbf{I}\mathbf{B}^{-1}\right)$$
$$= \det\left(\mathbf{B}\left(\mathbf{S} - \lambda\,\mathbf{I}\right)\mathbf{B}^{-1}\right)$$

To finish the calculation, recall that the determinant of a product is the product of the determinants, and that determinants of inverse matrices are reciprocals (Theorem 6.10 and Corollary 6.12 of Chapter 4). So

$$\det\left(\mathbf{T} - \lambda\,\mathbf{I}\right) = \det\left(\mathbf{B}\right)\det\left(\mathbf{S} - \lambda\,\mathbf{I}\right)\det\left(\mathbf{B}^{-1}\right)$$
$$= \det\left(\mathbf{B}\right)\det\left(\mathbf{B}^{-1}\right)\det\left(\mathbf{S} - \lambda\,\mathbf{I}\right)$$
$$= \det\left(\mathbf{S} - \lambda\,\mathbf{I}\right)$$

In short $\mathbf{T}$ and $\mathbf{S}$ (hence T and S) have the same characteristic polynomials. Since the characteristic polynomial determines both its roots (the eigenvalues), and their algebraic multiplicity, T and S have the same eigenvalues with the same algebraic multiplicities, as claimed.

To show they have the same *geometric* multiplicities, take any eigenvalue λ of $\mathbf{S}$, and let $\mathbf{v}$ be any λ-eigenvector of $\mathbf{S}$. Then $\mathbf{B}\mathbf{v}$ is an λ-eigenvalue of $\mathbf{T}$, as follows by rewriting (54) as $\mathbf{T}\mathbf{B} = \mathbf{B}\mathbf{S}$ and then computing

$$\mathbf{T}\mathbf{B}\mathbf{v} = \mathbf{B}\mathbf{S}\mathbf{v} = \mathbf{B}(\lambda\mathbf{v}) = \lambda\,\mathbf{B}\mathbf{v}$$

Thus, $\mathbf{T}(\mathbf{B}\mathbf{v}) = \lambda\,\mathbf{B}\mathbf{v}$, and $\mathbf{B}\mathbf{v}$ is an λ-eigenvector of $\mathbf{T}$. Applying this to a *basis* of λ-eigenvectors for $\mathbf{S}$, we see—by Exercise 346(b) and the invertibility of $\mathbf{B}$—that the dimension of the λ-eigenspace of $\mathbf{T}$ is at *least* as big as that of the λ-eigenspace of $\mathbf{S}$. The geometric multiplicity of λ in T is thus greater than or equal to that of λ in $\mathbf{S}$.

This whole argument is reversible, however, since $\mathbf{B}^{-1}$ similarly maps any λ-eigenvector of $\mathbf{T}$ to one for $\mathbf{S}$. So the inequality goes the other way too: the geometric multiplicity of λ in S is at greater than or equal to that of λ in $\mathbf{T}$.

Both inequalities can hold only if the geometric multiplicity of λ is the same for both operators, just as we hoped to prove. □

The contrapositive version of Proposition 5.14 tells us that operators with different spectra *cannot* be similar.

EXAMPLE 5.15. The operators represented by

$$\mathbf{A} = \begin{bmatrix} 11 & -4 \\ 12 & -3 \end{bmatrix} \quad \text{and} \quad \mathbf{B} = \begin{bmatrix} 4 & -6 \\ 18 & -7 \end{bmatrix}$$

are *not* similar, as we can check by computing their characteristic polynomials:

$$\det(\mathbf{A} - \lambda) = (11 - \lambda)(-3 - \lambda) + 4 \cdot 12 = \lambda^2 - 8\lambda + 15$$
$$\det(\mathbf{B} - \lambda) = (4 - \lambda)(-7 - \lambda) + 6 \cdot 18 = \lambda^2 + 7\lambda + 10$$

Different characteristic polynomials means different eigenvalues and/or algebraic multiplicities. That rules out similarity, by Proposition 5.14.

EXAMPLE 5.16. The operators represented by

$$\mathbf{A} = \begin{bmatrix} 2 & 1 & 0 \\ 0 & 2 & 1 \\ 0 & 0 & 2 \end{bmatrix} \quad \text{and} \quad \mathbf{B} = \begin{bmatrix} 2 & 1 & 1 \\ 0 & 2 & 0 \\ 0 & 0 & 2 \end{bmatrix}$$

are *not* similar, even though they have the *same* characteristic polynomial, namely $(\lambda - 2)^3$, as can be seen by inspection. Both $\mathbf{A}$ and $\mathbf{B}$ have the same single eigenvalue $\lambda = 2$, with algebraic multiplicity three in each case. But to check *geometric* multiplicities, we compute

$$\mathbf{A} - 2\mathbf{I} = \begin{bmatrix} 0 & 1 & 0 \\ 0 & 0 & 1 \\ 0 & 0 & 0 \end{bmatrix} \quad \text{and} \quad \mathbf{B} - 2\mathbf{I} = \begin{bmatrix} 0 & 1 & 1 \\ 0 & 0 & 0 \\ 0 & 0 & 0 \end{bmatrix}$$

We see that $\mathbf{A} - 2\mathbf{I}$ has *one* free column, while $\mathbf{B} - 2\mathbf{I}$ has *two*. This tells us the geometric multiplicities of $\lambda = 2$ for the two matrices are 1 and 2, respectively—they are *different*. The contrapositive of Proposition 5.14 now applies, and we conclude that $\mathbf{A} \not\sim \mathbf{B}$.

– **Practice** –

474. By conjugating with the matrix

$$\mathbf{S} = \mathbf{S}^{-1} = \begin{bmatrix} 0 & 1 \\ 1 & 0 \end{bmatrix}$$

Show that

$$\begin{bmatrix} 1 & 0 \\ 0 & -1 \end{bmatrix} \sim \begin{bmatrix} -1 & 0 \\ 0 & 1 \end{bmatrix}$$

475. By conjugating with the matrix

$$\mathbf{P} = \begin{bmatrix} 0 & 1 & 0 \\ 0 & 0 & 1 \\ 1 & 0 & 0 \end{bmatrix}$$

Show that

$$\begin{bmatrix} 1 & 0 & 0 \\ 0 & 2 & 0 \\ 0 & 0 & 3 \end{bmatrix} \sim \begin{bmatrix} 2 & 0 & 0 \\ 0 & 3 & 0 \\ 0 & 0 & 1 \end{bmatrix}$$

476. Find all eigenvalues, along with their algebraic and geometric multiplicities:

a) $\begin{bmatrix} 1 & 1 \\ 1 & 1 \end{bmatrix}$ b) $\begin{bmatrix} 1 & 1 \\ 0 & 1 \end{bmatrix}$ c) $\begin{bmatrix} 1 & 0 \\ 0 & 1 \end{bmatrix}$ d) $\begin{bmatrix} 1 & 1 \\ 0 & 0 \end{bmatrix}$

477. Find all eigenvalues, along with their algebraic and geometric multiplicities:

a) $\begin{bmatrix} 3 & 2 & 0 \\ 0 & 3 & 2 \\ 0 & 0 & 3 \end{bmatrix}$ b) $\begin{bmatrix} 3 & 2 & 0 \\ 0 & 3 & 0 \\ 0 & 0 & 3 \end{bmatrix}$ c) $\begin{bmatrix} 3 & 3 & 0 \\ 0 & 3 & 0 \\ 0 & 0 & 2 \end{bmatrix}$

478. Show that

$$\begin{bmatrix} 1 & 2 \\ 3 & 4 \end{bmatrix} \not\sim \begin{bmatrix} 1 & 2 \\ 4 & 3 \end{bmatrix}$$

479. Show that

$$\begin{bmatrix} 2 & 0 & 0 \\ 3 & -1 & 3 \\ 0 & 0 & 2 \end{bmatrix} \not\sim \begin{bmatrix} 2 & 0 & 0 \\ 3 & -1 & 0 \\ 0 & 0 & -1 \end{bmatrix}$$

480. Show that these two matrices have the same characteristic polynomial, but even so, are *not* similar.

$$\begin{bmatrix} 2 & 0 & -1 \\ 0 & 2 & -1 \\ 0 & 0 & -1 \end{bmatrix}, \quad \begin{bmatrix} 2 & -1 & 0 \\ 0 & 2 & -1 \\ 0 & 0 & -1 \end{bmatrix}$$

481. Complete the proof of Proposition 5.5 by showing that similarity is reflexive and transitive as explained in conclusions (a) and (c) of that Proposition.

482. Show that a non-trivial 2-by-2 symmetric matrix can never be similar to a non-trivial *skew*-symmetric matrix. [*Hint: Show they can never have the same characteristic equation.*]

483. A square matrix $\mathbf{A}$ and its transpose always have the same spectrum (compare Exercise 471).

   a) If $\lambda \in \mathrm{spec}(\mathbf{A})$, does λ have the same algebraic multiplicity in $\mathbf{A}$ and $\mathbf{A}^T$?

   b) f $\lambda \in \mathrm{spec}(\mathbf{A})$, does λ have the same geometric multiplicity in $\mathbf{A}$ and $\mathbf{A}^T$?

[*Hint: What do we know about determinants and ranks of transposes?*]

484. Prove Observation 5.8. Then find an $n \times n$ matrix $\mathbf{T}$ with $\mathrm{spec}(\mathbf{T}) = \{\lambda\} = \mathrm{spec}(\lambda \mathbf{I})$, but which is *not* similar to $\lambda \mathbf{I}$. [*Hint: Imitate Example 5.7.*]

485. (From the proof of Proposition 5.10.) Let $\mathbf{E}_{ij}$ be the matrix we get by changing the i, j entry of $\mathbf{0}_{n \times n}$ from a zero to a 1. Suppose $\mathbf{T} = [t_{ij}]$ is another $n \times n$ matrix, and for every $1 \leq i, j \leq n$, we have

(55) $$\mathbf{E}_{ij}\mathbf{T} - \mathbf{T}\mathbf{E}_{ij} = \mathbf{0}$$

Let $\mathbf{e}_i$ denote the ith standard basis vector in $\mathbf{R}^n$.

a) Show that for any $1 \leq i \leq n$, we have

$$\mathbf{TE}_{i1}\,\mathbf{e}_1 = \mathbf{T}\,\mathbf{e}_i$$

b) Show that for any $1 \leq i \leq n$, we have

$$\mathbf{E}_{i1}\mathbf{T}\,\mathbf{e}_1 = t_{11}\mathbf{e}_i$$

c) Use (a), (b), and (55) to deduce that $\mathbf{T} = t_{11}\mathbf{I}$, so that $\mathbf{T}$ represents a homothety.

[*Hint: Recall Theorem 5.6 of Chapter 1.*]

6. Diagonalizability and the Spectral Theorem

After homotheties, the simplest similarity classes are those that contain more general diagonal matrices. Operators in these classes are called *diagonalizable*.

DEFINITION 6.1. An operator is **diagonal** if it is represented by a diagonal matrix. It is **diagonalizable** if it is *similar* to a diagonal operator.   □

Most *diagonalizable* operators are not actually *diagonal*.

EXAMPLE 6.2. The matrix

$$\mathbf{A} = \begin{bmatrix} 2 & 3 \\ 5 & 0 \end{bmatrix}$$

is certainly not diagonal, but it *is* diagonalizable, for it is similar to

$$\mathbf{D} = \begin{bmatrix} 5 & 0 \\ 0 & -3 \end{bmatrix}$$

Indeed, $\mathbf{A} = \mathbf{BDB}^{-1}$, with

$$\mathbf{B} = \begin{bmatrix} 1 & -3 \\ 1 & 5 \end{bmatrix}, \quad \text{and} \quad \mathbf{B}^{-1} = \frac{1}{8}\begin{bmatrix} 5 & 3 \\ -1 & 1 \end{bmatrix}$$

Geometrically, the eigenspaces of $\mathbf{D}$ are just the x- and y-axes, so $\mathbf{D}$ magnifies along the x-axis by a factor of 5, while along the y-axis, it multiplies by -3.

Because $\mathbf{A}$ is *similar* to $\mathbf{D}$ it *also* multiplies by 5 and -3 in independent directions (those of the corresponding eigenvectors).   □

This example illustrates the geometric nature of diagonalizability, but how did we find the invertible matrix $\mathbf{B}$ that realized the precise algebraic similarity to a diagonal matrix? More generally, we face three obvious questions:

- *How can one tell whether an operator is diagonalizable?*
- *If it is diagonalizable, to what diagonal operator is it similar?*
- *Finally, how can we find an invertible matrix $\mathbf{B}$ that will realize the similarity?*

A single theorem settles all three questions:

THEOREM 6.3 (Main Diagonalizability test). *A linear operator T on $\mathbf{R}^n$ is diagonalizable if and only if it has n independent eigenvectors* $\mathbf{v}_1, \mathbf{v}_2, \ldots, \mathbf{v}_n$. *If so, it is similar to the diagonal matrix*

$$\mathbf{D} = \begin{bmatrix} \lambda_1 & 0 & \cdot & 0 \\ 0 & \lambda_2 & \cdots & 0 \\ \vdots & \vdots & \ddots & \vdots \\ 0 & 0 & \cdots & \lambda_n \end{bmatrix}$$

Each λ_i is the eigenvalue of the eigenvector $\mathbf{v}_i$, and we have the similarity

$$\mathbf{T} = \mathbf{B}\mathbf{D}\mathbf{B}^{-1} \quad (or \ \mathbf{D} = \mathbf{B}^{-1}\mathbf{T}\mathbf{B})$$

where $\mathbf{B}$ is the matrix whose jth column is $\mathbf{v}_j$:

$$\mathbf{B} = \begin{bmatrix} \mathbf{v}_1 & \mathbf{v}_2 & \cdots & \mathbf{v}_n \\ | & | & \cdots & | \\ | & | & \cdots & | \end{bmatrix}$$

PROOF. The proof is surprisingly short, and once one sees how it works, the result is almost obvious.

The key is to consider the effect of $\mathbf{T}$ on the matrix $\mathbf{B}$. The columns of $\mathbf{B}$ are eigenvectors, and matrix multiplication operates column-by-column, so we immediately have

$$\mathbf{T}\mathbf{B} = \begin{bmatrix} \lambda_1\mathbf{v}_1 & \lambda_2\mathbf{v}_2 & \cdots & \lambda_n\mathbf{v}_n \\ | & | & \cdots & | \\ | & | & \cdots & | \end{bmatrix}$$

On the other hand, it follows just as easily from the column description of matrix/vector multiplication and that of matrix multiplication that we get the same thing when we *right*-multiply $\mathbf{B}$ by the diagonal eigenvalue matrix $\mathbf{D}$:

$$
\mathbf{BD} =
\begin{bmatrix}
\mathbf{v}_1 & \mathbf{v}_2 & \cdots & \mathbf{v}_n \\
| & | & \cdots & | \\
& & \cdots &
\end{bmatrix}
\begin{bmatrix}
\lambda_1 & 0 & \cdot & 0 \\
0 & \lambda_2 & \cdots & 0 \\
\vdots & \vdots & \ddots & \vdots \\
0 & 0 & \cdots & \lambda_n
\end{bmatrix}
$$

$$
=
\begin{bmatrix}
\lambda_1\mathbf{v}_1 & \lambda_2\mathbf{v}_2 & \cdots & \lambda_n\mathbf{v}_n \\
| & | & \cdots & | \\
& & \cdots &
\end{bmatrix}
$$

In short, we easily get $\mathbf{TB} = \mathbf{BD}$, and if we right-multiply both sides of this equation by $\mathbf{B}^{-1}$, we get $\mathbf{T} = \mathbf{BDB}^{-1}$, just as the Theorem claims. □

EXAMPLE 6.4. The Main Diagonalizability result (just proven) shows how to get the similarity presented in Example 6.2. For if we compute the characteristic polynomial of the matrix

$$
\mathbf{A} =
\begin{bmatrix}
2 & 3 \\
5 & 0
\end{bmatrix}
$$

there, we get $\lambda^2 - 2\lambda - 15 = (\lambda - 5)(\lambda + 3)$, and hence $\mathrm{spec}(\mathbf{A}) = \{5, -3\}$.

Computing eigenspaces for these eigenvalues, one easily finds that $\mathbf{v}_1 = (1, 1)$ generates the 5-eigenspace, while $\mathbf{v}_2 = (-3, 5)$ generates the -3-eigenspace. As $\mathbf{v}_1$ and $\mathbf{v}_2$ are independent, they form a basis for $\mathbf{R}^2$. Theorem 6.3 immediately tells us (i) that $\mathbf{A}$ is diagonalizable, (ii) that it is similar to

$$
\mathbf{D} =
\begin{bmatrix}
5 & 0 \\
0 & -3
\end{bmatrix}
$$

and (iii) that $\mathbf{A} = \mathbf{BDB}^{-1}$, where

$$
\mathbf{B} =
\begin{bmatrix}
1 & -3 \\
1 & 5
\end{bmatrix}
$$

□

EXAMPLE 6.5. We can use Theorem 6.3 to conclude that *neither* matrix below is diagonalizable:

$$\mathbf{X} = \begin{bmatrix} 0 & 0 & -3 \\ 0 & 2 & 0 \\ 3 & 0 & 0 \end{bmatrix}, \qquad \mathbf{Y} = \begin{bmatrix} 3 & 0 & 1 \\ 0 & 3 & 0 \\ 0 & 0 & 3 \end{bmatrix}$$

To see that $\mathbf{X}$ isn't diagonalizable, compute its characteristic polynomial:

$$(\lambda^2 + 9)(2 - \lambda)$$

There is only one root: $\lambda = 2$. Subtracting $2\mathbf{I}$ from $\mathbf{X}$, we find that the $\lambda = 2$ eigenspace is spanned by $(0, 1, 0)$. Alas, we get only one independent eigenvector. Theorem 6.3 says that for diagonalizability, we need three.

The characteristic polynomial for $\mathbf{Y}$ is obvious, and more promising: $(\lambda - 3)^3$. There is only one eigenvalue, but it has algebraic multiplicity three, and we can hope for geometric multiplicity three as well—which would give us three independent eigenvectors. Unfortunately though, when we subtract 3 off the diagonal, we only get two free columns, hence a 2-dimensional eigenspace. As there is no other eigenspace, we again fall short of the three independent eigenvectors stipulated by Theorem 6.3, and hence $\mathbf{Y}$ is not diagonalizable. □

EXAMPLE 6.6. The vectors

$$\mathbf{v}_1 = (0, -1, 1), \ \mathbf{v}_2 = (-1, 1, 1), \ \text{and} \ \mathbf{v}_3 = (2, 1, 1)$$

are independent, as we can see by computing the determinant of the matrix having the $\mathbf{v}_j$'s as columns: it is $-6 \neq 0$.

They are also eigenvectors of the operator represented by

$$\mathbf{T} = \begin{bmatrix} -2 & 1 & 1 \\ 1 & 1 & -4 \\ 1 & -4 & 1 \end{bmatrix}$$

Indeed, one easily computes

$$\mathbf{T}\mathbf{v}_1 = (0, -5, 5) = 5\mathbf{v}_1$$
$$\mathbf{T}\mathbf{v}_2 = (4, -4, -4) = -4\mathbf{v}_2$$
$$\mathbf{T}\mathbf{v}_3 = (-2, -1, -1) = -\mathbf{v}_3$$

So, $\mathbf{T}$ has 3 independent eigenvectors, with eigenvalues $5, -4$, and -1, respectively. It is thus diagonalizable, and by Theorem 6.3, similar to

$$\mathbf{D} = \begin{bmatrix} 5 & 0 & 0 \\ 0 & -4 & 0 \\ 0 & 0 & -1 \end{bmatrix}$$

$\square$

6.7. "Distinct eigenvalue" test. Theorem 6.3 is a *sure* test for diagonalizability—an *if and only if* statement. It can be difficult to apply, however, because it can't certify diagonalizability of an operator on $\mathbf{R}^n$, until we know all its eigenspaces. For $n > 3$ (and sometimes even $n = 3$), finding all eigenspaces (or even just the eigenvalues) can be very hard.

We therefore present two weaker tests for diagonalizability. They are easier to apply, but not *equivalent* to diagonalizability: A matrix *is* diagonalizable if it passes either test, but it *may* be diagonalizable even when it fails both.

The first test saves us of the work of computing eigenspaces, though it asks for the full *spectrum* of an operator—knowledge that can be difficult to get, as we have seen.

THEOREM 6.8 (Distinct Eigenvalue test). *A linear operator on* $\mathbf{R}^n$ *with n **different** eigenvalues is diagonalizable.*

PROOF. This is really just a shortcut to Theorem 6.3. To prove it, we show that eigenvectors with different eigenvalues are always independent, so with n different eigenvalues, Theorem 6.3 applies.

Suppose to the contrary, that we had an $n \times n$ matrix $\mathbf{A}$ with a *dependent* set of non-zero eigenvectors

$$V = \{\mathbf{v}_1, \mathbf{v}_2, \ldots, \mathbf{v}_k\}$$

all having *different* eigenvalues $\lambda_1, \lambda_2, \ldots, \lambda_k$. Since V is dependent, some linear combination of its vectors must vanish, by conclusion (f) of Proposition 6.9 of Chapter 5. So for some scalars c_i, we have

$$c_1\mathbf{v}_1 + c_2\mathbf{v}_2 + \cdots + c_k\mathbf{v}_k = \mathbf{0}$$

We may assume each c_i is non-zero—if not, just throw away the ones that have zero coefficients, leaving a smaller dependent set, and call the new set V. In fact, we can assume $c_i = 1$ for each i. If not, just replace $\mathbf{v}_i$ by $c_i\mathbf{v}_i$. These are still eigenvectors, since eigenspaces are closed under scalar multiplication.

Our dependency equation now becomes

$$\mathbf{v}_1 + \mathbf{v}_2 + \cdots + \mathbf{v}_k = \mathbf{0}$$

If we multiply it by λ_1, we get

(56) $\lambda_1\mathbf{v}_1 + \lambda_1\mathbf{v}_2 + \cdots + \lambda_1\mathbf{v}_k = \mathbf{0}$

If instead, we multiply instead by $\mathbf{A}$, we get

(57) $\lambda_1\mathbf{v}_1 + \lambda_2\mathbf{v}_2 + \cdots + \lambda_k\mathbf{v}_k = \mathbf{0}$

Subtract the former from the latter. The $\mathbf{v}_1$ terms cancel, leaving

$$(\lambda_2 - \lambda_1)\mathbf{v}_2 + (\lambda_3 - \lambda_1)\mathbf{v}_3 \cdots + (\lambda_k - \lambda_1)\mathbf{v}_k = \mathbf{0}$$

By hypothesis, the λ_i's are all different, so none of the coefficients here vanish, and by conclusion (f) of Proposition 6.9 again, we now see that

$$V' = \{\mathbf{v}_2, \mathbf{v}_3, \ldots, \mathbf{v}_k\}$$

is dependent.

Now just repeat the argument $k-2$ more times. Each time, we produce a dependent set with one less eigenvector until finally, we conclude that $\{\mathbf{v}_k\}$, by itself, is dependent—absurd, since a single non-zero vector is never dependent. By assuming the Theorem false, we derived a contradiction, so it must be true. $\square$

EXAMPLE 6.9. The eigenvalues of a triangular matrix always equal its diagonal entries (do you see why?). So an $n \times n$ triangular matrix with *distinct* diagonal entries will have n distinct eigenvalues and hence must be diagonalizable by Theorem 6.8. For instance, we can immediately decide that

$$\begin{bmatrix} 1 & 2 & 3 \\ 0 & 4 & 5 \\ 0 & 0 & 6 \end{bmatrix} \quad \text{and} \quad \begin{bmatrix} 1 & 0 & 0 \\ 2 & 3 & 0 \\ 4 & 5 & 6 \end{bmatrix}$$

are diagonalizable by Theorem 6.8. The same Theorem, however, tells us nothing about

$$\begin{bmatrix} 1 & 2 & 3 \\ 0 & 1 & 2 \\ 0 & 0 & 1 \end{bmatrix} \quad \text{and} \quad \begin{bmatrix} 1 & 0 & 0 \\ 2 & 2 & 0 \\ 4 & 2 & 1 \end{bmatrix}$$

As it turns out, at least one of these *is* diagonalizable (Exercise 491), even though both fail to satisfy the hypothesis of Theorem 6.8. $\square$

6.10. The Spectral Theorem. Our final test applies only to symmetric matrices, but it is one of the crowning results of any first course in linear algebra, and one of the most important in all of mathematics.

Recall Definition 2.4 of Chapter 4: $\mathbf{A}$ is *symmetric* if an only if it equals its transpose:

$$\mathbf{A}^T = \mathbf{A}$$

DEFINITION 6.11. We call a linear operator **symmetric** when it is represented by a symmetric matrix.   □

Remarkably, *symmetry alone guarantees diagonalizability.* In fact, symmetry implies even *more* than diagonalizability—it implies *orthogonal* diagonalizability:

DEFINITION 6.12. A linear operator T on $\mathbf{R}^n$ is **orthogonally diagonalizable** when T has n *orthonormal* eigenvectors.

When T has n *independent* eigenvectors, it is diagonalizable by Theorem 6.3, but orthonormality is stronger than independence (Corollary 3.4 of Chapter 6). We shall see that orthogonal diagonalizability is stronger than diagonalizability too.

Recall that when an operator is diagonalizable, it magnifies or shrinks along n independent eigenvector directions. When it is *orthogonally* diagonalizable, these directions are mutually perpendicular—just like the standard coordinate axes. Since the standard axes are the lines along which actual *diagonal* operators do their rescaling, orthogonally diagonalizable operators resemble diagonal operators even more than merely diagonalizable operators do.

EXAMPLE 6.13. The matrices $\mathbf{A}$ and $\mathbf{A}'$ below are both diagonalizable, but only $\mathbf{A}$ is *orthogonally* diagonalizable.

$$\mathbf{A} = \begin{bmatrix} 1 & 2 \\ 2 & 1 \end{bmatrix}, \qquad \mathbf{A}' = \begin{bmatrix} 3 & 0 \\ 4 & -1 \end{bmatrix}$$

Indeed, it is easy to check that $(1,1)$ and $(1,-1)$ are eigenvectors of $\mathbf{A}$, with eigenvalues $\lambda = 3$ and $\lambda = -1$, respectively. Since $(1,1)$ and $(1,-1)$ are *orthogonal*, we can normalize them to get an *orthonormal* basis of eigenvectors, which then form the columns of an *orthogonal* diagonalizing matrix

$$\mathbf{B} = \frac{1}{\sqrt{2}} \begin{bmatrix} 1 & 1 \\ 1 & -1 \end{bmatrix}$$

Because **B** is orthogonal, we can invert it by simply transposing. That makes it easy to check the diagonalization formula for **A**. Indeed, $\mathbf{A} = \mathbf{BDB}^{-1}$ becomes

$$\mathbf{A} = \mathbf{BDB}^T \quad \text{where} \quad \mathbf{D} = \begin{bmatrix} 3 & 0 \\ 0 & -1 \end{bmatrix}$$

This means that **A** is just a "rotated" version of **D**. It stretches by a factor of 3 along the line spanned by $(1,1)$, while it flips the *orthogonal* line spanned by $(1,-1)$. The diagonal matrix **D** does exactly the same thing, except that it stretches by a factor of 3 along the x-axis, while flipping the perpendicular direction—the y-axis.

Our second matrix $\mathbf{A}'$ has the same (distinct) eigenvalues as **A**, ($\lambda = 3, -1$) so it too is similar to **D**.

But the eigenspaces of $\mathbf{A}'$ are spanned by the *non*-orthogonal vectors $(1,1)$ and $(1,0)$. So $\mathbf{A}'$ stretches by a factor of 3 along the line spanned by $(1,1)$ and flips along the line spanned by $(1,0)$, but these lines are *not* perpendicular—they form an angle of 45°.

In short, while **A** and $\mathbf{A}'$ are both similar to **D**, the former is "more similar" than the latter because like **D** (and every other diagonal matrix), *its eigenspaces are mutually perpendicular*. This is the meaning of *orthogonal diagonalizability* (see Figures 10 and 11).

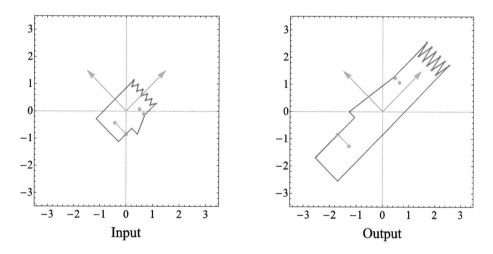

Input   Output

Figure 10. Orthogonal diagonalizability. The symmetric matrix **A** in Example 6.13 above stretches by a factor of 3 along the $(1,1)$ eigenvector, while flipping along the $(-1,1)$ eigenvector.

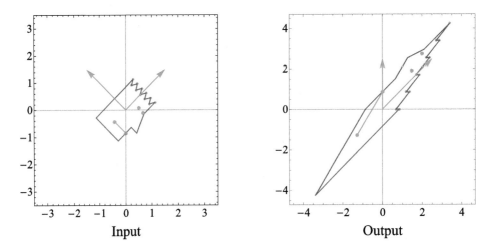

Figure 11. Non-orthogonal diagonalizability. The matrix $\mathbf{A}'$ in Example 6.13 also stretches by a factor of 3 along the $(1, 1)$ eigenvector, but (not being symmetric) it flips along a *non*-perpendicular direction (in this case, vertical), producing a more complicated type of distortion.

Remarkably, an operator is orthogonally diagonalizable *if and only if* it is symmetric. Simple but profound, this fact plays a key role in almost every mathematical subject. Formally,

THEOREM 6.14 (Spectral Theorem). *A linear operator is orthogonally diagonalizable if and only if it is symmetric.*

We can prove this using undergraduate mathematics only, but we defer the proof to Appendix B because it isn't fully linear-algebraic—it requires some basic multivariable calculus. We therefore content ourselves here with examples and discussion.

The Spectral Theorem is fundamental partly because symmetric matrices arise in so many applications of linear algebra. But much more broadly, we shall see in Section 7 that it yields decisive information about *all* linear transformations, even those represented by *non*-symmetric— even non-*square*—matrices. How? Because $\mathbf{A}^T\mathbf{A}$ is always symmetric, even when $\mathbf{A}$ itself is not, so we can always apply the Spectral Theorem to $\mathbf{A}^T\mathbf{A}$. We return to this point in Section 7. For now, we focus on its direct application to symmetric matrices by looking at some examples.

EXAMPLE 6.15. The Spectral Theorem 6.14 lets us see immediately that a complicated matrix like

$$\mathbf{S} = \begin{bmatrix} 1 & 4 & 3 & 0 & 7 \\ 4 & 5 & -2 & 3 & 0 \\ 3 & -2 & -1 & -2 & 3 \\ 0 & 3 & -2 & 2 & -5 \\ 7 & 0 & 3 & -5 & 4 \end{bmatrix}$$

is diagonalizable—even orthogonally diagonalizable—simply because it is symmetric. This is an amazing shortcut, given how difficult it would be to compute any of its eigenvalues precisely. By symmetry alone, however, we may conclude that $\mathbf{S}$ has 5 eigenvalues, and 5 mutually orthogonal directions in $\mathbf{R}^5$ along which it rescales its input by these eigenvalues. $\square$

EXAMPLE 6.16. The matrix

$$\mathbf{A} = \begin{bmatrix} 7 & 6 \\ 6 & -2 \end{bmatrix}$$

is symmetric, so the Spectral Theorem 6.14 predicts an orthogonal (or even orthonormal) basis of eigenvectors.

To find such a basis, we start as usual with the spectrum. The characteristic equation is clearly

$$\begin{aligned} \det\left(\mathbf{A} - \lambda \mathbf{I}\right) &= (7 - \lambda)(-2 - \lambda) - 36 \\ &= \lambda^2 - 5\lambda - 50 \\ &= (\lambda - 10)(\lambda + 5) \end{aligned}$$

Hence $\operatorname{spec}(\mathbf{A}) = \{-5, 10\}$, and we can proceed to find the corresponding eigenvectors:

$\boxed{\lambda = -5}$ To get an eigenvector for $\lambda = -5$, we row-reduce

$$\mathbf{A} + 5\mathbf{I} = \begin{bmatrix} 12 & 6 \\ 6 & 3 \end{bmatrix} \xrightarrow{\text{row-reduce}} \begin{bmatrix} 1 & 1/2 \\ 0 & 0 \end{bmatrix}$$

Reading off the homogeneous generator, we find that $\mathbf{v}_{-5} := (-1/2, 1)$ spans the $\lambda = -2$ eigenspace, or, clearing the denominator, we can take $\mathbf{v}_{-5} = (-1, 2)$.

$\boxed{\lambda = 10}$ Here, we row-reduce

$$\mathbf{A} - 10\mathbf{I} = \begin{bmatrix} -3 & 6 \\ 6 & -12 \end{bmatrix} \xrightarrow{\text{row-reduce}} \begin{bmatrix} 1 & -2 \\ 0 & 0 \end{bmatrix}$$

It follows that $\mathbf{v}_{10} = (2, 1)$ spans the $\lambda = 10$ eigenspace.

Combined, these calculations give us the eigenvector basis

$$\{\mathbf{v}_{-5}, \mathbf{v}_{10}\} - \left\{ \begin{pmatrix} -1 \\ 2 \end{pmatrix}, \begin{pmatrix} 2 \\ 1 \end{pmatrix} \right\}$$

which are orthogonal, as predicted by the Spectral Theorem 6.14. Given the eigenvalues, we also see that geometrically, $\mathbf{A}$ mimics the diagonal matrix

$$\mathbf{D} = \begin{bmatrix} -5 & 0 \\ 0 & 10 \end{bmatrix}$$

except that it rescales by by -5, and 10 along "rotated" axes spanned by $\mathbf{v}_{-5}$ and $\mathbf{v}_{10}$ instead of the standard coordinate axes.

Algebraically, we express this via the (orthogonal) diagonalization formula

$$\mathbf{A} = \mathbf{B}\mathbf{D}\mathbf{B}^T$$

where the columns of $\mathbf{B}$ form an *orthonormal* basis for $\mathbf{R}^2$. Though orthogonal, our vectors $\mathbf{v}_{-5}$ and $\mathbf{v}_{10}$ are not ortho*normal*, so we have to normalize—to divide each one by its length—to get $\mathbf{B}$. Since

$$|\mathbf{v}_{-5}| = |\mathbf{v}_{10}| = \sqrt{5}$$

this gives

$$\mathbf{B} = \frac{1}{\sqrt{5}} \begin{bmatrix} -1 & 2 \\ 2 & 1 \end{bmatrix}$$

Since $\mathbf{B}$ is orthogonal and *symmetric*, it is its own transpose, hence its own inverse, and we have

$$\mathbf{A} = \mathbf{B}\mathbf{D}\mathbf{B} = \frac{1}{5} \begin{bmatrix} -1 & 2 \\ 2 & 1 \end{bmatrix} \begin{bmatrix} -5 & 0 \\ 0 & 10 \end{bmatrix} \begin{bmatrix} -1 & 2 \\ 2 & 1 \end{bmatrix}$$

$\square$

EXAMPLE 6.17. The matrix

$$\mathbf{A} = \begin{bmatrix} 1 & 2 & 1 \\ 2 & 0 & 2 \\ 1 & 2 & 1 \end{bmatrix}$$

is symmetric, so the Spectral Theorem 6.14 predicts an orthogonal (or even orthonormal) basis of eigenvectors. Let us try to find such a basis.

The hardest part, as usual, is to find the spectrum. We leave the reader to show that the characteristic equation is

$$\det(\mathbf{A} - \lambda\mathbf{I}) = -\lambda^3 + 8\lambda + 2\lambda^2$$
$$= -\lambda(\lambda^2 - 2\lambda - 8)$$
$$= -\lambda(\lambda + 2)(\lambda - 4)$$

It follows that $\mathrm{spec}(\mathbf{A}) = \{-2, 0, 4\}$, and we can proceed to find the corresponding eigenvectors:

$\boxed{\lambda = -2}$ To get an eigenvector for $\lambda = -2$, we row-reduce

$$\mathbf{A} - \lambda\mathbf{I} = \mathbf{A} + 2\mathbf{I} = \begin{bmatrix} 3 & 2 & 1 \\ 2 & 2 & 2 \\ 1 & 2 & 3 \end{bmatrix} \xrightarrow[\cdots]{\text{row-reduce}} \begin{bmatrix} 1 & 0 & -1 \\ 0 & 1 & 2 \\ 0 & 0 & 0 \end{bmatrix}$$

Reading off the homogeneous generator, we see that $\mathbf{v}_{-2} := (1, -2, 1)$ spans the $\lambda = -2$ eigenspace.

$\boxed{\lambda = 0}$ The presence of 0 as an eigenvalue tells us that $\mathbf{A}$ has a non-trivial nullspace. To find it, we row-reduce $\mathbf{A}$ itself:

$$\mathbf{A} - \lambda\mathbf{I} = \mathbf{A} = \begin{bmatrix} 1 & 2 & 1 \\ 2 & 0 & 2 \\ 1 & 2 & 1 \end{bmatrix} \xrightarrow[\cdots]{\text{row-reduce}} \begin{bmatrix} 1 & 0 & 1 \\ 0 & 1 & 0 \\ 0 & 0 & 0 \end{bmatrix}$$

It follows that the homogeneous generator $\mathbf{v}_0 := (1, 0, -1)$ spans the $\lambda = 0$ eigenspace, that is, the kernel of the operator.

$\boxed{\lambda = 4}$ Calculating similarly for this last eigenvalue, we get

$$\mathbf{A} - \lambda\mathbf{I} = \mathbf{A} - 4\mathbf{I} = \begin{bmatrix} -3 & 2 & 1 \\ 2 & -4 & 2 \\ 1 & 2 & -3 \end{bmatrix} \xrightarrow[\cdots]{\text{row-reduce}} \begin{bmatrix} 1 & 0 & -1 \\ 0 & 1 & -1 \\ 0 & 0 & 0 \end{bmatrix}$$

and hence $\mathbf{v}_4 := (1, 1, 1)$.

In sum, we have the eigenvector basis

$$\{\mathbf{v}_{-2}, \mathbf{v}_0, \mathbf{v}_4\} = \left\{ \begin{pmatrix} 1 \\ -2 \\ 1 \end{pmatrix}, \begin{pmatrix} 1 \\ 0 \\ -1 \end{pmatrix}, \begin{pmatrix} 1 \\ 1 \\ 1 \end{pmatrix} \right\}$$

As predicted by the Spectral Theorem 6.14, each eigenvector is orthogonal to the other two. Geometrically, $\mathbf{A}$ then mimics the diagonal matrix

$$\mathbf{D} = \begin{bmatrix} -2 & 0 & 0 \\ 0 & 0 & 0 \\ 0 & 0 & 4 \end{bmatrix}$$

except that the scalings by $-2, 0$, and 4 take place along "rotated" axes spanned by $\mathbf{v}_{-2}$, $\mathbf{v}_0$, and $\mathbf{v}_4$ instead of the standard coordinate axes. Algebraically, we express this same fact via the (orthogonal) diagonalization formula

$$\mathbf{A} = \mathbf{BDB}^T$$

where the columns of $\mathbf{B}$ form an *orthonormal* basis for $\mathbf{R}^3$. Our vectors $\mathbf{v}_{-2}, \mathbf{v}_0$ and $\mathbf{v}_4$ above are orthogonal, but we must divide each one by its length to make them orthonormal and get $\mathbf{B}$. Since

$$|\mathbf{v}_{-2}| = \sqrt{6}, \quad |\mathbf{v}_0| = \sqrt{2}, \quad \text{and} \quad |\mathbf{v}_4| = \sqrt{3}$$

this gives

$$\mathbf{B} = \begin{bmatrix} \frac{1}{\sqrt{6}} & \frac{1}{\sqrt{2}} & \frac{1}{\sqrt{3}} \\ \frac{-2}{\sqrt{6}} & 0 & \frac{1}{\sqrt{3}} \\ \frac{1}{\sqrt{6}} & \frac{-1}{\sqrt{2}} & \frac{1}{\sqrt{3}} \end{bmatrix} = \frac{1}{\sqrt{6}} \begin{bmatrix} 1 & \sqrt{3} & \sqrt{2} \\ -2 & 0 & \sqrt{2} \\ 1 & -\sqrt{3} & \sqrt{2} \end{bmatrix}$$

□

Finally we emphasize again that when we have orthogonal diagonalizability—or equivalently, by the Spectral Theorem 6.14, symmetry of the representing matrix—we get an orthonormal basis of eigenvectors. As in the Examples above, we can make the eigenvector basis matrix $\mathbf{B}$ *orthogonal*, and we can invert it simply by transposing. The usual diagonalization equation

$$\mathbf{T} = \mathbf{BDB}^{-1}$$

then reduces to the much simpler $\mathbf{T} = \mathbf{BDB}^T$ when T is symmetric.

– Practice –

486. Determine which matrices below are diagonalizable without trying to diagonalize them. Justify each answer using a diagonalizability test.

a) $\begin{bmatrix} 1 & 2 \\ 3 & 2 \end{bmatrix}$ b) $\begin{bmatrix} 2 & 0 \\ 3 & 2 \end{bmatrix}$ c) $\begin{bmatrix} 2 & 1 \\ 3 & 2 \end{bmatrix}$ d) $\begin{bmatrix} 2 & 0 \\ 0 & 2 \end{bmatrix}$

487. Determine which matrices below are diagonalizable without trying to diagonalize them. Justify each answer using a diagonalizability test.

a) $\begin{bmatrix} 1 & 2 & 3 \\ 3 & 2 & 1 \\ 0 & 0 & 2 \end{bmatrix}$ b) $\begin{bmatrix} 1 & 2 & 3 \\ 3 & 2 & 1 \\ 0 & 0 & 4 \end{bmatrix}$ c) $\begin{bmatrix} 1 & 2 & 3 \\ 2 & 2 & 1 \\ 3 & 1 & 4 \end{bmatrix}$

488. Diagonalize each 2-by-2 matrix $\mathbf{A}$ by finding, in each case, a diagonal matrix $\mathbf{D}$, and an invertible matrix $\mathbf{B}$ such that $\mathbf{A} = \mathbf{BDB}^{-1}$.

a) $\mathbf{A} = \begin{bmatrix} 1 & 2 \\ 1 & 0 \end{bmatrix}$ b) $\mathbf{A} = \begin{bmatrix} 1 & 2 \\ 3 & 6 \end{bmatrix}$ c) $\mathbf{A} = \begin{bmatrix} 3 & 4 \\ 5 & 2 \end{bmatrix}$

489. The matrix $\mathbf{T}$ in Example 6.6 is symmetric. What orthogonal matrix diagonalizes it?

490. Diagonalize each 3-by-3 matrix $\mathbf{A}$ by finding, in each case, a diagonal matrix $\mathbf{D}$, and an invertible matrix $\mathbf{B}$ such that $\mathbf{A} = \mathbf{BDB}^{-1}$.

a) $\mathbf{A} = \begin{bmatrix} 1 & 2 & 0 \\ 0 & 2 & -1 \\ 0 & 0 & 3 \end{bmatrix}$ b) $\mathbf{A} = \begin{bmatrix} 1 & \sqrt{2} & 0 \\ \sqrt{2} & 1 & \sqrt{2} \\ 0 & \sqrt{2} & 1 \end{bmatrix}$

491. Determine which of these matrices from Example 6.9 are diagonalizable.

$\begin{bmatrix} 1 & 2 & 3 \\ 0 & 1 & 2 \\ 0 & 0 & 1 \end{bmatrix}$ and/or $\begin{bmatrix} 1 & 0 & 0 \\ 2 & 2 & 0 \\ 4 & 2 & 1 \end{bmatrix}$

492. A certain operator T on $\mathbf{R}^2$ has $(2,5)$ as an eigenvector, with eigenvalue $\lambda = -1$, and $(1,3)$ as an eigenvector, with eigenvalue $\lambda = 2$.

a) What diagonal matrix (or matrices) represent operators similar to T?

b) What matrix represents T itself?

493. A certain operator S on $\mathbf{R}^3$ has $(3, 2, 1)$ as an eigenvector, with eigenvalue $\lambda = 2$, and $(2, 2, 1)$ as an eigenvector, with eigenvalue $\lambda = 0$, and $(1, 1, 1)$ as an eigenvector with eigenvalue $\lambda = -2$.

    a) What diagonal matrix (or matrices) represent operators similar to S?

    b) What matrix represents S itself?

494. Determine whether each matrix is orthogonally diagonalizable; justify your conclusion. You shouldn't need any calculations:

$$
\text{a)} \begin{bmatrix} 4 & -2 & 7 \\ -2 & 0 & -11 \\ 7 & -11 & 0 \end{bmatrix}
\qquad
\text{b)} \begin{bmatrix} 0 & -1 & 0 \\ 1 & 0 & -1 \\ 0 & 1 & 0 \end{bmatrix}
$$

$$
\text{c)} \begin{bmatrix} 7 & -6 & 5 & -4 \\ -6 & 7 & -4 & 3 \\ 5 & -4 & 7 & 2 \\ -4 & 3 & 2 & 7 \end{bmatrix}
\qquad
\text{d)} \begin{bmatrix} -4 & 5 & -6 & 7 \\ 5 & -4 & 7 & 2 \\ 2 & 7 & -4 & 5 \\ 7 & -6 & 3 & -4 \end{bmatrix}
$$

495. Find an *orthogonal* 2-by-2 matrix that diagonalizes

$$
\mathbf{S} = \begin{bmatrix} 7 & 6 \\ 6 & -2 \end{bmatrix}
$$

496. Find an *orthogonal* 3-by-3 matrix that diagonalizes

$$
\mathbf{S} = \begin{bmatrix} 0 & -2 & 2 \\ -2 & -1 & 0 \\ 2 & 0 & 1 \end{bmatrix}
$$

497. The Spectral Theorem 6.14 says that symmetric matrices have orthogonal eigenvectors. But "almost" symmetric matrices can have eigenvectors that aren't "almost" orthogonal. Find the angle between the eigenvectors of this "almost" symmetric matrix.

$$
\mathbf{Z} = \begin{bmatrix} 1 & 0.01 \\ 0 & 1.01 \end{bmatrix}
$$

498. (Compare Exercise 456.) Show that the only eigenvalue a *skew-symmetric* matrix of any size can have is $\lambda = 0$. [*Hint: If* $\mathbf{Ax} = \lambda \mathbf{x}$, *what is* $\mathbf{Ax} \cdot \mathbf{x}$? *What happens if we apply Corollary 2.8 of Chapter 4 to* $\mathbf{Ax} \cdot \mathbf{x}$ *with* $\mathbf{A}$ *skew-symmetric?*]

499. In Exercise 410 we saw that every 2×2 orthogonal matrix takes one of these two forms for some scalar θ:

$$\begin{bmatrix} \cos\theta & -\sin\theta \\ \sin\theta & \cos\theta \end{bmatrix} \quad \text{or} \quad \begin{bmatrix} \cos\theta & \sin\theta \\ \sin\theta & -\cos\theta \end{bmatrix}$$

  a) Using inspection only, deduce that one of these is orthogonally diagonalizable. The other represents a rotation by θ.

  b) Show that the rotational form is diagonalizable if and only if θ is a multiple of π, when it is *orthogonally* diagonalizable.

  c) Compute the spectra of both forms (assume θ is *not* a multiple of π.

  d) Show that the non-rotational form represents reflection across the line spanned by $(\cos\theta/2, \sin\theta/2)$. (That is, it leaves that line fixed, while flipping lines perpendicular to it.)

500. Show that when $\mathbf{A}$ and $\mathbf{B}$ are operators on $\mathbf{R}^n$ with a common basis of eigenvectors $\{\mathbf{v}_1, \mathbf{v}_2, \dots, \mathbf{v}_n\} \subset \mathbf{R}^n$, they commute ($\mathbf{AB} = \mathbf{BA}$). [*Hint: Diagonalize.*]

501. Let $\mathbf{A}_{n \times m}$ be any matrix.

  a) Show that $\mathbf{A}^T\mathbf{A}$ and $\mathbf{AA}^T$ are always symmetric. What are their sizes?

  b) Show that the eigenvalues of $\mathbf{A}^T\mathbf{A}$ cannot be negative. [*Hint: Suppose* $\mathbf{x}$ *is an eigenvector of* $\mathbf{A}^T\mathbf{A}$, *and note that* $\mathbf{A}^T\mathbf{Ax} \cdot \mathbf{x} = \mathbf{Ax} \cdot \mathbf{Ax}$.]

  c) Show that $\mathrm{nul}(\mathbf{A}^T\mathbf{A}) = \mathrm{nul}(\mathbf{A})$.

502. Let $\mathbf{A}_{n \times m}$ be any matrix. Show that $\mathbf{A}^T\mathbf{A}$ and $\mathbf{AA}^T$ always have the same eigenvalues.

$- \star -$

7. Singular Value Decomposition

7.1. Overview. The Spectral Theorem 6.14 claims orthogonal di
agonalizability for *symmetric* matrices. Shortly after stating it, however, we claimed it would also have big implications for *all* linear transformations—symmetric or not.

Here we show why. The basic idea is beautifully simple:

No matter how badly $\mathbf{A}$ *fails to be symmetric, the product* $\mathbf{A}^T\mathbf{A}$ *is*
always *symmetric, and hence (by the Spectral Theorem 6.14)* $\mathbf{A}^T\mathbf{A}$ *is
always orthogonally diagonalizable, with eigenvalues and eigenvectors
that reveal the geometry of* $\mathbf{A}$ *itself.*

Because the transpose of a product is always the product of the transposes in reverse order (Proposition 2.7 of Chapter 4), we can easily
verify that $\mathbf{A}^T\mathbf{A}$ is its own transpose:

$$\left(\mathbf{A}^T\mathbf{A}\right)^T = \mathbf{A}^T\left(\mathbf{A}^T\right)^T = \mathbf{A}^T\mathbf{A}$$

By definition, this makes $\mathbf{A}^T\mathbf{A}$ symmetric, so we can apply the Spectral
Theorem 6.14 to $\mathbf{A}^T\mathbf{A}$, getting an orthonormal basis of eigenvectors,
which along with their eigenvalues, say much about $\mathbf{A}$. The precise
result is called the *Singular Value Decomposition (SVD)* of $\mathbf{A}$. We
summarize it in Theorem 7.4 below, the last major goal of our book.

We focus on theory here, but the Singular Value Decomposition enjoys
an enormous range pivotal applications too. Data compression algorithms and the "data-mining" technique called Principle Component
Analysis are among the most important. Readers interested in applications may want to start with a 2012 article[2] whose abstract includes
the following statement:

> *The SVD can be used to characterize political positions of
> congressmen, measure the growth rate of crystals in ig-
> neous rock, and examine entanglement in quantum com-
> putation. We also discuss higher-dimensional generaliza-
> tions of the SVD, which have become increasingly crucial
> with the newfound wealth of multidimensional data, and
> have launched new research initiatives in both theoretical
> and applied mathematics. With its bountiful theory and
> applications, the SVD is truly extraordinary.*

---

[2]Carla D. Martin & Mason A. Porter, The extraordinary SVD, *American Mathematical Monthly* **119** (2012), no. 10, 838–851.

To understand SVD, we first need to generalize our notion of "diagonal matrix" to the non-square setting.

DEFINITION 7.2. An $n \times m$ matrix $\mathbf{A} = [a_{ij}]$ is **quasi-diagonal** if its non-diagonal entries all vanish. In other words,

$$a_{ij} = 0 \quad \text{when } i \neq j$$

For square matrices, *quasi-diagonal* and *diagonal* are the same, but quasi-diagonal matrices need not be square.

When a linear map $T \colon \mathbf{R}^m \to \mathbf{R}^n$ is represented by a quasi-diagonal matrix, we call it a **quasi-diagonal transformation**. $\qquad\square$

These matrices, for instance, are all quasi-diagonal:

$$\begin{bmatrix} 3 & 0 \\ 0 & -2 \\ 0 & 0 \end{bmatrix}, \quad \begin{bmatrix} 9 & 0 \\ 0 & 0 \\ 0 & 0 \end{bmatrix}, \quad \text{and} \quad \begin{bmatrix} -1 & 0 & 0 & 0 & 0 \\ 0 & 4 & 0 & 0 & 0 \end{bmatrix}$$

As with diagonal matrices, zeros *on* the main diagonal are permitted, but zeros *off* it are *required*.

If we let d denote the *smaller* of m and n, we can understand any quasi-diagonal transformation $T \colon \mathbf{R}^m \to \mathbf{R}^n$ as a two-step operation:

- First, T *rescales* along the first d coordinate axes in the *domain* $\mathbf{R}^m$, multiplying each of of the first d input coordinates by a scalar (possibly negative or zero).

- Second, T places these first d rescaled axes into the first d slots of the output, zeroing out any other output coordinates.

We illustrate with two simple examples where $d = 2$.

EXAMPLES 7.3. The quasi-diagonal matrix

$$\mathbf{A} = \begin{bmatrix} 3 & 0 \\ 0 & -2 \\ 0 & 0 \end{bmatrix}$$

represents the linear map given by $T(x, y) = (3x, -2y, 0)$ (check this). In terms of our two-step process, T rescales the x- and y-axes in the domain $\mathbf{R}^2$ by -3 and 2 respectively, then places that rescaled copy of $\mathbf{R}^2$ into $\mathbf{R}^3$ as the $z = 0$ plane.

Similarly, the quasi-diagonal matrix

$$\mathbf{B} = \begin{bmatrix} -1 & 0 & 0 & 0 & 0 \\ 0 & 4 & 0 & 0 & 0 \end{bmatrix}$$

represents the linear map $T \colon \mathbf{R}^5 \to \mathbf{R}^2$ given by $T(x_1, x_2, x_3, x_4, x_5) = (-x_1, 4x_2)$. Here $n = 2 < 5 = m$, and thus $d = 2$. So T "collapses" all of $\mathbf{R}^5$ down onto the plane of its first two coordinate axes, rescales those two axes by -1 and 4 respectively, then maps that plane onto the range, $\mathbf{R}^2$. $\square$

We now state the main SVD result. We will devote the rest of the section to its meaning and proof.

THEOREM 7.4 (Singular Value Decomposition). *If* $\mathbf{A}$ *is any* $n \times m$ *matrix of rank* r, *then*
$$\mathbf{A} = \mathbf{V} \mathbf{D} \mathbf{U}^T$$
where $\mathbf{U}$ *and* $\mathbf{V}$ *are orthogonal, and* $\mathbf{D}$ *is quasi-diagonal. The main diagonal entries of* $\mathbf{D}$, *called the **singular values** of* $\mathbf{A}$, *are the square-roots of the eigenvalues of* $\mathbf{A}^T\mathbf{A}$.

REMARK 7.5. Lemma 7.10 below shows that the eigenvalues of $\mathbf{A}^T\mathbf{A}$ are always non-negative, so the square-roots in Theorem 7.4 always make sense. $\square$

REMARK 7.6. When $\mathbf{A}$ is *symmetric*, the SVD Theorem 7.4 reduces to the Spectral Theorem 6.14. For then we can take $\mathbf{V} = \mathbf{U}$, with the columns of $\mathbf{U}$ given by eigenvectors of $\mathbf{A}$, and the quasi-diagonal $\mathbf{D}$ becomes truly diagonal. *The SVD Theorem 7.4 thus generalizes the Spectral Theorem*, extending it to *all* matrices, not just symmetric ones. $\square$

The SVD Theorem 7.4 likely seems abstract and puzzling to most readers at first glance. But it has a striking and clear geometric interpretation, and we want to convey that before we explain the proof. This interpretation comes from

- The orthogonality of $\mathbf{U}$ and $\mathbf{V}$, and

- The quasi-diagonality of $\mathbf{D}$

The orthogonality of $\mathbf{U}$ and $\mathbf{V}$ means—by Proposition 3.14 of Chapter 6—that both $\mathbf{U}$ and $\mathbf{V}$ (and their transposes, which are also orthogonal by Observation 3.17 of Chapter 6) represent *distortion free* transformations.

As for $\mathbf{D}$, we saw in discussing quasi-diagonal transformations that *the distortion produced by a quasi-diagonal transformation is of the simplest kind: it merely "rescales" along certain coordinate axes, while collapsing any remaining axes to zero.*

So when the Singular Value Decomposition says we can factor $\mathbf{A}$ as $\mathbf{VDU}^T$, with $\mathbf{U}$ and $\mathbf{V}$ orthogonal, it is saying:

The transformation represented by an arbitrary matrix $\mathbf{A}$ can be accomplished in three steps, with no distortion except during the second step, where the distortion is very simple.

The three steps go like this: First, $\mathbf{U}^T$ does a distortion-free isomorphism that rotates or reflects the columns of $\mathbf{U}$ to the standard basis vectors. Then $\mathbf{D}$ rescales the first m or n (whichever is less) axes or $\mathbf{R}^m$ and sends them to the corresponding axes of $\mathbf{R}^n$. Finally, $\mathbf{V}$ makes a distortion-free rotation or reflection that takes the standard basis vectors in $\mathbf{R}^n$ to the columns of $\mathbf{V}$. When we put all three steps together, we get:

The factorization $\mathbf{A} = \mathbf{VDU}^T$ tells us that $\mathbf{A}$ rescales along each direction spanned by the (orthonormal) columns of $\mathbf{U}$, and then maps those directions, with no further distortion, to the orthogonal directions spanned by the columns of $\mathbf{V}$.

This last interpretation implies an illuminating corollary of the SVD Theorem 7.4:

COROLLARY 7.7. *Given any linear transformation $T\colon \mathbf{R}^m \to \mathbf{R}^n$, there is an orthonormal basis $\{\mathbf{u}_1, \mathbf{u}_2, \ldots, \mathbf{u}_m\}$ in the domain that T maps to an **orthogonal** set in the range. That is,*

$$T(\mathbf{u}_i) \cdot T(\mathbf{u}_j) = 0 \quad \text{if } i \neq j$$

PROOF. Let $\mathbf{A}$ be the matrix representing T, and use the SVD Theorem 7.4 to factor it as

$$\mathbf{A} = \mathbf{VDU}^T$$

Multiply both sides of the equation on the right by $\mathbf{U}$. Since $\mathbf{U}$ is orthogonal, this yields

(58) $$\mathbf{AU} = \mathbf{VD}$$

The columns of $\mathbf{U}$ form the orthonormal basis we seek. Indeed, let $\mathbf{u}_j$ denote the jth column of $\mathbf{U}$. Then $T(\mathbf{u}_j) = \mathbf{Au}_j$, and since matrix multiplication works column-by-column, (58) now tells us that $\mathbf{Au}_j = \mathbf{c}_j(\mathbf{VD})$ (the jth column of $\mathbf{VD}$). Using the column-by-column rule yet again, we now get more:

$$\mathbf{Au}_j = \mathbf{c}_j(\mathbf{VD})$$
$$= \mathbf{Vc}_j(\mathbf{D})$$
$$= \mathbf{V}\left(\sqrt{\lambda_j}\mathbf{e}_j\right)$$
$$= \sqrt{\lambda_j}\,\mathbf{Ve}_j$$
$$= \sqrt{\lambda_j}\,\mathbf{c}_j(\mathbf{V})$$

(For $\mathbf{D}$ is quasi-diagonal, and its jth column is $\sqrt{\lambda_j}\mathbf{e}_j$, where $\mathbf{e}_j$ as usual denotes the jth standard basis vector in $\mathbf{R}^n$.)

Since $\mathbf{V}$ is orthogonal, its columns are orthonormal. In particular, $\mathbf{c}_i(\mathbf{V}) \cdot \mathbf{c}_j(\mathbf{V}) = 0$ when $i \neq j$. This gives the desired result, since the $\mathbf{u}_i$'s are orthonormal, and we have now shown that

$$T(\mathbf{u}_i) \cdot T(\mathbf{u}_j) = \mathbf{Au}_i \cdot \mathbf{Au}_j = \sqrt{\lambda_i}\sqrt{\lambda_j}\,\mathbf{c}_i(\mathbf{V}) \cdot \mathbf{c}_j(\mathbf{V}) = 0$$

$\square$

We emphasize this Corollary in our final Examples 7.16 and 7.18.

7.8. Two lemmas. To fully understand and apply the SVD Theorem 7.4 requires familiarity with its proof, which amounts mostly to using the eigen*vectors* of $\mathbf{A}^T\mathbf{A}$ to locate the "right" orthonormal bases in the domain and range of T. These bases will form the columns of the matrices $\mathbf{U}$ and $\mathbf{V}$ in the SVD Theorem, while the eigen*values* of $\mathbf{A}^T\mathbf{A}$ determine $\mathbf{D}$.

To fill in the details, we need two easy lemmas that let us construct the bases mentioned above. With their help, the SVD Theorem 7.4 will become almost obvious.

The lemmas themselves are not difficult—the reader may already know them from Exercise 501.

LEMMA 7.9. *The nullspaces of* $\mathbf{A}$ *and* $\mathbf{A}^T\mathbf{A}$ *always agree. That is, for any matrix* $\mathbf{A}$, *we have*

$$\mathrm{nul}\!\left(\mathbf{A}^T\mathbf{A}\right) = \mathrm{nul}(\mathbf{A})$$

PROOF. We can prove this by showing that both

$$\mathrm{nul}\!\left(\mathbf{A}^T\mathbf{A}\right) \supset \mathrm{nul}(\mathbf{A}) \quad \text{and} \quad \mathrm{nul}\!\left(\mathbf{A}^T\mathbf{A}\right) \subset \mathrm{nul}(\mathbf{A})$$

The first containment is easy. For when $\mathbf{x} \in \mathrm{nul}(\mathbf{A})$, we have

$$(\mathbf{A}^T\mathbf{A})\mathbf{x} = \mathbf{A}^T(\mathbf{Ax}) = \mathbf{A}^T\mathbf{0} = \mathbf{0}$$

and hence $\mathbf{x} \in \mathrm{nul}\!\left(\mathbf{A}^T\mathbf{A}\right)$ too.

The reverse containment requires a bit more, namely the link between transpose and dot product of Corollary 2.8 in Chapter 4: *For any* $\mathbf{x}$ *and* $\mathbf{y}$ *in the appropriate dimensions, we have*

$$\mathbf{Ax} \cdot \mathbf{y} = \mathbf{x} \cdot \mathbf{A}^T \mathbf{y}$$

If we replace $\mathbf{y}$ by $\mathbf{Ax}$ in this identity, we get

$$\mathbf{Ax} \cdot \mathbf{Ax} = \mathbf{x} \cdot \mathbf{A}^T \mathbf{Ax}$$

When $\mathbf{x} \in \mathrm{nul}(\mathbf{A}^T\mathbf{A})$, the right side vanishes. The left must then vanish too, but it equals $|\mathbf{Ax}|^2$, which vanishes if and only if $\mathbf{Ax} = \mathbf{0}$. But that means $\mathbf{x} \in \mathrm{nul}(\mathbf{A})$. So $\mathbf{x} \in \mathrm{nul}(\mathbf{A}^T\mathbf{A})$ implies $\mathbf{x} \in \mathrm{nul}(\mathbf{A})$, and thus

$$\mathrm{nul}(\mathbf{A}^T\mathbf{A}) \subset \mathrm{nul}(\mathbf{A})$$

as needed. $\qquad\qquad\square$

The other fact we want assures us that, as mentioned earlier, eigenvalues of $\mathbf{A}^T\mathbf{A}$ can never be negative:

LEMMA 7.10. *Let* $\mathbf{A}$ *be any matrix. Then each eigenvalue of* $\mathbf{A}^T\mathbf{A}$ *is non-negative.*

PROOF. Every eigenvalue λ of $\mathbf{A}^T\mathbf{A}$ has an eigenvector $\mathbf{v} \neq \mathbf{0}$, so that

$$(\mathbf{A}^T\mathbf{A})\mathbf{v} = \lambda\,\mathbf{v}$$

Dot each side of this identity with $\mathbf{v}$ to get

$$(\mathbf{A}^T\mathbf{A})\mathbf{v} \cdot \mathbf{v} = \lambda\,\mathbf{v} \cdot \mathbf{v} = \lambda\,|\mathbf{v}|^2$$

At the same time, by Corollary 2.8 of Chapter 4 again (indeed, by the same calculation we used in the proof of Lemma 7.9 above) we have

$$|\mathbf{Av}|^2 = \mathbf{Av} \cdot \mathbf{Av} = (\mathbf{A}^T\mathbf{A})\mathbf{v} \cdot \mathbf{v}$$

Chaining these two calculations together yields

$$|\mathbf{Av}|^2 = \lambda\,|\mathbf{v}|^2$$

and hence (since $\mathbf{v} \neq \mathbf{0}$)

$$\lambda = \frac{|\mathbf{Av}|^2}{|\mathbf{v}|^2} \geq 0$$

which proves the Lemma. $\qquad\qquad\square$

We are now ready to construct the orthonormal bases we seek in the domain and range of the transformation represented by $\mathbf{A}$.

7.11. Domain basis. The domain of our transformation T is $\mathbf{R}^m$, so the representing matrix $\mathbf{A}$ has m columns, making $\mathbf{A}^T\mathbf{A}$ an $m \times m$ symmetric matrix. By the Spectral Theorem 6.14, $\mathbf{A}^T\mathbf{A}$ is orthogonally diagonalizable, with an orthonormal eigenvector basis

$$\{\mathbf{u}_1, \mathbf{u}_2, \ldots, \mathbf{u}_m\} \subset \mathbf{R}^m$$

Writing λ_i for the ith eigenvalue of $\mathbf{A}^T\mathbf{A}$, we then have

$$(\mathbf{A}^T\mathbf{A})\mathbf{u}_i = \lambda_i\mathbf{u}_i, \quad i = 1, 2, \ldots, m$$

and by Lemma 7.10, each eigenvalue λ_i is non-negative. Number the eigenvectors in decreasing eigenvalue order, so that

$$\lambda_1 \geq \lambda_2 \geq \lambda_3 \geq \cdots \geq \lambda_m \geq 0$$

Let r denote the rank of $\mathbf{A}$. If $r < m$, then $\mathbf{A}$—and hence $\mathbf{A}^T\mathbf{A}$, by Lemma 7.9—has a non-trivial nullspace, and the last $m - r$ eigenvalues must vanish. We need to treat the corresponding eigenvectors differently, so we rename these last $m - r$ eigenvectors—the $\mathbf{u}_i$'s that span $\mathrm{nul}(\mathbf{A}^T\mathbf{A})$—as

$$\mathbf{k}_1, \mathbf{k}_2, \ldots, \mathbf{k}_{m-r}$$

Our orthonormal eigenvector basis now looks like this:

$$\{\mathbf{u}_1, \mathbf{u}_2, \ldots, \mathbf{u}_r, \mathbf{k}_1, \mathbf{k}_2, \ldots, \mathbf{k}_{m-r}\} \subset \mathbf{R}^m$$

with corresponding eigenvalues

$$\lambda_1 \geq \lambda_2 \geq \cdots \geq \lambda_r > 0, \quad \text{and} \quad \lambda_{r+1} = \lambda_{r+2} = \cdots = \lambda_m = 0$$

The orthonormality of the basis makes the spans of the $\mathbf{u}_i$'s and the $\mathbf{k}_j$'s *orthogonal complements* in $\mathbf{R}^m$. Since the $\mathbf{k}_j$'s span $\mathrm{nul}(\mathbf{A})$, and $\mathrm{nul}(\mathbf{A})^\perp = \mathrm{row}(\mathbf{A})$, by Proposition 1.7 of Chapter 6, we have

LEMMA 7.12. *Let $\mathbf{A}$ be any $n \times m$ matrix. Then $\mathbf{R}^m$ has an orthonormal basis U composed of bases for $\mathrm{row}(\mathbf{A})$ and $\mathrm{nul}(\mathbf{A})$, respectively:*

$$U = \{\mathbf{u}_1, \mathbf{u}_2, \ldots, \mathbf{u}_r\} \cup \{\mathbf{k}_1, \mathbf{k}_2, \ldots, \mathbf{k}_{m-r}\}$$

where the $\mathbf{u}_i$'s are eigenvectors of $\mathbf{A}^T\mathbf{A}$ with strictly positive eigenvalues.

This Lemma gives us the special basis we need for the *domain* of any linear transformation $T: \mathbf{R}^m \to \mathbf{R}^n$. To understand the SVD Theorem 7.4, we also want a special basis for the *range* of T.

7.13. Range basis. Recall that $\mathrm{col}(\mathbf{A})$ and $\mathrm{nul}(\mathbf{A}^T)$ are ortho-complements in the range of T (Proposition 1.7 of Chapter 6 again). We construct a basis for $\mathbf{R}^n$—the range of T—by combining separate orthonormal bases for $\mathrm{col}(\mathbf{A})$ and $\mathrm{nul}(\mathbf{A}^T)$, just as we combined the $\mathbf{u}_i$'s and $\mathbf{k}_j$'s in Lemma 7.12 above.

We know that $\mathrm{rank}(\mathbf{A}^T) = \mathrm{rank}(\mathbf{A}) = r$ by Proposition 8.12 of Chapter 5. So the nullity of $\mathbf{A}^T$ is $n - r$, and any basis for $\mathrm{nul}(\mathbf{A}^T)$ must therefore contain exactly $n - r$ vectors. Accordingly, let

$$\left\{ \mathbf{k}_1^*, \mathbf{k}_2^*, \ldots, \mathbf{k}_{n-r}^* \right\}$$

be any orthonormal basis for $\mathrm{nul}(\mathbf{A}^T)$.

We need to be more particular about our basis for the image of T, which equals $\mathrm{col}(\mathbf{A})$ by Proposition 1.3. Any basis for $\mathrm{col}(\mathbf{A})$ will have $r = \mathrm{rank}(\mathbf{A})$ vectors, of course, since rank (by definition) equals dimension of column-space. So we seek r independent vectors

$$\mathbf{v}_1, \mathbf{v}_2, \ldots, \mathbf{v}_r \in \mathrm{col}(\mathbf{A})$$

Specifically, we will get each $\mathbf{v}_i$ from the corresponding $\mathbf{u}_i$ in the domain of T by defining

$$\mathbf{v}_i = \frac{\mathbf{A}\mathbf{u}_i}{\sqrt{\lambda_i}}, \quad i = 1, 2, \ldots, r$$

This definition certainly puts each $\mathbf{v}_i$ in $\mathrm{col}(\mathbf{A})$, since $\mathbf{A}\mathbf{x} \in \mathrm{col}(\mathbf{A})$ for *any* $\mathbf{x} \in \mathbf{R}^m$ (Observation 4.1 of Chapter 5). Note too that the square-roots here all make sense, because $\lambda_i > 0$ for each $1 \le i \le r$, by Lemma 7.12. Crucially, moreover, the $\mathbf{v}_i$'s, defined this way, form an *orthonormal* set, because

$$\mathbf{v}_i \cdot \mathbf{v}_j = \frac{\mathbf{A}\mathbf{u}_i \cdot \mathbf{A}\mathbf{u}_j}{\sqrt{\lambda_i}\sqrt{\lambda_j}} = \frac{\mathbf{A}^T\mathbf{A}\mathbf{u}_i \cdot \mathbf{u}_j}{\sqrt{\lambda_i\lambda_j}} = \frac{\lambda_i \mathbf{u}_i \cdot \mathbf{u}_j}{\sqrt{\lambda_i\lambda_j}} = \begin{cases} 0, & i \ne j \\ 1, & i = j \end{cases}$$

since the $\mathbf{u}_i$'s are orthonormal, and $\lambda_i/\sqrt{\lambda_i\lambda_j} = 1$ when $i = j$. As an orthonormal set, of course, the $\mathbf{v}_i$'s are independent by Corollary 3.4 of Chapter 6. Moreover, we have $r = \dim(\mathrm{col}(\mathbf{A}))$ of them. So they form a *basis* for $\mathrm{col}(\mathbf{A})$, by Corollary 8.8 of Chapter 5.

These facts give us a special basis for the *range* of T analogous to the domain basis of Lemma 7.12:

LEMMA 7.14. *Let* $\mathbf{A}$ *be any* $m \times n$ *matrix. Then* $\mathbf{R}^n$ *has an orthonormal basis* V *composed of bases for* $\mathrm{col}(\mathbf{A})$ *and* $\mathrm{nul}(\mathbf{A}^T)$, *respectively:*

$$V = \{\mathbf{v}_1,\, \mathbf{v}_2,\, \ldots,\, \mathbf{v}_r\} \;\cup\; \{\mathbf{k}_1^*,\, \mathbf{k}_2^*,\, \ldots,\, \mathbf{k}_{n-r}^*\}$$

where $\mathbf{A}\mathbf{u}_i = \sqrt{\lambda_i}\,\mathbf{v}_i$, *with the* $\mathbf{u}_i$'s *given by Lemma 7.12.*

We now have all we need to prove the SVD Theorem 7.4 in short order.

PROOF OF SVD THEOREM. The real work is already done. First, Lemma 7.12 gives an orthonormal basis of eigenvectors for $\mathbf{A}^T\mathbf{A}$ in the domain $\mathbf{R}^m$:

$$U = \{\mathbf{u}_1,\, \mathbf{u}_2,\, \ldots,\, \mathbf{u}_r, \mathbf{k}_1,\, \mathbf{k}_2,\, \ldots,\, \mathbf{k}_{m-r}\}$$

In particular, we have $\mathbf{A}^T\mathbf{A}\mathbf{u}_i = \lambda_i\mathbf{u}_i$, and $\mathbf{A}\mathbf{k}_j = \mathbf{0}$. The corresponding basis matrix, namely

$$\mathbf{U} = \begin{bmatrix} \mathbf{u}_1 & \mathbf{u}_2 & \cdots & \mathbf{u}_r & \mathbf{k}_1 & \cdots & \mathbf{k}_{m-r} \\ | & | & & | & | & & | \end{bmatrix}$$

is then orthogonal. This will serve as the matrix $\mathbf{U}$ in the statement of the SVD Theorem 7.4.

Similarly, Lemma 7.14 gives us an orthonormal basis in the range, and we take the corresponding basis matrix as the matrix $\mathbf{V}$ of the SVD Theorem 7.4:

$$\mathbf{V} = \begin{bmatrix} \mathbf{v}_1 & \mathbf{v}_2 & \cdots & \mathbf{v}_r & \mathbf{k}_1^* & \cdots & \mathbf{k}_{n-r}^* \\ | & | & & | & | & & | \end{bmatrix}$$

Lemma 7.14 says the first r columns of $\mathbf{U}$ and $\mathbf{V}$ are related: $\mathbf{A}\mathbf{u}_i = \sqrt{\lambda_i}\mathbf{v}_i$. Lemma 7.12 tells us that $\mathbf{A}\mathbf{k}_j = \mathbf{0}$ for each $\mathbf{k}_j$. So the columnwise description of matrix multiplication now implies

$$\mathbf{A}\mathbf{U} = \begin{bmatrix} \sqrt{\lambda_1}\,\mathbf{v}_1 & \sqrt{\lambda_2}\,\mathbf{v}_2 & \cdots & \sqrt{\lambda_r}\,\mathbf{v}_r & \mathbf{0} & \cdots & \mathbf{0} \\ | & | & & | & | & & | \end{bmatrix}$$

Computing column-by-column, and viewing matrix/vector multiplication as linear combination (our mantra again), it is easy to check that the matrix on the right above also equals $\mathbf{V}\mathbf{D}$, where $\mathbf{D}$ is this $n \times m$ quasi-diagonal matrix:

$$\mathbf{D} := \begin{bmatrix} \sqrt{\lambda_1} & 0 & 0 & \cdots & 0 & \cdots & 0 \\ 0 & \sqrt{\lambda_2} & 0 & \cdots & 0 & \cdots & 0 \\ \vdots & & \ddots & \vdots & \vdots & & \vdots \\ 0 & \cdots & 0 & \sqrt{\lambda_r} & 0 & \cdots & 0 \\ 0 & \cdots & \cdots & 0 & 0 & \cdots & 0 \\ \vdots & & \vdots & \vdots & & & \vdots \\ 0 & \cdots & \cdots & 0 & 0 & \cdots & 0 \end{bmatrix}$$

In sum, we have $\mathbf{AU} = \mathbf{VD}$, with $\mathbf{U}$ and $\mathbf{V}$ orthogonal, and $\mathbf{D}$ quasi-diagonal.

Now just multiply $\mathbf{AU} = \mathbf{VD}$ on the right by $\mathbf{U}^T$. Since $\mathbf{U}$ is orthogonal, $\mathbf{U}^T = \mathbf{U}^{-1}$, and we get the SVD Theorem 7.4:

$$\mathbf{A} = \mathbf{VDU}^T$$

As promised, $\mathbf{U}$ and $\mathbf{V}$ here are orthogonal, and the square-roots of the eigenvalues of $\mathbf{A}^T\mathbf{A}$ appear on the main diagonal of the quasi-diagonal matrix $\mathbf{D}$. □

REMARK 7.15. The proof of the SVD Theorem 7.4 is concrete: it doesn't just show that matrices $\mathbf{U}$, $\mathbf{V}$, and $\mathbf{D}$ with the required properties exist. Along with the Lemmas preceding it, it shows exactly how to construct those matrices. □

We conclude with two examples illustrating the construction of $\mathbf{U}$, $\mathbf{V}$, and $\mathbf{D}$, and the use of the SVD Theorem 7.4 for analyzing transformations that aren't symmetric.

EXAMPLE 7.16. Consider this square—but not symmetric—matrix:

$$\mathbf{A} = \begin{bmatrix} 2 & 2 \\ -2 & 1 \end{bmatrix}$$

The SVD Theorem 7.4 says we can write it as

$$\mathbf{A} = \mathbf{V}\,\mathbf{D}\,\mathbf{U}^T$$

where

$$\mathbf{D} = \begin{bmatrix} \sqrt{\lambda_1} & 0 \\ 0 & \sqrt{\lambda_2} \end{bmatrix}$$

and the λ_i's are given by the eigenvalues of

$$\mathbf{A}^T\mathbf{A} = \begin{bmatrix} 2 & -2 \\ 2 & 1 \end{bmatrix}\begin{bmatrix} 2 & 2 \\ -2 & 1 \end{bmatrix} = \begin{bmatrix} 8 & 2 \\ 2 & 5 \end{bmatrix}$$

The usual calculation finds the characteristic polynomial of the latter matrix to be

$$\lambda^2 - 13\lambda + 36 = (\lambda - 9)(\lambda - 4)$$

So $\lambda_1 = 9$ and $\lambda_2 = 4$. These eigenvalues already say much about **A**. For as explained above, the SVD Theorem 7.4 factors the transformation represented by **A** into three steps, *with no distortion except during the quasi-diagonal step.* Knowing the λ_i's, we can now see that the distortion step is represented by

$$\mathbf{D} = \begin{bmatrix} 3 & 0 \\ 0 & 2 \end{bmatrix}$$

Thus, except for distortion-free isomorphisms in the domain and range, **A** acts by magnifying its input along perpendicular lines by factors of roughly 3 and 2, respectively. It follows that no input gets magnified by less than 2, nor more than 3. This kind of information is often valuable, and can't be gotten from mere inspection of **A**.

We can delve deeper, however, by determining the "before and after" matrices **U** and **V**.

According to Lemma 7.12, the columns of **U** are the (normalized) eigenvectors of $\mathbf{A}^T\mathbf{A}$. Having computed $\lambda_1 = 9$ and $\lambda_2 = 4$ above, we can find these easily:

$\boxed{\lambda = 9:}$ We row-reduce

$$\mathbf{A}^T\mathbf{A} - 9\mathbf{I} = \begin{bmatrix} -1 & 2 \\ 2 & -4 \end{bmatrix} \xrightarrow[\cdots]{\text{row-reduce}} \begin{bmatrix} 1 & -2 \\ 0 & 0 \end{bmatrix}$$

So $\mathbf{w}_9 := (2, 1)$ spans the $\lambda = 9$ eigenspace.

$\boxed{\lambda = 4:}$ A similar calculation with $\lambda = 4$ yields

$$\mathbf{A}^T\mathbf{A} - 5\mathbf{I} = \begin{bmatrix} 4 & 2 \\ 2 & 1 \end{bmatrix} \xrightarrow[\cdots]{\text{row-reduce}} \begin{bmatrix} 1 & 1/2 \\ 0 & 0 \end{bmatrix}$$

So $\mathbf{w}_4 := (1, -2)$ spans the $\lambda = 4$ eigenspace.

Since $|\mathbf{w}_9| = |\mathbf{w}_4| = \sqrt{5}$, we now have **U**:

$$\mathbf{U} = \frac{1}{\sqrt{5}} \begin{bmatrix} \mathbf{w}_9 & \mathbf{w}_4 \\ | & | \end{bmatrix} = \frac{1}{\sqrt{5}} \begin{bmatrix} 2 & 1 \\ 1 & -2 \end{bmatrix}$$

Though we won't do so here, one easily finds that **U** has eigenvalues ± 1. It represents a (distortion-free) reflection across its $+1$ eigenspace.

Now let us find $\mathbf{V}$. Since $\mathbf{A}^T\mathbf{A}$ (and hence $\mathbf{A}$) has trivial nullspace, the formula at the end of Lemma 7.14 produces the columns of $\mathbf{V}$ from the corresponding columns $\mathbf{u}_1$ and $\mathbf{u}_2$ of $\mathbf{U}$:

$$\mathbf{v}_1 = \frac{\mathbf{A}\mathbf{u}_1}{\sqrt{\lambda_1}} = \frac{1}{3}\frac{(6,-3)}{\sqrt{5}} = \frac{(2,-1)}{\sqrt{5}}$$

$$\mathbf{v}_2 = \frac{\mathbf{A}\mathbf{u}_2}{\sqrt{\lambda_2}} = -\frac{1}{2}\frac{(2,4)}{\sqrt{5}} = -\frac{(1,2)}{\sqrt{5}}$$

So

$$\mathbf{V} = \frac{1}{\sqrt{5}}\begin{bmatrix} 2 & -1 \\ -1 & -2 \end{bmatrix}$$

One can show that $\mathbf{V}$, like $\mathbf{U}$, represents a (distortion-free) reflection. What does all this tell us? The SVD Theorem 7.4 says that $\mathbf{A} = \mathbf{V}\mathbf{D}\mathbf{U}^T$, and here, $\mathbf{U}$ is symmetric, so $\mathbf{U}^T = \mathbf{U}$, and we can check that

$$\mathbf{V}\mathbf{D}\mathbf{U} = \frac{1}{5}\begin{bmatrix} 2 & -1 \\ -1 & -2 \end{bmatrix}\begin{bmatrix} 3 & 0 \\ 0 & 2 \end{bmatrix}\begin{bmatrix} 2 & 1 \\ 1 & -2 \end{bmatrix} = \begin{bmatrix} 2 & 2 \\ -2 & 1 \end{bmatrix} = \mathbf{A}$$

as predicted.

Geometrically, the SVD Theorem 7.4 lets us visualize the transformation represented by $\mathbf{A}$ as follows: *If we take the columns of* $\mathbf{U}$ *as axes, any figure aligned with these axes will be magnified by 3 along the first axis, by 2 along the second, and then rotated to align with the axes spanned by the columns of* $\mathbf{V}$.

Figure 12 illustrates.

$\square$

REMARK 7.17. When $\mathbf{A}$ is square, as in the Example just given, we can interpret the SVD Theorem as an "almost diagonalizable" claim:

> *Any square matrix* $\mathbf{A}$ *is orthogonally similar to a product* $\mathbf{QD}$, *with* $\mathbf{Q}$ *orthogonal and* $\mathbf{D}$ *diagonal:*
>
> $$\mathbf{A} = \mathbf{U}(\mathbf{QD})\mathbf{U}^T$$
>
> *where* $\mathbf{U}$ *is orthogonal.*

For when $\mathbf{A}$ square, the quasi-diagonal matrix $\mathbf{D}$ in $\mathbf{A} = \mathbf{V}\mathbf{D}\mathbf{U}^T$ will be truly diagonal. Meanwhile, the orthogonality of $\mathbf{U}$ means $\mathbf{U}\mathbf{U}^T = \mathbf{I}$, so we can write

$$\mathbf{A} = \mathbf{V}\mathbf{D}\mathbf{U}^T = \left(\mathbf{U}\mathbf{U}^T\right)\mathbf{V}\mathbf{D}\mathbf{U}^T = \mathbf{U}(\mathbf{QD})\mathbf{U}^T$$

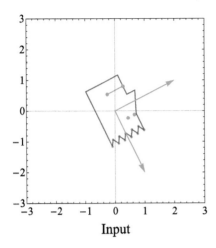
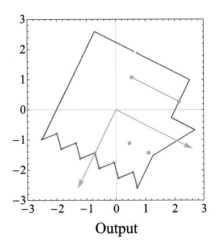

Input   Output

Figure 12. The SVD relation $\mathbf{A} = \mathbf{VDU}^T$ encodes the
geometry of $\mathbf{A}$. Any figure aligned with axes spanned by the
columns of $\mathbf{U}$ (like the cartoon aligned with the blue arrows
on left), will be magnified by factors of 2 and 3 along those
directions, then re-aligned with the directions spanned by the
columns of V (blue arrows on right).

where $\mathbf{Q} = \mathbf{U}^T\mathbf{V}$. This is orthogonal (the product of orthogonal matrices is always orthogonal by Proposition 3.18 of Chapter 6), making
$\mathbf{A}$ orthogonally similar to $\mathbf{QD}$, as claimed. $\square$

EXAMPLE 7.18. As a last example, we use the SVD Theorem 7.4 to
analyze the transformation $T\colon \mathbf{R}^2 \to \mathbf{R}^4$ represented by

$$\mathbf{A} = \begin{bmatrix} 1 & 3 \\ 6 & 2 \\ 5 & 7 \\ -2 & -2 \end{bmatrix}$$

As usual, we start by computing the symmetric matrix $\mathbf{A}^T\mathbf{A}$, which
turns out to be

$$\mathbf{A}^T\mathbf{A} = \begin{bmatrix} 66 & 54 \\ 54 & 66 \end{bmatrix}$$

We then find the characteristic polynomial to be

$$\lambda^2 - 132\lambda + 66^2 - 54^2 = \lambda^2 - 132\lambda + 1440$$
$$= (\lambda - 120)(\lambda - 12)$$

and hence $\operatorname{spec}(T) = \{12, 120\}$. It follows immediately that the quasi-
diagonal matrix $\mathbf{D}$ in the SVD decomposition of $\mathbf{A}$ is then

$$\mathbf{D} = \begin{bmatrix} \sqrt{12} & 0 \\ 0 & \sqrt{120} \\ 0 & 0 \\ 0 & 0 \end{bmatrix}$$

The input to T thus gets stretched by factors of $\sqrt{12}$ and $\sqrt{120}$ along perpendicular directions. By Lemma 7.12, those directions are spanned by the eigenvectors of $\mathbf{A}^T\mathbf{A}$, which, after normalization, also give the columns of $\mathbf{U}$. We find them the familiar way:

$\boxed{\lambda = 12:}$ Row-reduce

$$\mathbf{A}^T\mathbf{A} - 12\mathbf{I} = \begin{bmatrix} 54 & 54 \\ 54 & 54 \end{bmatrix} \longrightarrow \begin{bmatrix} 1 & 1 \\ 0 & 0 \end{bmatrix}$$

The homogeneous generator for the latter matrix, namely $\mathbf{w}_{12} := (1, -1)$, spans the $\lambda = 12$ eigenspace.

$\boxed{\lambda = 120:}$ Similarly,

$$\mathbf{A}^T\mathbf{A} - 120\mathbf{I} \longrightarrow \begin{bmatrix} 1 & -1 \\ 0 & 0 \end{bmatrix}$$

and the homogeneous generator $\mathbf{w}_{120} := (1, 1)$ thus spans the $\lambda = 120$ eigenspace of $\mathbf{A}^T\mathbf{A}$.

We then compute $|\mathbf{w}_{12}| = |\mathbf{w}_{120}| = \sqrt{2}$, and putting these facts together, can immediately write $\mathbf{U}$:

$$\mathbf{U} = \frac{1}{\sqrt{2}} \begin{bmatrix} \mathbf{w}_{12} & \mathbf{w}_{120} \\ | & | \end{bmatrix} = \frac{1}{\sqrt{2}} \begin{bmatrix} 1 & 1 \\ -1 & 1 \end{bmatrix}$$

As in the previous example, no eigenvalue $\mathbf{A}^T\mathbf{A}$ vanishes, so it—and by Lemma 7.9, $\mathbf{A}$ too—has rank 2. It follows that we can get the first two columns of $\mathbf{V}$ by normalizing the columns of

$$\mathbf{AU} = \frac{1}{\sqrt{2}} \begin{bmatrix} 1 & 3 \\ 6 & 2 \\ 5 & 7 \\ -2 & -2 \end{bmatrix} \begin{bmatrix} 1 & 1 \\ -1 & 1 \end{bmatrix} = \sqrt{2} \begin{bmatrix} -1 & 2 \\ 2 & 4 \\ -1 & 6 \\ 0 & -2 \end{bmatrix}$$

These columns are orthogonal, as expected, and after normalization, they give the first two columns of $\mathbf{V}$:

$$(59) \qquad \mathbf{v}_1 = \frac{1}{\sqrt{6}} \begin{pmatrix} -1 \\ 2 \\ -1 \\ 0 \end{pmatrix} \quad \text{and} \quad \mathbf{v}_2 = \frac{1}{\sqrt{15}} \begin{pmatrix} 1 \\ 2 \\ 3 \\ -1 \end{pmatrix}$$

According to Lemma 7.14, any orthonormal basis for $\mathrm{nul}(\mathbf{A}^T)$ will provide the last two columns of $\mathbf{V}$. We leave the reader to show (Exercise 512) that applying the Gram-Schmidt process to the homogeneous generators yields this orthonormal basis for $\mathrm{nul}(\mathbf{A}^T)$:

$$(60) \qquad \mathbf{k}_1^* = \frac{1}{\sqrt{21}} \begin{pmatrix} 2 \\ 1 \\ 0 \\ 4 \end{pmatrix} \quad \text{and} \quad \mathbf{k}_2^* = \frac{1}{\sqrt{210}} \begin{pmatrix} 11 \\ 2 \\ -7 \\ -6 \end{pmatrix}$$

For the record, this gives

$$\mathbf{V} = \begin{bmatrix} \mathbf{v}_1 & \mathbf{v}_2 & \mathbf{k}_1^* & \mathbf{k}_2^* \\ | & | & | & | \\ | & | & | & | \end{bmatrix}$$

with the $\mathbf{v}_i$'s from (59) and the $\mathbf{k}_i^*$'s above. But the actual contents of the last two columns here have little geometric interest. The geometry of the transformation shows up mainly in the columns of $\mathbf{U}$, the singular values $\sqrt{12}$ and $\sqrt{120}$ in $\mathbf{D}$, and the first two columns of $\mathbf{V}$. Specifically, these data tell us that T first stretches its input (in $\mathbf{R}^2$) by factors of $\sqrt{12}$ and $\sqrt{120}$, respectively, along the lines spanned by the first and second columns of $\mathbf{U}$—the lines generated by $(1, -1)$ and $(1, 1)$. The result is then placed—without distortion—onto the image plane in $\mathbf{R}^4$. That plane is spanned by the first two columns of $\mathbf{V}$ in $\mathbf{R}^4$. Figure 13 depicts the input and image planes for this process. $\qquad \square$

These examples only hint at the utility of the SVD Theorem 7.4. Exercises 513 and 514 explore another important application.

– Practice –

503. Suppose $a, b > 0$, and let

$$\mathbf{A} = \begin{bmatrix} 0 & a \\ -b & 0 \end{bmatrix}$$

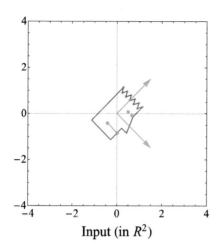

Input (in R^2)

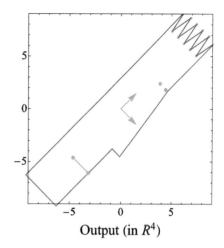

Output (in R^4)

Figure 13. The transformation analyzed in Example 7.18 maps the orthonormal basis (arrows) and input cartoon on the left to an orthonormal pair and output cartoon in the image plane, depicted on the right. The image plane lies in $\mathbf{R}^4$. Note that we shrunk the scale on the right to fit the page; the arrows there are actually equal in size to those on the left.

a) Show that $\mathbf{A}$ has no eigenvalues.

b) What are its singular values?

504. Compute the singular values of the operator represented by each matrix. Which operator has the largest distortion factor?

a) $\begin{bmatrix} -1 & -7 \\ 8 & -4 \end{bmatrix}$ b) $\begin{bmatrix} -7 & -9 \\ 6 & 2 \end{bmatrix}$ c) $\begin{bmatrix} 0 & -4 \\ 5 & -3 \end{bmatrix}$

505. Find the singular values of the transformation represented by

$$\mathbf{A} = \begin{bmatrix} 2 & 3 \\ 0 & 2 \end{bmatrix}$$

and the matrices $\mathbf{V}$, $\mathbf{D}$, and $\mathbf{U}$ in its singular value decomposition. What will $\mathbf{A}$ do to a unit square aligned with the columns of $\mathbf{U}$?

506. Let $T \colon \mathbf{R}^3 \to \mathbf{R}^2$ be the linear transformation represented by

$$\mathbf{A} = \begin{bmatrix} 3 & 2 & 4 \\ 1 & 4 & -2 \end{bmatrix}$$

Find the matrices $\mathbf{V}$, $\mathbf{D}$, and $\mathbf{U}$ in its singular value decomposition. T will map a unit cube aligned with the columns of $\mathbf{U}$ to a rectangle in $\mathbf{R}^2$. What will the side-lengths of that rectangle be? Its sides will be parallel to which lines through the origin?

507. Let $T\colon \mathbf{R}^3 \to \mathbf{R}^2$ be the linear transformation represented by

$$\mathbf{A} = \begin{bmatrix} 2 & 3 & -1 \\ 1 & 1 & 3 \end{bmatrix}$$

Find the matrices $\mathbf{V}$, $\mathbf{D}$, and $\mathbf{U}$ in its singular value decomposition. T will map a unit cube aligned with the columns of $\mathbf{U}$ to a rectangle in $\mathbf{R}^2$. What will the side-lengths of that rectangle be? Its sides will be parallel to which lines through the origin?

508. If a matrix has orthonormal columns,

   a) What are its singular values?

   b) How does the answer to (a) relate to the statement of Proposition 3.14 in Chapter 6?

509. Suppose a matrix has mutually orthogonal (but not necessarily orthonormal) columns. What are its singular values?

510. Show that for any matrix $\mathbf{A}$, the product $\mathbf{A}\mathbf{A}^T$ will (like $\mathbf{A}^T\mathbf{A}$) always be symmetric, and hence orthogonally diagonalizable.

511. Show that the $\mathbf{v}_i$'s of Lemma 7.14 are eigenvectors of $\mathbf{A}\mathbf{A}^T$, just as the $\mathbf{u}_i$'s of Lemma 7.12 are eigenvectors of $\mathbf{A}^T\mathbf{A}$.

512. Verify that when $\mathbf{A}$ is the matrix in Example 7.18, and we apply Gram–Schmidt to the homogeneous generators for its transpose, we get the basis displayed in Equation (60).

513. Suppose we have k "data points" $\mathbf{v}_i \in \mathbf{R}^n$, all lying on the line generated by a unit-vector $\mathbf{u}$, so that for some scalars $a_1, a_2, \ldots, a_k$, we have $\mathbf{v}_i = a_i \mathbf{u}$ for each $i = 1, \ldots, k$.

   a) If we make the $\mathbf{v}_i$'s *rows* of a matrix $\mathbf{A}$, what will $\mathbf{A}^T\mathbf{A}$ be?

   b) Show that in this case $\mathbf{A}$ will have only one non-zero singular value.

   c) Show that the eigenspace of $\mathbf{A}^T\mathbf{A}$ associated with the non-zero singular value will be the line generated by $\mathbf{u}$.

*The significance of Exercise 513 is that the eigenspace associated with the largest eigenvalue of $\mathbf{A}^T\mathbf{A}$ in such cases always picks out the line that **fits** the data.*

*More importantly, even when the data points do not all lie on one line, the eigenspace of largest eigenvalue picks out the subspace that **best fits** the data. This makes the SVD technique enormously useful for detecting "trends" in data sets. Exercise 514 gives a glimpse of this.*

514. (Calculator helpful) Regard the vectors $(2.9, 1.3), (-3.1, -0.7)$, and $(0.1, -0.3)$ as "data points" in the plane, and make them rows of a matrix

$$\mathbf{A} = \begin{bmatrix} 2.9 & 1.3 \\ -3.1 & -0.7 \\ 0.1 & -0.3 \end{bmatrix}$$

   a) Carefully plot the data on a standard set of axes.

   b) Compute $\mathbf{A}^T\mathbf{A}$.

   c) Verify that $\mathrm{spec}(\mathbf{A}^T\mathbf{A}) = \{20, 0.3\}$. (Subtract each eigenvalue off the diagonal and show that we get a non-trivial nullspace.)

   d) Verify that $(3, 1)$ and $(-1, 3)$ are eigenvectors of $\mathbf{A}^T\mathbf{A}$.

   e) On the axes from part (a) draw the line generated by the $\lambda = 20$ eigenvector, and decide whether you think any other line (through the origin) would better "fit" the data.

$-\star-$

APPENDIX A

Determinants

We return here to the *determinant* function introduced in §6 of Chapter 4. We stated various properties for the determinant there without proof. Now we provide the missing justification. We begin with a closer look at the "permutation formula" for determinants given in equation (33) of Chapter 4. Mastery of that formula yields accessible proofs for most of the determinant's key properties, including

- The basic determinant rules of Proposition 6.2, and

- The transposition-invariance of Proposition 6.15.

We then prove Theorem 6.10 from that same chapter: the product formula $\det(\mathbf{AB}) = \det \mathbf{A} \det \mathbf{B}$.

1. The Permutation Formula

DEFINITION 1.1. If n is a counting number, a **permutation** p of the set $\{1, 2, 3, \ldots, n\}$ is just a re-ordering of that set, with no repeats or omissions. Equivalently, it is a one-to-one, onto map

$$p : \{1, 2, 3, \ldots, n\} \to \{1, 2, 3, \ldots, n\}$$

where $p(i)$ gives the new position of i in the reordering. The set of *all* permutations of $\{1, 2, 3, \ldots, n\}$ is called the **symmetric group** S_n.
□

EXAMPLE 1.2. Suppose p is the permutation that re-orders the first five integers to produce $\{3, 5, 2, 1, 4\}$. We can describe p (or any permutation) with a 2-rowed matrix: inputs on the top row, outputs on the bottom. We enclose it in curly brackets to signal that the matrix describes a permutation. For this particular permutation, this gives

$$p: \quad \left\{ \begin{array}{ccccc} 1 & 2 & 3 & 4 & 5 \\ 3 & 5 & 2 & 1 & 4 \end{array} \right\}$$

Alternatively, we could just list $p(i)$ for each input, like this:

$$p(1) = 3, \ p(2) = 5, \ p(3) = 2, \ p(4) = 1, \ \text{and} \ p(5) = 4$$

$\square$

OBSERVATION 1.3. *There are exactly* $n!$ *distinct permutations of the set* $\{1, 2, 3, \ldots, n\}$. *The symmetric group* S_n *thus contains* $n!$ *elements.*

PROOF. If we want to construct a permutation of $\{1, 2, 3, \ldots, n\}$, we have n possible choices for $p(1)$. That leaves $n - 1$ choices for $p(2)$, and then $n - 2$ choices for $p(3)$ and so forth. The total number of permutations in S_n is thus

$$n \cdot (n - 1) \cdot (n - 2) \cdots 3 \cdot 2 \cdot 1 = n!$$

as claimed.

$\square$

Note that we always count the *identity* permutation $p(i) = i$ for each $i = 1, 2, \cdots, n$ as an element of S_n, even though no input actually moves.

OBSERVATION 1.4. *If* $p \in S_n$ *is any permutation* **except** *the identity, there are inputs* i *and* j *with* $p(i) > i$ *and* $p(j) < j$.

PROOF. The reader can probably supply a proof, but here's an elegant one. First note that

$$\big(p(1) + p(2) + p(3) + \cdots + p(n)\big) - \big(1 + 2 + 3 + \cdots + n\big) = 0$$

since both sums contain the same numbers, just differently ordered. Rearrange this as

$$(p(1) - 1) + (p(2) - 2) + (p(3) - 3) + \cdots + (p(n) - n) = 0$$

There are only two ways that this (or any) sum can vanish. The first is that *every term* vanishes. Here that would mean $p(i) = i$ for every i, making p the identity permutation. That would be consistent with our statement.

The only other possibility is for the sum to have both *positive* and *negative* terms. In that case, there are inputs i and j with $p(i)-i > 0$ and $p(j) - j < 0$. Equivalently $p(i) > i$ and $p(j) < j$, which proves our statement. □

EXAMPLE 1.5. The symmetric group S_2 contains just $2! = 2$ permutations: the identity and one other, namely

$$\left\{\begin{matrix} 1 & 2 \\ 2 & 1 \end{matrix}\right\}$$

Note that $p(1) > 1$ and $p(2) < 2$ in accordance with Observation 1.4.

Similarly, the reader can check that S_3 contains these $3! = 6$ permutations and no others (compare Exercise 31):

$$\left\{\begin{matrix} 1 & 2 & 3 \\ 1 & 2 & 3 \end{matrix}\right\}, \quad \left\{\begin{matrix} 1 & 2 & 3 \\ 2 & 3 & 1 \end{matrix}\right\}, \quad \left\{\begin{matrix} 1 & 2 & 3 \\ 3 & 1 & 2 \end{matrix}\right\}$$

$$\left\{\begin{matrix} 1 & 2 & 3 \\ 1 & 3 & 2 \end{matrix}\right\}, \quad \left\{\begin{matrix} 1 & 2 & 3 \\ 2 & 1 & 3 \end{matrix}\right\}, \quad \left\{\begin{matrix} 1 & 2 & 3 \\ 3 & 2 & 1 \end{matrix}\right\}$$

It is also easy to see that each one (except the identity) has inputs i and j, with $p(i) > i$ and $p(j) < j$. □

DEFINITION 1.6. Two outputs in a permutation form an **inversion** if they *reverse* the order of their inputs; that is, if $i < j$, but $p(i) > p(j)$.

We call p an **even** or **odd** permutation when it has an even or odd number of inversions, respectively. We then define the **sign** of p as

$$\text{sign}(p) = \begin{cases} +1 & p \text{ is even} \\ -1 & p \text{ is odd} \end{cases}$$

□

EXAMPLE 1.7. Let us count the number of inversions in the permutation

$$p: \quad \left\{\begin{matrix} 1 & 2 & 3 & 4 & 5 \\ 3 & 5 & 2 & 1 & 4 \end{matrix}\right\}$$

of Example 1.2 above, and compute its sign.

To do so, we examine each output in the second row of the matrix, scan to the right (where the *inputs* are bigger), and count the *outputs* that are *smaller*.

Start with the first output: $p(1) = 3$. Scanning to the right, we find two smaller outputs: 2 and 1. So the first output, $p(1) = 3$ contributes two inversions in p.

The second output, $p(2) = 5$ accounts for three *new* inversions, since the smaller outputs $2, 1$, and 4, all lie to the right of it.

So far, we have five inversions, and we now consider the third output, $p(3) = 2$. To find new inversions, we again scan to the right, and in this case find just one smaller output, namely 1.

That gives p a total of six inversions, and there are no more: the reader will verify that the last two outputs, (1, and 4) yield no new inversions.

Finally, we note that p has an even number of inversions (six), so it is an even permutation, and hence $\text{sign}(p) = +1$. $\qquad\square$

EXAMPLE 1.8. Along with the *identity* permutation

$$I_n = \left\{ \begin{array}{ccccc} 1 & 2 & 3 & \cdots & n \\ 1 & 2 & 3 & \cdots & n \end{array} \right\}$$

the symmetric group S_n always includes the *countdown* permutation

$$Q_n = \left\{ \begin{array}{cccccc} 1 & 2 & 3 & \cdots & n-2 & n-1 & n \\ n & n-1 & n-2 & \cdots & 3 & 2 & 1 \end{array} \right\}$$

The identity permutation obviously has *zero* inversions, so it is even, with $\text{sign}(I_n) = +1$ for every n.

The inversions in the countdown permutation Q_n are harder to count, but each output $Q_n(i)$ is clearly inverted with respect to *every* output $Q_n(j)$ to the right of it. So $Q_n(1) = n$ contributes $n-1$ inversions, $Q_n(2) = n-1$ contributes $n-2$, and so forth. The total number of inversions is thus

$$(n-1) + (n-2) + (n-3) + \cdots + 3 + 2 + 1$$

There is a general (and easily justified) formula for this sum, as one often sees in first-year calculus: it equals $n(n-1)/2$. This number is clearly even when (and only when) either n or $n-1$ is divisible by *four*. It follows that

$$\text{sign}(Q_n) = \begin{cases} -1 & n = 2, 3, 6, 7, 10, 11, 14\ldots \\ +1 & n = 4, 5, 8, 9, 12, 13, 16\ldots \end{cases}$$

The countdown permutation Q_n can thus be either even or odd, depending on n. $\qquad\square$

REMARK 1.9. It is easy to see that swapping the *last* two outputs of a permutation p adds or subtracts 1 from the total number of inversions in p, and hence changes sign(p). (More surprisingly, the same holds if we swap *any* two outputs—see Exercise 516 below.) So if we divide S_n into its even and odd subsets, the "swap-last-two" operation yields a one-to-one correspondence between the even and odd subsets. They consequently have exactly the same size: half the permutations in S_n are even, and half are odd. □

We need one more fact about the symmetric group:

PROPOSITION 1.10. *Every permutation* $p \in S_n$ *has exactly one **inverse** permutation* p^{-1} *that "undoes"* p, *and* p^{-1} *has the same sign as* p.

PROOF. If $p \in S_n$, then the outputs $p(i)$ list every integer in the set $\{1, 2, \ldots, n\}$ exactly once. So the requirement that p^{-1} "undo" p, that is, that

$$(61) \qquad\qquad p^{-1}\left(p(i)\right) = i$$

for each $i = 1, 2, \ldots, n$ uniquely determines the input/output matrix for p^{-1}, and hence determines p^{-1} itself.

It remains to show that sign(p^{-1}) = sign(p). Actually, more is true: p^{-1} has the same *number of inversions* as p. For p has an inversion if and only if there are inputs i and j with

$$i < j \quad \text{and} \quad p(i) > p(j)$$

Applying (61) to the first of these inequalities, however, and defining $i' = p(i)$ and $j' = p(j)$, we can rewrite this pair of statements like this:

$$p^{-1}(i') < p^{-1}(j') \quad \text{and} \quad i' > j'$$

Thus, p^{-1} inverts the inputs i' and j' if and only if p inverts i and j. The inversions in p therefore correspond one-to-one with those of p^{-1}, and having the same number of inversions, p and p^{-1} have the same sign as claimed. □

EXAMPLE 1.11. If we know the bracketed 2-rowed "symbol" for p, we can produce the symbol for p^{-1} simply by swapping the two rows, and then re-ordering the columns so that the top row increases from 1 to n.

For instance, if we have

$$p: \quad \left\{ \begin{array}{ccccc} 1 & 2 & 3 & 4 & 5 \\ 3 & 5 & 2 & 1 & 4 \end{array} \right\}$$

as in Example 1.2, we can get the the symbol for p^{-1} by first swapping the two rows, to get

$$\left\{ \begin{array}{ccccc} 3 & 5 & 2 & 1 & 4 \\ 1 & 2 & 3 & 4 & 5 \end{array} \right\}$$

Then re-order the columns to "standardize" the top row, and get

$$p^{-1}: \quad \left\{ \begin{array}{ccccc} 1 & 2 & 3 & 4 & 5 \\ 4 & 3 & 1 & 5 & 2 \end{array} \right\}$$

It's easy to check that this "worked" in the sense that $p^{-1}(p(i)) = i$ for each input i. $\square$

1.12. The Determinant formula. Now that we have the basic facts about permutations in hand, we can see more deeply into the general determinant formula (cf. (33) in Chapter 4). For an $n \times n$ matrix $\mathbf{A}$, that formula, again, is

$$(62) \qquad \det \mathbf{A} = \sum_{p \in S_n} \operatorname{sign}(p)\, a_{1p(1)} a_{2p(2)} a_{3p(3)} \cdots a_{np(n)}$$

Each summand here is (plus or minus) a product of n matrix entries a_{ij}. The row indices in these entries simply go from 1 to n, so there is one a_{ij} from each row. The *column* indices are the outputs of p, which *also* go from 1 to n, albeit in a different order. So each term in (62) also has exactly one a_{ij} from each *column*.

EXAMPLE 1.13. Let us verify that (62) gives the expected result for a 2×2 matrix, namely

$$(63) \qquad \begin{vmatrix} a_{11} & a_{12} \\ a_{21} & a_{22} \end{vmatrix} = a_{11}a_{22} - a_{12}a_{21}$$

To see that (62) yields (63), note that there are two permutations in S_2, as listed in Example 1.5, namely the identity permutation I_2, and the countdown permutation Q_2. These are even and odd, respectively, so when $n = 2$, the sum in (62) can be written out as

$$(64) \qquad + a_{1I(1)} a_{2I(2)} - a_{1Q(1)} a_{2Q(2)}$$

Since $I(1) = 1$ and $I(2) = 2$, while Q reverses these two outputs, we easily see that (64) agrees with (63), as hoped. $\square$

2. Basic Properties of the Determinant

The determinant formula (62) leads quickly to the basic determinant rules of Proposition 6.2, along with the transposition invariance $|\mathbf{A}^T| = |\mathbf{A}|$ (Proposition 6.15). For the sake of convenience, we restate each Proposition here before giving its proof.

PROPOSITION 6.2 (Determinant Rules) *The determinant function assigns a scalar $|\mathbf{A}|$ to every $n \times n$ matrix $\mathbf{A} = [a_{ij}]$. Among its properties are these:*

   a) *When $\mathbf{A}$ is triangular, $|\mathbf{A}| = a_{11}a_{22}\ldots a_{nn}$ (the diagonal product).*

   b) *Swapping two rows of $\mathbf{A}$ changes only the sign of $|\mathbf{A}|$.*

   c) *Multiplying any single row of $\mathbf{A}$ by a scalar λ multiplies $|\mathbf{A}|$ by λ as well.*

   d) *Adding a multiple of one row of $\mathbf{A}$ to another row of $\mathbf{A}$ does not change $|\mathbf{A}|$.*

PROOF. **(a)**: Suppose $\mathbf{A}$ is upper triangular. Then all entries *below* the main diagonal vanish: $a_{ij} = 0$ whenever $j > i$. It follows that in the determinant formula (62), $a_{ip(i)} = 0$ whenever $p(i) > i$. According to Observation 1.4, however, *every permutation except the identity has $p(i) > i$ for some i*. As a result, every *term* in (62) vanishes *except* the one we get when p is the identity. That one remaining term is the diagonal product $+a_{11}a_{22}a_{33}\cdots a_{nn}$, so (a) holds for upper triangular matrices.

Similarly, when $\mathbf{A}$ is *lower* triangular, all entries $a_{jp(j)}$ with $p(j) < j$ vanish. Observation 1.4 applies here too, and again kills every term in the determinant summation formula *except* the one given by the identity. So (a) holds for lower-triangular matrices too.

(b): Now we need to show that the determinant changes sign when we swap two rows. This depends the result of Exercise 516: *Swapping any two outputs of a permutation changes its sign.*

Given that, a little careful bookkeeping yields (b). For suppose we swap rows $r < s$ in $\mathbf{A} = [a_{ij}]$ to get a new matrix $\mathbf{A}' = [a'_{ij}]$. Also, given any $p \in S_n$, let p' denote the new permutation we get by swapping outputs $p(r)$ and $p(s)$. Then for each $j = 1, 2, \ldots, n$, we have

(65) $\quad \begin{cases} a_{ij} = a'_{ij} & (i \neq r, s) \\ a_{rj} = a'_{sj} \\ a_{sj} = a'_{rj} \end{cases}$ and $\begin{cases} p(i) = p'(i) & (i \neq r, s) \\ p(r) = p'(s) \\ p(s) = p'(r) \end{cases}$

Note too that the transformation $p \mapsto p'$ maps S_n to S_n in a one-to-one, onto fashion.

We now claim that the p-summand for $\det \mathbf{A}$ is *exactly the opposite* of the p'-summand for $\det \mathbf{A}'$. For, starting with the p-summand of $\det \mathbf{A}$ and applying the information in (65), we can deduce that

$$\begin{aligned} & \text{sign}(p) a_{1p(1)} a_{2p(2)} \cdots a_{rp(r)} \cdots a_{sp(s)} \cdots a_{np(n)} \\ = \ & \text{sign}(p) a'_{1p(1)} a'_{2p(2)} \cdots a'_{sp(r)} \cdots a'_{rp(s)} \cdots a'_{np(n)} \\ = \ & \text{sign}(p) a'_{1p'(1)} a'_{2p'(2)} \cdots a'_{sp'(s)} \cdots a'_{rp'(r)} \cdots a'_{np'(n)} \\ = \ & -\text{sign}(p') a'_{1p'(1)} a'_{2p'(2)} \cdots a'_{rp'(r)} \cdots a'_{sp'(s)} \cdots a'_{np'(n)} \end{aligned}$$

since (thanks to Exercise 516) $\text{sign}(p') = -\text{sign}(p)$. One the one hand, if we now sum the terms at the top of this chain of identities over all $p \in S_n$, we get $\det \mathbf{A}$. On the other, summing the terms at the bottom of the chain over all $p' \in S_n$, we get $-\det \mathbf{A}'$. So $\det \mathbf{A}' = -\det \mathbf{A}$, as claimed by (b).

(c): This is the easiest conclusion. If we multiply row r of $\mathbf{A}$ by a scalar λ, we get a new matrix $\mathbf{A}' = [a'_{ij}]$, with $a'_{ij} = a_{ij}$ unless $i = r$, in which case we have $a'_{ij} = \lambda a_{ij}$. The p-summand in the formula for $\det \mathbf{A}'$ thus becomes exactly λ times the p-summand in the formula for $\det \mathbf{A}$:

$$\begin{aligned} & \text{sign}(p) a'_{1p(1)} a'_{2p(2)} \cdots a'_{rp(r)} \cdots a'_{np(n)} \\ = \ & \text{sign}(p) a_{1p(1)} a_{2p(2)} \cdots \lambda \, a_{rp(r)} \cdots a_{np(n)} \\ = \ & \lambda \, \text{sign}(p) a_{1p(1)} a_{2p(2)} \cdots a_{rp(r)} \cdots a_{np(n)} \end{aligned}$$

Summing the terms on the left, we get $\det \mathbf{A}'$, while the final terms sum to $\lambda \det \mathbf{A}$. This proves (c).

(d): This argument goes much like those for (a), (b), and (c) above, so we sketch it more briefly. The interested reader will fill in missing details without difficulty.

Suppose we get $\mathbf{A}' = [a'_{ij}]$ from $\mathbf{A} = [a_{ij}]$ by adding λ times row $\mathbf{r}_s(\mathbf{A})$ to $\mathbf{r}_r(\mathbf{A})$. Then for any permutation p, we have

$$\begin{cases} a'_{ip(i)} = a_{ip(i)} & (i \neq r) \\ a'_{rp(r)} = a_{rp(r)} + \lambda\, a_{sp(r)} \end{cases}$$

This splits the p-summand in the determinant formula for $\det \mathbf{A}'$ in two:

$$\begin{aligned} \text{sign}(p)&a'_{1p(1)}a'_{2p(2)} \cdots a'_{rp(r)} \cdots a'_{np(n)} \\ = \quad & \text{sign}(p)a_{1p(1)}a_{2p(2)} \cdots a_{rp(r)} \cdots a_{np(n)} \\ & + \lambda\, \text{sign}(p)a_{1p(1)}a_{2p(2)} \cdots a_{sp(r)} \cdots a_{np(n)} \end{aligned}$$

The *first* of the two terms on the right above is exactly the p-summand for $\det \mathbf{A}$. So when we sum those, we get $\det \mathbf{A}$.

When we sum the *second* of the two terms on the right above, however, we get $\lambda \det \mathbf{A}''$, where $\mathbf{A}''$ is the matrix we get by *replacing* $\mathbf{r}_r(\mathbf{A})$ with $\mathbf{r}_s(\mathbf{A})$. So by summing on both sides of the identity above, we get

$$\det \mathbf{A}' = \det \mathbf{A} + \det \mathbf{A}''$$

But *rows r and s of $\mathbf{A}''$ are now the same, which forces* $\det \mathbf{A}'' = 0$. Indeed, the determinant of *any* matrix with two equal rows must vanish. We proved this in Corollary 6.8 of Chapter 4 using only Conclusion (b) above. Since we already proved (b), that fact is available here, and it completes the proof of (d).   □

PROPOSITION 6.15 (Transposition invariance) *For any square matrix* $\mathbf{A}$, *we have*

$$\left| \mathbf{A}^T \right| = |\mathbf{A}|$$

PROOF. If, as usual, we write a_{ij} for the entries of $\mathbf{A}$, and a'_{ij} for those of $\mathbf{A}^T$, then we have the key relation (compare (28) of Chapter 4)

$$a'_{ij} = a_{ji}$$

Applying the determinant formula (62), we then have

$$\begin{aligned} \det \mathbf{A}^T &= \sum_{p \in S_n} \text{sign}(p)a'_{1p(1)}a'_{2p(2)}a'_{3p(3)} \cdots a'_{np(n)} \\ &= \sum_{p \in S_n} \text{sign}(p)a_{p(1)1}a_{p(2)2}a_{p(3)3} \cdots a_{p(n)n} \end{aligned}$$

If we now define $i' = p(i)$ for each $i = 1, 2, \ldots, n$, however, and invoke
Proposition 1.10, which gives $p^{-1}(i') = i$ and $\text{sign}(p^{-1}) = \text{sign}(p)$, we
can rewrite the last expression above to deduce

$$\det \mathbf{A}^T = \sum_{p \in S_n} \text{sign}(p^{-1}) a_{1'p^{-1}(1')} a_{2'p^{-1}(2')} a_{3'p^{-1}(3')} \cdots a_{n'p^{-1}(n')}$$

Two final observations let us complete the proof. First, since the correspondence $p \leftrightarrow p^{-1}$ is one-to-one, summing over all $p \in S_n$ is the
same as summing over all $p^{-1} \in S_n$. So we can abbreviate $q := p^{-1}$
and rewrite the result above as

$$\det \mathbf{A}^T = \sum_{q \in S_n} \text{sign}(q) a_{1'q(1')} a_{2'q(2')} a_{3'q(3')} \cdots a_{n'q(n')}$$

Finally, we can replace each row index $1', 2', \ldots, n'$ above by its actual
value, and then (since scalar multiplication is commutative), re-order
the resulting $a_{iq(i)}$'s to put the row-indices in standard order. We then
get the usual formula for $\det \mathbf{A}$, albeit with q replacing the letter p.
Thus,

$$\begin{aligned}
\det \mathbf{A}^T &= \sum_{q \in S_n} \text{sign}(q) a_{1q(1)} a_{2q(2)} a_{3q(3)} \cdots a_{nq(n)} \\
&= \det \mathbf{A}
\end{aligned}$$

just as we sought to prove.   □

3. The Product Formula

We conclude by proving the product formula $|\mathbf{AB}| = |\mathbf{A}||\mathbf{B}|$. Our
proof here follows the clever argument sketched in Gilbert Strang's
text.[1] The following lemma will help.

LEMMA 3.1. *Doing an elementary row operation to a matrix* $\mathbf{A}$ *before
or after multiplying on the right by a matrix* $\mathbf{B}$ *gives the same product.*

PROOF. This follows easily from the fact that $\mathbf{r}_i(\mathbf{AB}) = \mathbf{r}_i(\mathbf{A})\mathbf{B}$
(Exercise 210 of Chapter 4). For example, by multiplying both sides
of this identity by a scalar k, we get

$$k\mathbf{r}_i(\mathbf{AB}) = k\mathbf{r}_i(\mathbf{A})\mathbf{B}$$

which says we get the same result whether we multiply row i by k
before, or after right-multiplying by $\mathbf{B}$, just as the Lemma predicts.

---

[1]G. Strang, *Introduction to Linear Algebra*, Wellesley-Cambridge Press, Boston.

Similar arguments prove the lemma for the other elementary row operations: swapping two rows and adding a multiple of one row to another. We leave them as an exercise.   □

THEOREM 3.2. *The determinant of a product is the product of the determinants:* $\det(\mathbf{AB}) = \det(\mathbf{A})\det(\mathbf{B})$.

PROOF. First observe that the theorem holds when $\mathbf{B}$ is singular. For then case $\mathbf{AB}$ is also singular,[2] and determinants of singular matrices vanish. In short, $|\mathbf{AB}| = |\mathbf{A}||\mathbf{B}|$ when $\mathbf{B}$ is singular because both sides equal zero.

If, on the other hand, $\mathbf{B}$ is *not* singular, then $|\mathbf{B}| \neq 0$, and we can divide by it to define a function f_B on the set of all square matrices $\mathbf{A}$ as follows:

$$f_B(\mathbf{A}) := \frac{|\mathbf{AB}|}{|\mathbf{B}|}.$$

Now consider how elementary row operations on $\mathbf{A}$ affect $f_B(\mathbf{A})$. By Lemma 3.1 above, a row operation on $\mathbf{A}$ effects the same operation on $\mathbf{AB}$. But the behavior of determinants with respect to row operations (Proposition 6.2) and the definition of f_B then show that such an operation affects $f_B(\mathbf{A})$ in exactly the same way as it affects $|\mathbf{AB}|$, and hence $|\mathbf{A}|$.

It follows that $f_B(\mathbf{A})$ obeys the same rules as $|\mathbf{A}|$ with regard to row-operations. But that lets us compute $f_B(\mathbf{A})$ much as we would compute $|\mathbf{A}|$, row-reducing until we reach the identity, while keeping track of the multiplicative scalar we generate at each step. When we complete the row-reduction, we must then have

(66) $$f_B(\mathbf{A}) = |\mathbf{A}| f_B(\mathbf{I})$$

because $|\mathbf{A}|$ *is* the product of the constants we extracted, following the determinant rules, while reducing $\mathbf{A}$ to $\mathbf{I}$. At the same time, however, we obviously have

$$f_B(\mathbf{I}) = \frac{|\mathbf{IB}|}{|\mathbf{B}|} = \frac{|\mathbf{B}|}{|\mathbf{B}|} = 1$$

So (66) reduces to $f_B(\mathbf{A}) = |\mathbf{A}| \cdot 1 = |\mathbf{A}|$, and hence

---

[2]Compare Exercise 261 of Chapter 4: when $\mathbf{B}$ is singular, $\mathbf{Bx} = \mathbf{0}$ has a non-trivial solution $\mathbf{h} \neq \mathbf{0}$. But then $(\mathbf{AB})\mathbf{h} = \mathbf{A}(\mathbf{Bh}) = \mathbf{A0} = \mathbf{0}$. So $\mathbf{AB}$ has a non-trivial nullspace, hence is singular too, by Theorem 5.7 of Chapter 4.

$$|\mathbf{A}| = \frac{|\mathbf{AB}|}{|\mathbf{B}|}$$

This is the same as $|\mathbf{AB}| = |\mathbf{A}||\mathbf{B}|$ —exactly what we wanted.   □

– Practice –

515. Determine the sign of each permutation belonging to S_2 and S_3 as listed in Example 1.5. Note that each of the last three permutations listed for S_3 can be gotten by swapping the last two outputs of the permutation above it (compare Remark 1.9).

516. Show that swapping any two outputs of a permutation reverses its sign.

517. Proceeding as in Example 1.13, show that (62) yields the formula (31) we gave in Chapter 4 for the determinant of a 3×3 matrix. Use the list of $n = 3$ permutations from Example 1.5, and check your result against the "trick" depicted in Figure 7 of Chapter 4 too.

518. Complete the proof of Lemma 3.1 by showing that it holds for the two operations not yet addressed in its proof: swapping two rows, and adding a multiple of one row to another.

APPENDIX B

Proof of the Spectral Theorem

Here we prove the Spectral Theorem (Theorem 6.14 of Chapter 7):

THEOREM 6.14.(Spectral Theorem) *A linear operator is orthogonally diagonalizable if and only if it is symmetric.*

To set up our proof, we need to make a simple definition, prove an easy lemma, and finally, recall a basic fact about continuous functions.[1]

DEFINITION 0.3. We say that a linear operator $T : \mathbf{R}^n \to \mathbf{R}^n$ **preserves a subspace** $S \subset \mathbf{R}^n$ when $T(S) \subset S$. That is, T maps any vector in S back into S:

$$\mathbf{x} \in S \quad \text{implies} \quad T(\mathbf{x}) \in S$$

$\square$

Recall from Definition 2.4 of Chapter 7 that an operator $T \colon \mathbf{R}^n \to \mathbf{R}^n$ is **symmetric** when it is represented by a symmetric matrix.

LEMMA 0.4. *If a symmetric $n \times n$ operator preserves a subspace $S \subset \mathbf{R}^n$, then it preserves $S^{\perp}$ too.*

PROOF. We have to show that $T(\mathbf{x}) \in S^{\perp}$ whenever $\mathbf{x} \in S^{\perp}$. That is, we must show that $T(\mathbf{x}) \cdot \mathbf{s} = 0$ whenever $\mathbf{x} \in S^{\perp}$ and $\mathbf{s} \in S$. This follows immediately from the symmetry of T. For when $\mathbf{x} \in S^{\perp}$ and $\mathbf{s} \in S$, we have

$$T(\mathbf{x}) \cdot \mathbf{s} = \mathbf{A}\mathbf{x} \cdot \mathbf{s} = \mathbf{x} \cdot \mathbf{A}\mathbf{s} = \mathbf{x} \cdot T(\mathbf{s}) = 0$$

because T preserves S, while $\mathbf{x} \in S^{\perp}$. $\square$

Recall from Example 1.3 of Chapter 5 that the **unit sphere** in $\mathbf{R}^n$ is the set of all *unit* vectors in $\mathbf{R}^n$: all $\mathbf{v} \in \mathbf{R}^n$ with $|\mathbf{v}| = 1$.

---

[1]We regret having to leave the realm of algebra and using a fact about continuity—a concept from Calculus—to prove the Spectral Theorem. A purely algebraic proof *is* possible, but we would need to move beyond real scalars and introduce the complex number system, an even bigger departure from the familiar.

The basic fact about continuous functions we need involves this sphere: *Any continuous function on the unit sphere attains a maximum.*[2]

PROOF OF SPECTRAL THEOREM. Let $T : \mathbf{R}^n \to \mathbf{R}^n$ be a symmetric operator, represented by a symmetric matrix $\mathbf{A}$. If $\mathbf{A}$ is orthogonally diagonalizable, it is easy to see that it must also be symmetric (cf. Exercise 520 below). The real content of the Spectral Theorem is the converse: that *when $\mathbf{A}$ is symmetric, it must be orthogonally diagonalizable.* To prove that, we must show that a symmetric $n \times n$ matrix $\mathbf{A}$ will always have n orthonormal eigenvectors. The matrix $\mathbf{B}$ having those eigenvectors as columns is then orthogonal and diagonalizes $\mathbf{A}$. That is, $\mathbf{A} = \mathbf{B}\mathbf{D}\mathbf{B}^T$ with $\mathbf{D}$ diagonal, as discussed in §6 of Chapter 7.

In short, it remains to show that symmetry of $\mathbf{A}$ forces the existence of n orthonormal eigenvectors.

To produce these eigenvectors, we study the function $Q : \mathbf{R}^n \to \mathbf{R}$ given by

$$Q(\mathbf{x}) := \mathbf{A}\mathbf{x} \cdot \mathbf{x}$$

If we write the entry in row i, column j of $\mathbf{A}$ as a_{ij} in the usual way, it is easy to verify that Q expands as a quadratic polynomial in the coordinates $(x_1, x_2, \ldots, x_n)$ of $\mathbf{x}$:

$$Q(\mathbf{x}) = \sum_{i,j=1}^{n} a_{ij} x_i x_j$$

All polynomials are continuous, so by the basic fact mentioned above, Q attains a max on the unit sphere. Let $\mathbf{v} \in \mathbf{S}^{n-1}$ be a vector where that max is attained, so that

$$Q(\mathbf{v}) \geq Q(\mathbf{x}) \quad \text{for all } \mathbf{x} \text{ with } |\mathbf{x}| = 1$$

This inequality will force $\mathbf{v}$ to be an eigenvector of $\mathbf{A}$.

To see how, we leave the reader to check (recalling that $|\mathbf{v}| = 1$) that when

(67) $$\lambda := \mathbf{A}\mathbf{v} \cdot \mathbf{v} \quad \text{and} \quad \mathbf{w} = \mathbf{A}\mathbf{v} - \lambda\mathbf{v}$$

---

[2]This is just a slightly more sophisticated version of a theorem one always encounters in first-year Calculus: *A continuous function on a closed interval always attains a maximum.* Our version can be found in most basic Analysis texts; see for instance Corollary 21.5 of *Elementary Analysis: The Theory of Calculus* by Kenneth A. Ross.

we have

$$\mathbf{A}\mathbf{v} = \lambda\mathbf{v} + \mathbf{w} \quad \text{and} \quad \mathbf{w} \cdot \mathbf{v} = 0 \tag{68}$$

Now consider the vector function $\mathbf{v}(t)$ we get by normalizing the vectors along the line through $\mathbf{v}$ in the $\mathbf{w}$ direction, mapping them all to the unit sphere:

$$\mathbf{v}(t) = \frac{\mathbf{v} + t\mathbf{w}}{\sqrt{1 + t^2|\mathbf{w}|^2}}$$

Note that $\mathbf{v}(0) = \mathbf{v}$, the vector where Q attains its max. So if we now define the 1-variable function

$$f(t) = Q\left(\mathbf{v}(t)\right)$$

This function must attain a max at $t = 0$. We will show this forces $\mathbf{w} = \mathbf{0}$, which in turn means $\mathbf{A}\mathbf{v} = \lambda\mathbf{v} + \mathbf{0} = \lambda\mathbf{v}$, and hence that $\mathbf{v}$ is an eigenvector, as promised.

To check this we expand $f(t)$ explicitly:

$$
\begin{aligned}
f(t) \\
= \quad & Q\left(\mathbf{v}(t)\right) \\
= \quad & \mathbf{A}\mathbf{v}(t) \cdot \mathbf{v}(t) \\
= \quad & \frac{\mathbf{A}\mathbf{v} \cdot \mathbf{v} + 2t(\mathbf{A}\mathbf{v} \cdot \mathbf{w}) + t^2 \mathbf{A}\mathbf{w} \cdot \mathbf{w}}{1 + t^2|\mathbf{w}|^2} \\
= \quad & \frac{Q(\mathbf{v}) + 2t(\mathbf{A}\mathbf{v} \cdot \mathbf{w}) + t^2 Q(\mathbf{w})}{1 + t^2|\mathbf{w}|^2}
\end{aligned}
$$

Note that *we used the symmetry of* $\mathbf{A}$ *here to rewrite* $\mathbf{v} \cdot \mathbf{A}\mathbf{w} = \mathbf{A}\mathbf{v} \cdot \mathbf{w}$. We also simplified using the definition of Q.

It is now easy to check that the derivative of the denominator above vanishes at $t = 0$, so when we apply the quotient rule and set $t = 0$, we simply get

$$0 = f'(0) = 2\,\mathbf{A}\mathbf{v} \cdot \mathbf{w}$$

Conclusion: $\mathbf{A}\mathbf{v} \cdot \mathbf{w} = 0$. By dotting the definition of $\mathbf{w}$ in (67) above with $\mathbf{w}$ itself, however, and using this conclusion along with our assumption that $\mathbf{w} \cdot \mathbf{v} = 0$ from (68), we now see that $|\mathbf{w}|^2 = 0$, so that $\mathbf{w} = \mathbf{0}$ as hoped. In short, $\mathbf{v}$ is an eigenvector:

$$\mathbf{A}\mathbf{v} = \lambda\mathbf{v}$$

Now that we have one eigenvector, it is straightforward to finish the proof recursively using Lemma 0.4. Indeed, since $\mathbf{v}$ is an eigenvector,

A preserves the 1-dimensional subspace (line) spanned by **v**, Lemma 0.4 shows that **A** preserves its orthocomplement, the hyperplane $\mathbf{v}^{\perp}$. We may thus repeat the argument above, maximizing Q on $\mathbf{v}^{\perp}$ to get a second eigenvector $\mathbf{v}_2$ in this hyperplane. Now **A** preserves the plane spanned by $\mathbf{v}_1$ and $\mathbf{v}_2$, so by Lemma 0.4 again, it also preserves the orthcomplement of this plane, where we can maximize Q a third time to get a third unit eigenvector $\mathbf{v}_3$ orthogonal to both $\mathbf{v}_1$ and $\mathbf{v}_2$.

Nothing prevents us from continuing this process, decreasing the dimension of the orthocomplement by one each time, until the $\mathbf{v}_i$'s span all of $\mathbf{R}^n$, forming the desired orthonormal basis of eigenvectors. $\qquad\square$

– **Practice** –

519. Show that the operator $T\colon \mathbf{R}^2 \to \mathbf{R}^2$ represented by

$$\mathbf{A} = \begin{bmatrix} 1 & 1 \\ 0 & 1 \end{bmatrix}$$

preserves the x-axis, but not its orthocomplement. Lemma 0.4 thus *fails* without the symmetry assumption.

520. Prove the "easy half" of the Spectral Theorem: *If an operator is orthogonally diagonalizable, then it is symmetric.*

521. Show that the identities in Equation (68) follow from those in Equation (67).

$$-\star-$$

APPENDIX C

Lexicon

- **Augmented matrix.** See Matrix.

- **Basis.** A **basis** for a subspace $S \subset \mathbf{R}^n$ is a set of vectors in S which *spans* S, and is *independent*.

- **Characteristic polynomial.** The **characteristic polynomial** of an $n \times n$ matrix $\mathbf{A}$ is the degree-n polynomial in the variable λ given by

$$\det(\mathbf{A} - \lambda \mathbf{I})$$

Its roots are the eigenvalues of $\mathbf{A}$. The characteristic polynomial of a linear operator T on $\mathbf{R}^n$ is simply that of the matrix that represents it.

- **Codimension.** The **codimension** of a subspace $S \subset \mathbf{R}^n$, denoted $\text{codim}(S)$, is the is the difference between the ambient dimension n and the dimension of S. In short, $\text{codim}(S) = n - \dim(S)$.

- **Coefficient matrix.** See Matrix.

- **Column space.** The **column space** of an $n \times m$ matrix $\mathbf{A}$ is the span of its columns. Denoted by $\text{col}(\mathbf{A})$, it lies in $\mathbf{R}^n$, and is always a subspace.

- **Conjugate.** (Same as **similar**) We say that square matrices $\mathbf{A}$ and $\bar{\mathbf{A}}$ are **conjugate** if there is an invertible matrix $\mathbf{B}$ such that

$$\bar{\mathbf{A}} = \mathbf{B}\mathbf{A}\mathbf{B}^{-1}$$

In this case, $\bar{\mathbf{A}}$ represents a linear transformation that affects the basis formed by the columns of $\mathbf{B}$ in the same way that $\mathbf{A}$ affects the standard basis.

- **Contrapositive.** The **contrapositive** of a statement $P \Rightarrow Q$ is the statement *not* $Q \Rightarrow$ *not* P. The contrapositive is *logically equivalent* to the original statement: Either both are true, or neither is.

- **Converse.** The **converse** of a statement $P \Rightarrow Q$ is the statement $Q \Rightarrow P$. The converse is *logically independent* of the original statement: The truth of one implies nothing about the truth of the other.

- **Dependent.** See Independent.

- **Diagonal matrix.** A **diagonal matrix** is a square matrix with zeros everywhere **except** for the main diagonal. All entries in row i of a diagonal matrix are zero except (possibly) for the entry in column i.

- **Diagonalize.** An $n \times n$ matrix $\mathbf{A}$ is **diagonalizable** when it is similar to (conjugate to) a diagonal matrix:

$$\mathbf{A} = \mathbf{B}\mathbf{D}\mathbf{B}^{-1}$$

with $\mathbf{D}$ diagonal. In this case we say that $\mathbf{B}$ (or the basis for $\mathbf{R}^n$ given by its columns) **diagonalizes** $\mathbf{A}$.

Not all square matrices are diagonalizable, but all symmetric matrices are. In fact, they are orthogonally diagonalizable, by the Spectral Theorem.

- **Dimension.** The **dimension** of a subspace $S \subset \mathbf{R}^n$, denoted $\dim(S)$, is the number of vectors in any basis for S.

If $\dim(S) \geq 1$, S has infinitely many different bases. One proves, however, that all bases for S have the same number of vectors, so that $\dim(S)$ as defined above is unambiguous (Theorem 8.2 of Chapter 5).

- **Domain.** The **domain** of a mapping $F : X \to Y$ is the set X containing all allowed inputs to F.

- **Dot product.** The dot product of two numeric vectors

$$\mathbf{v} = (v_1, v_2, \ldots, v_n) \quad \text{and} \quad \mathbf{w} = (w_1, w_2, \ldots w_n)$$

in $\mathbf{R}^n$ is defined to be

$$\mathbf{v} \cdot \mathbf{w} = v_1 w_1 + v_2 w_2 + \cdots + v_n w_n$$

- **Eigenspace.** For any scalar λ, the λ-**eigenspace** of a linear operator T on $\mathbf{R}^n$ is the kernel of $T - \lambda I$, or equivalently, the nullspace of $\mathbf{A} - \lambda \mathbf{I}$, where $\mathbf{A}$ is the matrix representing T.

For each $\lambda \in \mathbf{R}$, the λ-eigenspace is a subspace, though it contains only the origin for all but at most n values of λ.

- **Eigenvalue.** A scalar λ is an **eigenvalue** of a linear operator T on $\mathbf{R}^n$ (or of the matrix $\mathbf{A}$ that represents T) if and only if the λ-eigenspace of T has positive dimension (and thus if and only if $\mathbf{A} - \lambda\mathbf{I}$ has a free column).

  The eigenvalues of a linear operator T on $\mathbf{R}^n$ (or an $n \times n$ matrix $\mathbf{A}$) are precisely the roots of its characteristic polynomial, which has degree n. An operator on $\mathbf{R}^n$ thus has at most n different eigenvalues.

- **Eigenvector.** When λ is an eigenvalue of an operator T on $\mathbf{R}^n$ (or of the matrix $\mathbf{A}$ that represents T), any vector $\mathbf{v} \in \mathbf{R}^n$ with $T(\mathbf{v}) = \lambda\mathbf{v}$ (or $\mathbf{A}\mathbf{v} = \lambda\mathbf{v}$) is called a λ-**cigenvector** (or just **eigenvector**) of T (or of $\mathbf{A}$). Equivalently, the λ-eigenvectors are the members of the λ-eigenspace.

- **Explicit.** When a subspace $S \subset \mathbf{R}^n$ is described as the span of a set of vectors $\{\mathbf{u}_1, \mathbf{u}_2, \ldots, \mathbf{u}_k\}$, we say we have an **explicit** description of S. The term highlights the fact that a span description lets us easily and *explicitly* produce vectors in S as linear combinations of the $\mathbf{u}_i$'s.

- **Homogeneous/Inhomogeneous.** A linear system $\mathbf{A}\mathbf{x} = \mathbf{b}$ is **homogeneous** when $\mathbf{b} = \mathbf{0}$, and **inhomogeneous** otherwise.

- **Hyperplane.** By a **hyperplane in $\mathbf{R}^m$**, we mean the set of all vectors $\mathbf{v} = (x_1, x_2, \ldots, x_m) \in \mathbf{R}^m$ that solve some linear equation (having at least one non-zero coefficient) in the m variables x_i. Thus, a hyperplane is the set of all *solutions* to a non-trivial linear equation.

- **Identity matrix.** The $n \times n$ **identity matrix**, denoted by $\mathbf{I}_n$ (or simply $\mathbf{I}$), is the matrix whose entry in row i, column j is 0 unless $i = j$, when it equals 1:

$$\mathbf{I}_n = \begin{pmatrix} 1 & 0 & 0 & \cdots & 0 \\ 0 & 1 & 0 & \cdots & 0 \\ 0 & 0 & 1 & \cdots & 0 \\ & \vdots & & \vdots & \\ 0 & 0 & 0 & \cdots & 1 \end{pmatrix}$$

The key property of the identity matrix is that it satisfies $\mathbf{I}_n\mathbf{A} = \mathbf{A}\mathbf{I}_m = \mathbf{A}$ for any $n \times m$ matrix $\mathbf{A}$.

- **Image.** The **image** of a linear transformation $T : \mathbf{R}^m \to \mathbf{R}^n$, denoted $\mathrm{im}(T)$, is the set of all $\mathbf{w} \in \mathbf{R}^n$ that are *actual outputs* of T. Thus, a vector $\mathbf{w}$ in the range of T belongs to $\mathrm{im}(T)$ if and only if $\mathbf{w} = T(\mathbf{v})$ for some $\mathbf{v}$ in the domain of T.

  The image of a linear transformation is always a subspace of its range. We always have $\mathrm{im}(T) = \mathrm{col}(\mathbf{A})$, where $\mathbf{A}$ is the matrix representing T.

- **Implicit.** A subspace S is defined **implicitly** when it is described as the solution of a homogeneous linear system—the nullspace of a matrix. We call such a description *implicit* because it does not *explicitly* name any vectors in S—it just provides a test we can use to determine whether vectors belong to S or not.

  Every subspace description is either *implicit* or *explicit*.

  An implicit description is the same as a perp description of a subspace, since $\mathbf{x}$ solves the homogeneous system $\mathbf{A}\mathbf{x} = \mathbf{0}$ if and only if $\mathbf{x}$ is orthogonal to each row of $\mathbf{A}$.

- **Independent.** A set of U of vectors in $\mathbf{R}^n$ is **linearly independent** (or just **independent**) if no proper subset of U has the same span as all of U. A non-independent set is called **dependent**.

  Independence has many equivalent definitions (see Proposition 6.9 of Chapter 5). For instance, U is independent if and only if no $\mathbf{v}_i \in U$ can be expressed as a linear combination of the remaining $\mathbf{v}_j$'s. More practically, the columns of a matrix $\mathbf{A}$ are independent if and only if each column of $\mathrm{RRE}(\mathbf{A})$ has a pivot.

- **Inverse.** The **inverse** $\mathbf{B}$ of a matrix $\mathbf{A}$ is defined by the condition
  $$\mathbf{A}\mathbf{B} = \mathbf{I}_n = \mathbf{B}\mathbf{A}$$
  where $\mathbf{I}_n$ denotes the identity matrix. We typically denote the inverse $\mathbf{B}$ by $\mathbf{A}^{-1}$. Note that because $\mathbf{I}_n$ is square, this definition forces both $\mathbf{A}$ and $\mathbf{B}$ to be square—only square matrices have inverses. Not all square matrices have inverses, however.

  If we know that $\mathbf{A}$ and $\mathbf{B}$ are both square, the conditions $\mathbf{A}\mathbf{B} = \mathbf{I}_n$ and $\mathbf{B}\mathbf{A} = \mathbf{I}_n$ each *separately* suffice to guarantee the other, and hence that $\mathbf{B} = \mathbf{A}^{-1}$.

- **Invertible matrix.** A matrix $\mathbf{A}$ is **invertible** if it has an inverse.

- **Invertible transformation.** A linear transformation $T :$ $\mathbf{R}^n \to \mathbf{R}^n$ is **invertible** when it is both one-to-one and onto. Invertibility of T is equivalent to that of the matrix $\mathbf{A}$ that represents T.

- **Kernel.** The **kernel** of a linear transformation $T : \mathbf{R}^m \to$ $\mathbf{R}^n$, denoted by $\ker(T)$ is the set of all vectors $\mathbf{v} \in \mathbf{R}^m$ that T sends to the origin. Thus $\mathbf{v}$ belongs to $\ker(T)$ if and only if $T(\mathbf{v}) = \mathbf{0}$.

  The kernel of a linear transformation is always a subspace of its domain. It always equals $\mathrm{nul}(\mathbf{A})$, where $\mathbf{A}$ is the matrix representing T.

- **Leading one.** See Reduced Row-Echelon form.

- **Left nullspace.** The **left nullspace** of an $n \times m$ matrix $\mathbf{A}$ is the nullspace of the transpose of $\mathbf{A}$, that is $\mathrm{nul}(\mathbf{A}^T)$. As such, it is always a subspace of the range of T.

- **Length.** The **length** of a numeric vector $\mathbf{v} \in \mathbf{R}^n$ is defined as
$$|\mathbf{v}| := \sqrt{\mathbf{v} \cdot \mathbf{v}} = \sqrt{v_1^2 + v_2^2 + \cdots + v_n^2}$$
when $\mathbf{v} = (v_1, v_2, \ldots, v_n)$.

- **Linear combination.** A **linear combination** of k vectors $\mathbf{v}_1, \mathbf{v}_2, \ldots, \mathbf{v}_k$ in $\mathbf{R}^n$ is a sum of the form
$$c_1 \mathbf{v}_1 + c_2 \mathbf{v}_2 + \cdots + c_k \mathbf{v}_k$$
The coefficients c_i's can be any scalars.

  Since the $\mathbf{v}_i$'s are in $\mathbf{R}^n$ so is the linear combination. Note that k and n need not bear any relation to each other.

- **Linear equation.** A **linear equation** in the m variables $x_1, x_2, \ldots, x_m$ is an equation of the form
$$a_1 x_1 + a_2 x_2 + a_3 x_3 + \cdots + a_m x_m = b$$
The a_i's are called *coefficients*, while b is called the *right-hand side*.

  Some or all of the coefficients may vanish. When the right-hand side b vanishes, we say the equation is homogeneous. Otherwise, it is inhomogeneous.

- **Linear operator.** A linear transformation $T : \mathbf{R}^m \to \mathbf{R}^n$ is an **operator** when $m = n$, i.e., the domain and range are the same.

  Any matrix that represents a linear operator is necessarily square.

- **Linear system.** A **linear system** of n equations in m unknowns is a set of n linear equations, each in m variables. Such a system thus takes the form

$$
\begin{aligned}
a_{11}x_1 + a_{12}x_2 + a_{13}x_3 + \cdots + a_{1m} &= b_1 \\
a_{21}x_1 + a_{22}x_2 + a_{23}x_3 + \cdots + a_{2m} &= b_2 \\
\vdots \qquad\qquad\qquad \vdots &= \vdots \\
a_{n1}x_1 + a_{n2}x_2 + a_{n3}x_3 + \cdots + a_{nm} &= b_n
\end{aligned}
$$

- **Linear transformation.** A transformation (or mapping) from $\mathbf{R}^m$ to $\mathbf{R}^n$ is **linear** if it commutes with linear combinations, in the sense that for any scalars a_i and vectors $\mathbf{v}_i$ in $\mathbf{R}^m$, we have

$$
\begin{aligned}
T\left(a_1\mathbf{v}_1 + a_2\mathbf{v}_2 + \cdots + a_k\mathbf{v}_k\right) \\
= a_1\,T(\mathbf{v}_1) + a_2\,T(\mathbf{v}_2) + \cdots + a_k\,T(\mathbf{v}_k)
\end{aligned}
$$

  Equivalently, T is linear if and only if it commutes with addition and scalar multiplication separately, meaning that for any scalar k and any $\mathbf{v}, \mathbf{w} \in \mathbf{R}^m$ we have both

  i) $T(\mathbf{v} + \mathbf{w}) = T(\mathbf{v}) + T(\mathbf{w})$, and

  ii) $T(k\mathbf{v}) = k\,T(\mathbf{v})$

- **Matrix.** A **matrix** is a rectangular array of scalars. We say that a matrix is $n \times m$ if it has n (horizontal) rows and m (vertical) columns.

  When we extract the coefficients a_{ij} from a linear system and place them in a matrix with a_{ij} in row i, column j, we get the **coefficient matrix** of the system. When we add an additional column on the right with the right-hand side b_i of the ith equation in the ith position, we get the **augmented matrix** of the system.

- **Matrix transformation.** Multiplication by a $n \times m$ matrix $\mathbf{A}$ yields a linear transformation $T : \mathbf{R}^m \to \mathbf{R}^n$, given by $T(\mathbf{v}) = \mathbf{A}\mathbf{v}$. In this case, we call T a **matrix transformation**, and we say that $\mathbf{A}$ represents T.

  By Theorem 5.6 of Chapter 1, every linear transformation T is represented by some matrix $\mathbf{A}$. All linear transformations can thus be seen as matrix transformations.

- **Nullity.** The **nullity** of a linear transformation T is the dimension of its kernel. The nullity of a matrix $\mathbf{A}$ is then the nullity of the linear transformation it represents.

  The kernel of a linear transformation always equals the nullspace of the matrix that represents it, so the nullity of a matrix $\mathbf{A}$ is the dimension of the nullspace of $\mathbf{A}$.

- **Nullspace.** The **nullspace** of an $n \times m$ matrix $\mathbf{A}$, denoted $\mathrm{nul}(\mathbf{A})$, is the set of all solutions to $\mathbf{A}\mathbf{x} = \mathbf{0}$. It is always a subspace of $\mathbf{R}^m$.

- **One-to-one.** A mapping $F : X \to Y$ is **one-to-one** if each $\mathbf{y}$ in the range of F has at *most* one pre-image.

  Equivalently, F is one-to-one when $F(\mathbf{x}) = \mathbf{y}$ never has more than one solution $\mathbf{x} \in X$ for any fixed $\mathbf{y} \in Y$.

  This also means F is one-to-one if and only if unequal inputs $\mathbf{x}_1 \neq \mathbf{x}_2$ always map to unequal outputs $F(\mathbf{x}_1) \neq F(\mathbf{x}_2)$.

- **Onto.** A mapping $F : X \to Y$ is **onto** if every $\mathbf{y}$ in the range has at *least* one pre-image.

  Equivalently, F is onto when $F(\mathbf{x}) = \mathbf{y}$ has at least one solution for every $\mathbf{y} \in Y$.

  This also means F is onto if and only if its image equals its entire range.

- **Origin.** The numeric vector $(0, 0, \ldots, 0) \in \mathbf{R}^n$ is called the **origin** of $\mathbf{R}^n$. We often denote it by $\mathbf{0}$.

- **Orthocomplement.** Same as orthogonal complement.

- **Orthogonal** (matrix). An $n \times n$ matrix $\mathbf{A}$ is **orthogonal** if and only $\mathbf{A}^{-1} = \mathbf{A}^T$. Equivalently, $\mathbf{A}^T \mathbf{A} = \mathbf{A}\mathbf{A}^T = \mathbf{I}$, which is in turn equivalent to saying that the rows of $\mathbf{A}$, and the columns of $\mathbf{A}$, *both* form orthonormal bases for $\mathbf{R}^n$.

- **Orthogonal** (vectors). We say that non-zero vectors $\mathbf{v}, \mathbf{w} \in \mathbf{R}^n$ are **orthogonal** when their dot product vanishes: $\mathbf{v} \cdot \mathbf{w} = 0$.

- **Orthogonal complement.** The **orthogonal complement** (or simply **orthocomplement**) of a subspace S in $\mathbf{R}^n$ is the set of all $\mathbf{w} \in \mathbf{R}^n$ that are orthogonal to every $\mathbf{v} \in S$. We denote the orthogonal complement of S by $S^\perp$.

  The orthocomplement of a subspace S is itself always a subspace.

- **Orthogonally diagonalizable.** A (necessarily square) matrix $\mathbf{A}$ is **orthogonally diagonalizable** if it is conjugate to a diagonal matrix via an orthogonal matrix $\mathbf{B}$. That is, when

$$\mathbf{A} = \mathbf{B}\mathbf{D}\mathbf{B}^T$$

  with $\mathbf{B}$ orthogonal and $\mathbf{D}$ diagonal. Orthogonally diagonalizable matrices are always diagonalizable, but diagonalizable matrices need not be *orthogonally* diagonalizable.

- **Orthonormal.** We call a set of vectors $\{\mathbf{u}_1, \mathbf{u}_2, \ldots, \mathbf{u}_k\}$ **orthonormal** if each $\mathbf{u}_i$ is a unit vector, and is orthogonal to all other vectors in the set.

- **Perp.** The **perp** of a set U of vectors in $\mathbf{R}^n$ (denoted $U^\perp$) is the set of all vectors $\mathbf{x} \in \mathbf{R}^n$ that are orthogonal to each vector in U.

- **Pivot.** See Reduced Row-Echelon form.

- **Pre-image.** The **pre-image** of a point $\mathbf{y}$ in the range of a mapping $F : X \to Y$ is the (possibly empty) set of all inputs $\mathbf{x} \in X$ that F sends to $\mathbf{y}$. The pre-image of $\mathbf{y}$ thus comprises all solutions $\mathbf{x}$ of the equation $F(\mathbf{x}) = \mathbf{y}$. We often denote the pre-image of $\mathbf{y}$ by $F^{-1}(\mathbf{y})$.

- **Range.** The **range** of any mapping $F : X \to Y$ is the set Y containing all potential outputs of the mapping.

- **Rank.** The **rank** of a matrix $\mathbf{A}$ is the dimension of its column space. Computationally, it equals the number of pivots in $\mathrm{RRE}(\mathbf{A})$.

  The **rank** of a linear transformation T is the dimension of its image. Since the image of a linear transformation coincides with the column space of the matrix that represents it, the rank of T is the same as the rank of its representing matrix.

- **Reduced Row-Echelon (RRE) form.** We say that a matrix **A** has **Reduced Row-Echelon** form if and only if it satisfies all four of the following conditions:

  - **Leading 1's:** *The leftmost non-zero entry in every row is a "1". These entries are called **leading ones**, or just **pivots**.*

  - **Southeast:** *Each pivot lies to the right of all pivots above it. ("Pivots head southeast.")*

  - **Pivot columns:** *If a column has a pivot, all its other entries vanish.*

  - **0-rows sink:** *Any row populated entirely by zeros lies at the bottom of the matrix—all non-zero entries lie above it.*

- **Represent.** We say that a matrix **A represents** a linear transformation T if $T(\mathbf{v}) = \mathbf{A}\mathbf{v}$ for every input vector $\mathbf{v}$.

- **Row-space.** The **row-space** of an $n \times m$ matrix **A** is the span of the rows of **A**. Equivalently, it is the column space of the transpose: $\mathrm{row}(\mathbf{A}) = \mathrm{col}(\mathbf{A}^T)$. It is always subspace of $\mathbf{R}^m$.

- **R** and **$\mathbf{R}^n$.** We denote the set of all real numbers—the points of the number line—by **R**. Note that **R** forms the set of all allowed scalars for this text. The **numeric vector space $\mathbf{R}^n$** is the set of **all** n-entried vectors $\mathbf{x} = (x_1, x_2, \ldots, x_n)$, with each x_i belonging to **R**.

  For instance, $(1, 2, 2, 1) \in \mathbf{R}^4$, and if we write $\mathbf{v} \in \mathbf{R}^3$, we mean that $\mathbf{v}$ has three coordinates: $\mathbf{v} = (x, y, z)$, for example.

- **Scalar.** The preferred term for *number* in linear algebra.

  In this text, scalars are always *real* numbers, but in more advanced work they may belong to other number systems, such as the complex numbers.

- **Span** (noun). The **span** of a set of vectors $\{\mathbf{u}_1, \mathbf{u}_2, \ldots, \mathbf{u}_k\} \subset \mathbf{R}^n$ is denoted by $\mathrm{span}\{\mathbf{u}_1, \mathbf{u}_2, \ldots, \mathbf{u}_k\}$. It is the family of all linear combinations

$$c_1\mathbf{u}_1 + c_2\mathbf{u}_2 + \cdots + c_k\mathbf{u}_k$$

The span of a set in $\mathbf{R}^n$ is always a subspace of $\mathbf{R}^n$.

- **Span** (verb). When a subspace $S \subset \mathbf{R}^n$ is the span of a set of vectors, we say those vectors **span** S. In particular, any set of vectors "spans its span."

- **Spectrum.** The **spectrum** of a linear operator T on $\mathbf{R}^n$ is the set of all the eigenvalues $\{\lambda_1, \lambda_2, \ldots, \lambda_k\}$ of T, of which there are at most n. We denote the spectrum by $\mathrm{spec}(T)$.

- **Standard basis.** The **standard basis** for $\mathbf{R}^n$ is comprised of the n vectors

$$
\begin{aligned}
\mathbf{e}_1 &= (1,\ 0,\ 0,\ \ldots,\ 0,\ 0) \\
\mathbf{e}_2 &= (0,\ 1,\ 0,\ \ldots,\ 0,\ 0) \\
\mathbf{e}_3 &= (0,\ 0,\ 1,\ \ldots,\ 0,\ 0) \\
&\ \ \vdots \\
\mathbf{e}_n &= (0,\ 0,\ 0,\ \ldots,\ 0,\ 1)
\end{aligned}
$$

Note that these are also the rows (and columns) of the $n \times n$ identity matrix.

- **Subspace.** A **subspace** is a subset $S \subset \mathbf{R}^n$ that is closed under both vector addition and scalar multiplication. Closure under addition (or scalar multiplication) means that the sum of any two vectors in S and all scalar multiples of vectors in S remain in S.

- **Symmetric.** An $n \times n$ matrix $\mathbf{A}$ is **symmetric** when it equals its own transpose: $\mathbf{A}^T = \mathbf{A}$. All symmetric matrices are consequently square, though square matrices are not generally symmetric.

- **Transpose.** The **transpose** of an $n \times m$ matrix $\mathbf{A}$, denoted $\mathbf{A}^T$, is the $m \times n$ matrix whose ith row is the ith *column* of $\mathbf{A}$. Equivalently, it is the matrix whose (i, j) entry is the (j, i) entry of $\mathbf{A}$.

- **Unit vector.** A vector $\mathbf{v} \in \mathbf{R}^n$ is a **unit vector** when it has length one ("unit length"): $\mathbf{v} \cdot \mathbf{v} = 1$.

- **Vector** (numeric). A **numeric vector** (or simply **vector** when "numeric" is clear from the context) is an element of $\mathbf{R}^n$, i.e., an ordered n-tuple of scalars. The individual scalars in a vector are called its *entries*, *components*, or *coordinates*. We use bold lowercase to denote vectors, as in $\mathbf{v} = (v_1, v_2, \ldots, v_n)$ or $\mathbf{a} = (2, -1, 3, 0)$.

The entries of a numeric vector may be written horizontally or vertically. The following, for instance, represent the same numeric vector:

$$(2, -3, 5, 7) \qquad \text{and} \qquad \begin{pmatrix} 2 \\ -3 \\ 5 \\ 7 \end{pmatrix}$$

- **Vector** (geometric). A **geometric vector** (or simply **vector** when "geometric" is clear from the context) is an arrow emanating from the origin in the plane or in space. We indicate geometric vectors by putting an arrow above a lowercase letter, as in $\vec{v}$.

  If one puts axes through the origin and introduces units of measurement along them, each geometric vector $\vec{v}$ in the plane or in space can be identified with a numeric vector $\mathbf{v}$ that lists its coordinates relative to those axes.

Index